DICTIONARY OF GENETICS

ROBERT C. KING
PROFESSOR OF BIOLOGY
NORTHWESTERN UNIVERSITY

WILLIAM D. STANSFIELD
PROFESSOR OF BIOLOGY
CALIFORNIA POLYTECHNIC STATE UNIVERSITY

New York   Oxford
OXFORD UNIVERSITY PRESS
1985

Oxford University Press

Oxford   London   New York   Toronto
Delhi   Bombay   Calcutta   Madras   Karachi
Kuala Lumpur   Singapore   Hong Kong   Tokyo
Nairobi   Dar es Salaam   Cape Town
Melbourne   Auckland

and associated companies in
Beirut   Berlin   Ibadan   Mexico City   Nicosia

Published by Oxford University Press, Inc.,
200 Madison Avenue, New York, New York 10016

Library of Congress Cataloging in Publication Data

King, Robert C.
   A dictionary of genetics.

   Rev. ed. of: A dictionary of genetics / Robert C.
King. 2nd ed., rev. 1974
   1. Genetics—Dictionaries.   I. Stansfield,
William D., 1930–   .   II. King, Robert C., 1928–
Dictionary of genetics.   III. Title.
QH427.K55   1985   575.1'0321   84-29575
ISBN 0-19-503494-5
ISBN 0-19-503495-3 (pbk.)

Printing (last digit):   9 8 7 6 5 4 3

Printed in the United States of America

PREFACE

What has happened in the field of genetics since the publication of the Second Edition of this Dictionary is nothing short of spectacular. The first restriction enzyme that cuts DNA at specific sites was discovered in 1970, and by 1972 the first recombinant DNA molecules had been generated. The next year, genetically engineered plasmids had been functionally cloned in a bacterium. The possibilities for manipulating heredity and perhaps creating potentially dangerous novel microorganisms by this new technology caused concerned scientists to call for a worldwide moratorium (at the famous Asilomar conference of 1975) on certain classes of such experiments. In 1977, recombinant DNA technology went commercial, eukaryotic genes were found to contain noncoding regions (introns), and procedures for rapid DNA sequencing became available.

This rapid evolution of scientific knowledge and technical applications has created a burgeoning vocabulary of new terminology, most of which is not to be found in standard collegiate dictionaries or even dictionaries of biology. One can hardly read a few lines of current journal articles in genetics or related fields before encountering one or more new words from the jargon of molecular genetics. This deeper understanding of genetics has had ramifications into all of the disciplines in the biological sciences, especially those of microbiology, ecology, systematics, and evolution, as well as into the applied disciplines of medicine and agriculture. Furthermore, one must be well versed in physics and biochemistry (especially nucleic acid enzymology) in order to appreciate the kinds of techniques and instrumentation utilized in the synthesis, purification, and analysis of nucleic acids and proteins. Since it is difficult for the nonspecialist to keep abreast of the advances in all of these fields, the scope of this Third Edition has been expanded to reflect the broadly interdisciplinary nature of modern genetics. The use of acronyms has become very popular as a means of rapid communication, both orally and in scientific writings. Therefore, an attempt has been made to identify some of the more commonly employed acronyms (e.g., **HAT medium, HLA antigens, kbp, PAGE,** etc.).

Sometimes research reports and other technical references cite scientific names of species and genera that are experimentally manipulated by geneticists without specifying whether the organism is a grass, a wasp, or a rodent. To aid the nonspecialist in this regard, this Dictionary lists those species and genera that have been extensively studied either because of their economic importance or because they provide certain useful advantages for studying specific genetic problems. In the later case, a brief description of these advantages is given. The reader can find where a species fits into the classification of living organisms by referring to Appendix A and can find the scientific names of common domesticated species in Appendix B.

The Third Edition has grown considerably. The number of definitions has increased from 3,800 in the Second Edition to 5,920, and the number of illustrations from 172 to 225. The classification presented as Appendix A has been revised to follow Whittaker's five kingdom systems. Appendix B contains the common and scientific names of about 200 domesticated species. In Appendix C a Chronology of events important to the development of genetics begins with the invention of the compound microscope in 1590 and ends with the Nobel Prize laureates of 1984. The number of entries in the Chronology (Appendix C) has grown from 360 to 560, and an index has been added containing the names of over 800 scientists cited. A bibliography of major sources for the Chronology is also part of Appendix C. A list of periodicals cited in the literature of genetics, cytology, and molecular biology (Appendix D) has increased from 323 to 393 entries.

Certain rules have been followed regarding the arrangement of definition entries. Each is placed in alphabetical order using the bold face title letter by letter, while ignoring spaces between words. Thus **S phase** is placed between *Sphaerocarpus* and **spherocytic anemia.** Identical alphabetical listings are entered so that lower case letters precede upper case letters. Thus the **r** entry comes before the **R** entry. In titles beginning with a Greek letter, the letter is spelled out. Thus λ **phage** is found under **lambda phage.** In titles starting with a number or containing numbers, the numbers are ignored in the alphabetical placement. Thus **M5 technique** is treated as **M technique** and **T24 oncogene** as **T oncogene.** However, numbers are used to determine the order in a series, *e.g.,* **4S, 5S,** and **5.8S RNA.** When looking for a two word term, if you don't find the definition listed under the first word, try the second. For example, definitions for **genetic recombination** and **beta thalassemia** occur under **recombination** and **thalassemias,** respectively.

The authors would appreciate being informed of errors of commission and/or omission, and welcome suggestions for improvements so that these might be included in any subsequent printings or editions.

Evanston, Ill.   R.C.K.
San Luis Obispo, Calif.   W.D.S.
February 1985

Acknowledgments: Drs. Roger Melvold and Pamela Mulligan were especially generous with advice and provided many useful definitions. They also made valuable suggestions as to discoveries worth including in the chronology, and Drs. Lynn Margulis, Hampton Carson, Yataro Tazima, and Susanne Gollin were also helpful in this regard. Dr. Joseph Gall kindly provided the light micrograph of salamander oocyte chromosomes used on the cover and again in the text.

APPENDICES

A DICTIONARY OF GENETICS

A

A **1.** mass number of an atom; **2.** haploid set of autosomes; **3.** ampere; **4.** adenine or adenosine.

Å Angstrom unit.

A₂ *See* **hemoglobin.**

A 23187 *See* **ionophore.**

AA-AMP amino acid adenylate.

A, B antigens mucopolysaccharides responsible for the ABO blood group system. The A and B antigens reside on the surface of erythrocytes, and differ only in the sugar attached to the penultimate monosaccharide unit of the carbohydrate chain. This minor chemical difference makes the macromolecule differentially active antigenically. The I^A and I^B genes presumably control the formation or functioning of the enzymes that add the specific sugar units to the carbohydrate chains in a preformed mucopolysaccharide molecule. The I° allele is inactive in this regard, and when homozygous results in the O phenotype. Glyco-proteins with properties antigenically identical to the A, B antigens are ubiquitous, having been isolated from bacteria and plants. Every human being more than 6 months old pos-sesses those antibodies of the A, B system that are not directed against its own blood-group antigens. These "pre-existing natural" antibodies probably result from immunization by the ubiquitous antigens mentioned above. The I alleles reside on autosome 9. *See* **blood group, H substance.**

aberrations *See* **chromosomal aberration, radiation-induced chromosomal aberration.**

ABM paper aminobenzyloxy methyl cellulose paper, which when chemically activated, reacts covalently with single-stranded nucleic acids.

ABO blood group system a system of alleles residing on human chromosome 9 that spec-ifies certain red cell antigens. *See* **AB antigens, blood groups, Bombay blood group.**

abortive transduction failure of a transducing exogenote to become integrated into the host chromosome, but rather existing as a non-replicating particle in only one cell of a clone. *See* **transduction.**

abortus a dead fetus born prematurely, whether the abortion was artificially induced or spontaneous.

3

absolute plating efficiency the percentage of individual cells that give rise to colonies when inoculated into culture vessels. *See* **relative plating efficiency.**

absorbance (*also* **absorbency**) a measure of the loss of intensity of radiation passing through an absorbing medium. It is defined in spectrophotometry by the relation: log (I_o/I), where I_o = the intensity of the radiation entering the medium and the I = the intensity after traversing the medium.

abundance in molecular biology, the average number of molecules of a specific mRNA in a given cell, also termed *representation*. The abundance, $A = NRf/M$, where N = Avogadro's number, R = the RNA content of the cell in grams, f = the fraction the specific RNA represents of the total RNA, and M = the molecular weight of the specific RNA in daltons.

acaricide miticide (*q.v.*).

acatalasemia the hereditary absence of catalase (*q.v.*) in man; inherited as an autosomal recessive.

acatalasia synonym for acatalasemia (*q.v*).

acceleration *See* **heterochrony.**

accelerator an apparatus that imparts kinetic energy to charged subatomic particles to produce a high-energy particle stream for analyzing the atomic nucleus.

acceptor stem the double-stranded branch of a tRNA molecule to which an amino acid is attached (at the 3′, CCA terminus) by a specific aminoacyl-tRNA synthetase.

accessory chromosomes *See* **B chromosomes.**

accessory nuclei bodies resembling small nuclei that occur in the oocytes of most Hymenoptera and those of some Hemiptera, Coleoptera, Lepidoptera, and Diptera. Accessory nuclei are covered by a double membrane possessing annulate pores. They are originally derived from the oocyte nucleus, but they subsequently form by the amitotic division of other accessory nuclei.

***Ac, Ds* system** Activator–Dissociation system (*q.v.*).

ace *See* **symbols used in human cytogenetics.**

acentric designating a chromatid or a chromosome that lacks a centromere. *See* **chromosome bridge.**

Acer the genus of maple trees. *A. rubrum,* the red maple, and *A. saccharum,* the sugar maple, are studied genetically because of their commercial importance.

Acetabularia a genus of large, unicellular green algae. Grafting experiments between species of this genus have provided information on the nuclear control of cytoplasmic differentiation.

Acetobacter a genus of aerobic bacilli which secure energy by oxidizing alcohol to acetic acid.

aceto-carmine a stain used in the preparation of chromosome squashes consisting of a 5% solution of carmine in 45% acetic acid. Largely supplanted by aceto-orcein.

aceto-orcein a fluid consisting of 1% orcein (*q.v.*) dissolved in 45% acetic acid, used in making squash preparations of chromosomes. *See* **salivary gland squash preparation.**

acetylcholine a compound playing an important role in the transmission of nerve impulses across synapses and from nerve endings to the muscles innervated.

4

acetylcholine

$$H_3C-\overset{\overset{\displaystyle O}{\parallel}}{C}-O-CH_2-CH_2-\overset{+}{N}\overset{\displaystyle \diagup CH_3}{\underset{\diagdown CH_3}{-CH_3}}$$

acetyl-coenzyme A *See* **coenzyme A.**

acetyl serine *See* **N acetyl serine.**

achiasmate referring to meiosis without chiasmata. In those species in which crossing over is limited to one sex, the achiasmate meiosis generally occurs in the heterogametic sex.

achondroplasia a hereditary dwarfism due to retarded growth of the long bones. In humans it is inherited as an autosomal dominant. Homozygotes die at an early age. *See* **bovine achondroplasia, fowl achondroplasia.**

achromatic figure the mitotic apparatus (*q.v.*).

A chromosomes *See* **B chromosomes.**

acid fuchsin an acidic dye used in cytochemistry.

acidic amino acid an amino acid (*q.v.*) having a net negative charge at neutral *p*H. Those universally found in proteins are aspartic acid and glutamic acid, which bear negatively charged side chains in the *p*H range generally found in living systems.

acidic dye an organic anion that binds to and stains positively charged macromolecules.

Acoelomata a subdivision of the Protostomia containing species in which the space between the epidermis and the digestive tube is occupied by a cellular parenchyma. *See* **classification.**

acquired characteristics, inheritance of inheritance by offspring of characteristics that arose in their parents as responses to environmental influences and are not the result of gene action. *See* **Lamarckism.**

Acraniata a subphylum of Chordata containing species without a true skull. *See* **classification.**

acrasin a chemotactic agent produced by *Dictyostelium discoideum* that is responsible for the aggregation of the cells. Acrasin has been shown to be cyclic AMP (*q.v.*).

Acrasiomycota the phylum containing the cellular slime molds. These are protoctists that pass through a unicellular stage of amobae that feed on bacteria. Subsequently these amoebae aggregate to form a fruiting structure that produces spores. The two most extensively studied species from this phylum are *Dictyostelium discoideum* and *Polysphondylium pallidum.*

acridine dye any of a class of organic molecules that bind to DNA and in bacteriophages act as mutagenic agents by causing additions or deletions in the base sequences.

5

acridine orange   an acridine dye that functions both as a fluorochrome and a mutagen.

acriflavin   an acridine dye that produces reading frame shifts (*q.v.*).

acritarchs   the oldest known fossils, thought to represent the earliest eukaryotic cells, estimated to be about 1.6–1.4 billion years old. Most acritarchs were probably thick-walled, cyst-forming, unicellular, photosynthetic algae of some kind. The name is a compound term from the Greek meaning "of uncertain origin."

acrocentric   designating a chromosome or chromatid with a nearly terminal centromere. *See* **telocentric chromosome.**

acromycin   *See* **tetracycline.**

acronym   a word formed by the initial letters of a compound term. Laser, LET, and rep (*q.v.*) would be examples.

acrosome   an apical organelle in the sperm head secreted by the Golgi material, which digests the egg coatings to permit fertilization.

acrostical hairs   one or more rows of small bristles along the dorsal surface of the thorax of *Drosophila.*

acrosyndesis   telomeric pairing by homologues during meiosis.

acrotrophic   *See* **meroistic.**

acrylamide   *See* **polyacrylamide gel.**

ACTH   adrenocorticotropic hormone (*q.v.*).

actidione   cycloheximide.

actin   a family of proteins making up 5%–10% of the total protein of cells. Cytoplasmic actins from diverse species closely resemble each other and the actins from muscle tissues, so it is clear that actin has been conserved during evolution. Under the electron microscope actin fibers are seen as microfilaments 5–7 nm in diameter. Cytochalasin B (*q.v.*) inhibits the rate of elongation of actin filaments. *Compare with* **intermediate filaments.** *See* **stress fibers.**

actinins   muscle proteins; these have been separated into alpha and beta classes of 200 kd and 60 kd molecular weights, respectively.

actinomycin D   an antibiotic produced by *Streptomyces chrysomallus* that prevents the transcription of messenger RNA. *See* **RNA polymerase.**

actinomycin D

```
                    H                    H
                    |                    |
        O    CH₃-C-CH₃        CH₃-C-CH₃    O
        ‖       |                |         ‖
     ┌─ C ──────── C-H        H-C ──────── C ─┐
     │            |            |             │
     │            N-CH₃        N-CH₃         │
     │            |            |             │
     │         sarcosine    sarcosine       O
     │            |            |             │
     O         L-proline    L-proline        │
     │            |            |             │
     │         D-valine     D-valine         │
     │            |            |             │
     │            C=O          C=O           │
     │            |            |             │
 CH₃-HC ──────── C-H        H-C ──────── CH-CH₃
     │            |            |
     │            NH           NH
     │            |            |
     │            C=O          C=O
     │            |            |
               C      N      C     NH₂
       H-C    C    C    C
       HC   C    C    C
          C   O  H  C     O
          |            |
          CH₃          CH₃
```

activated macrophage   a macrophage which has been stimulated (usually by a lympho-kine) to enlarge, to increase its enzymatic content, and to increase its non-specific phago-cytic activity.

activated water   a transient chemically reactive state produced in water by absorbed ion-izing radiations.

activating enzyme   an enzyme that catalyzes a reaction involving ATP and a specific amino acid. The product is an activated complex that subsequently reacts with a specific transfer RNA.

activation analysis   a method of extremely sensitive analysis based on the detection of characteristic radionuclides produced by neutron activation.

activation energy   the energy required for a chemical reaction to proceed. Enzymes com-bine transiently with a reactant to produce a new complex that has a lower activation energy. Under these circumstances the reaction can take place at the prevailing tempera-ture of the biological system. Once the product is formed, the enzyme is released unchanged.

activator   a molecule that converts a repressor into a stimulator of operon transcription; e.g., the repressor of a bacterial arabinose operon becomes an activator when combined with the substrate.

***Activator-Dissociation* system**   controlling elements discovered by Barbara McClintock in maize. *Ac* is an *autonomous* element that is inherently unstable. It has the ability to excise itself from one chromosomal site and to transpose to another. *Ac* is detected by its acti-vation of *Ds*. *Ds* is *nonautonomous* and is not capable of excision or transposition by itself. *Ac* need not be adjacent to *Ds* or even on the same chromosome in order to activate *Ds*. When *Ds* is so activated, it can alter the level of expression of neighboring genes, the struc-ture of the gene product, or the time of development when the gene expresses itself, as a consequence of nucleotide changes inside or outside of a given cistron. An activated *Ds* can also cause chromosome breakage, which may yield deletions or generate a bridge-breakage-fusion-bridge cycle (*q.v.*). *See* **transposable elements.**

active immunity immunity conferred on an organism by its own exposure and response to antigen. In the case of immunity to disease-causing agents, the antigenic pathogens may be administered in a dead or attenuated form. *See also* **passive immunity.**

active site that portion(s) of a protein that must be maintained in a specific shape and amino acid content to be functional. Examples: **1.** in an enzyme, the substrate-binding region; **2.** in histones or repressors, the parts that bind to DNA; **3.** in an antibody, the part that binds antigen; **4.** in a hormone, the portion that recognizes the cell receptor.

actomyosin the basic contractile element in muscle fibers; a molecular complex of the proteins actin and myosin.

acute transfection infection of cells with DNA for a short period of time.

adaptation **1.** the process by which organisms undergo modification so as to function more perfectly in a given environment. **2.** any developmental, behavioral, anatomical, or physiological characteristic of an organism which, in its environment, improves its chances for survival and of leaving descendants.

adaptive enzyme inducible enzyme.

adaptive landscape a three-dimensional graph that shows the frequencies of two genes, each present in two allelic forms (aA and bB in the illustration) plotted against average fitness for a given set of environmental conditions, or a comparable conceptual plot in multidimensional space to accommodate more than two loci.

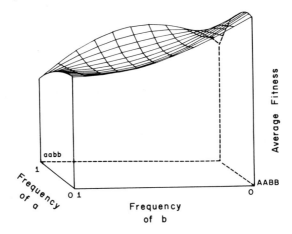

adaptive norm the array of genotypes (compatible with the demands of the environment) possessed by a given population of a species.

adaptive peak a high point (perhaps one of several) on an adaptive landscape (*q.v.*), from which movement in any planar direction (changed gene frequencies) results in lower average fitness.

adaptive radiation evolution from a generalized, primitive species of diverse, specialized species, each adapted to a distinct mode of life.

adaptive surface synonymn for adaptive landscape (*q.v.*).

adaptive value the property of a given genotype when compared with other genotypes that confers fitness (*q.v.*) to an organism in a given environment.

adaptor   a short, synthetic DNA segment containing a restriction site that is coupled to both ends of a blunt-ended restriction fragment. The adaptor is used to join one molecule with blunt ends to a second molecule with cohesive ends. The restriction site of the adaptor is made identical to that of the other molecule so that when cleaved by the same restriction enzyme both DNAs will contain mutually complementary cohesive ends.

adaptor hypothesis   In 1958 F. H. C. Crick suggested that during translation a given amino acid is carried to the template by an adaptor molecule containing nucleotides and that the adaptor is the part which actually fits on the RNA template. Crick thus predicted the discovery of transfer RNA.

adaptor RNA   transfer RNA.

ADCC   antibody-dependent cellular cytotoxicity; also known as antibody-dependent cell-mediated cytotoxicity. Cell-mediated cytotoxicity requiring prior binding of antibody to target cells for killing to occur. It does not involve the complement cascade. *See* **K cells.**

additive factor   one of a group of non-allelic genes affecting the same phenotypic characteristics and each enchancing the effect of the other in the phenotype. *See* **quantitative inheritance.**

additive gene action   **1.** a form of allelic interaction in which dominance is absent; the heterozygote is intermediate in phenotype between homozygotes for the alternative alleles. **2.** the cumulative contribution made by all loci (of the kind described above) to a polygenic trait.

additive genetic variance   genetic variance attributed to the average effects of substituting one allele for another at a given locus, or at the multiple loci governing a polygenic trait.

adenine   a purine base found in nucleic acids.

adenine deoxyriboside   *See* **nucleoside.**

adenohypophysis   the anterior, intermediate, and tuberal portions of the hypophysis, which originate from the buccal lining in the embryo.

adenosine   *See* **nucleoside.**

adenosine phosphate   any of three compounds in which the nucleoside adenosine is attached through its ribose group to one, two, or three phosphoric acid molecules. They are:

9

AMP, ADP, and ATP are interconvertible. ATP upon hydrolysis yields the energy used to drive a multitude of biological processes (muscle contraction, photosynthesis, bioluminescence, and the biosynthesis of proteins, nucleic acids, polysaccharides and lipids).

adenovirus   any of a group of spherical DNA viruses characterized by a shell containing 252 capsomeres. Adenoviruses infect a number of mammalian species including man; some are oncogenic.

adenylcyclase   the enzyme that catalyzes the conversion of adenosine triphosphate (ATP) into cyclic adenosine monophosphate (AMP).

adenylic acid   *See* **nucleotide.**

adhesion plaques   *See* **vincullin.**

adjacent disjunction   *See* **translocation heterozygote.**

adjacent segregation   *See* **translocation heterozygotic.**

adjuvant   a substance administered with an antigen which enhances the stimulation of the immune response to the antigen.

adoptive immunity   *See* **adoptive transfer.**

adoptive transfer   the transfer of an immune function from one organism to another through the transfer of immunologically active or competent cells.

ADP   adenosine diphosphate.   *See* **adenosine phosphate.**

adrenocorticotropic hormone   a single-chain peptide hormone (39 amino acids long) stimulating secretion by the adrenal cortex. It is produced by the adenohypophysis of vertebrates. Abbreviated ACTH. Also called *corticotropin.*

advanced   in systematics, the later or derived stages or conditions within a lineage that exhibits an evolutionary advance; the opposite of primitive.

adventitious   designating a structure arising in an unusual place. For example, roots growing from leaves are *adventitious.*

adventitious embryony   the production by mitotic divisions of an embryonic sporophyte from the tissues of another sporophyte without a gametophytic generation intervening.

Aedes   a genus containing over 700 species, several of which transmit important human diseases. *A. aegypti,* the vector of yellow fever and dengue fever, has a low chromosome number and is easily reared in the laboratory. The genetics of insecticide resistance has been intensively studied.

Aegilops   a genus of grasses including several species of genetic interest especially *A. umbellulata,* a wild Mediterranean species resistant to leaf rust. A gene for rust resistance has been transferred from *A. umbellulata* to *Triticum vulgare.*

aerobe   a cell that lives in air and utilizes oxygen. A strictly aerobic cell cannot live in the absence of oxygen.

aestivate   to pass through a hot, dry season in a torpid condition. *See also* **hibernate.**

afferent   leading toward the organ or cell involved. In immunology, the events or stages involved in activating the immune system.

affinity in immunology, the inate binding power of an antibody combining site with a single antigen binding site.

affinity chromatography a technique for separating molecules by their affinity to bind to ligands (e.g., antibodies) attached to an insoluble matrix (e.g., Sepharose). The bound molecules can subsequently be eluted in a relatively pure state.

afibrinogenemia an inherited disorder of the human blood clotting system characterized by the inability to synthesize fibrinogen; inherited as an autosomal recessive.

agammaglobulinemia the inability in humans to synthesize certain immunoglobulins. The most common form is inherited as an X-linked recessive, and plasma cells are absent. *See* **antibody.**

agamospermy the formation of seeds without fertilization. The male gametes, if present, serve only to stimulate division of the zygote. *See* **apomixis.**

Agapornis a genus of small parrots. The nest building of various species and their hybrids has provided information on the genetic control of behavior patterns.

agar a polysaccharide extract of certain seaweeds used as a solidifying agent in culture media.

agar plate count the number of bacterial colonies that develop on an agar-containing medium in a petri dish seeded with a known amount of inoculum. From the count the concentration of bacteria per unit volume of inoculum can be determined.

agglutination the clumping of viruses or cellular components in the presence of a specific immune serum.

agglutinin any antibody capable of causing clumping of erythrocytes, or more rarely other types of cells.

agglutinogen an antigen that stimulates the production of agglutinins.

aggregation chimera a mammalian chimera made through the mingling of cells of two embryos. The resulting composite embryo is then transferred into the uterus of a surrogate mother where it comes to term.

agonistic behavior any social interaction between members of the same species that involves aggression or threat and conciliation or retreat.

agouti the grizzled color of the fur of mammals resulting from alternating bands of light (phaeomelanin) and dark (eumelanin) pigments in the individual hairs. It is also the name given to the genes (usually dominant to the dark color genes) responsible for the insertion of the light pigment bands into the hairs. *See* **melanin.**

agranular reticulum endoplasmic reticulum devoid of attached ribosomes.

agranulocytes white blood cells (leukocytes) whose cytoplasm contains few, if any, granules, and possessing an unlobed nucleus (mononuclear leukocytes); includes lymphocytes and monocytes

agriculturally important species *See* **Appendix B.**

Agrobacterium tumefaciens the bacterium responsible for crown gall disease in a wide range of dicotyledonous plants. The bacterium enters only dead, broken plant cells and then may transmit a tumor-inducing plasmid into adjacent living plant cells. This infective process is a natural form of genetic engineering. Strains of *A. tumefaciens* carrying the plasmid may be artificially genetically engineered to introduce foreign genes of choice into plant cells, and then by growing the cells in tissue culture, whole plants can be regenerated, every cell of which contains the foreign gene. *See* **Ti plasmid.**

Agropyron elongatum a weed related to crabgrass noted for its resistance to stem rust. Genes conferring rust resistance have been transferred from this species to *Triticum aestivum.*

AHG antihemophilic globulin. *See* **blood clotting.**

AI, AID, AIH *See* **artificial insemination.**

AIA anti-immunoglobulin antibodies, produced in response to foreign antibodies introduced into an experimental animal.

akinetic acentric (*q.v.*).

ala alanine. *See* **amino acid.**

alanine *See* **amino acid.**

alate having wings.

albinism 1. deficiency of chromoplasts in plants. 2. the inability to form melanin (*q.v.*) in the eyes, skin and hair, due to a tyrosine deficiency. In humans, the condition is inherited as an autosomal recessive. Amelanic melanocytes are present in the skin of albinos. In a wide variety of animals, the albino gene has pleiotropic effects on the visual pathways, resulting in nystagmus and crossed eyes. *See* **ocular albinism.**

albino 1. an animal lacking pigmentation. 2. a plant lacking chromoplasts.

albomaculatus referring to a variegation consisting of irregularly distributed white and green regions on plants resulting from the mitotic segregation of genes or plastids.

albumin a 70 kd protein which represents 40–50% of the plasma protein in adult mammals. It is important both as an osmotic and a *p*H buffer and also functions in the transport of metal ions and various small organic molecules. Albumin is synthesized and secreted by the liver. In the mouse the albumin gene resides on chromosome 5, separated from the alpha fetoprotein gene by a DNA segment about 13.5 kb long. In humans these two genes are in the long arm of chromosome 4. *See* **alpha fetoprotein.**

alcaptonuria alkaptonuria (*q.v.*).

aldehyde any of a class of organic compounds having the general formula $C_nH_{2n}O$ and containing a terminal $-C\!\!\overset{\displaystyle H}{\underset{\displaystyle O}{}}$ group.

aldosterone a hormone from the adrenal cortex influencing electrolyte balance.

aleurone the outer layer of the endosperm of a seed.

aleurone grain a granule of protein occurring in the aleurone.

Aleutian mink an autosomal recessive mutation in *Mustela vison* producing diluted pigmentation of the fur and eyes. The homozygotes show a lysosomal defect similar to the Chédiak-Steinbrinck Higashi syndrome of man.

alga (*plural* **algae**) any of a large group of aquatic, chlorophyll-bearing organisms ranging from single cells to giant seaweeds. *See* **classification;** Cyanobacteria, Dinoflagellata, Euglenophyta, Xanthophyta, Bacillariophyta, Phaecophyta, Rhodophyta, Gamophyta, Chlorophyta.

alien addition monosomic   a genome that contains a single chromosome from another species in addition to the normal complement of chromosomes.

aliphatic   designating molecules made up of linear chains of carbon atoms. Aliphatic compounds are one of three structural types of organic compounds, the others being aromatic (*q.v.*) and heterocyclic (*q.v.*).

aliquot   a part, such as a representative sample, that divides the whole without a remainder. Two is an aliquot of six because it is contained exactly three times. Loosely used for any fraction or portion.

alkali metal   any of five elements in Group IA of the periodic table: lithium (Li), sodium (Na), potassium (K), rubidium (Rb), and cesium (Cs).

alkaline earth   any element of Group IIA of the periodic table: beryllium (Be), magnesium (Mg), calcium (Ca), strontium (Sr), barium (Ba), and radium (Ra).

alkaline phosphatase   an enzyme that removes 5'-P termini of DNA and leaves 5'-OH groups. Treatment of a plasmid with alkaline phosphatase renders it unable to be reconstituted by ligase, unless DNA from another source is inserted.

alkaloids   a group of nitrogen-containing organic substances found in plants; many are pharmacologically active (e.g., caffeine, cocaine, nicotine).

alkapton   2,5-dihydroxyphenylacetic acid. *See* **homogentistic acid.**

alkaptonuria (*also* **alcaptonuria**)   a relatively benign hereditary disese in man due to an autosomal recessive gene. Alkaptonurics cannot make the liver enzyme homogentisic acid oxidase. Therefore homogentisic acid (*q.v.*) is not broken down to simpler compounds but is excreted in the urine. Since the colorless homogentisic acid is readily oxidized to a black pigment, the urine of alkaptonurics darkens when exposed to air. This disease enjoys the historic distinction of being the first metabolic disease studied by A. E. Garrod. *See* **Inborn Errors of Metabolism.**

alkylating agent   a compound causing the substitution of an alkyl group for an active hydrogen atom in an organic compound. According to the number of reactive groups they contain, alkylating agents are classified as mono-, bi-, or polyfunctional. Many chemical mutagens are alkylating agents. *See* **busulfan, chlorambucil, cyclophosphamide, epoxide, ethylmethane sulfonate, melphalan, Myleran, nitrogen mustard, sulfur mustard, TEM, Thiotepa, triethylenethiophosphoramide.**

alkyl group   a univalent radical having the general formula C_nH_{2n+1} derived from a saturated aliphatic hydrocarbon by removal of one atom of hydrogen. Named by replacing the ending -*ane* of the hydrocarbon with -*yl* (e.g., meth*ane* becomes meth*yl*).

allantois   a sac-like outgrowth of the ventral side of the hindgut present in the embryos of reptiles, birds, and mammals. The allantois represents a large and precocious development of the urinary bladder.

allatum hormones   hormones synthesized by the insect corpus allatum. The titer of allatum hormones influences the qualitative properties of each molt in holometabolous insects. At high concentrations, larval development ensues; at lower levels, the insect undergoes pupal metamorphosis, and in the absence of the allatum hormones adult differentiation takes place. The allatum hormones thus have a juvenilizing action and for this reason have also been called juvenile hormones (JHs). The structural formulas for three of the juvenile hormones are shown below. In adult females, the allatum hormone is required for vitellogenesis. The JH analogue, ZR515 (*q.v.*), is often used as a substitute for natural JHs in *Drosophila* experiments. For example, mutations are known that block vitellogenesis. Female flies homozygous for certain of these mutants synthesize yolky oocytes when they receive a topical application of ZR515.

JH 1

$$CH_3-CH_2-\underset{\underset{O}{\diagdown\diagup}}{\overset{\overset{CH_3}{|}}{C}}-CH-CH_2-CH_2-\underset{\underset{CH_2}{|}}{\overset{}{C}}=CH-CH_2-CH_2-\underset{}{\overset{\overset{CH_3}{|}}{C}}=CH-\overset{\overset{O}{\|}}{C}-O-CH_3$$
with CH_2 then CH_3 below

JH 2

$$CH_3-CH_2-\underset{\underset{O}{\diagdown\diagup}}{\overset{\overset{CH_3}{|}}{C}}-CH-CH_2-CH_2-\underset{\underset{CH_3}{|}}{\overset{}{C}}=CH-CH_2-CH_2-\overset{\overset{CH_3}{|}}{C}=CH-\overset{\overset{O}{\|}}{C}-O-CH_3$$

JH 3

$$CH_3-\underset{\underset{O}{\diagdown\diagup}}{\overset{\overset{CH_3}{|}}{C}}-CH-CH_2-CH_2-\underset{\underset{CH_3}{|}}{\overset{}{C}}=CH-CH_2-CH_2-\overset{\overset{CH_3}{|}}{C}=CH-\overset{\overset{O}{\|}}{C}-O-CH_3$$

ZR 515

$$CH_3-\underset{\underset{O-CH_3}{|}}{\overset{\overset{CH_3}{|}}{C}}-CH_2-CH_2-CH_2-\underset{\underset{CH_3}{|}}{\overset{}{C}}H-CH_2-CH=CH-\overset{\overset{CH_3}{|}}{C}=CH-\overset{\overset{O}{\|}}{C}-O-CH\diagup\diagdown CH_3, CH_3$$

allele a shorthand form of **allelomorph**; one of a series of possible alternative forms of a given gene (cistron, *q.v.*), differing in DNA sequence, and affecting the functioning of a single product (RNA and/or protein). If more than two alleles have been identified in a population, the locus is said to show *multiple allelism. See* **heteroallele, homoallele.**

allelic complementation the production of nearly normal phenotype in an organism carrying two different mutant alleles in trans configuration. Such complementation is sometimes caused by the reconstruction in the cytoplasm of a functional protein from the inactive products of the two alleles. When such a phenomenon can be demonstrated by mixing extracts from individuals homozygous for each allele, the term *in vitro complementation* is used.

allelic exclusion in a clone of plasma cells heterozygous for an immunoglobulin gene, the detection of only one allelic product. The mechanism that excludes activation of one allele is presently unknown.

allelic frequency the percentage of all alleles at a given locus in a population gene pool represented by a particular allele. For example, in a population containing 20 *AA*, 10 *Aa*, and 5 *aa*, the frequency of the *A* allele is $[2(20) + 1(10)]/2(35) = 5/7 = 0.714$. *See* **gene frequency.**

allelism test complementation test.

allelomorph commonly shortened to allele (*q.v.*).

allelotype the frequency of alleles in a breeding population.

Allen's rule the generalization that the extended body parts of a warm-blooded species (tail, ears, and limbs) are relatively shorter in the colder regions of the species range than in the warmer.

allergen a substance inducing hypersensitivity.

allergy an immune hypersensitivity response to an agent which is non-antigenic to most of the individuals in a population.

allesthetic trait any individual characteristic that has an adaptive function only via the nervous systems of other organisms. For example, odors, display of color patterns, mating calls, etc., that are important components of courtship in various species.

Allium the genus that includes *A. cepa,* the onion; *A. porrum,* the leek; *A. sativum,* the garlic; and *A. schoenoprasum,* the chive—all classic subjects for cytological studies of mitotic chromosomes.

alloantigen an antigen (*q.v.*) that elicits an immune response (*q.v.*) when introduced into a genetically different individual of the same species. Antibodies produced in response to alloantigens are called *alloantibodies.*

allochromacy the formation of other coloring agents from a given dye which is unstable in solution. Nile blue (*q.v.*) exhibits allochromacy.

allocycly a term referring to differences in the coiling behavior shown by chromosomal segments or whole chromosomes. Allocyclic behavior characterizes the pericentric heterochromatin, the nucleolus organizer, and in some species entire sex chromosomes. If a chromosome or chromosomal segment is tightly condensed in comparison with the rest of the chromosomal complement, the chromosome or chromosomal segment is said to show *positive heteropycnosis* (*q.v.*). Allocycly is also used to describe asynchronous separation of bivalents during the first anaphase in meiosis. In man, for example, the X and Y chromosomes segregate ahead of the autosomes and are said to show *positive allocycly.*

allogeneic referring to genetically different members of the same species, especially with regard to alloantigens (*q.v.*).

allogeneic disease *See* **graft-versus-host reaction.**

allograft a graft of tissue from a donor of one genotype to a host of a different genotype but of the same species.

allolactose *See* **lactose.**

allometry the relation between the rate of growth of a part of an individual and the growth rate of the whole or of another part. In the case of isometry, the relative proportions of the body parts remain constant as the individual grows; in all other cases, the relative proportions change as total body size increases. *See* **heterauxesis.**

allomone any chemical secreted by an organism that influences behavior in a member of another species, benefiting only the producer. If both species benefit, it is a *synamone.* If only the receiver benefits, it is a *kairomone.*

alloparapatric speciation a mode of gradual speciation in which new species originate through populations that are initially allopatric, but later become parapatric before completely effective reproductive isolation has evolved. Natural selection may enhance incipient reproductive isolating mechanisms in the zone of contact by character displacement (*q.v.*), and other mechanisms. *Compare with* **parapatric speciation.**

allopatric occupying distinctly separate geographic ranges.

allopatric speciation the development of distinct species through differentiation of populations in geographic isolation.

allophene a phenotype not due to the mutant genetic constitution of the cells of the tissue in question. Such a tissue will develop a normal phenotype if transplanted to a wild-type host. *See* **autophene.**

allophenic mice chimeric mice produced by removing cleaving mouse eggs, fusing the blastomeres *in vitro,* and reimplanting the fused embryos into the uterus of another mouse to permit embryogenesis to continue. Viable mice containing cells derived from two or more embryos have been obtained and used in cell lineage studies.

allopolyploid (*also* **alloploid***)* a polyploid organism arising from the combination of genetically distinct chromosome sets.

allosteric effect the result of the reversible interaction of a small molecule with a protein molecule, which leads to changes in the shape of the protein and a consequent alteration of the interaction of that protein with a third molecule.

allosteric effectors small molecules that reversibly bind to allosteric proteins at a site different from the active site, causing an allosteric effect.

allosteric enzyme a regulatory enzyme whose catalytic activity is modified by the noncovalent attachment of a specific metabolite to a site on the enzyme other than the catalytic site.

allosteric protein a protein showing allosteric effects.

allosyndesis the pairing of homoeologous chromosomes (*q.v.*) in an allopolyploid (*q.v.*). Thus if the genetic composition of an alloploid is given by AABB, where AA represent the chromosomes derived from one parent species and BB the chromsomes derived from the other parent species, then during meiotic prophase, A undergoes allosyndetic pairing with B. Such pairing indicates that the A and B chromosomes have some segments that are homologous, presumably because the two parent species have a common ancestry. In the case of *autosyndesis,* A pairs only with A, and B with B. *Segmental alloploids* form both bivalents and multivalents during meiosis because of allosyndesis.

allotetraploid an organism that is diploid for two genomes, each from a different species; synonymous with amphidiploid.

allotypes antigenic markers on immunoglobulin chains or other serum proteins that are not common to all normal members of a species. For example, in humans 3 genetic variants of the kappa light chains are known that result from amino acid substitution at positions 153 and/or 191. *Compare with* **idiotypes, isotypes.**

allotype suppresion the systematic and long term suppression of the expression of an immunoglobulin allotype in an animal induced by treatment with antibodies against the allotype.

allotypic differentiation *See in vivo* **culturing of imaginal discs.**

allozygote an individual homozygous at a given locus, whose two homologous genes are of independent origin, as far as can be determined from pedigree information. *See* **autozygote.**

allozymes allelic forms of an enzyme that can be distinguished by electrophoresis, as opposed to the more general term *isozyme* (*q.v.*).

alpha amanitin a compound derived from the mushroom *Aminita phalloides* that inhibits transcription by eukaryotic RNA polymerase II, but does not inhibit mitochondrial RNA synthesis.

alpha chain one of the two polypeptides found in adult and fetal hemoglobin (*q.v.*).

alpha fetoprotein the major plasma protein of fetal mammals. AFP is a 70 kd glycoprotein that is synthesized and secreted by the liver and the yolk sac. The genes encoding AFP and serum albumen arose in evolution as the result of a duplication of an ancestral gene $(3–5) \times 10^8$ years ago. *See* **albumen.**

alpha galactosidase an enzyme that catalyzes the hydrolysis of substrates that contain α-galactosidic residues, including glycosphingolipids and glycoproteins. In humans, α-galactosidase exists in two forms, A and B. The A form is encoded by a gene on the X chromosome. Fabry disease (*q.v.*) is caused by mutations at this locus. The B form is encoded by a gene on chromosome 22.

alpha helix a characteristic helical secondary structure seen in many proteins. The alpha helix allows maximum intramolecular hydrogen bond formation between C=O and H−N groups. One turn of the helix occurs for each 3.6 amino acid residues. *See* **protein structure.**

alpha particle a helium nucleus, consisting of two protons and two neutrons, and having a double positive charge.

alpha ray a stream of alpha particles.

alpha tocopherol vitamin E. (*q.v.*)

alteration enzyme a protein of phage T4 that is injected into a host bacterium along with the phage DNA; this protein modifies host RNA polymerase by linking it to ADP-ribose. RNA polymerase modified in this way renders it incapable of binding to sigma factor and thus unable to initiate transcription at host promoters. *See* **RNA polymerase.**

alternate disjunction *See* **translocation heterozygote.**

alternate segregation *See* **translocation heterozygote.**

alternation of generations reproductive cycles in which a haploid phase alternates with a diploid phase. In mosses and vascular plants the haploid phase is the gametophyte, the diploid the sporophyte.

altricial referring to the type of ontogeny seen in vertebrate species characterized by large litters, short gestations, and the birth of relatively undeveloped, helpless young. *Compare with* **precocial.**

altruism behavior of an individual that benefits others; to the extent that the "others" are related to the altruist (the one exhibiting altruistic behavior), such actions may actually be an expression of fitness. *See* **inclusive fitness.**

Alu family the most common dispersed repeated DNA sequence in the human genome, consisting of about 300 base pairs in perhaps half a million copies, accounting for about 5% of human DNA. Each segment apparently is made up of two 140 base-pair sequences joined head to tail with a 31 base-pair insert in the right-hand monomer. These sequences appear to be readily transposable. The family name is derived from the fact that these sequences are cleaved by the restriction endonuclease *Alu I.*

alula a small lobe at the base of the *Drosophila* wing just above the haltere.

Am abbreviation for an amber suppressor (*q.v.*). For example, if mutation 35 in the *leu* gene is a suppressor-sensitive Am type, it would be written *leu*35 (Am).

amaurosis blindness occurring without an obvious lesion in the eye, as from a disease of the optic nerve or brain.

amber codon the mRNA triplet UAG that causes termination of protein translation; one of three "stop" codons. The terms *amber* and *ochre* (*q.v.*) originated from a private laboratory joke and have nothing to do with colors.

Amberlite trade name for a family of ion-exchange resins.

amber mutation   a mutation in which a polypeptide chain is terminated prematurely. Amber mutations are the result of a base substitution that converts a codon specifying an amino acid into UAG, which signals chain termination. In certain strains of *E. coli* amber mutations are suppressed. These strains contain a tRNA with a AUC anticodon, which inserts an amino acid at the UAG site and hence permits translation to continue. *See* **ochre mutation, nonsense mutation.**

amber suppressor   any mutant gene coding for a tRNA whose anticodon can respond to the UAG stop codon by the insertion of an amino acid that renders the gene product at least partially functional. For example, a mutant tyrosine-tRNA anticodon 3′AUC would recognize 5′UAG, tyrosine would be inserted, and chain growth would continue.

Ambystoma mexicanum   the Mexican axolotl, a widely used laboratory species. The urodele for which the most genetic information is available.

amensalism   a species interaction in which one is adversely affected and the other is unaffected.

Ames test   a bioassay for detecting mutagenic and possibly carcinogenic compounds, developed by Bruce N. Ames in 1974. Reverse mutants to histidine independence are scored by growing *his⁻ Salmonella typhimurium* on plates deficient in histidine in the presence of the chemical (test) and in its absence (control).

amethopterin   methotrexate (*q.v.*).

amino acid   any of the subunits that are polymerized to form proteins. The twenty amino acids universally found in proteins are listed below together with their abbreviations and their messenger RNA code designations; their formulas appear on page 19.

AMINO ACID	ONE LETTER SYMBOL	THREE LETTER SYMBOL	mRNA CODE DESIGNATION
alanine	A	ala	GCU, GCC, GCA, GCG
arginine	R	arg	CGU, CGC, CGA, CGG, AGA, AGG
asparagine	P	asn	AAU, AAC
aspartic acid	D	asp	GAU, GAC
cysteine	C	cys	UGU, UGC
glutamic acid	E	glu	GAA, GAG
glutamine	Q	gln	CAA, CAG
glycine	G	gly	GGU, GGC, GGA, GGG
histidine	H	his	CAU, CAC
isoleucine	I	ile	AUU, AUC, AUA
leucine	L	leu	UUA, UUG, CUU, CUC, CUA, CUG
lysine	K	lys	AAA, AAG
methionine	M	met	AUG
phenylalanine	F	phe	UUU, UUC
proline	P	pro	CCU, CCC, CCA, CCG
serine	S	ser	UCU, UCC, UCA, UCG, AGU, AGC
threonine	T	thr	ACU, ACC, ACA, ACG
tryptophan	W	trp	UGG
tyrosine	Y	tyr	UAU, UAC
valine	V	val	GUU, GUC, GUA, GUG

amino acid activation   a coupled reaction catalyzed by a specific aminoacyl synthetase which attaches a specific amino acid (AA) to a specific transfer RNA (tRNA) in preparation for translation (*q.v.*).

$$AA + ATP \rightarrow AA\text{-}AMP + 2P$$
$$AA\text{-}AMP + tRNA \rightarrow AA\text{-}tRNA + AMP$$

NH$_2$—CH$_2$—COOH	NH$_2$—CH—COOH 　　　$\|$ 　　　CH$_3$	NH$_2$—CH—COOH 　　　$\|$ 　　　CH 　　$\diagup$　$\diagdown$ 　CH$_3$　　CH$_3$	NH$_2$—CH—COOH 　　　$\|$ 　H—C—CH$_3$ 　　　$\|$ 　　　CH$_2$ 　　　$\|$ 　　　CH$_3$	NH$_2$—CH—COOH 　　　$\|$ 　　　CH$_2$ 　　　$\|$ 　　　CH 　　$\diagup$　$\diagdown$ 　CH$_3$　　CH$_3$
Glycine	Alanine	Valine*	Isoleucine*	Leucine*

ALIPHATIC, MONOAMINO, MONOCARBOXYLIC ACIDS

NH$_2$—CH—COOH 　　$\|$ 　　CH$_2$ 　　$\|$ 　　CH$_2$ 　　$\|$ 　　CH$_2$ 　　$\|$ 　　CH$_2$ 　　$\|$ 　　NH$_2$	NH$_2$—CH—COOH 　　$\|$ 　　CH$_2$ 　　$\|$ 　　CH$_2$ 　　$\|$ 　　CH$_2$ 　　$\|$ 　　N—H 　　$\|$ 　　C=NH 　　$\|$ 　　NH$_2$	NH$_2$—CH—COOH 　　$\|$ 　　CH$_2$ 　　$\|$ 　　SH	NH$_2$—CH—COOH 　　$\|$ 　　CH$_2$ 　　$\|$ 　　CH$_2$ 　　$\|$ 　　S 　　$\|$ 　　CH$_3$	H $\|$ H—N—C—COOH H$_2$C$\diagdown$　$\diagup$CH$_2$ 　　　C 　　$\diagup$　$\diagdown$ 　　H　　H
Lysine*	Arginine	Cysteine	Methionine*	Proline
ALIPHATIC, DIAMINO		ALIPHATIC, SULFUR-CONTAINING		

NH$_2$—CH—COOH 　　$\|$ 　　CH$_2$ 　　$\|$ 　　COOH	NH$_2$—CH—COOH 　　$\|$ 　　CH$_2$ 　　$\|$ 　　CH$_2$ 　　$\|$ 　　COOH	NH$_2$—CH—COOH 　　$\|$ 　　CH$_2$ 　　$\|$ 　　C=O 　　$\|$ 　　NH$_2$	NH$_2$—CH—COOH 　　$\|$ 　　CH$_2$ 　　$\|$ 　　CH$_2$ 　　$\|$ 　　C=O 　　$\|$ 　　NH$_2$	NH$_2$—CH—COOH 　　$\|$ 　　CH$_2$　H 　　$\|$　$\diagup$ 　　C=C 　　　　$\diagdown$NH 　　　$\diagup$ 　　C=C 　HC$\diagup$　$\diagdown$CH 　　HC—CH
Aspartic Acid	Glutamic Acid	Asparagine	Glutamine	Tryptophan*
ALIPHATIC, DICARBOXYLIC		ALIPHATIC AMIDES		

NH$_2$—CH—COOH 　　$\|$ 　　CH$_2$ 　　$\|$ 　　OH	NH$_2$—CH—COOH 　　$\|$ 　H—C—OH 　　$\|$ 　　CH$_3$	NH$_2$—CH—COOH 　　$\|$ 　　CH$_2$ 　　$\|$ 　　C 　$\diagup$　$\diagdown$ H—C　　C—H 　$\|\|$ H—C　　C—H 　$\diagdown$　$\diagup$ 　　C 　　$\|$ 　　H	NH$_2$—CH—COOH 　　$\|$ 　　CH$_2$ 　　$\|$ 　　C 　$\diagup$　$\diagdown$ HC　　C—H 　$\|\|$ HC　　C—H 　$\diagdown$　$\diagup$ 　　C 　　$\|$ 　　OH	NH$_2$—CH—COOH 　　$\|$ 　　CH$_2$ 　　$\|$ 　　C=C—H 　　$\|$　$\|$ 　　N　NH 　　$\diagdown\!\!\diagup$ 　　C 　　$\|$ 　　H
Serine	Threonine*	Phenylalanine*	Tyrosine	Histidine
ALIPHATIC, HYDROXYL-CONTAINING		AROMATIC		HETEROCYCLIC

*required in the diet of mammals.

amino acid sequence　the linear order of the amino acids in a peptide or protein.

amino acid side chain　a group attached to an amino acid, represented by **R** in the general formula for an amino acid:

$$\text{NH}_2\text{—CH—COOH}$$
$$|$$
$$\textbf{R}$$

aminoaciduria   the presence of one or more amino acids in the urine in abnormal quantities because of a metabolic defect.

aminoacyl adenylate   the activated compound that is an intermediate in the formation of a covalent bond between an amino acid and its specific transfer RNA; abbreviated AA-AMP. *See* **AMP.**

aminoacyl-tRNA   an aminoacyl ester of a transfer RNA molecule.

aminoacyl-tRNA binding site   *See* **translation.**

aminoacyl-tRNA synthetases   enzymes that activate amino acids and attach each activated amino acid to its own species of tRNA. These enzymes catalyze: (1) the reaction of a specific amino acid (AA) with adenosine triphosphate (ATP) to form AA-AMP, and (2) the transfer of the AA-AMP complex to a specific transfer RNA, forming AA-tRNA and free AMP (adenosine monophosphate). *See* **adenosine phosphates.**

amino group   a chemical group ($-NH_2$) which with the addition of a proton can form $-NH_3^+$.

aminopeptidase   an enzyme (in both prokaryotes and eukaryotes) that removes the formulated methionine (fMet) or methionine from the NH_2 terminus of growing or completed polypeptide chains.

aminopterin   a folic acid analogue. *See* **methotrexate.**

aminopurine   a base analogue producing transitions (*q.v.*).

amino terminal end   the end of a polypeptide chain that has a free amino group.

amniocentesis   sampling of amniotic fluid for the prenatal diagnosis of fetal disorders. During the procedure, a hollow needle is inserted through the skin and muscle of the mother's abdomen, through the uterus, and into the amniotic sac that surrounds the fetus. Cells that have sloughed from the fetus are suspended in the fluid. Cells in the sample are cultured for about three weeks to raise their numbers to the point where chromosomal and biochemical analyses can be made. Amniocentesis cannot be done until about the 16th week of pregnancy, since the sac containing the embryo is not large enough to permit safe withdrawal of the fluid until this time. *See* **chorionic villi sampling.**

amniocytes   cells obtained by amniocentesis (*q.v.*).

amnion   A fluid-filled sac within which the embryos of reptiles, birds, and mammals develop. The wall of this sac has a two-layered epithelium. The inner epithelium of the wall is the amnion, although the term is sometimes applied to the whole sac. The outer epithelium is usually called the chorion. Amniotic fluid within the sac provides a liquid environment for the embryo.

amniote   a land-living vertebrate (reptile, bird, or mammal) whose embryos have an amnion and allantois.

Amoeba proteus   a common species of rhizopod; a giant protozoan used for microsurgical nuclear transplantations.

amoeboid movement   cellular motility involving cytoplasmic streaming into cellular extensions called *pseudopodia*.

amorph   a mutant gene having no effect compared with that of the wild-type allele.

AMP   adenosine monophosphate.

amphidiploid   an organism that is diploid for two genomes, each from a different species; synonymous with allotetraploid.

amphimixis   sexual reproduction. *See* **apomixis.**

Amphioxus   See **Branchiostoma.**

amphipathic   descriptive of a molecule that has distinct polar and nonpolar segments (e.g., membrane phospholipids).

amphiprotic   pertaining to a compound (like water) which can donate or accept a proton.

amphoteric compound (*also* **ampholyte**)   a substance that can act both as an acid and a base. Thus a protein is amphoteric because it tends to lose protons on the more alkaline side of its isoelectric point and to gain protons on the acid side of its isoelectric point.

amphotropic virus   a virus that can replicate in cells from its host species or in cells from other species. *See* **host range;** *compare with* **ecotropic virus.**

ampicillin   a penicillin derivative that kills growing bacteria by interfering with cell wall synthesis.

amplification   *See* **gene amplification.**

ampoule (*also* **ampule**)   a sealed glass container, often used to contain a measured quantity of a volatile compound or of a solution of measured strength.

amylase   an enzyme that hydrolyzes glucosidic bonds in polyglucosans such as glycogen.

amyloplast   a starch-rich plastid.

anabolism   metabolic synthesis of complex molecules from simpler percursors, usually requiring energy expenditure and specific anabolic enzymes. *See* **catabolism.**

anaerobe   a cell that can live without oxygen. A strict anaerobe cannot live in the presence of oxygen.

anagenesis   phyletic evolution within a single lineage without subdivision or splitting; the opposite of cladogenesis.

analog computer   *See* **computer.**

analogous   referring to structures or processes that have evolved convergently, as opposed to the term *homologous* (*q.v.*). Analogous structures have similar functions but are different in evolutionary origin: e.g., the wing of a butterfly and of a bat.

analogous structures   structures that have the same function, but are of different evolutionary origin.

analogue   a compound related to, but slightly different structurally from, a biologically significant molecule, such as an amino acid (*see* **azaserine**), a pyrimidine or purine (*see* **base analogues**), or a hormone (*see* **ZR515**).

analysis of variance   a statistical technique that allows the partitioning of the total variation observed in an experiment among several statistically independent possible causes of the variation. Among such causes are treatment effects, grouping effects, and experimental errors. Checking the absence of an effect due to the treatment is often the purpose of the inquiry. The statistical test of the hypothesis that the treatment had no effect, is the F test or variance-ratio test. If the ratio of the mean square for treatments to the mean square for error exceeds a certain constant that depends on the respective degrees of freedom of the two mean squares at a chosen significance level, then the treatments are inferred to have been effective. Analysis of variance is particularly useful in judging which sources of uncontrolled variation in an experiment need to be allowed for in testing treatment effects.

anamnestic response   *See* **immune response, immunological memory.**

anaphase   *See* **mitosis.**

anaphase lag   delay in the movement of one or more chromosomes from the metaphase plate during anaphase, often resulting in chromosome loss (*q.v.*).

anaphylaxis   a systemic allergic or hypersensitivity response leading to immediate respiratory and/or vascular difficulties.

Anas platyrhyncha   the mallard duck, ancestor to the domestic or Pekin duck, *A.p. domestica.*

anastomosis   the joining of two or more tubular vessels to form a branching system.

anastral mitosis   the type of mitosis characteristically found in higher plants. A spindle forms, but no centrioles or asters are observed.

anautogenous insect   an adult female insect that must feed for egg maturation. *See also* **autogenous insect.**

Anderson disease   a hereditary glycogen storage disease of humans arising from a deficiency of the enzyme amylo-(1,4:1,6)-transglucosidase. Inherited as an autosomal recessive. Prevalence 1/500,000.

androdioecy   a sexual dimorphism in plants having bisexual and separate male individuals.

androecious   referring to plants having only male flowers.

androecium   the aggregate of the stamens in a flower.

androgen   any substance with male sex hormone activity. *See* **testosterone.**

androgenesis   development from a fertilized egg followed by disintegration of the maternal nucleus prior to syngamy so that the resulting individual possesses only paternal chromosomes and is haploid.

androgenic gland   a gland found in most crustaceans belonging to the subclass Malacostraca. When implanted into maturing females, the gland brings about masculinization of primary and secondary sex characters.

androgenital syndrome   masculinization of genitals in humans with XX chromosomal constitution, caused by overactivity of the adrenal cortex; inherited as an autosomal recessive trait.

anemia   a disorder characterized by a decrease in hemoglobin per unit volume of blood. In the case of *hemolytic* anemia there is a destruction of red blood cells. In the case of *hypochromic* anemia there is a reduction in the hemoglobin content of the erythrocyte.

anemophily   pollination by the wind.

anergy   the lack of an expected immune response.

aneucentric   referring to an aberration generating a chromosome with more than one centromere.

aneuploidy   a condition in which the chromosome number of an individual is not an exact multiple of the typical haploid set for the species. The suffex -somic is used in the nomenclature (e.g., monosomic, double trisomic, tetrasomic, etc.).

aneurin   vitamin B_1.

aneusomatic   referring to organisms whose cells contain variable numbers of individual chromosomes.

angiosperm   a flowering plant. *See* **classification.**

Angstrom unit   a unit of length equal to one ten-thousandth of a micron (10^{-4}; a micron being 10^{-6} meter); convenient for describing atomic dimensions; also equivalent to 10^{-1} millimicrons ($m\mu$) or 10^{-10} meter. Abbreviated A, A°, Å, Å.U., or A.U. Named in honor of the Swedish physicist Anders Jonas Ångstrom.

Animalia   the kingdom containing the multicellular animals. *See* **classification, Eukaryotes.**

animal pole   that pole of an egg which contains the most cytoplasm and the least yolk.

anion   a negatively charged ion.

anisogamy   that mode of sexual reproduction in which one of the sex cells, the egg, is large and nonmobile, whereas the other (the sperm) is small and motile. *See* **isogamy.**

anisotropy   a directional property of crystals and fibers having a high degree of molecular orientation. Anisotropic substances have different physical properties when tested in different directions. When a ray of plane polarized light passes through anisotropic material it is split into two rays polarized in mutually perpendicular planes. This property of anisotropic material is called *birefringence.* Muscle fibers and the metaphase spindle are examples of living materials exhibiting birefringence. Materials showing no birefringence are said to be *isotropic. See* **polarization microscope.**

ankylosing spondylitis   an arthritic disease resulting in a stiffening and bending of the spine; inherited as an autosomal dominant with reduced penetrance. Over 90% of patients with this disease carry the B27 HLA antigen. *See* **histocompatiblity.**

anlage   the embryonic primordium from which a specific part of the organism develops.

anneal   to subject first to heating then to cooling. In molecular genetics experiments, annealing is used to produce hybrid nucleic acid molecules containing paired strands, each from a different source. Heating results in the separation of the individual strands of any double stranded, nucleic acid helix, and cooling leads to the pairing of any molecules that have segments with complementary base pairs.

annidation   the phenomenon where a mutant is maintained in a population because it can flourish in an available ecological niche which the parent organisms cannot utilize. A wingless mutant insect, for example, might be poorly adapted in its ancestral habitat but able to live in tunnels and crevices that a winged form could not occupy.

annulate lamellae   paired membranes arranged in stacks and possessing annuli resembling those of the nuclear membrane. Annulate lamellae may serve to transfer nuclear material to the cytoplasm by the replication of the nuclear envelop and may be a mechanism for storing gene-derived information to be used for cytoplasmic differentiation during early embryogenesis. During insect oogenesis, annulate lamellae occur alongside nurse cell nuclei and germinal vesicles, and they are abundant in the ooplasm.

annulus   a ring. Applied to any of a number of ring-shaped parts of animals and plants. Used in cytology to refer to the ring-shaped nuclear pores.

anode   the positive electrode; the electrode to which negative ions are attracted.

anodontia   the congenital absence of teeth. Hypodontia is currently the preferred term.

Anopheles   a genus containing about 150 species of mosquitoes, many of which are of medical importance. *A. quadrimaculatus, A. atroparvus, A. albimanus, A. stephensi,* and *A. pharoensis* are malaria vectors. The cytotaxonomy of this genus has been intensively studied. Polytene chromosomes occur in larval salivary gland cells and in ovarian nurse cells of adults.

Anser anser   the Gray Lag goose, a favorite experimental organism for students of animal behavior and its hereditary components.

antagonist   a molecule that bears sufficient structural similarity to a second molecule to compete with that molecule for binding sites on a third molecule. *See* **competition.**

antenatal   before birth; during pregnancy.

antennae   the first paired appendages on the head of arthropods.

antepenultimate   third from the end.

anterior   toward the head end of an animal.

anther   the terminal portion of a stamen bearing pollen sacs.

antheridium   the male gametangium of algae, fungi, bryophytes, and pteridophytes.

anthesis   the time of flowering.

anthocyan   a red, violet, or blue glycosidic pigment occurring in solution in the cell-sap of flowers, fruits, stems, and leaves. These pigments are responsible for some autumn coloring of leaves and the tinting of young shoots and buds in the spring. *See* **pelargonin.**

anthropocentrism   also called *anthropomorphism.* **1.** explanation of natural phenomena or processes in terms of human values. **2.** assuming humans to be of central importance in the universe or ultimate end of creation. **3.** ascribing human characteristics to a nonhuman organism.

anthropoid   designating the great apes of the family Pongidae, including the gibbons, orangutans, gorillas, and chimpanzees.

anthropometry   the science that deals with the measurements of the human body and its parts.

antiauxin   a molecule that competes with an auxin (*q.v.*) for auxin receptor sites. A well known antiauxin is 2,6-dichlorophenoxyacetic acid.

antibiotic   a bacteriocidal or bacteriostatic substance produced by certain microorganisms. *See* **actinomycin D, ampillicin, chloramphenicol, cyclohexamide, erythromycin, neomycin, novobiocin, penicillin, puromycin, streptomycin,** and **tetracycline.**

antibody   a protein produced by lymphoid cells (plasma cells) in response to foreign substances (antigens) and capable of coupling specifically with its homologous antigen (the one that stimulated the immune response) or with substances that are chemically very similar to that same antigen. *See* **immunoglobulin.**

anticlinal   referring to a layer of cells cutting the circumference of a cylindrical plant organ at right angles. *Compare* **periclinal.**

anticoding strand   *See* **sense strand.**

anticodon   the triplet of nucleotides in a transfer RNA molecule which associates by complementary base pairing with a specific triplet (codon) in the messenger RNA molecule during its translation in the ribosome.

anti-diuretic hormone   *See* **vasopressin.**

antigen   **1.** a foreign substance that, upon introduction into a vertebrate animal, stimulates the production of homologous antibodies; a complete antigen or immunogen. A complex antigenic molecule may carry several antigenically distinct sites (determinants). **2.** a substance that is chemically similar to certain parts of an immunogen and can react specifically with its homologous antibody, but is too small to stimulate antibody synthesis by itself; an incomplete antigen or hapten (*q.v.*).

antigenic conversion   **1.** appearance of a specific antigen(s) on cells as a consequence of virus infection. **2.** antibody-induced shift in certain protozoans or parasites to a new cell-surface antigen and cessation of expression of another antigen; resulting from switches in gene activity; synonymous in this sense with *serotype transformation; compare with* **antigenic modulation.**

antigenic determinant   a small chemical complex (relative to the size of the macromolecule or cell of which it is a part) that determines the specificity of an antigen-antibody interaction; an epitope. *See* **paratope.**

antigenic modulation   suppression of cell-surface antigens in the presence of homologous antibodies.

antihemophilic globulin   *See* **hemophilia A.**

antilogarithm   the number corresponding to a given logarithm. Example: 2 is the logarithm of 100 (to the base 10), and therefore 100 is the antilogarithm of 2.

antimetabolite   in general, a molecule that functions as an antagonist or metabolic poison.

antimitotic agent   any compound that suppresses the mitotic activity of a population of cells.

antimongolism   a syndrome of congenital defects accompanying hypoploidy for chromosome 21. Such children carry a normal chromosome 21 and one which contains a large deficiency. From the standpoint of chromosome balance, mongolism and anti-mongolism represent reciprocal phenomena.

antimorph   **1.** enantiomer. **2.** a mutant allele that acts in a direction opposite to the normal allele.

antimutagen   a compound (generally a purine nucleoside) that antagonizes the action of mutagenic agents (generally purines or purine derivatives) on bacteria. Some anti-mutagens reduce the spontaneous mutation rate.

antiparallel **1.** the opposite strand orientations with which all nucleic acid duplexes (DNA-DNA, DNA-RNA, or RNA-RNA) associate; if one strand is oriented left to right 5′ to 3′, the complementary strand is oriented left to right 3′ to 5′ antiparallel to it. **2.** two segments of a polypeptide chain may lie antiparallel with one end N-terminal to C-terminal and the other end C-terminal to N-terminal, respectively.

antipodal one of a group of three haploid nuclei formed during megasporogenesis. In corn these divide mitotically and eventually form a group of 20 to 40 antipodal cells, which aid in nourishing the young embryo.

antirepressor in lambda phage, the dimeric protein product of gene *cro* (control of *repres*sor and *o*ther things) that prevents the synthesis of repressor and inhibits the expression of early genes (those involved in replication of the phage genome), allowing the phage to enter the late stage of development leading to cell lysis.

Antirrhinum majus the snapdragon; a classic species for the study of multiple alleles.

antisense strand *See* **coding strand.**

antiserum a serum containing antibodies.

antisigma factor a protein that is synthesized during infection of *Escherichia coli* by T4 phage and prevents recognition of initiation sites by the sigma factor of RNA polymerase.

antispecificity factor *See* **antisigma factor.**

anti-termination factor a protein that allows an RNA polymerase to ignore signals to stop translation at particular sites in DNA molecules.

antitoxin an antibody that neutralizes a specific toxin.

anucleate without a nucleus.

anucleolate without a nucleolus.

anucleolate mutation a mutation that lacks the nucleolus organizer. Such mutations have been observed in *Drosophila, Chironomus,* and *Xenopus.*

AP endonuclease any enzyme that cleaves DNA on the 5′ side of either apurinic or apyrimidinic (AP) sites. For example, uracil-*N*-glycosylate removes uracil bases from DNA; AP endonuclease breaks the phosphodiester backbone, thereby allowing the excision-repair pathway to operate. *See* **apurinic acid.**

aphasic lethal a lethal mutation in which death may occur randomly throughout development.

apholate an aziridine mutagen (*q.v.*) also used as an insect chemosterilant.

apiary a place where colonies of bees are kept in hives for the production of honey and beeswax and for the pollination of fruit trees.

Apis mellifera   the domesticated honeybee. There are about 25 races recognized, but the ones most commonly used are the Italian, Caucasian, Carniolan, and German races. Reproduction is by arrhenotokous parthenogenesis (*q.v.*).

aplasia   the failure of development of an organ.

apnea, drug-induced   a hereditary disease in man arising from a deficiency of the enzyme pseudocholinesterase.

apodeme   an invagination of the insect exoskeleton which serves for muscle attachment.

apoenzyme   the protein portion of a holoenzyme (*q.v.*), which requires a specific coenzyme for its functioning.

apoferritin   *See* **ferritin.**

apoinducer   a protein that, when bound to DNA, activates transcription by RNA polymerase.

apomict   an organism produced by apomixis.

apomixis   asexual reproduction in plants through agamospermy (*q.v.*) or vegetative reproduction. *See* **amphimixis.**

apomorphic character   one derived from an ancestral character. For example, in flowering plants, whorled floral parts are considered to be an apomorphic character derived from the more primitive spiral arrangement. *Contrast with* **plesiomorphic character.**

aporepressor   a regulatory protein that, when bound to its repressor, inhibits transcription of genes in an operon (*q.v.*).

aposematic coloration   warning coloration (*q.v.*).

apospory   the development in plants of a diploid embryo sac by the somatic division of a nucellus or integument cell; a form of agamospermy (*q.v.*).

apostatic selection   frequency-dependent selection for the rarer morph (e.g., in Batesian mimicry, the rarer the mimic, the greater its selective advantage). *See* **mimicry.**

a posteriori   descriptive of biometric or statistical tests in which the comparisons of interest are unplanned and become evident only after experimental results are obtained.

apposition   a process of increasing volume by adding new material to old at the periphery.

a priori   descriptive of biometric or statistical tests in which the comparisons of interest are determined on theoretical grounds in advance of experimentation.

aptation   any character currently subject to selection whether its origin can be ascribed to selective processes (adaptation) or to processes other than selection or selection for a different function (exaptation).

apterous   without wings.

apterygote   any of the primitive insects belonging to the cohort Apterygota which contains the orders Thysanura (silverfish) and Diplura (campodeids and japigids). The species are wingless and possess abdominal styli. *See* **classification.**

aptitude   the specific physiological state of a lysogenic bacterium during which it can produce infectious bacteriophages when exposed to an inducing agent.

apurinic acid (AP)   a polynucleotide from which purine bases have been removed by hydrolysis of the *N*-glycosidic bonds, leaving the sugar-phosphate backbone where the purines once were. The symbol AP represents either apurinic or apyrimidinic. *See* **AP endonuclease.**

apyrene sperm *See* **sperm polymorphism.**

A.R. analytical reagent.

Arabidopsis thaliana an herb in which gene dosage and interaction have been studied extensively.

Arachnida a class of Arthropoda containing the spiders, scorpions, opilionids, and mites. *See* **classification.**

Araldite trademark for a plastic commonly used for embedding tissues for electron microscopy.

Araneida the subclass of Arthropoda containing the spiders.

arbovirus a virus which replicates in both an arthropod and a vertebrate host.

archegonium the female sex organ of liverworts, mosses, ferns, and most gymnosperms.

archenteron the primitive digestive cavity of any metazoan embryo. It is formed by gastrulation.

Archeopteryx a genus of Jurassic fossil birds related to *Archeornis.*

Archeornis a genus of Jurassic birds. Fossils show teeth, three free digits with claws on the wings, and a long feathered tail containing numerous vertebrae. In many respects *Archeornis* is a feathered reptile.

Archeozoic the older of the two subdivisions of the Precambrian. *See* **Proterozoic;** *see also* **geologic time divisions.**

arcuate arched.

arg arginine. *See* **amino acid.**

arginine-urea cycle *See* **ornithine cycle.**

arithmetic mean an average; the number found by dividing the sum of a series by the number of items in the series.

arithmetic progression a series of numbers in which each number is larger (or smaller) than the number that precedes it by a constant value.

aromatic designating a compound having a closed ring of carbon atoms. *See* **aliphatic, heterocyclic.**

aromatic amino acid an amino acid whose side chain includes a derivative of a phenyl group (phenylalanine, tyrosine, or tryptophan).

Arrhenius plot a plot of the logarithm of the growth rate versus the reciprocal of the absolute temperature. *See* **generation time.**

arrhenotokous parthenogenesis the phenomenon by which unfertilized eggs produce haploid males and fertilized eggs produce diploid females.

arrhenotoky arrhenotokous parthenogenesis.

arthritis an autoimmune disease of the joints involving an attack by the body's immune system upon the synovial membranes.

Arthropoda the largest phylum in the animal kingdom in number of species. It contains the Chelicerata and Mandibulata. *See* **classification.**

Arthus reaction localized inflammation and swelling caused by a hypersensitivity reaction. This type of reaction was first described in 1903 by N. M. Arthus and M. Breton.

Articulata a division of the invertebrates containing segmented, coelomate protostomes. *See* **classification.**

artifact any structure which is not typical of the actual specimen, but which results from cytological processing, post mortem changes, etc.

artificial insemination (AI) the placement of sperm into a female reproductive tract or the mixing of male and female gametes by other than natural means. In AID, the sperm are provided by a donor other than the woman's husband; in AIH, the sperm are those of her husband.

artificial parthenogenesis induction of the development of an unfertilized egg by chemical or physical stimulation.

artificial selection the choosing by man of the genotypes contributing to the gene pool of succeeding generations of a given organism.

Ascaris megalocephala *Parascaris equorum* (*q.v.*) (the preferred nomenclature).

ascertainment bias *See* **proband method.**

Aschelminthes a phylum containing the pseudocoelomate species with an anterior mouth, posterior anus, and straight digestive tube. *See* **classification.**

Ascobolus immersus an ascomycete fungus that is convenient for tetrad analysis (*q.v.*). *See* **classification.**

ascogenous hypha a hypha that develops from the surface of an ascogonium after plasmogamy (*q.v.*). The hypha therefore contains nuclei of both mating types. After crozier formation (*q.v.*) such hyphae produce asci.

ascogonium in fungi, a female cell that receives haploid nuclei from an antheridium.

ascomycete a fungus of class Ascomycetes which produce ascospores. *See* **classification.**

ascorbic acid vitamin *C* (*q.v.*). Scurvy is a deficiency disease resulting from inadequate ascorbic acid in the diet.

ascospore a meiospore (*q.v.*) contained within an ascus sac.

ascus a sac containing ascospores. The fact that all the products of a meiotic division are contained in an ascus in some fungi makes tetrad analysis (*q.v.*) possible.

asexual reproduction reproduction without sexual processes; vegetative propagation.

Ashkenazi Jews *See* **Tay-Sachs disease.**

A site *See* **translation.**

A-site-P-site model *See* **translation.**

asn asparagine. *See* **amino acid.**

asp aspartate; aspartic acid. *See* **amino acid.**

asparagine *See* **amino acid.**

Asparagus officinalis the asparagus.

aspartic acid *See* **amino acid.**

Aspergillus nidulans a fungal species belonging to the Deuteromycota (*see* **classification**). Parasexuality (*q.v.*) was discovered in this species.

association the joint occurrence of two genetically determined characteristics in a population at a frequency that is greater than expected according to the product of their independent frequencies.

associative overdominance linkage of a neutral locus to a selectively maintained (balanced) polymorphism that increases heterozygosity at the neutral locus, provided that linkage disequilibrium exists between the loci in question. *Compare with* **hitchhiking effect.**

associative recognition requirement for the initiation of an immune response of the simultaneous recognition by T lymphocytes of the antigen in association with another structure, normally a cell surface alloantigen encoded within the major histocompatibility complex.

assortative mating sexual reproduction in which the pairing of male and female is not random, but involves a tendency for males of a particular kind to breed with females of a particular kind. If the two parents of each pair tend to be more (less) alike than is to be expected by chance, then positive (negative) assortative mating is occurring.

assortment the random distribution to the gametes of different combinations of chromosomes. Each 2N individual has a paternal and a maternal set of chromosomes forming N homologous pairs. At anaphase of the first meiotic division, one member of each chromosome pair passes to each pole, and thus the gametes will contain one chromosome of each type, but this chromosome may be of either paternal or maternal origin.

aster *See* **mitotic apparatus.**

asthenic frail.

asynapsis the failure of homologous chromosomes to pair during meiosis. *Contrast with* **desynapsis.**

atavism the reappearance of a character after several generations, the character being caused by a recessive gene or by complementary genes. The aberrant individual is sometimes termed a "throw-back."

Atebrin a trade name for quinacrine (*q.v.*). Also spelled Atabrine.

ateliosis retarded growth resulting in a human of greatly reduced size but normal proportions. Such midgets generally show a marked deficiency in pituitary growth hormone. *See* **pituitary dwarfism.**

(A+T)/(G+C) ratio the ratio between the amount of adenine-thymine pairs and the amount of guanine-cytosine pairs in a given DNA sample.

atom the smallest particle of an element that is capable of undergoing a chemical reaction. *See* **chemical element.**

atomic mass the mass of a neutral atom of a nuclide, usually expressed in terms of atomic mass units.

atomic mass unit one-twelfth the weight of a ^{12}C atom; equivalent to 1.67×10^{-24} g.

atomic number the number of protons in the nucleus, or the number of positive charges on the nucleus; symbolized by Z. It also represents the number of orbital electrons surrounding the nucelus of a neutral atom.

atomic weight the weighted mean of the masses of the neutral atoms of an element expressed in atomic weight units.

ATP adenosine triphosphate. *See* **adenosine phosphate.**

atresia congenital absence of a normal passageway.

atrichia hairlessness. In the domestic dog the condition is inherited as an autosomal dominant. Homozygotes are stillborn.

attached X-chromosome in *Drosophila,* an aberrant compound X-chromosome (*q.v.*) possessing a median centromere. *See* **detached X, double X, matrocliny.**

attachment point (ap) a hypothetical analog of the centromere on chloroplast DNA in *Chlamydomonas.* Chloroplast genes may be mapped with respect to the attachment point.

attenuate **1.** in bacteriology, to weaken or reduce in virulence. **2.** in radiation physics, to reduce the intensity of a beam of radiation by passing it through some shielding material.

attenuation **1.** in molecular genetics, a mechanism for regulating the action of certain bacterial operons involving premature termination of transcription. The ribosome functions as the equivalent of a positive regulator protein. When starved for an amino acid (e.g., tryptophan), the ribosome stalls in the leader and allows formation of a hairpin secondary structure that does not interfere with RNA polymerase activity. When tryptophan is not limiting, ribosomes do not stall in the leader and an alternative hairpin forms in the leader that causes RNA polymerase to dissociate. The corresponding amino acyl-tRNA(s) provide the equivalent of a corepressor signal by preventing ribosome stalling. *See* **leader sequence.** **2.** in immunology, the process whereby a pathogenic microorganism loses its virulence. Pathogens or their toxic products have been successfully attenuated for use as immunogens by aging, heating, drying, passage through unnatural hosts, chemical treatments, etc.

attenuator region a region within the leader sequence (*q.v.*) of a gene where most RNA polymerase molecules stop elongation of the transcript unless a specific molecular signal (antitermination factor, *q.v.*) is received.

att sites loci on a phage and the chromosome of its bacterial host where recombination integrates the phage into or excises it from the bacterial chromosome.

audiogenic seizure convulsions induced by sound. Certain strains of mice, rats, and rabbits are especially prone to such seizures.

aureomycin *See* **tetracycline.**

Australian one of the six primary biogeographic realms (*q.v.*), comprising Australia, the Celebes, New Guinea, Tasmania, New Zealand, and the oceanic islands of the South Pacific.

Australopithecine designating a member of a group of man-like apes, fossils of which have been found in South Africa. These primates had an average brain volume of 500 cc, and they lived in the early Pleistocene, some 700,000 years ago. *See* **Olduvai Gorge.**

Australopithecus afarensis a group of 3–4 million year old fossil hominids discovered in the Afar region of Ethopia and represented by the famous fossil "Lucy" (a 40% complete skeleton found by Donald Johanson in 1977). Some anthropologists argue that this group should be included in the species *A. africanus;* others argue that "Lucy" represents the earliest appearance of the genus *Homo.*

Australopithecus africanus an extinct species of man-like apes from the early Pleistocene epoch of Africa.

autapomorphic character **1.** a derived character evolved from a plesiomorphic character state in the immediate ancestor of a single species **2.** uniquely derived characters shared by several synapomorphous taxa. For example, hair was an autapomorphy of the first mammalian species and is also a synapomorphy of all mammals.

autarchic genes in mosaic organisms these genes are not prevented from manifesting their phenotypic effects by gene products diffusing from genetically different neighboring tissues, whereas *hyparchic* genes are so inhibited.

autocatalysis the promotion of a reaction by its product.

autochthonous pertaining to a species that has evolved within the region where it is native.

autoclave a chamber for sterilizing equipment or materials with high pressure steam.

autogamy in *Paramecium,* a process of self-fertilization resulting in homozygosis. In single individuals, the two micronuclei each undergo meiosis and seven of the eight resulting nuclei degenerate. The remaining haploid nucleus divides mitotically and the two identical nuclei fuse. The fusion nucleus gives rise to the micro- and macronuclei of the paramecium and its progeny. Autogamy is therefore a mode of nuclear reorganization that constitutes an extreme form of inbreeding. In *P. aurelia,* autogamy occurs spontaneously at regular intervals.

autogenous (autogeneic) self-derived; autogenous mosquitoes can produce eggs without a blood meal. *See* **anautogenous insect.**

autogenous control regulation of gene expression by its own product either inhibiting (negative autogenous control) or enhancing (positive autogenous control) its activity. For example in *E. coli,* AraC protein (the regulatory factor for the arabinose operon) controls its own synthesis by binding to the AraC promoter.

autograft the transplantation of a living piece of tissue from one site to another on the same animal.

autoimmune disease any pathological condition resulting from an individual's immune response to its own cells or tissues; may involve humoral immunity (antibodies) and/or cell-mediated immunity (T cell response).

autoimmunity the condition produced by an autoimmune disease.

autologous referring to a graft from one region to another on the same animal.

automimic a palatable individual which is an automatic mimic of members of the same species that are unpalatable to predators. *See* **automimicry.**

automimicry the phenomenon, seen for example in the Monarch butterfly (*q.v.*), in which the species has a polymorphism in terms of its palatability to predators. The polymorphism arises from the different food plants chosen by the ovipositing female. Most members of the species are rendered unpalatable because they feed as larvae upon plant species rich in substances toxic to birds. Those insects that feed on plants containing no toxins are palatable. However, such insects mimic perfectly the more abundant, unpalatable members of the species.

automutagen any mutagenic chemical formed as a metabolic product by an organism.

autonomous controlling element a controlling element, apparently having both receptor and regulator functions, that enters a gene and makes it an unstable mutant.

autophagic vacuole an enlarged lysosome containing mitochondria and other cellular organelles in process of being digested.

autophene a phenotype due to the genetic constitution of the cells showing it. Transplantation of such mutant cells to a wild-type host does not modify their mutant phenotype.

autopolyploid a polyploid that originates by the multiplication of one basic set of chromosomes.

autoradiograph a photographic picture showing the position of radioactive substances in tissues, obtained by coating a squash preparation or a section with a photographic emulsion in the dark, and subsequently developing the latent image produced by the decay radiations. In the case of colony hybridization (*q.v.*) a filter containing the radioactive chimeric vectors is taken to a dark room, placed in an x-ray film holder, and covered with a sheet of x-ray film. The film is then left to expose for several hours or a few days before it is processed. The position of the silver grains on the film marks the location of the colonies of interest. *Also see* **DNA fiber autoradiography.**

autoradiographic efficiency the number of activated silver grains (produced in a photographic emulsion coating a section) per 100 radioactive disintegrations occurring within the tissue section during the exposure interval.

autoradiography a technique for localizing radioactively-labeled molecules by applying a photosensitive emulsion to the surface of a radioactive specimen.

autoselection the process whereby a genetic element tends to increase in frequency by virtue of the nature of its transmission, even though it has no effect on the viability, fecundity, or fertility of the individual that bears it.

autosexing the use of sex-linked genes with obvious phenotypic effects to permit the identification by external inspection of the sex of immature organisms (larval silkworms or chicks, for example) before sexual dimorphic traits become obvious.

autosome a chromosome other than a sex chromosome.

autosyndesis *See* **allosyndesis.**

autotetraploid an autopolyploid with four similar genomes. If a given gene exists in two allelic forms *A* and *a*, then five genotypic classes can be formed in an autotetraploid: *AAAA* (quadruplex), *AAAa* (triplex), *AAaa* (duplex), *Aaaa* (simplex), and *aaaa* (nulliplex).

autotroph a microorganism that can build its own macromolecules and obtain energy from very simple, inorganic molecules, such as ammonia and carbon dioxide.

autozygote an individual homozygous at a given locus whose two homologous genes are identical by descent, in that both are derived from the same gene in a common ancestor. *See* **allozygote.**

auxesis growth in size by increase in cell volume without cell division.

auxin a phytohormone promoting longitudinal growth. Natural auxins are indole derivatives biosynthesized from tryptophan. The most common auxin is indole acetic acid (*q.v.*). *See* **antiauxin.**

auxocyte a cell whose nucleus is destined to enter meiotic prophase; a primary oocyte, primary spermatocyte, megasporocyte, or microsporocyte.

auxotroph a mutant microorganism that can be grown only upon minimal medium that has been supplemented with growth factors not required by wild-type strains.

Avena the genus to which the various species of oats belong. The most commonly cultivated oat is *A. sativa.*

Avena test a technique using the curvature of *Avena* coleoptiles as a bioassay for auxins.

average life the average of the individual lives of all the atoms of a particular radioactive substance. It is 1.443 times the radioactive half-life (*q.v.*).

avian leukosis *See* **leukemia.**

avian myeloblastosis virus an oncogenic RNA virus.

avidity the total combining power of an antibody with an antigen. It involves both the affinity of each binding site and the number of binding sites per antibody and antigen molecule.

Avogadro's number the number of atoms (6.025×10^{23}) in one gram atomic weight of an element; also the number of molecules in the gram molecular weight of a compound.

awn a stiff, bristle-like appendage occurring on the flowering glumes of grasses and cereals.

axenic growth of organisms of a given species in the complete absence of members of any other species.

axil the angle between the upper surface of a leaf and the stem that bears it.

axis one of the principal imaginary lines which divides a body or figure symmetrically.

axolotl the aquatic larval stage of salamanders of the genus *Ambystoma* which does not metamorphose but breeds while a larva.

axon the long process of a nerve cell, normally conducting impulses away from the nerve cell body.

axoneme a shaft of microtubules extending the length of a cilium, flagellum or pseudopod. Axonemes from all cilia and flagella (including sperm tails) contain the same "9 + 2" arrangement of microtubules. In the center of each axoneme are two singlet microtubules that run the length of the shaft. The central tubules are surrounded by a circle of doublet microtubules, each consisting of an A and B subfiber. Each A subfiber has longitudinally repeating pairs of arm-like projections that contain dynein (*q.v.*). *See* **flagellum.**

axoplasm the cytoplasm contained in axons.

axopodia rigid, linear cellular projections composed mostly of microtubules found in species belonging to the Actinopoda. *See* **classification,** Protoctista.

5-azacytidine an analogue of cytidine in which a nitrogen atom is substituted for a carbon in the number 5 position of cytosine (*q.v.*). The analogue is incorporated into newly synthesized DNA, and such DNA is undermethylated. Since a reduction in the number of methyl groups attached to genes is associated with an increase in their transcriptional activities, 5-azacytidine can switch on certain genes. For example, patients given the drug may start making fetal hemoglobin, which implies that their gamma genes have been switched on. *See* **hemoglobin, hemoglobin genes.**

azaguanine a purine analogue.

azaserine an antibiotic that inhibits purine biosynthesis and produces chromosomal aberrations.

$$NH{=}CH{-}\underset{\underset{O}{\|}}{C}{-}O{-}CH_2{-}\underset{\underset{NH_2}{|}}{CH}{-}COOH$$

azathioprene an immunosuppressive drug (*q.v.*).

34

azauracil   a pyrimidine analogue.

aziridine mutagen   a mutagenic alkylating agent containing aziridinyl groups ($-N\overset{\displaystyle CH_2}{\underset{\displaystyle CH_2}{\big|}}$).

See **apholate, ethylenimine, hemel, hempa, TEM, tepa, tetramine.**

azoospermia   absence of motile sperm in the ejaculate.

Azotobacter   a genus of nonsymbiotic nitrogen-fixing bacteria. *See* **Rhizobium.**

azure B   a metachromatic, basic dye used in cytochemistry.

B

B   symbol for boron.

B₁, B₂, B₃, etc.   the first, second, third, etc., backcross generations. The first backcross is made by mating an individual with one of its parents or with an individual of that identical genotype. The offspring produced belong to the B_1 generation. The second backcross is made by crossing B_1 individuals again with individuals of genotype identical to the parent referred to in the first backcross, etc.

Bacillus   a genus of rod-shaped bacteria. *B. subtilis* is a Gram-positive, spore-forming, soil bacillus that grows readily in a chemically defined medium and undergoes genetic exchange by transformation and transduction (*q.v.*). While it has been studied in the most detail, considerable work has also been done on *B. cereus* and *B. megatherium*.

backbone   the supporting structure of atoms in a polymer from which the side chains project.

backcross   a cross between an offspring and one of its parents or an individual genetically identical to one of its parents.

backcross parent   that parent of a hybrid with which it is again crossed or with which it is repeatedly crossed. Backcross may involved individuals of genotype identical to the parent rather than the parent itself.

background constitutive synthesis   the occasional transcription of genes in a repressed operon due to a momentary dissociation of the repressor that allows a molecule of RNA polymerase to bind to its promoter and initiate transcription. Sometimes called "sneak synthesis."

background genotype   the genotype of the organism in addition to the genetic loci primarily responsible for the phenotype under study.

background radiation   ionizing radiation arising from sources other than that under study. Background radiation due to cosmic rays and natural radioactivity is always present, and there may also be background radiation due to man-made contaminating radiation.

back mutation   reverse mutation.

bacteriochlorophyll   a compound related to chlorophyll found in a few autotrophic bacteria.

bacteriocins   proteins synthesized by various bacterial species that are toxic when

36

absorbed by bacteria belonging to sensitive strains. Resistance to and the ability to synthesize bacteriocins are controlled by plasmids. *Escherichia coli* strains produce bacteriocins called colicins (*q.v.*). Bacteriocins from *Pseudomonas aeruginosa* are called pycocins.

bacteriophage a virus whose host is a bacterium; commonly called *phage*. Below are listed some common bacteria and their viral parasites:

Eschericheria coli	P_1, P_2, P_4, λ, Mu 1, N_4, ϕX174, R_{17}, T_1 through T_6.
Salmonella typhimurium	P_{22}
Corynebacterium diphtheriae	β
Bacillus subtilis	SP_{82}, ϕ 29
Shigella dysenteriae	P_1, P_2, P_4

Bacterial viruses show extreme variations in complexity. For example, the RNA phage R_{17} has a genome size of 1.1×10^6 daltons, whereas the DNA of the T_4 phage weighs 130×10^6 daltons. *See* **plaque, temperate phage, virulent phage.**

bacteriophage packaging insertion of recombinant bacteriophage lambda DNA into *E. coli* for replication and encapsidation into plaque-forming bacteriophage particles.

bacteriostatic agent a substance that prevents the growth of bacteria without killing them.

balanced lethal system a strain of organisms bearing nonallelic recessive lethal genes, each in a different homologous chromosome. When interbred such organisms appear to breed true, because one-half of the progeny are homozygous for one or the other lethal gene and die prior to their detection. The surviving progeny, like their parents, are heterozygous for the lethal genes.

balanced polymorphism genetic polymorphism maintained in a population because the heterozygotes for the alleles under consideration have a higher adaptive value than either homozygote.

balanced selection selection favoring heterozygotes that produces a balanced polymorphism (*q.v.*).

balanced stock a genetic stock which, though heterozygous, can be maintained generation after generation without selection. Such stocks may contain balanced lethal genes; or a recessive lethal gene, which kills hemizygous males, combined with a nonallelic recessive gene, which confers sterility on homozygous females.

balancer haltere.

Balbiani chromosome a polytene chromosome. So called because such banded chromosomes were first discovered by E. G. Balbiani in *Chironomus* larvae in 1881.

Balbiani ring a giant RNA puff present on a polytene chromosome of a salivary gland cell during a significant portion of larval development. The largest and most extensively studied Balbiani rings are on chromosome 4 of *Chironomus tentans*. See illustration on page 38.
The transcription product of one of these puffs (BR 2) is a 75S RNA, which encodes the message for a giant polypeptide of the saliva. Balbiani rings contain thousands of DNA loops upon which mRNAs are being transcribed. These combine with proteins to form RNP granules (Balbiani ring granules) that eventually pass into the cytoplasm through nuclear pores.

baldness in humans, early baldness is thought to be due to an autosomal gene behaving as a dominant in males, but as a recessive in females (a sex-influenced trait).

Bal 31 exonuclease an exonuclease from *Brevibacterium albidum* that digests both 5′ and 3′ strands of linear DNA from each end; used to introduce deletions into cloned DNA molecules.

Balbiani ring

BR 1

BR 2

BR 3

Chromosome 4 in the salivary glands of *Chironomus tentans* with the BR 2 band in an unpuffed *(above)* and in a puffed stage *(below)*. The position of the BR 2 band is indicated by an arrow.

Balzer freeze fracture apparatus   *See* **freeze etching.**

Bam HI   *See* **restriction endonuclease.**

band   a vertical stripe on a polytene chromosome resulting from the specific association of a large number of homologous chromomeres at the same level in the somatically paired bundle of chromosomes.

BAP   6-benzylaminopurine (*q.v.*).

Bar   a sex-linked mutation in *Drosophila melanogaster* (symbolized by *B*) due to a tandem duplication (*q.v.*) of segment 16A. Its genetic locus is 57.0 The mutation results in a reduction in the number of facets in the compound eye. *Bar* reverts to wild type by unequal crossing over. In analyzing the *Bar* phenomenon Sturtevant discovered position effects (*q.v.*).

barley   *See* ***Hordeum vulgare.***

barn   a unit of area of the atomic nucleus, numerically equal to 10^{-24}cm^2, used as a measure of the probability of capture in a nuclear reaction. *See* **capture cross section.**

Barr body   the condensed single X-chromosome seen in the nuclei of somatic cells of female mammals; named for M. Barr its discover. *See* **dosage compensation, late replicating X-chromosome, Lyon hypothesis, sex chromatin.**

basal body (granule)   a structure generally composed of a ring of nine triplet microtubules surrounding a central cavity, found at the base of cilia. The structure of flagella is similar, but generally composed of a ring of nine doublet microtubules surrounding two singlet microtubules ("9 + 2" arrangement).

base analogue   a purine or pyrimidine base that differs slightly in structure from the normal base. Some analogues may be incorporated into nucleic acids in place of the normal constitutent. *See* **aminopurine, azaguanine, azauracil, mercaptopurine.** Analogues of nucleosides behave similarly. *See* **5-bromodeoxyuridine.**

basement membrane   a delicate acellular membrane that underlies most animal epithelia.

base pair   a pair of hydrogen-bonded nitrogenous bases (one purine and one pyrimidine) that join the component strands of the DNA double helix. *See* **deoxyribonucleic acid.**

base-pairing rules   the rule that adenine forms a base pair with thymine (or uracil) and guanine with cytosine in a double stranded nucleic acid molecule.

base-pair ratio   *See* **$(A+T)/(G+C)$ ratio.**

base-pair substitution   a type of lesion in a DNA molecule that results in a mutation. There are two subtypes. In the case of *transitions* one purine is substituted by the other or one pyrimidine by the other, and so the purine-pyrimidine axis is preserved. In the case of *transversions* a purine is substituted by a pyrimidine or *vice versa,* and the purine-pyrimidine axis is reversed.

basic amino acids   amino acids that have a net positive charge at neutral *p*H. Lysine and arginine bear positively charged side chains under most conditions.

basic dye   any organic cation that binds to and stains negatively charged macromolecules, such as nucleic acids.

basic number   the lowest haploid chromosome number in a polyploid series (symbolized by *x*). The monoploid number.

basikaryotype   the karyotype corresponding to the monoploid.

Basilarchia archippus   the palatable Viceroy butterfly, which mimics the Monarch. *See* **mimicry.**

basitarsus   metatarsus (*q.v.*).

basophilic   designating an acidic compound that readily stains with basic dyes.

Batesian mimicry   *See* **mimicry.**

B cell   B lymphocyte. *See* **lymphocyte.**

B chromosomes   supernumeray chromosomes that are not duplicates of any of the members of the basic complement of the normal or "A" chromosomes. B chromosomes are devoid of structural genes. During meiosis, B chromosomes never pair with A chromosomes, and B's show an irregular and nonmendelian pattern of inheritance. B chromosomes are extremely widespread among flowering plants and have been extensively studied in rye and maize. B chromosomes are believed to perpetuate and spread themselves in populations because they replicate faster than A chromosomes.

bee dances   circling and waggling movements (*Rundtanzen* and *Schwanzeltanzen*) performed by worker bees to give their hive mates information as to the location of a new source of food.

Beer-Lambert law   the law that the absorption of light by a solution is a function of the concentration of solute. It yields a relation commonly used in photometry: $E = \log_{10} I_0/I = \text{kcb}$, where E = optical density; I_0 = the intensity of the incident monochromatic light; I = the intensity of the transmitted light; k = a constant determined by the solvent, wavelength, and temperature; c = the concentration of absorbing material in moles per liter; and b = the thickness in cm of the solution traversed by the light.

behavioral isolation   a prezygotic (premating) isolating mechanism in which two allopatric species refuse to mate because of differences in courtship behavior; ethological isolation.

behavior genetics   a branch of genetics that concerns the inheritance of forms of behavior such as courtship displays, nest building, etc., in lower animals and intelligence and personality traits in humans. Many traits that are of interest to behavioral scientists are quantitative characters (*q.v.*).

Bellevalia romana   a lily used in cytology because it possesses a small number (N = 4) of large chromosomes.

Belling's hypothesis   a hypothesis which assumes that crossing over does not require breakage and reunion. According to this theory the genes replicate first, and next the intergenic connections are formed between the adjacent newly synthesized genes. On rare occasions the interlinks form between genes, one of maternal and one of paternal origin. Crossovers arising by this mechanism would occur only between newly synthesized genes. Belling's hypothesis is a foreruner of the copy choice hypothesis (*q.v.*). *Contrast with* **breakage and reunion.**

Bence-Jones protein   a highly homogeneous urinary protein consisting of immunoglobulin light chains (or fragments thereof) produced by malignancies of antibody secreting cells (plasmacytomas or myelomas).

benign   *See* **neoplasm.**

benign subtertian malaria   *See* **malaria.**

benzene   the simplest of the aromatic organic compounds.

6-benzylaminopurine   a synthetic cytokinin.

Bergmann's rule   the generalization that geographical races of a warm-blooded species possess smaller body sizes in the warmer parts of the range and larger body sizes in the colder parts of the range.

beta chain   one of the two polypeptides found in adult hemoglobin.

beta endorphin   *See* **endorphin.**

beta galactosidase   *See* **lac operon.**

beta-2 microglobulin   *See* **MHC molecules.**

beta particle   a high-energy electron emitted from an atomic nucleus undergoing radioactive decay.

beta ray   a stream of beta particles.

betatron   an apparatus for accelerating electrons by means of electromagnetic induction.

bidirectional replication   a mechanism of DNA replication involving two replication forks moving in opposite directions away from the same origin.

biennial   designating a plant that requires two years to complete its life cycle, from seed germination to seed production and death. The plant develops vegetatively during the first growing season and flowers in the second.

bifid   forked.

bilateral symmetry   a form of symmetry in which the body can be divided by a longitudinal plane into two parts that are mirror images of each other.

Bilateria   animals with bilateral symmetry. *See* **classification.**

bilharziasis   *See* **schistosomiasis.**

bilirubin   an orange pigment formed as a breakdown product of the heme component of hemeproteins, especially the hemoglobin released during the normal destruction of erythrocytes by the reticuloendothelial system. Bilirubin released into the circulation by the reticuloendothelial system is taken up by the liver and excreted into the bile. The accumulation of bilirubin in plasma and tissues results in jaundice. *See* **Crigler Najjar syndrome.**

bimodal population   a population in which the measurements of a given character are clustered around two values.

binary fission   an amitotic, asexual division process by which a parent prokaryote cell splits transversely into daughter cells of approximately equal size.

Binet-Simon classification   *See* **intelligence quotient classification.**

binomial distribution   a probability function so named because the probabilities that an event will or will not occur, $n, n - 1, n - 2, \ldots, 0$ times are given by the successive coefficients in the binomial expansion $(a + b)^n$. Since a and b are the probabilities of occurrence and non-occurrence, respectively, their sum equals 1. The coefficients in a given binomial expansion can be found by referring to Pascal's pyramid. (*See* page 42).

binomial distribution

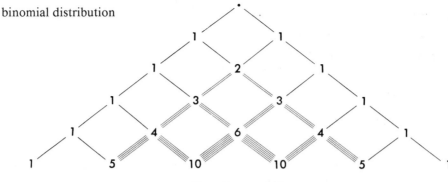

Pascal's pyramid

Here each horizontal row consists of the coefficients in question for consecutive values of n. The expansions for n equaling 1, 2, 3, 4, or 5 are shown below:

$(a + b)^1 = 1a + 1b$
$(a + b)^2 = 1a^2 + 2ab + 1b^2$
$(a + b)^3 = 1a^3 + 3a^2b + 3ab^2 + 1b^3$
$(a + b)^4 = 1a^4 + 4a^3b + 6a^2b^2 + 4ab^3 + 1b^4$
$(a + b)^5 = 1a^5 + 5a^4b + 10a^3b^2 + 10a^2b^3 + 5ab^4 + 1b^5$

Note that each term of the triangle is obtained by adding together the numbers to the immediate left and right on the line above. Such a binomial distribution can be used for calculating the frequency of families in which a certain proportion of individuals show a given phenotype. If we ask, for example, what will be the distribution of girls and boys in families numbering four children and let the frequency of boys = a, and that of girls = b; then using the formula $(a + b)^4$, we conclude that the distribution would be 1/16 all boys; 4/16 3 boys and 1 girl; 6/16 2 boys and 2 girls; 4/16 1 boy and 3 girls; and 1/16 all girls.

binomial nomenclature the current method of scientifically naming species of animals and plants. The name is in two parts: (1) the generic name designating the genus to which it belongs, and (2) the specific name peculiar to the new species. The capitalized generic name is written first; the specific second, usually with a lower case letter; italic type is used (broader classifications are capitalized and written in ordinary roman letters). The name of the author responsible for naming and describing the species should follow (*Drosophila melanogaster* Meigen), but except in taxonomic publications the author's name is usually omitted.

bioassay determination of the relative potency or effectiveness of a substance (e.g., a drug or a hormone) by comparing its effect on a group of test organisms or cells (e.g., tissue culture) using appropriate controls.

biocoenosis a group of plant and animal species living together as a community in a particular habitat.

biogenesis the production of a living cell from a parent cell. *Compare with* **biopoesis.**

biogeographic realms the divisions of the land masses of the world according to their distinctive floras and faunas. See map on page 43.

biological clock **1.** any mechanism that allows expression of specific genes at periodic intervals. **2.** any physiological factor that regulates body rhythms.

biological evolution *See* **evolution.**

biological species groups of naturally interbreeding populations that are reproductively isolated from other such species.

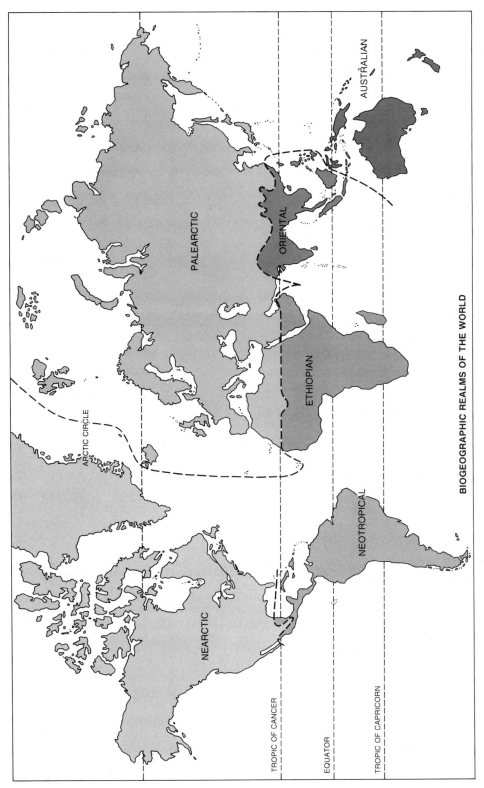

BIOGEOGRAPHIC REALMS OF THE WORLD

Redrawn from *General Zoology* by T. I. Storer and R. L. Usinger. 4th ed. Copyright © 1965 by McGraw-Hill, Inc. Used by permission of McGraw-Hill Book Company.

43

biolumniescence   the emission of light by living organisms.

biomass   the totality of living material in the region or trophic level under discussion.

biome   a community of living organisms in a major terrestrial ecological region (e.g., tropical rain forest biome, desert biome, tundra biome, coniferous forest biome, etc.).

biometry   the application of statistics to biological problems.

Biomphalaria glabrata   *See* **schistosomiasis.**

biopoesis   the origin of a living system from nonliving material; spontaneous generation.

biopterin   a precursor of the drosopterins (*q.v.*).

$$H_2N-C(=N)-C=C-CH$$
$$HN-C(=O)...N=C-CH-CH-CH_3$$
$$OH\ OH$$

biosphere   all of the ecosystems of the earth.

biosynthesis   the production of a chemical compound by a living organism.

biotechnology   the collection of industrial processes that involve the use of biological systems. For some industries, these processes involve the use of genetically engineered microorganisms.

biotic potential   *See* **reproductive potential.**

biotin   a vitamin functioning as a coenzyme (*q.v.*).

$$O$$
$$\|$$
$$C$$
$$HN\quad NH$$
$$HC-CH$$
$$H_2C\quad CH-(CH_2)_4-COOH$$
$$S$$

biotron   a group of rooms designed for the control of environmental factors, singly and in combinations. Biotrons are used for producing uniform experimental organisms, and for providing controlled conditions, for experiments.

biparental zygote   **1.** the common state for nuclear genes in diploid zygotes to contain equal genetic contributions from male and female parents. **2.** the rare state for cytoplasmic genes in diploid zygotes to contain DNA from both parents (e.g., chloroplast DNA in *Chlamydomonas*).

biparous   producing two individuals at one birth.

birefringence   *See* **anisotropy.**

birth defect   **1.** any morphological abnormality present at birth (congenital); such abnormalities may have a genetic basis or they may be environmentally induced (*see* **phenocopy**). **2.** any biochemical or physiological abnormality present at birth; such abnormalities usually have a genetic basis and have been called "inborn errors of metabolism" (*q.v.*).

bisexual   **1.** pertaining to a species made up of individuals of both sexes. **2.** pertaining to an animal having both ovaries and testes, or to a flower having both stamens and pistils.

Bison   a genus including the American bison, *B. bison,* and the European bison, *B. bonasus,* the latter of which has been used in studies of the effects of inbreeding.

Biston betularia   the peppered moth, a species showing industrial melanism (*q.v.*). A black form of the species, called *carbonaria,* spread through the industrial areas of England beginning about 1880. The melanic form is due to a dominant gene, and moths of this phenotype were less conspicuous in soot-darkened habitats. Thus, those individuals with genotypes that darkened their color tended to avoid bird predation and survived to reproduce.

The ancestral (left) and melanic (right) form of the peppered moth shown (in the upper drawing) resting upon lichen-covered bark in a nonpolluted woods and (in the lower drawing) resting upon soot-darkened bark.

Bittner's mouse milk virus   *See* **mammary tumor agent.**

bivalent   a pair of homologous, synapsed chromosomes. *See* **meiosis.**

blackwater fever   a name for malaria, with reference to the urinary excretion of heme. The malaria parasite uses only the globin fraction of hemoglobin for its metabolism and discards the heme which is excreted, darkening the urine.

blast cell transformation the differentiation, when antigenically stimulated, of a T lymphocyte to a larger, cytoplasm-rich lymphoblast.

blastema a small protuberance composed of competent cells from which an animal organ or appendage begins its regeneration.

blastocyst the mammalian embryo at the time of its implantation into the uterine wall.

blastoderm the layer of cells in an insect embryo that completely surrounds an internal yolk mass. The cellular blastoderm developes from a syncytial blastoderm by the partitioning of the cleavage nuclei with membranes derived from infoldings of the oolemma.

blastodisc a disc-shaped superficial layer of cells formed by the cleavage of a large yolky egg such as that of a bird or reptile. Mitosis within the blastodisc products the embryo.

blastomere one of the cells into which the egg divides during cleavage. When blastomeres differ in size the term *macromere* and *micromere* are often used.

blastoporal lip the dorsal rim of the amphibian blastopore which functions as the organizer inducing the formation of the neural tube. *See* **chordamesoderm.**

blastula an early embryonic stage consisting of a hollow sphere of cells.

Blatella germanica the German cockroach, the hemimetabolous insect for which the most genetic information is available.

blending inheritance **1.** an obsolete theory of heredity proposing that certain traits of an offspring are an average of those of its parents because of the blending of their fluid-like germinal influences; hereditary characters transmitted in this way would not segregate in later generations. **2.** a term incorrectly applied to condominant traits, to genes lacking dominance, or to additive gene action.

blepharoplast the basal granule of flagellates.

blood clotting a cascade of enzymatic reactions in blood plasma that produces strands of fibrin to stop bleeding. Fibrinogen, a protein found in the blood plasma, is acted upon by the enzyme thrombin. As a result a negatively charged peptide is split off the fibrinogen molecule, leaving monomeric fibrin which is capable of rapid polymerization to produce a clot. Active thrombin is formed from an inactive precursor prothrombin, also found in blood plasma. The conversion of prothrombin into thrombin is a very complex process which requires a number of factors, including a lipoprotein factor liberated from rupturing blood platelets, a protein plasma thromboplastin component (PTC), antihemophilic globulin (AHG), calcium ions, and others.

blood coagulation blood clotting (*q.v.*).

blood group a type in a system of classification of blood, based on the occurrence of agglutination of the red blood cells when bloods from incompatible groups are mixed. The classical human blood groupings were A, B, AB, and O. However, a multitude of more recently identified groups exist. *See* **A, B antigens, Kell-Cellano, Kidd, Lewis, Lutheran, MN, P,** and **Rh.**

blood group chimerism the phenomenon in which dizygotic twins exchange hematopoietic stem cells while *in utero* and continue to form blood cells of both types after birth. *See also* **radiation chimera.**

bloodline in domesticated animals, a line of direct ancestors.

blood plasma the straw-color fluid remaining when the suspended corpuscles have been removed from blood.

blood relation   a nontechnical term referring to an individual genetically related to another.

blood typing   determination of antigens on red blood cells, usually for the purpose of matching donor and recipient for blood transfusion. Conventionally only antigens of the ABO and Rh systems are typed for this purpose.

Bloom syndrome   a type of dwarfism inherited as an autosomal recessive. Patients have decreased immunity and show sun sensitivity. This is one of the hereditary syndromes characterized by chromosome fragility. *See* **Franconi anemia.**

blotting   the general name given to methods by which electrophoretically or chromatographically resolved RNAs, DNAs, or proteins can be transferred from the support medium (e.g., gels) to an immobilizing paper or membrane matrix. Blotting can be performed by two major methods: (1) capillary blotting involves transfer of molecules by capillary action (e.g., Southern blotting, northern blotting, *q.v.*), and (2) electro-blotting, which involves transfer of molecules by electrophoresis.

blunt end ligation   the use of a DNA ligase (*q.v.*) to join blunt ended restriction fragments. *Compare with* **cohesive end ligation.**

blunt ends   *See* **restriction endonuclease.**

B lymphocyte   *See* **lymphocyte.**

boar   an uncastrated adult male swine, especially an undomesticated one.

bobbed   a gene (*bb*) in *Drosophila melanogaster* producing a small bristle phenotype. The locus of *bb* is very near the centromere, and *bb* is the only gene known to have alleles on both the X and Y chromosomes. The plus allele of *bb* is the nucleolus organizer, and the various hypomorphic alleles may represent partial deletions of ribosomal RNA cistrons resulting from unequal crossing over (*q.v.*).

Bombardia lunata   an ascomycete commonly used in tetrad analysis.

Bombay blood group   a rare human variant of the ABO blood group system (first discovered in Bombay, India) that does not have A, B, or O antigens. Individuals homozygous for an autosomal recessive allele (*h/h*) cannot make the precursor H substance (*q.v.*) from which the A and B antigens are formed. This is a classical case of recessive epistasis in human genetics, because without the product of allele *H,* the products of the ABO locus cannot be formed. Bombay bloods appear to be group O when routinely tested by antibodies against the A or B antigens, but an individual with the Bombay phenotype may be carrying unexpressed genes for the A and/or B antigens. However, they make anti-H that is not found in individuals of groups A, B, or O. Therefore it is possible for a child of group A or B to be produced from parents that appear to be group O, if one of them is a Bombay phenotype and carries the genes for antigens A or B or both. *See* **A, B antigens.**

Bombus   the genus containing the bumblebee.

Bombyx mori   the commercial silk moth, first domesticated for the purpose of silk production in China over 4,000 years ago. Next to *Drosophila melanogaster,* it is the insect whose genetics is best understood. *See* **silk.**

bond energy   the energy required to break a given chemical bond. For example, 58.6 kilogram calories per mol are required to break a carbon to carbon (C—C) bond.

border cell   in *Drosophila* oogenesis, one of the migratory follicle cells that form the micropylar apparatus.

Bos   the genus that includes the domestic cow, *B. taurus,* the Brahman, *B. indicus,* and the yak, *B. grunniens.*

bottleneck effect   fluctuations in gene frequencies occurring when a large population passes through a contracted stage and then expands again with an altered gene pool (usually one with reduced variability) as a consequence of genetic drift (*q.v.*).

botulism   poisoning by an exotoxin synthesized by *Clostridium botulinum,* found in certain foods during decay.

bouquet configuration   *See* **meiosis.**

bovine   referring to members of the cattle family, especially to those of the domestic cattle species *Bos taurus.*

bovine achondroplasia   hereditary chondrodystrophy seen in "bull-dog" calves of the Dexter breed. The condition is inherited as an autosomal recessive.

bp   abbreviation for "base pairs."

brachydactyly   abnormal shortness of fingers or toes or both.

brachyury   a short-tailed mutant phenotype in the mouse governed by a gene (T). It was through this mutant that the T complex was discovered.

Bracon hebetor   *See* ***Microbracon hebetor*** (earlier called *Habrobracon juglandis*).

bract   a modified leaf in whose axil a flower or other reproductive organ arises.

bradyauxesis   *See* **heterauxesis.**

bradytelic   used to refer to a lower than average rate of evolution.

bradytelic evolution   *See* **evolutionary rates.**

Brahman   a breed of humped domestic cattle *(Bos indicus).*

brain hormone   prothoracicotropic hormone (*q.v.*).

Branchiostoma   a genus of lancelets, commonly called *Amphioxus. Branchiostoma lanceolatum* is the sole living representative of the Cephalochordata. *See* **classification.**

branch migration   *See* **Holliday model.**

breakage and reunion   the classical and generally accepted model of crossing over by physical breakage and crossways reunion of broken chromatids during meiosis.

breakage-reunion enzymes   enzymes that use continuous stretches of DNA molecules, rather than preexisting termini, as substrates. The DNA duplex is broken and rejoined. The energy released by breakage is stored in a covalent enzyme-DNA intermediate and utilized in rejoining the molecules.

breakthrough   an individual that escapes the deleterious action of its genotype. In a population of individuals homozygous for a given recessive lethal gene, almost all will die at a defined developmental stage. Those that develop past this stage are called "breakthroughs," or "escapers."

breathing   in molecular genetics, the periodic, localized openings of a DNA duplex molecule to produce single-stranded "bubbles."

breed   an artificial mating group derived from a common ancestor for genetic study and domestication.

breeding   the controlled propagation of plants and animals.

breeding size   the number of individuals in a population that are actually involved in reproduction during that generation.

breeding true   to produce offspring of phenotype identical to the parents; said of homozygotes.

bridge-breakage-fusion-bridge cycle   a cycle which begins with a dicentric chromosome forming a bridge as it is pulled toward both poles at once during anaphase. Such dicentric chromosomes may arise from an exchange within a paracentric inversion or may be radiation-induced. Once the dicentric breaks, the broken ends remain sticky, and these fuse subsequent to duplication. The result is another dicentric which breaks at anaphase, and so the cycle continues, with the chromosomes being broken anew at every mitosis (see diagram below). Since each subsequent break is likely to be at a different place than the previous ones, there will be a repeated regrouping of the genetic loci to produce duplications and deficiencies. *See* **chromosome bridge.**

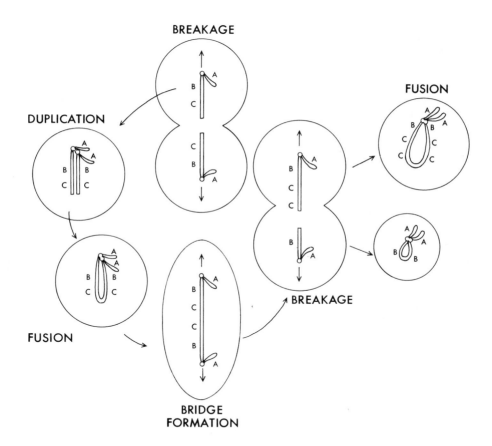

bridge migration   synonomous with branch migration. *See* **Holliday model.**

bridging cross   a mating made to transfer one or more genes between two reproductively isolated species by first transferring them to an intermediate species that is sexually compatible with the other two species.

Briggsian logarithms   *See* **logarithms.**

brilliant cresyl blue   a basic dye used in cytochemistry.

bristle mutant   one of a group of about 100 nonallelic mutations that affect bristle pattern or bristle morphology in *Drosophila melanogaster*.

bristle organ   Each insect bristle is an organ consisting of four cells: the cell that secretes the bristle, the socket cell that secretes the ring which encloses the bristle, a sensory nerve cell whose process ends near the base of the bristle, and the sheath cell that surrounds the nerve axon. *See* **trichogen.**

broad bean   *Vicia faba*. This is the European plant to which the term *bean* was originally applied.

broad heritability   the proportion of the total phenotypic variance (for a polygenic trait in a given population) that is attributed to the total genetic variance (including additive, dominance, epistatic, and other types of gene action); symbolized H^2. *See* **heritability.**

5-bromodeoxyuridine   a thymidine analogue that can be incorporated into DNA during its replication. This substitution profoundly affects that structure of the DNA. When both strands are substituted with BUDR, a chromatid stains less intensely than when only one strand is so substituted. Thus when cells are grown in the presence of BUDR for two replication cycles, the two sister chromatids stain differentially and therefore are called harlequin chromosomes. Consequently the BUDR labeling method can be used to detect sister chromatide exchanges. BUDR causes breakage in chromosomal regions rich in heterochromatin.

5-bromouracil   a mutagenically active pyrimidine analogue.

brood the offspring from a single birth or from a single clutch of eggs.

broodiness the tendency of female birds to incubate eggs.

Bruce effect blockage of implantation of early mouse embryos due to exposure of the mother to an unfamiliar male (or to its urine), particularly one from a strain different than the stud male; named for its discoverer H. M. Bruce. *See* **pheromone.**

Bryophyllum a genus of succulent plants studied in terms of the genetic control of the photoperiodic flowering response. *See* **phytochrome.**

Bryophyta the phylum containing mosses, liverworts, and hornworts. Bryophytes have no vascular system. *See* **classification.**

bubble a structure formed in a duplex DNA at the site of initiation and strand separation during replication.

buccal pertaining to the mouth.

buck an adult male of various animals including the rabbit, deer, mouse, or rat.

bud **1.** a daughter cell produced by mitosis in yeast; one daughter cell retains the cell wall of the parent cell, whereas the daughter cell produced by budding generates a new celll wall. **2.** an underdeveloped plant shoot, consisting of a short stem bearing crowded, overlapping immature leaves.

budding **1.** in yeast and plants, the process by which a bud (*q.v.*) is produced. **2.** in enveloped viruses such as influenza virus and Sindbis virus, a mode of release from the host cell in which a portion of the cell membrane forms an envelope around the nucleocapsid. The envelope contains viral proteins, but no cellular proteins.

BUDR 5-bromodeoxyuridine (*q.v.*).

buffer a compound that, in solution, tends to prevent or resist rapid changes in pH upon the addition of small quantities of acid or base. *See* **cacodylate buffer, tris buffer, veronal buffer.**

buffering the resistance of a system to change by outside forces.

Bufo a genus of toads. Wild populations of species of this genus have been extensively studied by population geneticists.

bulb a modified shoot consisting of a very much shortened underground stem enclosed by fleshy scale-like leaves. It serves as an organ of vegetative reproduction. The onion, daffodil, tulip, and hyacinth produce bulbs.

bull the adult male of various animals including domesticated cattle, elephants, moose, or elk.

bull-dog calf *See* **bovine achondroplasia.**

buoyant density a measure of the weight of a molecule obtained by centrifuging in a density gradient. *See* **centrifugation separation.**

Burkitt's lymphoma a monoclonal malignant proliferation of B lymphocytes primarily affecting the jaw and associated facial bones. The cancer is named after Denis Burkitt, who first described it in central African children in 1958. Most Burkitt tumors occurring in Africans contain Epstein-Barr virus (*q.v.*), and this virus is believed to be mosquito borne. Burkitt lymphomas from United States and European patients lack EBV. Burkitt's lymphoma cells always contain a reciprocal translocation involving the long arm of chromosome 8 and chromosome 14, or less frequently 22 or 2. The break point on chromosome 8 is always near the *myc* oncogene (*q.v.*). The break point on the other chromosome is always near an immunoglobulin gene, namely, 14 (heavy chains), 22 (lambda light chains), or 2 (kappa light chains). In its translocated state *myc* is activated and the cancer ensues. *See* **immunoglobulin chains.**

bursa of Fabricius a sac-like structure connected to the posterior alimentary canal in birds. The bursa is the major site where B lymphocytes become mature immunoglobulin (antibody)-secreting plasma cells. The equivalent organ in mammals has not been definitely identified. Most evidence suggests the bone marrow.

bursicon an insect hormone that appears in the blood after molting and is required for the tanning and hardening of new cuticle.

burst size the average number of bacteriophages released from a lysed host.

busulfan a mutagenic, alkylating agent.

$$CH_3 - \overset{\overset{\displaystyle O}{\|}}{\underset{\underset{\displaystyle O}{\|}}{S}} - O - (CH_2 - CH_2)_2 - O - \overset{\overset{\displaystyle O}{\|}}{\underset{\underset{\displaystyle O}{\|}}{S}} - CH_3$$

C

c   curie (radiation unit).

C   **1.** Celsius (also Centigrade), **2.** carbon. **3.** the haploid amount of DNA. **4.** cytosine or cytidine.

¹⁴C   a radioactive isotope of normal carbon (^{12}C) emitting a weak beta particle. The half-life of ^{14}C is 5700 years.

Є   Cambrian.

Ca   calcium.

CAAT box   part of a conserved DNA sequence of about 75 bp upstream from the initiator for eukaryotic transcription; possibly involved in binding RNA polymerase II. *See* **Hogness box.**

cacodylate buffer   a buffer containing an organic arsenical salt commonly used in preparing fixatives for electron microscopy.

Caenobacter taenospiralis   *See* **killer paramecia.**

Caenorhabditis elegans   a small nematode whose developmental genetics has been extensively investigated. The worm is about 1 mm in length, and its life cycle, when reared at 20°C, is 3.5 days. Its transparent cuticle allows the visualization of every cell. The adult has 816 somatic cells, of which 302 are neurons. The complete lineage history and fate of every cell is known. *C. elegans* normally reproduces as a self-fertilizing hermaphrodite, which has two X chromosomes per cell, plus five pairs of autosomes. Loss of an X by meiotic nondisjunction leads to the production of males. These arise spontaneously among the progeny of hermaphrodites at a frequency of about 0.2%. The mating of hermaphrodites with males provides a method of genetic analysis, and over 300 genes have been defined and mapped. *See* **cell lineage, cell lineage mutants,** *Panagrellus redivivus, Turbatrix aceti.*

caffeine   an alkaloid stimulant found in tea and coffee. Caffeine is a purine analogue which is mutagenically active in microbial systems. Its structure is given on page 54.

Cairns molecule   *See* **theta structure.**

cal   calorie.

calciferol   vitamin D.

caffeine

$$
\begin{array}{c}
\quad\; O \quad\quad CH_3 \\
\quad\; \| \quad\quad\; | \\
\quad\; C \qu\quad N \\
CH_3-N \quad C \quad\quad CH \\
\quad\; | \quad\quad | \quad\quad \| \\
\quad\; C \quad\quad C - N \\
O \quad\quad N \\
\quad\quad | \\
\quad\quad CH_3
\end{array}
$$

calcium, Ca an element universally found in small amounts in tissues. Atomic number 20; atomic weight 40.08; valence 2 +; most abundant isotope ^{40}Ca; radioisotope ^{45}Ca, half-life 164d, radiation emitted—beta particles.

calico cat a tortoiseshell cat (*q.v.*) with white spots, which add a third color to the coat color pattern.

Calliphora erythrocephala a large fly in which polytene chromosomes occur in the ovarian nurse cells in certain inbred lines. The banding pattern of these giant chromosomes have been compared with those of pupal trichogen cells (*q.v.*).

callus the cluster of plant cells that results from tissue culturing a single plant cell.

calorie the amount of heat required to raise the temperature of one gram of water from 14.5°C to 15.5°C at a pressure of one atmosphere. The kilocalorie (kcal) is equal to 1000 calories. The calorie is sometimes referred to as a small calorie and the kilocalorie as a large calorie.

calyx the sterile, outer whorl of floral parts composed of sepals.

cambium the lateral meristem of vascular plants.

Cambrian the most ancient period in the Paleozoic era. *See* **geologic time divisions.**

Camelus the genus of camels including *C. bactrianus,* the two-humped camel; and *C. dromedarius,* the one-humped camel, also called the dromedary.

cAMP *See* **cyclic AMP.**

Campbell model of λ integration a hypothesis devised by A. Campbell to explain the mechanism of integration of phage lambda into the *E. coli* host chromosome. According to the model, linear lambda DNA is first circularized. Then prophage integration occurs as a physical breakage and reunion of phage and host DNA molecules precisely between the bacterial DNA site for phage attachment and a corresponding site in the phage DNA.

canalization the existence of developmental pathways that lead to a standard phenotype in spite of genetic or environmental disturbances.

canalized character a trait whose variability is restricted within narrow boundaries even when the organisms are subjected to disturbing environments or mutations.

canalizing selection elimination of genotypes that render developing individuals sensitive to environmental fluctuations.

cancer a class of diseases characterized by uncontrolled cellular growth. *See* **neoplasm.**

Canis familiaris the dog. More than a hundred breeds of the one species are distinguished, and extensive studies have been made of the genetic control of their social behavior. *See* **dog.**

canonical sequence a sequence that describes the nucleotides most often present in a

DNA segment of interest. For example, in the Pribnow box and the Hogness box, the canonical sequences are $T_{89} A_{89} T_{50} A_{65} A_{65} T_{100}$ and $T_{83} A_{97} T_{93} A_{85} A_{63} A_{83} A_{50}$, respectively. The subscript indicates the percent occurrence of the most frequently found base.

cap   *See* **methylated cap.**

CAP   catabolite activating protein (*q.v.*).

capillarity   the action by which the surface of a liquid where it is in contact with a hydrophilic solid is elevated because of the attraction of the molecules of the liquid for those of the solid.

capillary attraction   the force of adhesion between a solid and a liquid in capillarity.

capon   a castrated domestic fowl.

capped 5′ ends   the 5′ ends of eukaryotic mRNAs containing methylated caps (*q.v.*).

capping   1. addition of a cap (*q.v.*) to mRNA molecules. 2. redistribution of cell surface structures to one region of the cell, usually mediated by cross-linkage of antigen-antibody complexes.

Capsicum   a genus that includes red pappers and pimentos, *C. annum,* and green pepper, *C. frutescens.*

capsid   the protein coat of a virus particle.

capsomere   one of the subunits from which a virus shell is constructed. Capsomeres may contain several different polypeptide chains. The virus shell is formed by assembling capsomeres about the nucleic acid core in a precise geometrical pattern.

capture cross section   the probability that an atomic nucleus will capture an incident particle. The barn (*q.v.*) is the unit used to measure nuclear cross sections.

Carassius auratus   the aquarium goldfish, bred for ornament from time immemorial.

carbohydrate   a compound, having the general formula, $C_x H_{2x} O_x$. Common examples of carbohydrates are glucose, cellulose, glycogen, and starches (*q.v.*).

carbon   the third most abundant of the biologically important elements. Atomic number 6; atomic weight 12.01115; valence 4; most abundant isotope ^{12}C; radioisotope ^{14}C (*q.v.*).

carbonaria   a melanic variety of the Peppered Moth, commonly found in soot-polluted woods surrounding industrial cities in Great Britain. It is an example of industrial melanism (*q.v.*).

3′ carbon atom end   nucleic acids are conventionally written with the 3′ carbon of the pentose to the right. Transcription or translation from a nucleic acid proceeds from 5′ to 3′ carbon.

5′ carbon atom end   nucleic acids are conventionally written with the end of the pentose containing the 5′ carbon to the left. *See* **deoxyribonucleic acid.**

carbon dioxide sensitivy   *See* **sigma virus.**

carbonyl group   a doubly bonded carbon-oxygen group (C=O). The secondary structure of a polypeptide chain involves hydrogen bonds between the carboneal group of one residue (amino acid) and the imino (NH) group of the fourth residue down the chain. *See* **alpha helix.**

Carboniferous   a period of the Paleozoic era. *See* **geologic time divisions.**

carboxyl group   a chemical group (COOH) which is acidic because it can become negatively charged ($-C-O^-$) if a proton dissociates from its hydroxyl group.

$$-C-O^-$$
with $\parallel$ over O, i.e.

$$
\begin{array}{c}
\parallel \\
O
\end{array}
$$

carboxyl terminal   C-terminus.

carboxypeptidases   two pancreatic enzymes (A and B) that hydrolyze protein chains beginning at the carboxyl terminal end of the chain and liberating amino acids one at a time. These enzymes are useful for amino acid sequence studies.

carcinogen   a physical or chemical agent that induces cancer.

carcinogenic   capable of inducing cancer.

carcinoma   a cancer of epithelial tissues (e.g., skin cancer); *adenocarcinoma* is a cancer of gland epithelia.

carnivore   a meat-eating animal. Also applied to a few insectivorous plants.

carotenoids   lipid soluble pigments ranging in color from yellow to red. The carotenes whose structures appear below are plant carotenoids.

α-Carotene

β-Carotene

carpel   the part of a flower which encloses the ovules and extends into a compound pistil.

carrier   1. an individual heterozygous for a single recessive gene. 2. a stable isotope of an element mixed with a radioisotope of that element to give a total quantity sufficient to allow chemical operations. 3. an immunogenic molecule (e.g., a foreign protein) to which a hapten (*q.v.*) is coupled, thus rendering the hapten capable of inducing an immune response.

carrier-free radioisotope   a radioisotope essentially undiluted with a stable contaminating isotope.

carrying capacity   the size or density of a population that can be supported in stable equilibrium with the other biota of a community; symbolized K.

cartilage   a skeletal connective tissue formed by groups of cells that secrete into the intercellular space a ground substance containing a protein, collagen, and a polysaccharide, chondroitin sulfuric acid.

Carya a genus that includes *C. ovata,* the shagbark hickory; and *C. pecan,* the pecan.

caryonide a lineage of paramecia that derive their macronuclei from a single macronuclear primordium. Such paramecia are generally immediate descendants of the exconjugants.

caryopsis a dry indehiscent multiple-seeded fruit derived from a compound ovary. The corn ear is an example.

cassettes eukaryotic DNA sequences of related loci that can be substituted for genetically identical or different sequences by transposition. For example, the cassette model for mating type in yeast proposes that substituting a silent cassette (from either of two different loci) into the mating-type locus results in activation of that cassette and a specific change in mating type from α to a.

caste a class of structurally and functionally specialized individuals within a colony of social insects.

castrate to remove the testes from an animal.

cat any of a number of domesticated breeds of the species *Felis catus.* Popular breeds include: SHORT-HAIRED BREEDS: Domestic Shorthair, Siamese, Burmese, Abyssinian, Russian Blue, Havana Brown, Manx, and Rex. LONG-HAIRED BREEDS: Persian, Angora, and Himalayan.

catabolism metabolic breakdown of complex molecules to simpler products, often requiring catabolic enzymes and accompanied by the release of energy.

catabolite a compound generated by the breakdown of food molecules.

catabolite activating protein (CAP) a constitutively produced, dimeric, positive regulator protein in bacteria which, when bound to a promoter region and cAMP, facilitates transcription by RNA polymerase of certain catabolite-sensitive adjacent genes in inducible and glucose-sensitive operons (such as the *lac* operon of *E. coli*). Also known as cyclic AMP receptor protein (CRP) or catabolite gene activator (CGA) protein.

catabolite repression the reduction or cessation of synthesis of enzymes involved in catabolism of sugars such as lactose, arabinose, etc., when bacteria are grown in the presence of glucose. The enzyme adenyl cyclase is inhibited by glucose from converting ATP to cyclic adenosine monophosphate (cAMP); cAMP must complex with catabolite activator protein (CAP) in order for RNA polymerase to bind to promoters of genes responsible for enzymes capable of catabolizing sugars other than glucose.

catalase an enzyme that splits hydrogen peroxide into water and oxygen. *See* **acatalasemia.**

catalyst a substance that increases the rate of a chemical reaction without being consumed. Enzymes are biological catalysts.

catarrhine pertaining to Old World monkeys.

catastrophism a geological theory proposing that the earth has been shaped by violent events of great magnitude (e.g., worldwide floods, near collisions with comets, etc.); the opposite of uniformitarianism (*q.v.*).

catatonia a clinical subdivision of schizophrenia (*q.v.*). In the active stage of catatonia the patient displays impulsive, belligerent behavior and resists any direction or manipulation; whereas in the passive stage he remains immobile, mute, refuses food, and may be insensitive to pain.

cat cry syndrome   a syndrome of multiple congenital malformations in humans with a deficiency in the short arm of chromosome 5. Infants with this condition produce a peculiar cry that sounds like a cat mewing. Also known as the *cri du chat* syndrome.

category   a rank in a taxonomic hierarchy to which one or more taxa may be assigned: e.g., phylum, class, order, family, genus, species.

catenane   a structure made up of two or more interlocking rings.

catenate   to convert two or more rings into a system of interlocking rings.

cathepsin   any of certain proteolytic enzymes thought to reside in lysosomes (*q.v.*). Such enzymes are abundant, for example, in metamorphosing tadpoles during the resorption of the tail.

cathode   the negative electrode to which positive ions are attracted.

cation   a positively charged ion so named because it is attracted to the negatively charged cathode.

Cattanach's translocation   a translocation in the mouse discovered by B. M. Cattanach. The aberration involves an X chromosome into which a segment of autosome 7 has been translocated. The insertion carries the wild-type alleles of three autosomal genes that control the color of the fur. Studies on mice heterozygous for Cattanach's translocation have shown that during X-chromosomal inactivation in somatic cells, the genes in the inserted autosomal segment are turned off sequentially in order of their distances from the X chromosomal element. Thus the X inactivation spreads into the attached autosomal segment, but does not travel unabated to the end of the segment.

cattle   any of a number of domesticated breeds of the species *Bos taurus*. Popular breeds include: BEEF CATTLE: Hereford, Shorthorn, Aberdeen-Angus, and Santa Gertrudis. DAIRY CATTLE: Holstein-Friesian, Jersey, Guernsey, Ayrshire, and Brown Swiss.

Cavia porcellus   the domestic guinea pig, a species of rodent used as a laboratory animal. Numerous mutants are known affecting coat color and texture.

C banding   a method for producing stained regions around centromeres. *See* **chromosomal banding techniques.**

cDNA   complementary DNA produced from a RNA template by the action of RNA-dependent DNA polymerase (reverse transcriptase). If the RNA template has been processed to remove introns, the cDNA will not be identical to the gene from which the RNA was transcribed. Also called *copy DNA*.

cDNA clone   a duplex DNA sequence complementary to an RNA molecule of interest, carried in a cloning vector.

cecidogen   a gall-forming substance.

cell   the smallest, membrane-bound protoplasmic body capable of independent reproduction.

cell affinity   a property of eukaryotic cells of the same type to adhere to one another but not to those of a different type; this property is lost when the cell transforms to the cancerous state.

cell culture   a term used to denote the growing of cells *in vitro,* including the culturing of single cells. In cell cultures the cells are not organized into tissues.

cell cycle   the sequence of events between one mitotic division and another in a eukaryotic cell. Mitosis (M phase) is followed by a growth (G_1) phase, then by DNA synthesis (S phase), then by another growth (G_2) phase, and finally by another mitosis. In HeLa cells (*q.v.*), for example, the G_1, S, G_2, and M phases take 8.2, 6.2, 4.6, and 0.6 hours, respectively.

cell determination   an event in embryogenesis that specifies the developmental pathway that a cell will follow.

cell differentiation   the process whereby descendants of a single cell achieve and maintain specializations of structure and function. Differentiation presumably is the result of differential transcriptions.

cell division   the process (binary fission in prokaryotes, mitosis in eukaryotes) by which two daughter cells are produced from one parent cell.

cell fractionation   the separation of the various components of cells after homogenization of a tissue and differential centrifugation. Four fractions are generally obtained: (1) the nuclear fraction, (2) the mitochondrial fraction, (3) the microsomal fraction, and (4) the soluble fraction or cytosol.

cell-free extract   a fluid obtained by rupturing cells and removing the particulate material, membranes, and remaining intact cells. The extract contains most of the soluble molecules of the cell.

cell fusion   the experimental formation of a single hybrid cell with nuclei and cytoplasm from different somatic cells. The cells that are fused may come from tissue cultures derived from different species. Such fusions are facilitated by the adsorption of certain viruses by the cells. *See* **Sendai virus, polyethylene glycol, Zimmermann cell fusion.**

cell hybridization   the production of viable hybrid somatic cells following experimentally induced cell fusion (*q.v.*). In the case of interspecific hybrids, there is a selective elimination of chromosomes belonging to one species during subsequent mitoses. Eventually cell lines can be produced containing a complete set of chromosomes from one species and a single chromosome from the other. By studying the new gene products synthesized by the hybrid cell line, genes residing in the single chromosome can be identified. *See* **syntenic genes.**

cell interaction genes   a term sometimes used to refer to some genes in the I region of the mouse H-2 complex which influence the ability of various cellular components of the immune system to cooperate effectively in an immune response.

cell line   a heterogeneous group of cells derived from a primary culture (*q.v.*) at the time of the first transfer.

cell lineage   a pedigree of the cells produced from an ancestral cell by binary fission in prokaryotes or mitotic division in eukaryotes. *Caenorhabditis elegans* (*q.v.*) is the only multicellular eukaryote for which the complete pattern of cell divisions from single-celled zygote to mature adult has been elucidated. Cell lineage diagrams are available which detail each cell or nuclear division and the fate of each cell produced by a terminal division.

cell lineage mutants   mutations that affect the division of cells or the fates of their progeny cells. Cell lineage mutants generally fall into two broad classes. The first contains mutations that affect general cellular processes, such as cell division or DNA replication.

Mutants perturbing the cell division cycle have been analyzed most extensively in *Saccharomyces cerevisae.* The second class of mutations show a striking specificity in their effects. For example, cell lineage mutants are known in *Caenorhabditis elegans* where particular cells are transformed to generate lineages or to adopt differentiated fates characteristic of cells normally found in different positions, at different times, or in the opposite sex. Some of these mutants result from transformations in cell fates. For example, a particular cell "A" will adopt the fate of another cell "B", and this results in the loss of the cells normally generated by A and the duplication of cells normally generated by B. Such transformations resemble the homeotic mutations (*q.v.*) of *Drosophila.* In *Caenorhabditis,* mutations of this type are generally symbolized by *lin. See* **developmental control genes, heterochronic mutations.**

cell lysis   disruption of the cell membrane, allowing the dissolution of the cell and exposure of its contents to the environment. Examples: bacteria undergo *bacteriolysis,* red blood cells experience *hemolysis.*

cell-mediated immunity   immune responses produced by T lymphocytes rather than by immunoglobulins (humoral- or antibody-mediated immunity); abbreviated CMI.

cell-mediated lympholysis   the killing of "target" cells by activated T lymphocytes through direct cell-cell contact. Often used as an *in vitro* test of cell-mediated immunity.

cell plate   a semi-solid structure formed by the coalescence of droplets that are laid down between the daughter nuclei following mitosis in plants. The cell plate is the precursor of the cell walls, and it is synthesized by the phragmoplast (*q.v.*).

cell strain   cells derived from a primary culture or cell line by the selection and cloning of cells having specific properties or markers. The properties or markers must persist during subsequent cultivation. *See* **in vitro markers, in vivo markers.**

cell theory   the basic theory (put forward by Schleiden and Schwann in 1838) that all animals and plants are made up of cells, and that growth and reproduction are due to division of cells.

cellular immunity   immune responses carried out by active cells rather than by antibodies.

cellulase   an enzyme that degrades cellulose to glucose.

cellulifugal   moving away from the center of the cell.

cellulose   a complex structural polysaccharide that makes up the greater part of the walls of plant cells. Cellulose is composed of a linear array of β-D-glucose molecules.

cell wall   a rigid structure secreted external to the plasma membrane. In plants it contains cellulose and lignin; in fungi it contains chitin; and in bacteria it contains peptidoglycans.

Celsius   *See* **temperature.**

cen   *See* **symbols used in human cytogenetics.**

cenospecies   a group of species which, when intercrossed, produce partially fertile hybrids.

Cenozoic the era extending from 65 million years ago to the present; the age of mammals. *See* **geologic time divisions.**

center of origin an area from which a given taxonomic group of organisms has originated and spread.

centi- a prefix meaning one-hundredth, applied to units of metric measurement.

Centigrade *See* **temperature.**

centimorgan *See* **morgan unit.**

central dogma the concept describing the functional interrelations between DNA, RNA, and protein; that is, DNA serves as a template for its own replication and for the transcription of RNA which, in turn, is translated into protein. Thus the direction of the transmission of genetic information is DNA → RNA → protein.

centric fusion breakage in the very short arms of two acrocentric chromosomes, followed by fusion of the long parts into a single chromosome; the two small fragments are usually lost; also termed a *Robertsonian translocation* or *whole arm fusion.*

LOSS

centrifugal acting in a direction away from the center.

centrifugal selection *See* **disruptive selection.**

centrifugation separation any of various methods of separation dispersions by the application of centrifugal force. In the case of *density gradient equilibrium centrifugation* a gradient of densities is established in a centrifuge tube by adding a low molecular weight salt such as cesium chloride. The mixture of molecules to be studied is layered in the surface of the gradient and then centrifuged until each molecule reaches the layer in the gradient with a buoyant density equal to its own. In the case of *density gradient zonal centrifugation* the macromolecules are characterized by their velocities of sedimentation through a preformed sucrose gradient. In this case the sedimentation velocity is determined by molecular size and shape.

centrifuge an apparatus used for the separation of substances by the application of centrifugal force generated by whirling at a high rate of rotation a vessel containing a fluid in which the substances are suspended. *See also* **ultracentrifuge.**

centriole a self-reproducing cellular organelle generally consisting of short cylinders containing nine groups of peripheral microtubules disposed about a central cavity. When a centriole reproduces, it forms a "daughter" centriole at right angles to itself. The daughter centriole grows outwards away from the "mother" until it reaches its mature size. Centrioles are capable of movement and always come to lie at the polar regions of the spindle apparatus in dividing animal cells. The behavior of the centrioles is illustrated in the **meiosis** entry. An organelle that is ultrastructurally identical to the centriole forms the basal body of a cilium. Centrioles do not occur in the cells of higher plants.

centripetal   acting in a direction toward the center.

centripetal selection   *See* **stabilizing selection.**

centrolecithal egg   one having centrally placed yolk.

centromere   a region of a chromosome to which spindle traction fibers attach during mitosis. The position of the centromere determines whether the chromosome will appear as a rod, a J, or a V during its poleward migration at anaphase. In a very few species the traction fibers seem to attach along the length of the chromosome. Such chromosomes are said to be *polycentric* or to have a *diffuse centromere.* A replicated chromosome consists of two chromatids joined at the centromere region. Late in prophase, kinetochores develop on the two faces of the centromere that point toward the spindle poles. The microtubules of the traction fiber attach to the kinetochores as shown below.

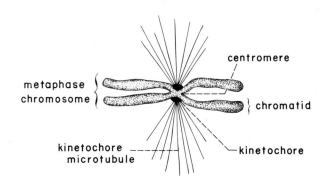

In the older literature, the terms centromere and kinetochore are used synonomously. The centromere of metaphase chromosomes is narrower than the regions distal to it, and therefore it is called the *primary chromosomal constriction.* The centromere is generally bordered by heterochromatin containing repetitious DNA (*q.v.*) and it is late to replicate. *See* **Luzula, meiosis, mitosis.**

centromere interference   the inhibitory effect of the centromere upon crossing over in adjacent chromosomal regions.

centromere misdivision   *See* **isochromosome.**

centromeric index   the percentage of the total length of a chromosome encompassed by its shorter arm. For example, in human somatic cells during metaphase, chromosomes 1 and 13 have centromeric indexes of 48 and 17, respectively. Therefore chromosome 1 is metacentric with its short arm occupying 48% of the total length of the chromosome, and chromosome 13 is acrocentric with a short arm that only occupies 17% of the total length.

centrosome   a region of differentiated cytoplasm containing the centriole.

Cepaea   a genus of land snails belonging to the family Helicidae. *C. hortenses* and *C. memoralis* exhibit extensive variation in color and ornamentation of the shell with longitudinal bands. These species have been extensively studied in the field and in laboratory colonies by population geneticists.

cephalic   designating the head or the anterior end of an animal.

cereal   a cultivated grass, whose seeds are used as food; for example, wheat, oats, barley, rye, maize, etc.

cerebroside a molecule composed of sphingosine, a fatty acid, and a sugar; abundant in the myelin sheaths of nerve cells.

ceruloplasmin a blue copper protein present among the α_2 globulins of the plasma. Approximately 95% of the circulating copper of human beings is bound to ceruloplasmin. Ceruloplasmin is made up of eight subunits, each of molecular weight 18,000.

cesium-137 a radioisotope of cesium with a half life of about 30 years generated during the explosion of certain nuclear weapons. It is one of the major sources of radiation contamination from fallout.

cesium chloride gradient centrifugation *See* **centrifugation separation.**

C genes genes that code for the constant region of immunoglobulin protein chains. *See* **immunoglobulin.**

chaeta a bristle, especially of an insect.

chaetotaxy the taxonomic study of the bristle pattern of insects.

Chagas disease a disease in man produced by *Trypanosoma cruzi* transmitted by blood-sucking bugs in the genera *Rhodnius* and *Triatoma*. Darwin is thought to have contracted Chagas disease while in South America, and as a result spent the remainder of his life as a semi-invalid.

chain reaction a biological, molecular, or atomic process in which some of the products of the process, or energies released by the process, are instrumental in the continuation or magnification of the process.

chain termination codon *See* **stop codon.**

chain terminator a molecule that stops the extension of a DNA chain during replication. *See* **2′, 3′-dideoxyribonucleoside triphosphates.**

character any detectable phenotypic property of an organism.

character displacement the exaggeration of species markers (visual clues, scents, mating calls, courtship rituals, etc.) or adaptions (anatomical, physiological, or behavioral) in sympatric populations relative to allopatric populations of related species. This phenomenon is attributed to the direct effects of natural selection intensifying allesthetic traits useful for species discrimination or for utilizing different parts of an ecological niche (thereby avoiding direct competition).

character states a suite of different expressions of a character in different organisms. These different states are said to be homologues. A character may have a minimum of two states (present/absent or primitive/derived) or have many states.

Chargaff's rules for the DNA of any species, the number of adenine residues equals the number of thymine residues; likewise, the number of guanines equals the number of cytosines; the number of purines (A + G) equals the number of pyrimidines (T + C).

charged tRNA a transfer RNA molecule to which an amino acid is attached; also termed *aminoacylated tRNA.*

charon phages a set of 16 derivatives of bacteriophage lambda that are designed as cloning vectors. They were named by their originators (F. R. Blattner and 11 colleagues) after the old ferryman of Greek mythology who conveyed the spirits of the dead across the River Styx.

chase *See* **pulse-chase experiment.**

chasmogamous   designating a plant in which fertilization takes place after the opening of the flower. *See* **cleistogamous.**

Chédiak-Steinbrinck-Higashi syndrome   a defect in lysosome formation involving leukocytes and melanocytes; inherited as an autosomal recessive. Patients show decreased pigmentation of hair and eyes, abnormal susceptibility to infections, and tendencies to develop malignant lymphomas. A similar syndrome occurs in mice, cattle, and mink. In the mouse, a homologous mutation results in an impairment in the functioning of natural killer cells (*q.v.*). *See* **Aleutian mink.**

chelating agent   a molecule capable of reacting with metallic ions and forming a stable, inert, water soluble complex. Chelating agents are used to remove metals which occur in minute amounts even in distilled water. *See* **EDTA.**

Chelicerata   a subphylum of arthropods containing the species that have no antennae and possess pincer-like chelicerae as the first pair of appendages. *See* **classification.**

chemical elements   Listed alphabetically by their symbols. *See* **periodic table.**

Ac	Actinium	Ge	Germanium	Pr	Praseodymium
Ag	Silver	H	Hydrogen	Pt	Platinum
Al	Aluminum	He	Helium	Pu	Plutonium
Am	Americium	Hf	Hafnium	Ra	Radium
Ar	Argon	Hg	Mercury	Rb	Rubidium
As	Arsenic	Ho	Holmium	Re	Rhenium
At	Astatine	I	Iodine	Rh	Rhodium
Au	Gold	In	Indium	Rn	Radon
B	Boron	Ir	Iridium	Ru	Ruthenium
Ba	Barium	K	Potassium	S	Sulfur
Be	Beryllium	Kr	Krypton	Sb	Antimony
Bi	Bismuth	La	Lanthanum	Sc	Scandium
Bk	Berkelium	Li	Lithium	Se	Selenium
Br	Bromine	Lr	Lawrencium	Si	Silicon
C	Carbon	Lu	Lutetium	Sm	Samarium
Ca	Calcium	Md	Mendelevium	Sn	Tin
Cd	Cadmium	Mg	Magnesium	Sr	Strontium
Ce	Cerium	Mn	Manganese	Ta	Tantalum
Cf	Californium	Mo	Molybdenum	Tb	Terbium
Cl	Chlorine	N	Nitrogen	Tc	Technetium
Cm	Curium	Na	Sodium	Te	Tellurium
Co	Cobalt	Nb	Niobium	Th	Thorium
Cr	Chromium	Nd	Neodymium	Ti	Titanium
Cs	Cesium	Ne	Neon	Tl	Thallium
Cu	Copper	Ni	Nickel	Tm	Thulium
Dy	Dysprosium	No	Nobelium	U	Uranium
Er	Erbium	Np	Neptunium	V	Vanadium
Es	Einsteinium	O	Oxygen	W	Tungsten
Eu	Europium	Os	Osmium	Xe	Xenon
F	Fluorine	P	Phosphorus	Y	Yttrium
Fe	Iron	Pa	Protactinium	Yb	Ytterbium
Fm	Fermium	Pb	Lead	Zn	Zinc
Fr	Francium	Pd	Palladium	Zr	Zirconium
Ga	Gallium	Pm	Promethium		
Gd	Gadolinium	Po	Polonium		

chemiosmotic theory the proposition that hydrogen ion concentration gradients across the mitochondrial membrane are responsible for the coupling of electron transfer and ATP formation.

chemostat an apparatus allowing the continuous cultivation of bacterial populations in a constant, competitive environment. Bacteria compete for a limiting nutrient in the medium. The medium is slowly added to the culture, and used medium plus bacteria are siphoned off at the same rate. The concentration of the limiting nutrient in the fresh medium determines the density of the steady state population, and the rate at which the medium is pumped into the chemostat determines the bacterial growth rate. In chemostat experiments, environmental variables can be changed, one by one, to ascertain how these effect natural selection, or the environment can be held constant and the differential fitness of two mutations can be evaluated.

chemotaxis the attraction or repulsion of organisms by a diffusing substance.

chemotherapy the treatment of a disease with drugs of known chemical composition which are assumed to be directly toxic to the pathogenic microorganisms.

chiasma (*pl.* **chiasmata**) the cytological manifestation of crossing-over; the cross-shaped points of junction between nonsister chromatids first seen in diplotene tetrads.

chiasma interference the more frequent (in the case of negative chiasma interference) or less frequent (in the case of positive interference) occurrence of more than one chiasma in a bivalent segment than expected by chance.

chiasmata (*sing.* **chiasma**) chromosomal sites where crossing-over produces an exchange of homologous parts betwen nonsister chromatids. *See* **meiosis.**

chiasmatype theory the theory that crossing-over between nonsister chromatids results in chiasma formation.

chicken *See **Gallus domesticus.***

chimera an individual composed of a mixture of genetically different cells. In plant chimeras, the mixture may involve cells of identical nuclear genotypes, but containing different plastid types. In more recent definitions, chimeras are distinguished from mosaics (*q.v.*) by requiring that the genetically different cells of chimeras be derived from genetically different zygotes. *See also* **aggregation chimera, mericlinal chimera, periclinal chimera, radiation chimera.**

Chinchilla lanigera a rodent bred on commercial ranches for its pelt. Numerous coat color mutants are known.

Chironomus a genus of delicate, primitive, gnat-like flies that spend their larval stage in ponds and slow streams. Nuclei from various larval tissues contain giant polytene chromosomes. The salivary gland chromosomes of *C. thummi* and *C. tentans* have been mapped, and the transcription processes going on in certain Balbiani rings (*q.v.*) have been studied extensively.

chi sequence an octomeric sequence in *E. coli* DNA, occurring about once every 10 kb, acting as a "hotspot" for RecA-mediated genetic recombination.

chi-square test (χ^2) a statistical procedure that enables the investigator to determine how closely an experimentally obtained set of values fits a given theoretical expectation. See page 480.

chi structure a structure resembling the Greek letter χ, formed by cleaving a dimeric circle with a restriction endonuclease that cuts each DNA circle only once. The parental monomeric duplex DNA molecules remain connected by a region of heteroduplex DNA at the point where crossing-over occurred. Thus, the identification of such chi structures provides evidence for cross-over events taking place between circular DNA molecules.

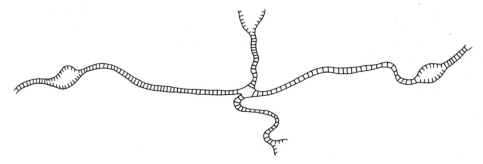

chitin a polymer of high molecular weight composed of N-acetylglucosamine residues joined together by beta glycosidic linkages between carbon atoms 1 and 4. Chitin is a component of the exoskeletons of arthropods.

Chlamydomonas reinhardi a species of green algae belonging to the Phylum Chlorophyta (*see* **classification**, Protoctista). The interaction of nuclear and cytoplasmic units of heredity has been extensively studied in this species.

chlorambucil a mutagenic, alkylating agent.

$$HOOC-CH_2-CH_2-CH_2-C$$

chloramphenicol an antibiotic produced by *Streptomyces venezuelae*. Chloramphenicol is a potent inhibitor of protein synthesis. It attaches to the 50 S subunit of the ribosome and prevents the addition of an amino acid to the growing polypeptide chain. It is also used to increase the proportion of plasmid DNA relative to that of "chromosomal" DNA in bacteria. *See* **ribosome, translation.**

Chlorella a genus of green algae extensively used in studies of photosynthesis and its genetic control.

chlorenchyma tissue possessing chloroplasts.

chlorine an element universally found in small amounts in tissues. Atomic number 17; atomic weight 35.453; valence $1-$; most abundant isotope ^{35}Cl; radioisotopes ^{38}Cl, half-life 37 minutes, ^{39}Cl, half-life 55 minutes, radiation emitted—beta particles.

chlorolabe *See* **color blindness.**

chloromycetin chloramphenicol.

chlorophyll the green photosynthetic pigment generally localized in intra-cellular organelles called chloroplasts. Chlorophyll is found in almost all members of the plant kingdom with the exception of the fungi. Chlorophyll is a mixture of two pigments: chlorophyll *a* ($C_{55} H_{72} O_5 N_4 Mg$) and chlorophyll *b* ($C_{55} H_{70} O_6 N_4 Mg$). The structural formula for the chlorophyll *a* molecule and the approximate dimensions of the porphyrin ring and phytol chain are given below:

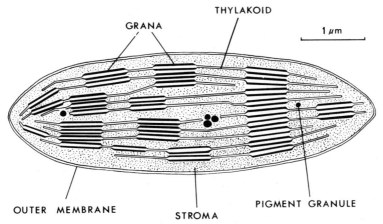

Porphyrin Ring

Phytol Chain

chloroplast the chlorophyll-containing, photosynthesizing organelle of plants. Chloroplasts are thought to be the descendants of endosymbiotic cyanobacteria. A typical chloroplast is shown below. Each is surrounded by a double membrane and contains a system of internal thylakoid membranes. These form stacks of flattened discs called grana in which chlorophyll molecules are embedded.

Chloroplasts contain DNA and can multiply. Replication of chloroplast DNA occurs throughout the cell cycle. Chloroplast DNAs are circular, like those of mitochondria, but many times larger. There are 40 to 80 DNA molecules per organelle. The DNA molecules form clusters in the stroma and are thought to attach to the inner membrane. The DNAs are devoid of histones. Chloroplasts contain 70S ribosomes and in this respect resemble bacterial ribosomes rather than those of the plant cytoplasm (*see* **ribosomes**). The reproduction and functioning of chloroplasts is under the control of both nuclear genes and those of the organelle. Chloroplasts develop from protoplastids. These are small organelles surrounded by a double membrane. The inner one gives rise to a sparce internal membrane system from which the thylakoids develop.

chlorosis   failure of chlorophyll development.

cholecystokinen   a hormone secreted by the duodenum which causes gall bladder contraction.

cholesterol   the quantitatively predominant steroid of man. A 27-carbon compound made up of a fused ring system.

cholinesterase   an enzyme that breaks down acetylcholine (*q.v.*).

chondriome   all the mitochondria of the cell referred to collectively.

chondriosome   a mitochondrion.

chondroitin sulfuric acid   a mucopolysaccharide commonly found in cartilage.

chordamesoderm   a layer of cells derived from the blastoporal lip which later in embryogenesis forms the mesoderm and notochord of the vertebrate embryo. The chordamesoderm acts as an organizer upon the overlying ectoderm to induce differentiation of neural structures.

Chordata   the phylum of animals with a notochord, a hollow dorsal nerve cord, and gill slits at some developmental stage. *See* **classification.**

chorea a nervous disorder, characterized by irregular and involuntary actions of the muscles of the extremities and face.

chorio-allantoic grafting the grafting of pieces of avian or mammalian embryos upon the allantois of the chick embryo under the inner shell-membrane. The implant becomes vascularized from the allantoic circulation and continues development.

chorion 1. the insect egg shell. 2. *see* **amnion.**

chorionic appendages the anterior, dorsal projections of the *Drosophila* egg shell which serve as breathing tubes when the egg is submerged.

chorionic gonadotropin (CG) a hormone produced by the placenta that continues to stimulate the production of progesterone by the corpus luteum and thus maintains the uterine wall in a glandular condition. This hormone is the one for which most pregnancy tests assay. For humans it is abbreviated HCG.

chorionic villi sampling the harvesting of chorionic cells by introducing a catheter through the vagina and into the uterus until it touches the chorion which surrounds the human embryo. Chorionic cells are the same genotype as that of the embryo, and therefore these can be used to detect enzymatic and karyotypic defects. Chorionic villi sampling can be performed between the 8th and 9th week of pregnancy. The technique has the advantage over amniocentesis (*q.v.*) of allowing fetal disorders to be detected earlier. Thus, when necessary, the pregnancy can be terminated at an earlier stage with less risk to the mother.

Christmas disease *See* **hemophilia B.**

chromatid conversion a form of gene conversion (*q.v.*) rendered evident by identical sister-spore pairs in a fungal octad that exhibits a non-Mendelian ratio. For example, if an ordered octad of ascospores from a cross of $+ \times$ m $= (++) (++) (++)$ (mm), one chromatid of the m parental chromosome appears to have been converted to $+$. In *half-chromatid conversion,* an octad of $(++) (++) (+m)$ (mm) indicates that one chromatid of the m parental chromosome was "half-converted." *See* **ordered tetrad.**

chromatid interference a deviation from the expected 1:2:1 ratio for the frequencies of 2-, 3-, and 4-strand double crossovers signaling a nonrandom participation of the chromatids of a tetrad in successive crossovers.

chromatids the two daughter strands of a duplicated chromosome which are joined by a single centromere. Upon the division of the centromere the sister chromatids become separate chromosomes. *See* **meiosis, mitosis.**

chromatin the complex of nucleic acids (DNA and RNA) and proteins (histones and non-histones) comprising eukaryotic chromosomes.

chromatin-negative an individual (normally a male) whose cell nuclei lack sex chromatin.

chromatin-positive an individual (normally a female) whose cell nuclei contain sex chromatin. *See* **Barr body, sex chromatin.**

chromatograph the record produced by chromatography (*q.v.*). Generally applied to a filter paper sheet containing a grouping of spots which represent the compounds separated.

chromatography a technique used for separating and identifying the components from mixtures of molecules having similar chemical and physical properties. The population of different molecules is dissolved in an organic solvent miscible in water, and the solution is allowed to migrate through a stationary phase. Since the molecules migrate at slightly different rates they are eventually separated. In *paper chromatography* filter paper serves as the stationary phase. In *column chromatography* (*q.v.*) the stationary phase is packed into a cylinder. In *thin layer chromatography* the stationary phase is a thin layer of absorbent silica gel or alumina spread on a flat glass plate. *See also* **gas chromatography.**

chromatophore submicroscopic particles isolated from photosynthetic bacteria that contain the photosynthetic pigments.

chromocenter a central aggregation of heterochromatic chromosomal elements in the *Drosophila* larval salivary-gland cell nucleus. The euchromatic chromosome arms extend from the chromocenter.

chromomere one of the serially aligned beads or granules of an eukaryotic chromosome, resulting from local coiling of a continuous DNA thread; best seen when most of the rest of the chromosome is relatively uncoiled as in the leptotene and zygotene stages of meiosis *(q.v.)*. In polytene chromosomes *(q.v.)*, the chromomeres lie in register and give the chromosome its banded appearance.

chromonema (*plural* **chromonemata**) the chromosome thread.

chromoneme the DNA thread of bacteria and their viruses.

chromoplast a carotenoid-containing plastid that colors ripe fruits and flowers.

chromosomal aberration an abnormal chromosomal complement resulting from the loss, duplication, or rearrangement of genetic material. *Intrachromosomal* or *homosomal* aberrations involve changes which occur in but one chromosome. Such aberrations include *deficiencies* and *duplications* which result in a reduction or increase in the number of loci borne by the chromosome. *Inversions* and *shifts* involve changes in the arrangement of the loci, but not in their number. In the case of an inversion a chromosomal segment has been deleted, turned through 180°, and reinserted at the same position on a chromosome, with the result that the gene sequence for the segment is reversed with respect to that of the rest of the chromosome. In the case of shift a chromosomal segment has been removed from its normal position and inserted (in the normal or reversed sequence) into another region of the same chromosome. *Interchromosomal* or *heterosomal* aberrations arise from situations in which non-homologous chromosomes are broken, and interchange occurs between the resulting fragments producing a *translocation. See* **radiation-induced chromosomal aberrations.**

chromosomal mutation *See* **chromosomal aberration.**

chromosomal polymorphism the existence within a population of two or more different structural arrangements of chromosomal material (e.g., inversions, translocations, duplications, etc.).

chromosomal puff a localized swelling of a specific region of a polytene chromosome due to localized synthesis of DNA or RNA. Extremely large RNA puffs are called *Balbiani rings (q.v.). See* **heat shock puffs.**

chromosomal RNA ribonucleic acid molecules associated with chromosomes during either division (e.g., primer, *q.v.*) or interphase (e.g., incomplete transcripts).

chromosomal sterility sterility from the lack of homology between the parental chromosomes in a hybrid.

chromosomal tubules microtubules of the spindle apparatus that originate at kinetochores of centromeres. The chromosomal tubules interpenetrate on the spindle with polar tubules *(q.v.)* and are hypothesized to slide by one another during anaphase as a consequence of making and breaking of cross bridges between them.

chromosome **1.** in prokaryotes, the circular DNA molecule containing the entire set of genetic instructions essential for life of the cell. **2.** in the eukaryotic nucleus, one of the threadlike structures consisting of chromatin (*q.v.*) and carrying genetic information arranged in a linear sequence.

chromosome arms   the two major segments of a chromosome, whose length is determined by the position of the centromere. *See* **metacentric, submetacentric, acrocentric, telocentric.**

chromosome banding techniques   there are four popular methods for staining human chromosomes. To produce *G-banding,* chromosomes are usually treated with trypsin and then stained with Giemsa. Most euchromatin stains lightly, and most heterochromatin stains darkly under these conditions. *C-bands* are produced by treating chromosomes with alkali and controlling the hydrolysis in a buffered salt solution. C-banding is particularly useful for staining and highlighting centromeres and polymorphic bands (especially those of meiotic chromosomes). With *Q-banding,* chromosomes are stained with a fluorochrome dye, usually quinacrine mustard or quinacrine dihydrochloride, and are viewed under untraviolet light. The bright bands correspond to the dark G-bands (with the exception of some of the polymorphic bands). Q-banding is especially useful for identifying the Y chromosome and polymorphisms that are not easily demonstrated by the G-banding procedure. *R-bands* are produced by treating chromosomes with heat in a phosphate buffer. They can then be stained with Giemsa to produce a pattern that is the reverse (hence the R in the term) of G-bands, thereby allowing the evaluation of terminal bands that are light after G-banding. Alternatively, chromosomes can be heated in buffer and then stained with acridine orange. When viewed under ultraviolet light, the bands appear in shades of red, orange, yellow, and green. They can also be photographed in color, but printed in black and white to reveal more distinctive R-bands.

chromosome bridge   a bridge formed between the separating groups of anaphase chromosomes because the two centromeres of a dicentric chromosome are being drawn to opposite poles.

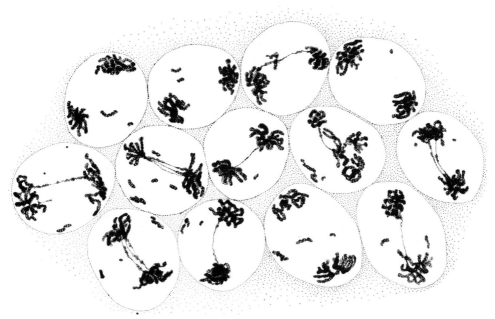

Such bridges may form as the result of single- or three-strand double-exchanges within the reverse loop of a paracentric inversion heterozygote (*see* **inversion**). They may also arise as radiation-induced chromosome aberrations (*q.v.*). Such chromosome bridges are always accompanied by acentric chromosome fragments. *See* **bridge-breakage-fusion-bridge cycle.**

chromosome congression   the movement of chromosomes to the spindle equator during mitosis.

chromosome diminution or elimination   the elimination during embryogenesis of certain chromosomes from cells that form somatic tissues. Germ cells, however, retain these chromosomes.

chromosome jumping   *See* **chromosome walking.**

chromosome loss   failure of a chromosome to be included in a daughter nucleus during cell division. *See* **anaphase lag.**

chromosome map   *See* **cytogenetic map, genetic map.**

chromosome polymorphism   the presence in the same interbreeding population of one or more chromosomes in two or more alternative structural forms.

chromosome rearrangement   a chromosomal aberration involving new juxtapositions of chromosomal segments; e.g., inversions, translocations.

chromosome scaffold   when histones are removed from isolated metaphase chromosomes and these are centrifuged onto electron microscope grids, extremely long loops of DNA can be seen to project from an irregular mass whose dimensions are similar to the original intact chromosome. The nature of the "scaffold" is unclear, and it may represent a precipitation artifact.

chromosome set   a group of chromosomes representing a genome (*q.v.*), consisting of one representative from each of the pairs characteristic of the somatic cells in a diploid species.

chromosome substitution   replacement by a suitable crossing program of one or more chromosomes by homologous or homoeologous chromosomes from another source. This may be a different strain of the same species or a related species that allows hybridization. *See* **homoeologous chromosomes, homologous chromosomes.**

chromosome theory of heredity   the theory put forth by W. S. Sutton in 1902 that chromosomes are the carriers of genes and that their meiotic behavior is the basis for Mendel's laws (*q.v.*).

chromosome walking   the sequential isolation of clones carrying overlapping restriction fragments to span a segment of chromosome that is larger than can be carried in a phage or a cosmid vector. The technique is generally needed to isolate a locus of interest for which no probe is available but that is known to be linked to a gene which has been identified and cloned. This probe is used to screen a genome library. As a result, all fragments containing the marker gene can be selected and sequenced. The fragments are then aligned, and those segments farthest from the marker gene in both directions are subcloned for the next step. These probes are used to rescreen the genome library to select new collections of overlapping sequences. As the process is repeated, the nucleotide sequences of areas farther and farther away from the marker gene are identified, and eventually the locus of interest will be encountered. If a chromosomal aberration is available which shifts a particular gene that can serve as a molecular marker to another position on the chromosome or to another chromosome, then the chromosome walk can be shifted to another position in the genome. The use of chromosome aberrations in experiments of this type is referred to as *chromosome jumping.*

chromotrope   a substance capable of altering the color of a metachromatic dye (*q.v.*).

chronic exposure   radiation exposure of long duration. Applied to experimental conditions in which the organism is given either a continuous low level exposure or a fractionated dose.

chronocline   in paleontology, a character gradient in the time dimension.

chronospecies   *See* **paleospecies.**

chrysalis   the pupa of a lepidopteron that makes no cocoon.

chymotrypsin   a proteoyltic enzyme from the pancreas which hydrolyzes peptide chains internally at peptide bonds on the carboxyl side of various amino acids, especially phenylalanine, tyrosine, and tryptophan.

cilia (*singular,* **cilium**)   populations of thin, motile processes found covering the surface of ciliates or the free surface of the cells making up a ciliated epithelium. Each cilium arises from a basal granule in the superficial layer of cytoplasm. The movement of cilia propels ciliates through the liquid in which they live. The movement of cilia on a ciliated epithelium serves to propel a surface layer of mucus or fluid. *See* **axoneme.**

circadian rhythm   a biological rhythm with a period of about 24 hours.

circular linkage map   the linkage-map characteristic of *Escherichia coli.* In preparation for genetic transfer during conjugation the ring-shaped chromosome of an *Hfr* bacterium breaks in such a way that when the ring opens the *F* factor is left attached to the region of the chromosome destined to enter the F^- cell last. Circular linkage maps have been constructed for several other bacteria and for certain viruses. *See also* **linkage map.**

circular overlap   the phenomenon in which a chain of continuous and intergrading populations of one species curves back until the terminal links overlap each other. Individuals from the terminal populations are then found to be reproductively isolated from each other; that is, they behave as if they belonged to separate species. A ring of races so formed is referred to as a *Rassenkreis.*

cis-acting locus   a genetic region affecting the activity of genes on that same DNA molecule. Such a locus usually does not code for a protein. *Contrast with* **trans-acting locus.**

cis dominance   the ability of a genetic locus to influence the expression of one or more adjacent loci in the same chromosome, as occurs in *lac* operator mutants of *E. coli.*

cisterna   a flattened, fluid-filled reservoir enclosed by a membrane.

cis, trans configurations   terminology which is currently used in the description of pseudoallelism. In the *cis* configuration both mutant recons are on one homologue and both wild-type recons are on the other ($a^1 \, a^2/+ \, +$). The phenotype observed is wild type. In the *trans* configuration each homologue has a mutant and a nonmutant recon ($a^1 \, +/+ \, a^2$), and the mutant phenotype is observed. In the case of pseudoallelic genes the terms *cis* and *trans* configurations correspond to the *coupling* and *repulsion* terminology used to refer to nonallelic genes. *See* **transvection effect.**

cis-trans test   a test for functional allelism to determine whether two mutants affecting the same character lie within the same or different cistrons; also known as a *complementation test.* Two mutants in cis position ($a^1 \, a^2/+ \, +$) will produce a wild phenotype whether they are alleles or pseudoalleles. The two mutants in trans position ($a^1 \, +/+ \, a^2$) will produce wild phenotype if they are pseudoalleles (representing different cistrons) but mutant phenotype if they are true alleles (representing the same cistron).

cistron   the section of the DNA molecule that specifies the formation of a particular polypeptide chain. The test that decides that two mutants reside in different cistrons demands that the wild phenotype be produced when the defective cistrons are in either the *cis* or *trans* position. *See* **gene.**

citric acid cycle (*also* **citrate cycle**)   a cycle of enzyme-controlled reactions by which the acetyl-coenzyme A produced by the catabolism of fats, proteins, and carbohydrates is oxidized, and the energy released used to form ATP from ADP. The condensation of acetyl-coenzyme A with a four-carbon compound (oxaloacetic acid) yields the six-carbon compound citric acid for which the cycle is named. Through a further series of oxidative decarboxylations, citric acid is broken down to oxaloacetic acid, and so the cycle is completed. As a result, one molecule of activated acetate is converted to two CO_2 molecules, eight H atoms are taken up and oxidized to water, and twelve molecules of ATP are formed concurrently. The enzymes are localized in mitochondria (*q.v.*) *See* **glycolysis, cytochrome system.** (See diagram on p. 75.)

citrullinemia   a hereditary disease in humans arising from a deficiency of the enzyme argi-nosuccinate synthetase. It is caused by a recessive gene on autosome 9.

citrullinuria   synonomous with citrullinemia (*q.v.*). *See* **ornithine cycle.**

Citrus   a genus of fruit-bearing trees including *C. aurantium,* the bitter orange; *C. aurantifolia,* the lime; *C. paradisi,* the grapefruit; *C. limon,* the lemon; *C. reticulata,* the tangerine; *C. sinensis,* the sweet orange. The navel orange is a cultivar (*q.v.*) of the last-named species.

clade   a taxon or other group consisting of a single species and its descendents; a monophyletic group; a set of species representing a distinct branch on a phylogenetic tree.

cladistic classification   synonomous with **phylogenetic classification** (*q.v.*).

cladistics   a method of classification that attempts to reconstruct phylogenies in terms of branching sequences of ancestor-descendant lineages. A species is delimited by successive speciation events, and thus is represented by the distance between successive branching points on a cladogram.

cladogenesis   branching evolution. *See* **dendritic evolution.**

cladogram   synonomous with phylogenetic tree (*q.v.*).

class   a grouping used in the classification of organisms. It is a subdivision of a phylum, and it is in turn subdivided into orders.

classification   *See* **Appendix A.**

class switching   *See* **heavy chain class switching.**

clastogen   a substance that causes chromosomal abnormalities.

clathrin   *See* **endocytotic vesicle.**

***ClB* technique**   a technique used for detecting sex-linked lethal and viable mutations in *Drosophila melanogaster.* The name is derived from the X-chromosome used. It contains a *C*rossover suppressor (an inversion), a *l*ethal, and the dominant marker *B*ar eye.

cleavage   the processes by which a dividing egg-cell gives rise to all the cells of the organism. Cleavage in some species follows a definite pattern, where it is said to be determinate (permitting the tracing of *cell lineages*); in other species the pattern is lost after the first few cell divisions.

cleidoic egg   an egg enclosed within a shell that is permeable only to gases.

cleistogamous   designating a self-pollinated plant which produces inconspicuous flowers that never open. *See* **chasmogamous.**

cline   a gradient of phenotypic and/or gene frequency change along a geographical transect of the population's range. *See* **isophenes.**

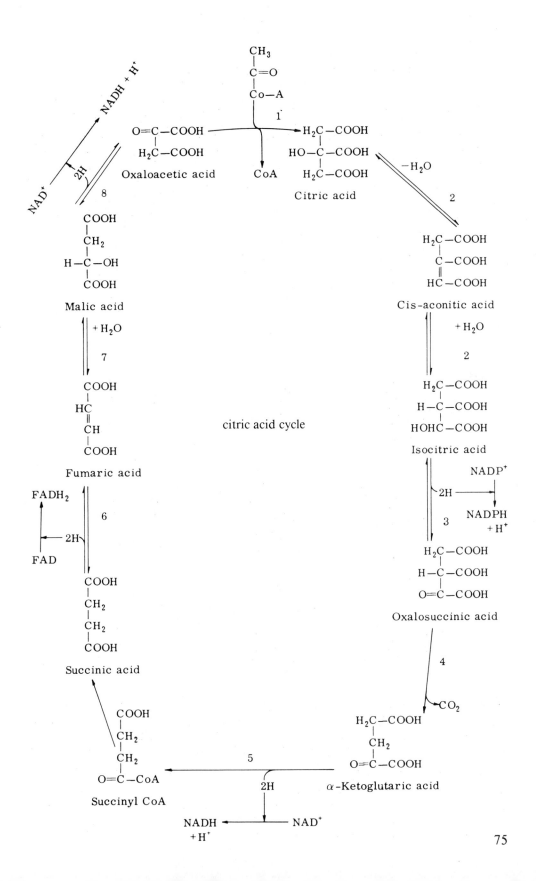

clock mutants   *Drosophila* mutants that show a disruption of the normal 24-hour circadian periodicity. Arrhythmic mutants are known, as well as mutants with short-periods ($\sim$ 19 hours) or long-periods ($\sim$ 28 hours).

clomiphene   any chemical that stimulate ovaries to release eggs.

clonal analysis   the use of mosaics generated by genetic methods or surgical operations to investigate cell autonomous and cell non-autonomous developmental processes. Cell autonomous genetic markers are used to label cells at an early stage in development and the subsequent morphological fates of the offspring of these cells can then be followed. *See* **compartmentalization.**

clonal selection theory   a hypothesis for explaining the specific nature of the immune response in which the diversity among various cells for the recognition of specific antigens exists prior to their exposure to the antigen. Subsequent exposure to a particular antigen causes the appropriate cells to undergo a clonal proliferation.

clone   **1.** a group of genetically identical cells or organisms all descended from a single common ancestral cell or organism by mitosis in eukaryotes or by binary fission in prokaryotes. **2.** genetically engineered replicas of DNA sequences.

cloned DNA   any DNA fragment that passively replicates in the host organism after it has been joined to a vector plasmid. Also called *passenger DNA.*

cloned library   a collection of cloned DNA sequences representative of the genome of the organism under study.

cloning   formation of clones or exact genetic replicas.

cloning vector   *See* **plasmid cloning vector.**

cloning vehicle   plasmid cloning vector (*q.v.*).

clonotype   the phenotype or homogenous product of a clone of cells.

closed reading frame   a segment of mRNA containing one or more termination codons that prevent its translation into protein. *Compare with* **open reading frame.**

close pollination   the pollination of a flower with pollen of another flower on the same plant or of the same clone.

Clostridium   a genus of bacilli to which the species causing tetanus, gangrene, and botulism belong.

cloud chamber   a device for observing the paths of ionizing particles. It uses the principle that super-saturated vapor condenses on ions. By photographing the condensed droplets which form about ions it is possible to compare the distribution of ions along the path of an ionizing radiation or particle.

club wheat   *Triticum compactum* (N = 21). *See* **wheat.**

clutch   a nest of eggs.

C-meiosis   colchicine-blocked meiosis. *See* **metaphase arrest.**

C-metaphase   colchicine-blocked metaphase. *See* **metaphase arrest.**

C-mitosis   colchicine-blocked mitosis. *See* **metaphase arrest.**

CNS   central nervous system.

coacervation   the production of an aggregation of viscous droplets in a continuous liquid phase caused by a coagulation of hydrophilic colloids.

coadaptation the selection process that tends to accumulate harmoniously interacting genes in the gene pool of a population.

coancestry coefficient of consanguinity (*q.v.*).

coated pit, vesicle *See* **endocytotic vesicle.**

coat protein the structural protein making up the external covering of a virus.

cobalamin vitamin B_{12}, a coenzyme for a variety of enzymes. It is essential for deoxyribose synthesis.

cobalt a biological trace element. Atomic number 27; atomic weight 58.9332; valence 2, 3^+; most abundant isotope Co^{59}; radioisotope Co^{60}, half-life 5.2 years, radiation emitted—beta and gamma rays (used extensively as a gamma-ray source).

cockerel a male chicken less than a year old. *See* **pullet.**

coconversion the concurrent correction of two sites during gene conversion.

cocoon **1.** a covering of silky strands secreted by the larva of many insects in which it develops as a pupa. Commercial silk is derived from the cocoon of *Bombyx.* **2.** a protective covering of mucus secreted about the eggs of some species.

code for the genetic code dictionary, *see* **amino acids, start codon, stop codon.**

code degeneracy *See* **degenerate code.**

coding strand the strand of duplex DNA having the same nucleotide sequence as mRNA, excepting that T substitutes in DNA for U in RNA. The coding strand is not the template for mRNA synthesis and is therefore the antisense strand.

coding triplet   codon. *See* **amino acid.**

codominant   designating genes when both alleles of a pair are fully expressed in the heterozygote. For example, the human being of AB blood group is showing the phenotypic effect of both I^A and I^B codominant genes.

codon (*also* **coding triplet**)   the nucleotide triplet in messenger RNA which specifies the amino acid to be inserted in a specific position in the forming polypeptide during translation. A complementary codon resides in the cistron specifying the mRNA in question. For the codon designations. *See* **amino acid, genetic code.**

coefficient of coincidence   *See* **coincidence.**

coefficient of consanguinity   the probability that two homologous genes drawn at random, one from each of the two parents, will be identical and therefore be homozygous in an offspring. The inbreeding coefficient of an individual is the same as the coefficient of consanguinity of its parents. *See* **Wright's inbreeding coefficient.**

coefficient of inbreeding   Wright's inbreeding coefficient.

coefficient of kinship   coefficient of consanguinity (*q.v.*).

coefficient of parentage   coefficient of consanguinity (*q.v.*).

coefficient of relationship (r)   the proportion of alleles in any two individuals that are replicas inherited from a common ancestor.

coefficient of selection (s)   *See* **selection coefficient.**

coelacanth   a member of an ancient group of bony fishes, belonging to the Crossopterygii, allies of the lungfishes, and near the ancestry of the amphibians. *See* **Latimeria chalumnae.**

coelenterate   an animal belonging to the phylum Cnidaria consisting mainly of marine organisms characterized by radial symmetry, a simple diploblastic body, a gastrovascular cavity, and usually showing metagenesis.

Coelomata   a superphylum of animals having a coelom or body cavity formed in and surrounded by mesoderm. *See* **classification.**

coenobium   a colony of unicellular organisms surrounded by a common membrane.

coenzyme   an organic molecule that must be loosely associated with a given enzyme in order for it to function. Generally, the coenzyme acts as a donor or acceptor of groups of atoms that have been added or removed from the substrate. Well-known coenzymes include NAD, NADP, ATP, FAD, FMN, and coenzyme A. Some coenzymes are derivatives of vitamins.

coenzyme A   a coenzyme made up of adenosine diphosphate, pantothenic acid, and mercaptoethylamine. The acetylated form of the coenzyme plays a central role in the citric acid cycle (*q.v.*). *See* formula on p. 79.

coenzyme Q   a molecule that functions as a hydrogen acceptor and donor in the electron transport chain (*q.v.*) Its formula is on p. 79. *See* **cytochrome system (Q and QH₂).**

coevolution   the evolution of one or more species in synchrony with another species as a consequence of their interdependence within the community; e.g., changes in a prey species are usually accompanied by adaptational changes in one or more of its predator species.

coenzyme A

*R = H in coenzyme A

$$R = \overset{O}{\overset{\|}{C}} - CH_3 \text{ in acetylcoenzyme A}$$

n = 6 – 10

coenzyme Q

cofactor a factor such as a coenzyme or a metallic ion required in addition to a protein enzyme for a given reaction to take place.

cognate tRNAs transfer RNA molecules that can be recognized by aminoacyl-tRNA synthetase enzymes.

cohesive end ligation the use of DNA ligase to join double-stranded DNA molecules with complementary cohesive termini that base pair with one another and bring together 3'-OH and 5'-P termini. *Compare with* **blunt end ligation.**

cohesive ends *See* **restriction endonuclease, sticky ends.**

cohort a group of individuals of similar age within a population.

coincidence, coefficient of an experimental value equal to the observed number of double crossovers divided by the expected number.

coincidental evolution *See* **concerted evolution.**

cointegrate structure the circular molecule formed by fusing two replicons, one possessing a transposon, the other lacking it. The cointegrate structure has two copies of the transposon located at both replicon junctions, oriented as direct repeats. The formation of a cointegrate structure is thought to be an obligatory intermediate in the transposition process. The donor molecule (containing the transposon) is nicked in opposite strands at the ends of the transposon by a site-specific enzyme. The recipient molecule is nicked at staggered sites. Donor and recipient strands are then ligated at the nicks. Each end of the transposon is connected to one of the single strands protruding from the target site, thereby generating two replication forks. When replication is completed, a cointegrate structure is formed, which contains two copies of the transposon oriented as direct repeats. An enzyme required in the formation of the cointegrate structure is called a *transposase*. The cointegrate can be separated into donor and recipient units each of which contains a copy of the transposon. This process is called *resolution* of the cointegrate, and it is accomplished by recombination between the transposon copies. The enzyme involved in resolution is called a *resolvase*.

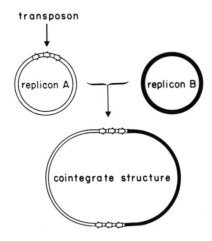

coisogenic   designating inbred strains of organisms which differ from one another only by a single gene as the result of mutation. *Contrast with* **cogenic.**

coitus   copulation; sexual intercourse.

Colcemid   trademark for a colchicine derivative extensively used as a mitotic poison.

colchicine   an alkaloid that inhibits the formation of the spindle and delays the division of centromeres. Colchicine is used to produce polyploid varieties of horticulturally important species. It is also used in medicine in the treatment of gout. Also used to stop mitosis at metaphase (when chromosomes are maximally condensed) for preparation of karyotypes. *See* **Colchicum.**

R = NHCOCH$_3$ for colchicine.

R = NHCH$_3$ for Colcemid.

Colchicum   the genus of crocuses, including: *C. autumnale,* the autumn crocus, source of colchicine; *C. aureus,* the Dutch crocus; *C. sativus,* the Saffron crocus.

cold-sensitive mutant   a gene that is defective at low temperature but functional at normal temperature.

80

coleoptile   the first leaf formed during the germination of monocotyledons.

col **factors**   bacterial plasmids that allow the cell to produce colicins (*q.v.*).

colicin   any of a group of proteins produced by certain strains of *Escherichia coli* and related species which have bactericidal effects. Colicinogenic bacteria are immune to the lethal effects of their own colicins.

colicinogenic   pertaining to a colicin-producing bacterium.

coliform   designating a Gram-negative, lactose-fermenting rod related to *Escherichia coli.*

colinearity   correspondence between the location of mutant sites within a bacterial cistron and the location of amino acid substitutions in its translational product. This correspondence is not complete in most eukaryotic genes because of the presence of nontranslated introns (*q.v.*).

coliphage   a bacteriophage (*q.v.*) that parasitizes *Escherichia coli.*

collagen   the most abundant of all proteins in mammals; the major fibrous element of skin, bone, tendon, cartilage, and teeth, representing one-quarter of the body's protein. It is the longest protein known and consists of a triple helix 3,000 angstroms long and 15 angstroms across. Five types of collagen are known, differing in amino acid sequence of the three polypeptide chains. In some molecules of collagen, the three chains are identical; in others, two of the chains are identical and the third contains a different amino acid sequence. The individual polypeptide chains are translated as longer precursors to which hydroxyl groups and sugars are attached. A triple helix is formed and secreted into the space between cells. Specific enzymes trim the ends of each helix. A mutation that produces exceptionally stretchable skin and loose jointedness (Ehlers-Danlos syndrome) is near one end of the gene for the alpha-2-polypeptide chain and results in failure of that end to be trimmed off. *See* **cartilage.**

collagenase   an enzyme that digests collagen.

collenchyma   plant tissue composed of cells that fit closely together and have thickened walls; in metazoans, undifferentiated mesenchyme cells lying in a gelatinous matrix.

Collinsia   a genus of annual herbaceous plants containing about twenty species. The cytotaxonomy of the genus and the cytogenetics of interspecific hybrids have been extensively studied.

collision, inelastic   the event occurring when a moving electron passes close to an atom. There is repulsion between the moving electron and an atomic electron sufficient to dislodge the latter from its nucleus.

colony, bacterial   a contiguous group of single cells derived from a single ancestor and growing on a solid surface.

colony bank   *See* **gene library.**

colony hybridization   an *in situ* hybridization (*q.v.*) technique used to identify bacteria carrying chimeric vectors whose inserted DNA is homologous with the sequence in question. Colony hybridization is accomplished by transferring bacteria from a Petri plate to a nitrocellulose filter. The colonies on the filter are then lysed, and the liberated DNA is fixed to the filter by raising the temperature to 80°C. After hybridization with a ^{32}P-labeled probe, the position of the colonies containing the sequence under study is determined by autoradiography (*q.v.*).

color blindness   defective color vision in man due to an absence or reduced amount of one of the three visual pigments. The pigments *chlorolabe, erythrolabe,* and *cyanolabe* absorb green, red, and blue light, respectively. The pigments are made of three different opsins combined with vitamin A aldehyde. The green-blind individual suffers from *deuteranopia* and cannot make chlorolabe. The red-blind individual suffers from *protanopia* and cannot make erythrolabe. The blue-blind individual suffers *tritanopia* and cannot make cyanolabe. *Deuteranomaly, protanomaly,* and *tritanomaly* are conditions caused by reduced amounts of the visual pigments in question, rather than their complete absence. The *protan* and *deutan* genes for red- and green-blindness, respectively, occupy different loci on the X chromosome. Blue blindness is quite rare and appears to be due to an autosomal gene.

colt a male horse less than four years old.

Columba livia the wild dove, ancestor to the domestic pigeon, *C. l. domestica.*

column chromatography the separation of organic compounds by percolating a liquid containing the compounds through a porous material in a cylinder. The porous material may be an ion exchange resin. *See* **chromatography.**

combinatorial association within a pool of immunoglobulin molecules, the association of molecules from any class of heavy chain with molecules from any type of light chain. Within a given immunoglobulin molecule, however, there is only one class of heavy and one type of light chain. *See* **immunoglobulin.**

combinatorial translocation for an immunoglobulin chain (heavy or light), the association of any variable region gene with any constant region gene within that same multigene family (*q.v.*). The two genes are brought together by interstitial translocation that may involve deletion of intervening genetic material. *See* **imunoglobulin.**

combining ability **1.** *general:* average performance of a strain in a series of crosses. **2.** *specific:* deviation of a strain's performance in a given cross from that predicted on the basis of its general combining ability.

comb shape a character influenced by two non-allelic gene pairs, *Rr* and *Pp,* in the chicken. One of the early examples of gene interaction (*q.v.*).

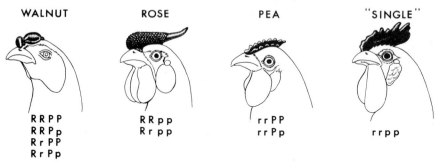

WALNUT	ROSE	PEA	"SINGLE"
RRPP	RRpp	rrPP	
RRPp	Rrpp	rrPp	
RrPP			rrpp
RrPp			

commaless genetic code successive codons that are contiguous and not separated by noncoding bases or groups of bases. Bacterial genes do not have introns (*q.v.*), and therefore the sequences of amino acids in polypeptides and of codons in the gene are colinear. Most eukaryotic genes contain coding regions called exons (specifying amino acids) interrupted by noncoding regions called introns, and in these situations the code is said to contain commas.

commensalism the association of organisms of different species without either receiving benefits essential or highly significant to survival. The species may live in the same shell or burrow or one may be attached to or live within the other. *See* **parasite, symbiosis.**

community all of the species that occupy a given region or a distinctive range of habitat conditions.

compartmentalization   a phenomenon discovered in *Drosophila* by Garcia-Bellido during investigations of the distribution of genetically marked clones of cells during the development of imaginal discs. When such clones are analyzed it is found that they do not randomly overlap any area of the disc, but are confined to compartments and never cross the borders between compartments. A compartment contains all the descendants of a small number of founder cells called a *polyclone*. As development proceeds, large compartments are progressively split into smaller ones. The founder cells are related to each other by position, not ancestry; the progeny of the founder cells form the compartment under consideration, and no other cells contribute to it. The developmental pattern of each compartment is controlled by a selector gene. When selector genes mutate, the cells in a compartment may develop a pattern of cell types appropriate for another compartment. The homeotic mutants (*q.v.*) are examples of mutated selector genes.

compensator genes   sex-linked genes in *Drosophila* which when present in the female (in double dose) reduce the activity of her two doses of a given "primary sex-linked gene" so as to make the total phenotypic effect equivalent to that seen in the male (which has one dose of both the primary gene and the compensator gene).

competence   the state of a part of an embryo that enables it to react to a given morphogenetic stimulus by determination and subsequent differentiation in a given direction. In a bacterium, competence refers to a stage of its life cycle during which the cell can naturally bind and internalize exogeneous DNA molecules, thereby allowing transformation (*q.v.*).

competition   **1.** the mutually exclusive use of the same limited resources (for example, food or a place to live, to hide, or to breed) by two or more organisms; **2.** the mutually exclusive binding of two different molecules to the same site on a third molecule. For example, folic acid and aminopterin compete for combination sites on various folic acid-dependent enzymes.

competitive exclusion principle   the assumption that two species with identical ecological requirements cannot coexist in the same niche in the same location indefinitely. One species will eventually supplant the other, unless they evolve adaptations that allow them to partition the niche and thereby reduce competition. Also known as *Gause's principle*.

complement   a group of at least nine proteins (C1, C2, . . . C9) normally found in vertebrate blood serum that can be activated immunologically (by antibodies of immunoglobulin classes IgG or IgM) or nonimmunologically (by bacterial lipopolysaccharides and other substances) through an alternate (properdin) pathway. Activation of the system involves sequential conversion of proenzymes to enzymes in a manner analogous to the formation of fibrin through the blood-clotting sequence. Some activated complement components enhance phagocytosis (opsonic activity), some make antigen-antibody-complement complexes "sticky" and cause them to become affixed to endothelial tissues or blood cells (serological adhesion or immune adherence), some cause release of vasoactive amines from blood basophils or tissue mast cells (anaphylotoxins), and some cause dissolution of bacterial cells (bacteriolysis).

complementary base sequence   a sequence of polynucleotides related by the base-pairing rules. For example, in DNA a sequence A-G-T in one strand is complementary to T-C-A in the other strand. A given sequence defines the complementary sequence. *See* **deoxyribonucleic acid.**

complementary DNA   *See* **cDNA.**

complementary factors   complementary genes.

complementary genes   nonallelic genes that complement one another. In the case of dominant complementarity the dominant alleles of two or more genes are required for the expression of some trait. In the case of recessive complementarity the dominant allele of either gene suppresses the expression of some trait (i.e. only the homozygous double recessive shows the trait).

complementary interaction the production by two interacting genes of effects distinct from those produced by either one separately.

complementary RNA (cRNA) synthetic transcripts of a specific DNA molecule or fragment, made by an *in vitro* transcription system. This cRNA can be labeled with radioactive uracil and then used as a probe (*q.v.*).

complementation appearance of wild-type phenotype in an organism or cell containing two different mutations combined in a hybrid diploid or a heterokaryon. *See* **complementation test.**

complementation group mutants lying within the same cistron; more properly called a *noncomplementation group.*

complementation map a diagrammatic representation of the complementation pattern of a series of mutants occupying a short chormosomal segment. Mutually complementing mutants are drawn as nonoverlapping lines, and noncomplementing mutants are represented by overlapping, continuous lines. Complementation maps are generally linear, and the positions of mutants on the complementation and genetic maps usually agree. A complementation map is thought to show sites where lesions have been introduced into the polypeptides coded for by the DNA segment under study.

complementation test the introduction of two mutant chromosomes into the same cell to see if the mutations in question occurred in the same gene. If the mutations are nonallelic, the genotype of the hybrid may be symbolized (a $+/+$ b). The wild phenotype will be expressed, since each chromosome "makes up for" or "complements" the defect in the other. *See also* **allelic complementation, cis-trans test.**

complete linkage a condition in which two genes on the same chromosome fail to be recombined and therefore are always transmitted together in the same gamete.

complete medium in microbiology a minimal medium supplemented with nutrients (such as yeast extract, casein hydrolysate, etc.) upon which nutritional mutants can grow and reproduce.

complete penetrance the situation in which a dominant gene always produces a phenotypic effect or a recessive gene in the homozygous state always produces a detectable effect.

complete sex linkage *See* **sex linkage.**

complexity in molecular biology, the total length of different sequences of DNA present in a given preparation as determined from reassociation kinetics; usually expressed in base pairs, but the value may also be given in daltons or any other mass unit.

complex locus a closely linked cluster of functionally related genes: e.g., the human hemoglobin gene complex or the bithorax locus in *Drosophila. See* **pseudoalleles.**

complex RNA that population of RNA molecules consisting of individual mRNA species, each present in only a few copies per cell; also known as *scarce mRNA.*

composite a plant of the immense family Compositae, regarded as comprising the most highly developed flowering plants. It includes species such as the daisy and sunflower.

composite transposon a DNA segment flanked on each end by insertion sequences (*q.v.*), either or both of which allow the entire element to transpose.

compositionism the philosophical thesis that higher levels of organization possess attributes that cannot be predicted from knowledge of simpler levels; also known as *holism* (*q.v.*) or *organicism;* the opposite of reductionism (*q.v.*).

compound   a molecule formed by a union of two or more atoms in a definite proportion.

compound eye   the multifaceted kind of eye of insects. In *Drosophila* each compound eye contains nearly 800 ommatidia.

compound X   designating metacentric, attached X chromosomes or double X chromosomes.

Compton effect   an attenuation process observed for x- or gamma-radiation. An incident photon interacts with an orbital electron of an atom to eject an electron and a scattered photon of less energy than the incident photon.

computer   a system, usually electronic, capable of receiving information, applying pre-scribed processes to it, and supplying the results. The *digital computer* receives informa-tion in separate "bits" and processes it in a series of discrete steps. The *analog computer* receives information in the form of continuous variables and processes it by means of analogous physical variables.

concanavalin A   a lectin (*q.v.*) derived from *Canavalia ensiformis,* the jack bean; abbre-viated conA. The compound stimulates T lymphocytes to enter mitosis. *Compare with* **pokeweed mitogen.**

concatemer   the structure formed by concatenation of unit sized components.

concatenation   linking of multiple subunits into a tandem series or chain, as occurs during replication of genomic subunits of phage lambda.

concerted evolution   the ability of two or more related genes to evolve together as though they occupied the same locus. As a consequence of concerted evolution, variation among the individual members of a family of repeated DNA sequences may accumulate between species, but with little or no variation among members of that same family within a spe-cies. *See* **molecular drive.**

concordant   twins are said to be concordant for a given trait if both exhibit the trait.

conditional mutation   a mutation that exhibits wild phenotype under certain (permissive) environmental conditions, but exhibits a mutant phenotype under other (restrictive) con-ditions. Some bacterial mutants are conditional lethals that cannot grow above 45°C, but grow well at 37°C.

conditioned dominance   referring to an allele which may or may not express itself depend-ing upon environmental factors or the residual genotype. Thus under different conditions the gene may behave as a dominant or as a recessive in heterozygotes.

cone   **1.** one of the elongate, unicellular photoreceptors in the vertebrate eye involved with vision in bright light and color recognition. **2.** ovule- or pollen-bearing scales, characteristic of conifers.

confusing coloration   a form of protective coloration which tends to confuse the predator by having a different appearance according to whether its possessor is at rest or in motion.

congenic strains   strains that differ from one another only with respect to a small chro-mosomal segment. Several congenic mouse strains were developed by G. D. Snell. These differ only in the major or minor histocompatibility loci they contain.

congenital   existing at birth. Congenital defects usually exist before birth.

congression   *See* **chromosome congression.**

conidium an asexual haploid spore borne on an aerial hypha. In *Neurospora* two types of conidia are found: oval *macroconidia,* which are multinucleate, and *microconidia,* which are smaller, spherical, and uninucleate. When incubated upon suitable medium, a conidium will germinate and form a new mycelium.

conifer a tree or shrub that bears cones, including pines, firs, spruces, and sequoias.

conjugation a temporary union of two single-celled organisms with at least one of them receiving genetic material from the other. (1) In bacteria the exchange is unidirectional with the "male" cell extruding all or a portion of one of its chromosomes into the recipient "female." (*See* **F factor**). (2) In protozoa, as shown below, entire nuclei are exchanged.

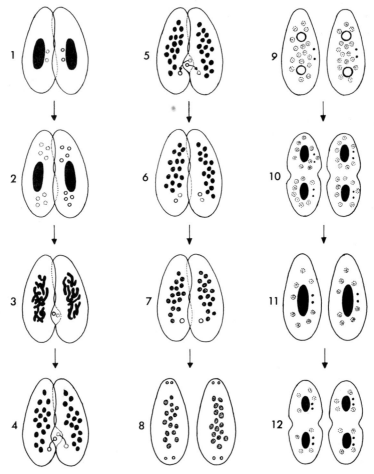

Nuclear changes accompanying conjugation in *Paramecium aurelia.* (1) Two parental animals, each with one macronucleus and two diploid micronuclei. (2) Formation of eight haploid nuclei from the micronuclei of each conjugant. (3) Seven nuclei in each conjugant disappear; the remaining nucleus resides in the paroral cone; the macronucleus breaks up into fragments. (4–7) The nuclei in the paroral cones divide mitotically, forming "male" and "female" gamete nuclei. The female nuclei pass into the interior of the parental animals, while the male nuclei are transferred to the partners. Male and female haploid nuclei fuse. (8) Each fusion nucleus divides twice mitotically. (9) Two of the four nuclei so formed differentiate into macronuclear anlagen (white circles) while the other two become micronuclei.(10–12) Each micronucleus divides, and transverse fission of the exconjugants produces four animals (two of which are shown in 11). Fragments of the old macronucleus are gradually lost. In (12) transverse fission begins in the animals seen previously in (11). (Redrawn from *Genetics of Paramecium aurelia* by G. H. Beale, 1954, published by Cambridge University Press.)

(3) In fungi conjugation also occurs between hyphae of opposite mating type to produce heterokaryons (*q.v.*).

conjugation tube *See* **pilus.**

conjugon a genetic element essential for bacterial conjugation. *See* **fertility factor.**

consanguinity genetic relationship. Consanguineous individuals have at least one common ancestor in the preceding few generations.

consecutive sexuality the phenomenon in which most individuals of a species experience a functional male phase when young and later change through a transitional stage to a functional female phase. A situation common in some molluscs.

consensus sequence synonomous with canonical sequence (*q.v.*).

conservative recombination breakage and reunion of preexisting strands of DNA in the absence of DNA synthesis.

conservative replication an obsolete model of DNA replication in which both strands of the daughter double helix are newly synthesized and both template strands of the parental molecule retain their original composition. *Compare with* **semiconservative replication.**

conservative substitution replacement of an amino acid in a polypeptide by one with similar characteristics; such substitutions are not likely to change the shape of the polypeptide chain, e.g., substituting one hydrophobic amino acid for another.

conspecific belonging to the same species.

constant region *See* **immunoglobulin.**

constitutive enzyme an enzyme that is always produced irrespective of environmental conditions.

constitutive gene a gene whose activity depends only on the efficiency of its promoter in binding RNA polymerase.

constitutive heterochromatin *See* **heterochromatin.**

constitutive mutation mutation that results in an increased constitutive synthesis by a bacterium of several functionally related, inducible enzymes. Such a mutation either modifies an operator gene so that the repressor cannot combine with it or modifies the regulator gene so that the repressor is not formed. *See* **regulator gene.**

constriction an unspiralized region of a metaphase chromosome. Kinetochores and nucleolus organizers are located in such regions.

contact inhibition the cessation of cell movement on contact with another cell. It is often observed when freely growing cells tissue cultured on a Petri plate come into physical contact with each other.

continental drift the concept advanced by A. Wegener that the world's continents once formed a part of a single mass and have since drifted into their present positions. The modern concept of plate tectonics (*q.v.*) has refined the theory of continental drift by placing the continents on larger sections of the earth's crust (tectonic plates) that are in motion. It is now accepted that the Americas, India, Australia, and Antarctica were joined to Africa and Europe, forming a vast land mass, and that the drifting apart of the continents began about 200 million years ago.

continental island an island assumed to have been once connected to a neighboring continent. *Contrast with* **oceanic island.**

continuous distribution a collection of data that yield a continuous spectrum of values.

For example, measurements such as height of plant or weight of fruit, carried out one or more decimal places. *See* **discontinuous distribution.**

continuous fibers   the microtubules that connect the two poles of the mitotic apparatus, as distinct from traction fibers and astral fibers. *See* **mitotic apparatus.**

continuous variation   an example of continuous variation would be the variation in a population of multicellular organisms for the characteristics of body weight, height, intelligence, etc. When measurements are collected from the population the values form a spectrum which is continuous from one extreme to the other. *See* **discontinuous variation.**

contractile ring   a transitory organelle which, during late anaphase and telophase, assumes the form of a continuous equatorial annulus beneath the plasma membrane of the cleavage furrow. The ring is composed of an array of fine cytoplasmic microfilaments (each 4 to 7 mμ in diameter) aligned circumferentially along the equator of the cell. These filaments are intimately associated both with each other and with the plasmalemma. The closing down of the contractile ring is responsible for the cleavage furrow.

control   a standard of comparison; a test or experiment established as a check of other experiments, performed by maintaining identical conditions except for the one varied factor, whose causal significance can thus be inferred.

controlled pollination   a common practice in plant hybridization of bagging the pistillate flowers to protect them from undesired pollen. When the pistillate flowers are in a receptive condition, they are dusted with pollen of a specified type.

controlling elements   a class of genetic elements that renders target genes unstably hypermutable, as in the Dissociation-Activator system of corn (*q.v.*). They include receptors and regulators. The receptor element is a mobile genetic element that, when inserted into the target gene, causes it to become inactivated. The regulator gene maintains the mutational instability of the target gene, presumably by its capacity to release the receptor element from the target gene and thus return that locus to its normal function. *See* **transposable elements.**

controlling gene   one which can switch cistrons on and off. *See* **regulator gene.**

convergence   the evolution of unrelated species occupying similar adaptive zones, resulting in structures bearing a superficial resemblance (for example, the wings of birds and insects).

convergent evolution   *See* **convergence.**

conversion   *See* **gene conversion.**

conversion factors   *See* **metric equivalents.**

Cooley's anemia   thalassemia (*q.v.*).

coordinated enzymes   enzymes whose rates of production vary together. For example, the addition of lactose to the medium causes the coordinated induction of beta-galactosidase *and* beta-galactoside permease in *Escherichia coli*. Such enzymes are produced by cistrons of the same operon. *See* **regulator genes.**

Cope's "law of the unspecialized"   a theory that the evolutionary novelities associated with new major taxa are more likely to originate from a generalized, rather than from a specialized member of an ancestral taxon.

Cope's rule   the general tendency for animals to increase in body size during the course of phyletic evolution (*q.v.*).

copia elements transposable elements of *Drosophila* existing as a family of closely related base sequences that code for abundant mRNAs. There are usually between 20 and 60 copia elements per genome. The actual number depends on the *Drosophila* strain employed. The copia elements are widely dispersed among the chromosomes, and the sites occupied vary between strains. Each copia element ranges in size from 5–9 kb, and it carries direct terminal repeats about 280 bp long. Several *Drosophila* mutations have been found that result from the insertion of copia-like elements.

copolymer a polymeric molecule containing more than one kind of monomeric unit. For example, copolymers of uridylic and cytidylic acids (poly UC) are used as synthetic messengers.

copper a biological trace element. Atomic number 29; atomic weight 63.45; valance 1^+, 2^+; most abundant isotope ^{63}Cu; radioisotope ^{64}Cu, half-life 12.8 hours, radiations emitted—gamma rays, electrons, and positrons.

Coprinus radiatus a basidiomycete fungus that is of considerable genetic interest because of the existence of dikaryotic and diploid strains.

copy-choice hypothesis an explanation of genetic recombination based on the hypothesis that the new strand of DNA alternates between the paternal and maternal strands of DNA during its replication. *See* **Belling's hypothesis.**

copy DNA synonomous with complementary DNA. *See* **cDNA.**

copy error a mutation resulting from a mistake during DNA replication.

cordycepin 3′-deoxyadenosine, an inhibitor of the polyadenylation of RNA.

core 1. the region of a nuclear reactor containing the fissionable material. 2. synaptonemal complex.

core DNA the DNA (146 bp) on a nucleosomal core particle.

core granule RNP granules in the ommatidia of *Drosophila.* The xanthomatins and drosopterins are normally bound to these granules.

core particle a structual unit of eukaryotic chromosomes revealed by digestion with micrococcal nuclease to consist of a histone octamer and a 146 bp core DNA. *See* **nucleosome.**

corepressor in repressible genetic systems, the small effector molecule (usually an end product of a metabolic pathway) that inhibits the transcription of genes in an operon by binding to a regulator protein (aporepressor). Also called a *repressing metabolite.*

corm a swollen vertical underground stem base containing food material and bearing buds. It can function as an organ of vegetative reproduction. The crocus and gladiolus have corms.

corn *Zea mays,* the most valuable crop plant grown in the United States, it ranks along with wheat, rice, and potatoes, as one of the four most important crops in the world. Corn is generally classified in five commercial varieties on the basis of kernel morphology: (1) Dent corn (var. *indentata*), the most common variety of field corn; the kernel is indented, from drying and shrinkage of the starch in the summit of the grain. (2) Flint corn (var. *indurata*), the kernel is completely enclosed by a horny layer and the grain is therefore smooth and hard; flint corn is the fastest to mature. (3) Sweet corn (var. *saccharata*), which is grown for human consumption, is picked when it is filled with a milky fluid, before the grain hardens, (4) Popcorn (var. *everta*), the ear is covered with small kernels that are enclosed by a tough coat; when heated the contained moisture is turned to steam and the kernel explodes. (5) Flour corn (var. *amylacea*), the ear contains soft, starchy kernels; it requires a long growing season and is therefore grown primarily in the tropics. *See* **double cross, hybrid corn.**

corpus allatum an endocrine organ in insects which synthesizes the allatum hormone (*q.v.*). In the larvae of cyclorrhaphous diptera the corpus allatum forms part of the ring gland (*q.v.*).

corpus cardiacum an endocrine organ in insects consisting of a central bundle of axons enveloped by cortical cells. Axons from the corpus cardiacum enter the corpus allatum. Most axons associated with the corpus cardiacum have their cell bodies in the pars intercerebralis. The cortical cells and many of the axons contain numerous neurosecretory spheres (*q.v.*).

corpus luteum a mass of yellowish tissue that fills the cavity left after the rupture of the mature ovum from the mammalian ovary.

correction in a hybrid DNA sequence, the replacement (i.e., by excision and repair) of illegitimate nucleotide base pairs by bases that pair properly.

correlated response the change in one character occurring as an incidental consequence of the selection for a seemingly independent character. For example, reduced fertility may accompany selection for increased bristle number in *Drosophila*.

correlation the degree to which statistical variables vary together. It is measured by the *correlation coefficient (r)* which has a value from zero (no correlation) to -1 or $+1$ (perfect negative or positive correlation, respectively).

corridor a migration route allowing easy dispersal for certain species.

cortex the peripheral region of a cell, tissue, or organ. *Compare with* **medulla.**

corticosterone one of a family of adrenal cortical hormones influencing glucose metabolism.

corticotropin *See* **adrenocorticotropic hormone.**

Corynebacterium diphtheriae the bacterium causing diptheria.

COS cells a monkey cell line that has been transformed by an SV40 viral genome containing a defective origin of viral replication. When introduced into COS cells, recombinant RNAs containing the SV40 origin and a foreign gene should replicate many copies.

cosmic rays high-energy particulate and electromagnetic radiations originating outside the earth's atmosphere.

cosmid plasmid vectors designed for cloning large fragments of eukaryotic DNA. The term signifies that the vector is a plas*mid* into which phage lambda *cos* sites (*q.v.*) have been inserted. As a result the plasmid DNA can be packaged in a phage coat *in vitro*.

cos sites *co*hesive end sites, nucleotide sequences that are recognized for packaging a phage DNA molecule into its protein capsule.

cot the point (symbolized by $C_0t_{1/2}$) in a reannealing experiment where half of the DNA is present as double-stranded fragments; also called the *half reaction time*. If the DNA fragments contain only unique DNA sequences and are similar in length, then $C_0t_{1/2}$ varies directly with DNA complexity (*q.v.*). *See* **reassociation kinetics.**

cotransduction the simultaneous transduction of two or more genes because the transduced element contains more than one locus.

cotransformation the simultaneous transformation of two or more bacterial genes; the genes cotransformed are inferred to be closely linked because transforming DNA fragments are usually small. Also called *double transformation.*

cotton *See Gossypium.*

Coturnix coturnix japonica the Japanese quail, a small bird used as a laboratory animal.

cot value *See* **cot.**

cotyledon the leaf-forming part of the embryo in a seed. Cotyledons may function as storage organs from which the seedling draws food, or they may absorb and pass on to the seedling nutrients stored in the endosperm. Once the cotyledon is exposed to light it develops chlorophyll and functions photosynthetically as the first leaf.

countercurrent distribution apparatus an automated apparatus used for separating mixtures. The method takes advantage of differences in the solubilities of the components of the mixture in two immiscible solvents. An example of the usefulness of the technique is in the separation of different transfer RNA molecules.

counterselection a technique used in bacterial conjugation experiments to allow recovery of recombinant F⁻ cells, while at the same time selecting against (preventing growth of) Hfr donor cells. For example, suppose the Hfr donor strain is susceptible to an antibiotic (such as streptomycin) and can synthesize histidine; the streptomycin locus must be so far from the origin of chromosome transfer that the mating pairs, which inevitably break apart, have separated before the *str* locus has been transferred. Suppose further that the recipient F⁻ cell cannot make histidine (His⁻) but is resistant to the antibiotic (Str⁻ʳ). Only His⁺Str⁻ʳ recombinants can survive on a medium lacking histidine and containing streptomycin. The desired gene (*his⁺* in this case) is called a *selected marker;* the gene that prevents growth of the male (*str⁻ʳ* in this case) is called the *counterselective marker.*

coupled reactions chemical reactions having a common intermediate and therefore a means by which energy can be transferred from one to the other. In the following pair of enzyme catalyzed reactions, glucose-1-phosphate is the common intermediate which is formed in the first reaction and used up in the second:

$$ATP + glucose \rightarrow ADP + glucose\text{-}1\text{-}phosphate$$
$$glucose\text{-}1\text{-}phosphate + fructose \rightarrow sucrose + phosphate$$

A molecule of sucrose is synthesized from glucose and fructose at the expense of the energy stored in ATP and transferred by glucose-1-phosphate.

coupled transcription-translation a characteristic of prokaryotes wherein translation begins on mRNA molecules before they have been completely transcribed.

coupling, repulsion configurations when both nonallelic mutants are present on one homologue and the other homologous chromosome carries the plus alleles (*a b/ + +*) the genes are said to be in the *coupling* configuration. The *repulsion* configuration refers to a situation in which each homologue contains a mutant and a wild-type gene (*a+/+b*). *See* **cis, trans configurations.**

courtship ritual a characteristic genetically determined behavioral pattern involving the production and reception of an elaborate sequence of visual, auditory, and chemical stimuli by the male and female prior to mating. Such rituals are interpreted as insuring that mating will occur only between individuals of opposite sex and the same species.

cousin the son or daughter of one's uncle or aunt. The children of siblings are *first* cousins. Children of *first* cousins are *second* cousins. Children of second cousins are third cousins, etc. The child of a first cousin is a *cousin once removed* of his father's or mother's first cousin.

covalent bond a valence bond formed by shared electron between the atoms in a covalent compound.

covariance a statistic employed in the computation of the correlation coefficient between two variables; the covariance is the sum of $(x-\bar{x})(y-\bar{y})$ over all pairs of values for the variables x and y, where $\bar{x}$ is the mean of the x values and $\bar{y}$ is the mean of all y values.

cover crop a crop grown between orchard trees or on fields between cropping seasons to protect the land from leaching and erosion.

cow the adult female of domestic cattle, elephants, and certain other animals.

cp chemically pure.

cp DNA chloroplast DNA. *See* **mtDNA.**

Craniata the subphylum of the Chordata containing the species with a true skull. The vertebrates.

Crassostrea virginica *See* **Pelecypoda.**

Creeper a gene in the chicken generally causing embryonic death when homozygous. Those few individuals surviving show extreme malformations of the limbs.

crenulate wavy, or with small scallops.

Crepis a genus of weedy herbs much studied with respect to the evolution of karyotype (*q.v.*).

Cretaceous the most recent Mesozoic period. *See* **geologic time divisions.**

cretinism a stunting of bodily growth and mental development in humans due to a deficiency of thyroid hormones. Hereditary cretinism, which is often accompanied by goiter (*q.v.*) and deafness, consists of a group of metabolic disorders that results in a failure in the formation of sufficient thyroxine and triiodothyronine. The defects include the inability of the thyroid gland to accumulate sufficient iodine, to convert it into organically bound iodine, and to couple iodotyrosines to form iodothyronines. All hereditary defects in thyroid hormonogenesis are inherited as autosomal recessives. *See* **thyroid hormones.**

Cricetulus griseus the Chinese hamster. The rodent is a favorite for cytogenetic studies because of its small chromosome number (N = 11).

cri du chat syndrome *See* **cat cry syndrome.**

Crigler-Najjar syndrome a rare hereditary defect in bilirubin metabolism inherited as an autosomal recessive. Patients lack hepatic bilirubin UDP-glucuronyl transferase. This enzyme functions to conjugate bilirubin with glucuronic acid prior to biliary excretion. In the absence of the enzyme, excess bilirubin builds up in all tissues causing jaundice, brain damage, and death.

crisis period the time interval of a primary cell culture, following a number of cell divisions, during which most secondary progeny die even though culture conditions are adequate to initiate a new primary culture of low cell density from a fresh isolate. *See* **Hayflick limit, tissue culture.**

criss-cross inheritance referring to the passage of sex-linked traits from mother to son and from father to daughter.

cristae elaborate invaginations of the inner mitochondrial membrane.

CRM cross-reacting material (*q.v.*).

cRNA *See* **complementary RNA.**

Cro-Magnon man *Homo sapiens sapiens* living in the upper Pleistocene. Cro-Magnon replaced the Neanderthal (*q.v.*) throughout its range.

cross in higher organisms, a mating between genetically different individuals of opposite sex. In microorganisms genetic crosses are often achieved by allowing individuals of different mating types to conjugate. In viruses genetic crossing requires infecting the host cells with viral particles of different genotypes. The usual purpose of an experimental cross is to generate offspring with new combinations of parental genes. *See* **backcross, conjugation, dihybrid, E_1, F_1, I_1, monohybrid cross, P_1, parasexuality, testcross.**

cross agglutination test one of a series of tests commonly employed in blood typing in which erythrocytes from a donor of unknown type are mixed with sera of known types.

crossbreeding outbreeding (*q.v.*).

cross hybridization (molecular) hybridization of a probe (*q.v.*) to a nucleotide sequence that is less than 100% complementary.

cross induction the induction of vegetative phage replication in lysogenic bacteria in response to compounds transferred from UV-irradiated F^+ to non-irradiated F^- bacteria during conjugation.

crossing over the exchange of genetic material between homologous chromosomes. Crossing over is characterized by positive interference and by reciprocality; that is, wild-type and double mutants are produced simultaneously in the same tetrad from a *trans* heterozygote. *See* **intragenic recombination.**

crossing over within an inversion *See* **inversion.**

cross linking formation of covalent bonds between a base in one strand of DNA and an opposite base in the complementary strand by mitotic poisons such as the antibiotic mitomycin C or the nitrite ion.

cross matching. *See* **cross agglutination test.**

crossopterygian a lobe-finned bony fish; one group of which was ancestral to the amphibians.

crossover fixation the spreading of a mutation in one member of a tandem gene cluster through the entire cluster as a consequence of unequal crossing-over.

crossover region the segment of a chromosome lying between any two specified marker genes.

crossover suppressor a gene, or an inversion (*q.v.*), that prevents crossing over in a pair of chromosomes. *Gowen's crossover suppressor* gene on chromosome 3 of *Drosophila* prevents the formation of synaptonemal complexes.

crossover unit a 1% crossover value between a pair of linked genes.

cross pollination the pollination of a flower with pollen from a flower of a different genotype.

cross-reacting material (CRM) any nonfunctional protein reactive with antibodies directed against its functional counterpart.

cross-reaction, serological union of an antibody with an antigen other than the one used to stimulate formation of that antibody; such cross-reactions usually involve antigens that are stereochemically similar or those that share antigenic determinants.

cross reactivation *See* **multiplicity reactivation.**

cross section *See* **transverse section.**

cross section, nuclear *See* **barn, capture cross section.**

crown gall disease *See Agrobacterium tumefaciens.*

crozier the hook formed by an ascogenous hypha of *Neurospora* or related fungi previous to ascus development. The hook is formed when the tip cell of an ascogenous hypha grows back upon itself. Within the arched portion of the hypha, cell walls are subsequently laid down in such a way that three cells are formed. The terminal cell of the branchlet is uninucleate, the penultimate one is binucleate, and the antipenultimate one is uninucleate. Fusion of the haploid nuclei of different mating types in the penultimate cell occurs, and it enlarges to form the ascus in which meiosis occurs.

CRP cyclic AMP receptor protein. *See* **catabolite activating protein (CAP).**

cruciform structure a cross-shaped configuration of DNA produced by complementary inverted repeats (palindromes) pairing with one another on the same strand instead of with its normal partner on the other strand.

cryostat a device designed to provide low-temperature environments in which operations (like sectioning frozen tissues) may be carried out under controlled conditions.

cryptic satellite a satellite DNA sequence that cannot be separated from the main-band DNA by density gradient ultracentrifugation. Cryptic satellite DNA can be isolated from the main-band DNA by its unique features (e.g., by the more rapid reannealing of the highly repeated segments that constitute the satellite).

cryptic species phenotypically similar species which never form hybrids in nature. *See* **sibling species.**

cryptoendomitosis somatic polyploidization taking place within an intact nuclear envelope. No stages comparable to the normal mitotic cycle are observed, but the DNA content increases in multiples of the haploid value.

cryptogam a spore-bearing rather than a seed-bearing plant. In older taxonomy a member of the Cryptogamia, including the ferns, mosses, algae, and fungi. *See* **phanerogam.**

crystal violet a basic dye used in cytochemistry.

c-src a cellular gene, present in various vertebrates, that hybridizes with *src*, the oncogene of the Rous sarcoma virus (*q.v.*). The *c-src* genes code for *pp60c-src* proteins that resemble *pp60v-src* proteins in their enzymatic properties.

C-terminus that end of the peptide chain which carries the free alpha carboxyl group of the last amino acid. By convention, the structural formula of a peptide chain is written with the C-terminus to the right. *See* **translation.**

"C"-type particles a group of RNA viruses with similar morphologies under the electron microscope, having a centrally placed, spherical RNA-containing nucleoid. These viruses are associated with many sarcomas and leukemias. The "C" represents "cancer."

Cucumis a genus of nearly 40 species including several of considerable economic importance, such as the cucumber *(C. sativus)* and the muskmelon *(C. melo)*. Considerable genetic information is available for both these species.

Cucurbita a genus of about 27 species including 5 that are extensively cultivated: *C. pepo,* summer squash; *C. mixta,* cushaws; *C. moschata,* winter squash; *C. maxima,* Hubbard squash; and *C. ficifolia,* Malabar gourds. Most genetic information is available for *C. pepo* and *C. maxima.*

Culex pipiens the most widely distributed species of mosquito in the world. The genetics of insecticide resistance has been intensively studied in this species. Giant polytene chromosomes occur in the salivary gland and Malpighian tubule cells of larvae.

cull to pick out and discard inferior animals from a breeding stock.

cultivar a variety of plant produced through selective breeding by humans and maintained by cultivation.

curie the quantity of a radioactive nuclide disintegrating at the rate of 3.700×10^{10} atoms per second.

cut a double-strand incision in a duplex DNA molecule. *Compare with* **nick.**

cut and patch repair repair of damaged DNA molecules by the enzymatic excision of the defective single-stranded segments and the subsequent synthesis of new segments. Using the complementary strand as a template the correct bases are inserted and are interlinked by a DNA polymerase. A DNA ligase joins the two ends of the "patch" to the broken strand to complete the repair.

cuticle the chitinous, acellular outer covering of insects.

cuvette a small glass or quartz cell used to hold fluids to be measured in a spectrometer.

C value the amount of DNA (e.g., expressed in picograms per cell) comprising the haploid genome for a given species.

C value paradox the paradox that there is often no correlation between the C values of species and their evolutionary complexity. For example, the C values for mammals fall into a narrow range (between 2 and 3 pg). By contrast, the C values for amphibia vary from 1 to 100 pg. However, the minimum C values reported for species from each class of eukaryotes does increase with evolutionary complexity. It follows that in species with C values above the expected range, a large fraction of the DNA must have functions other than coding for proteins, such as acting as controlling elements, or else it is functionless, *selfish DNA (q.v.).*

CVS chorionic villi sampling *(q.v.).*

cyanocobalamin cobalamin.

cyanogen bromide Br$-$C$\equiv$N, a reagent used for splitting polypeptides at methionine residues; commonly used in studies of protein structure and the determination of amino acid sequences.

cyanolabe *See* **color blindness.**

cyanosis a bluish coloration on the lips and on the ends of the extremities, usually from a cardiovascular condition.

cyclically permuted sequences DNA sequences of the same length containing genes in the same linear order, but starting and ending at different positions, as in a circle. In T4 DNA, each phage contains a different cyclically permuted sequence that is also terminally redundant. Cyclic permutation is a property of a population of phage DNA molecules, whereas terminal redundancy is a property of an individual phage DNA molecule. *See* **headful mechanism, terminal redundancy.**

cyclical selection   selection in one direction followed by selection in the opposite direction resulting from cyclical environmental fluctuations, such as seasonal temperature changes. If the generation time is short relative to the environmental cycle, different genotypes will be selected at different times, and the population will remain genetically inhomogeneous.

cyclic AMP   adenosine monophosphate with the phosphate group bonded internally to form a cyclic molecule; generated from ATP by the enzyme adenylcyclase; abbreviated cAMP. Likewise, guanosine monophosphate (GMP) can become a cyclic molecule by a phosphodiester bond between 3′ and 5′ atoms. Cyclic AMP has been shown to function as an acrasin in slime molds and to be active in the regulation of gene expression in both prokaryotes and eukaryotes. In *Escherichia coli,* cyclic AMP is required for the transcription of certain operons. *See* **adenylcyclase, catabolite repression.**

cycloheximide   an inhibitor of translation on 80S ribosomes; an antibiotic synthesized by *Streptomyces griseus.*

cyclophosphamide   an immunosuppressive drug (*q.v.*).

cyclorrhaphous diptera.   flies belonging to the suborder Cyclorrhapha, which contains the most highly developed flies. It includes the hover flies, the drosophilids, house flies, blow flies, etc.

cyclotron *See* **accelerator.**

cys cysteine. *See* **amino acid.**

cysteine *See* **amino acid.**

cystic fibrosis one of the most common hereditary diseases of white children, inherited as an autosomal recessive. The syndrome is due to the production of a unique glycoprotein that results in a mucous of abnormally high viscosity. In Caucasians in the United States the prevalence is 1/2,000, with a carrier rate of 1/20.

cystoblast a mitotically active germarial cell that forms the interconnected cystocytes of the *Drosophila* ovary. *See* **nurse cells.**

cystocyte a daughter cell formed by the division of a cystoblast in the *Drosophila* germarium. *See* **nurse cells, pro-oocyte.**

cytidine *See* **nucleoside.**

cytidylic acid *See* **nucleotide.**

cytochalasin B a mold metabolite that selectively inhibits cytokinesis by decomposing the contractile ring (*q.v.*).

cytochrome system electron transport chain (*q.v.*). The electrons produced during the oxidations occurring in the citric acid cycle (*q.v.*) are transferred via a series of coupled oxidations to the final hydrogen and electron acceptor, oxygen, to form water. The molecules involved in this electron transport chain are NAD (*q.v.*), FAD (*q.v.*), coenzyme Q (*q.v.*), and cytochromes b, c, a, and a_3. The cytochromes are heme-containing proteins, and electron transport depends upon the continual oxidation and reduction of the iron atom contained in heme. The sequence of reactions is diagrammed below.

electrons from citric
acid cycle

cytogenetic map   a map showing the locations of genes on a chromosome.

cytogenetics   the science that combines the methods and findings of cytology and genetics.

cytohet   an eukaryotic cell containing two genetically different types of a specific organelle; the term is an abbreviation for *cyto*plasmically *het*erozygous. For example, in the single-celled alga *Chlamydomonas,* the frequency of rare cytohets (containing chloroplasts from both parents) can be greatly increased by treatment of one parent (mating type +) with ultraviolet light.

cytokinesis   cytoplasmic division as opposed to karyokinesis (*q.v.*).

cytokinin   one of a group of N-substituted adenines which promote the division of plant cells. *See* **kinetin, zeatin.**

cytological hybridization   synonomous with *in situ* hybridization (*q.v.*).

cytological map   a diagramatic representation of the physical location of genes at specific sites, generally on dipteran giant polytene chromosomes or on human mitotic chromosomes.

cytology   the branch of biology dealing with the structure, function, and life history of the cell.

cytolysis   the dissolution of cells.

cytophotometry   quantitative studies of the localization within cells of various organic compounds using microspectrophotometry. Cytophotometric techniques are employed, for example, to determine changes in the DNA contents of cells throughout their life cycle. *See* **microspectrophotometer.**

cytoplasm   the protoplasm exclusive of that within the nucleus (which is called nucleoplasm).

cytoplasmic inheritance   non-Mendelian heredity involving replication and transmission of extrachromosomal genetic information found in organelles such as mitochondria and chloroplasts or in intracellular parasites such as viruses.

cytoplasmic male sterility   pollen abortion due to cytoplasmic factors which are maternally transmitted, but which act only in the absence of pollen-restoring genes. Such sterility can also be transmitted by grafting. This finding implicates a plant virus as the agent responsible for pollen abortion.

cytoplasmic matrix   *See* **microtrabecular lattice.**

cytoplast   the structural and functional unit of an eukaryotic cell formed by a lattice of cytoskeletal proteins to which are linked the nucleus and the cytoplasmic organelles.

cytosine   a pyrimidine base found in DNA and RNA.

cytosine deoxyriboside   *See* **nucleoside.**

cytoskeleton an internal skeleton that gives the eukaryotic cell its ability to move, to assume a characteristic shape, to divide, to undergo pinocytosis, to arrange its organelles, and to transport them from one location to another. The cytoskeleton contains microtubules, microfilaments, and intermediate filaments.

cytosol the fluid portion of the cytoplasm exclusive of organelles; synonomous with hyaloplasm. *See* **cell fractionation.**

cytostatic referring to any agent that suppresses cell multiplication and growth.

cytostatic agent an agent that inhibits cell growth.

cytotaxis the ordering and arranging of new cell structure under the influence of preexisting cell structure. The information controlling the three-dimensional architecture of the eukaryotic cell is thought to reside in the structure of the cytoplasmic ground substance. Evidence for this comes from microsurgical experiments on *Paramecium*. Cortical segments reimplanted with inverted polarity result in a changed pattern that is inherited through hundreds of generations. *See* **microtrabecular lattice.**

cytotoxic T lymphocytes *See* **K cells.**

D

d **1.** dextrorotatory. **2.** the dalton unit.

D Devonian.

2,4D 2,4 dichlorophenoxyacetic acid (*q.v.*).

(dA)$_{3000}$ a homopolymer chain containing about 3000 deoxyriboadenylate residues.

dalton a unit equal to the mass of the hydrogen atom (1.67×10^{-24} g). Abbreviated d.

dam the term used in the breeding of economically important mammals to refer to the female parent.

Danaus plexippus the Monarch butterfly (*q.v.*); the model mimicked by the Viceroy. *See* **Batesian mimicry.**

dark-field microscope a microscope designed so that the entering center light rays are blacked out and the peripheral rays are directed against the object from the side. As a result the object being viewed appears bright upon a dark background.

dark reactivation repair of mutagen-induced genetic damage by enzymes that do not require light photons for their action. *See* **photoreactivating enzyme.**

Darwinian evolution *See* **Darwinism.**

Darwinian fitness synonomous with adaptive value (*q.v.*).

Darwinian selection synonomous with natural selection (*q.v.*).

Darwinism the theory that the mechanism of biological evolution involves natural selection of adaptive variations.

Darwin's finches *See* **Geospizinae.**

Datura stramonium the Jimson weed, a species in which classic studies on polysomy have been done.

dauermodification an environmentally induced phenotypic change in a cell that survives in the generative or vegetative descendants of the cell in the absence of the original stimulus. However, with time the trait weakens and eventually disappears.

daughter cells (nuclei) the two cells (nuclei) resulting from division of a single cell (nucleus).

100

DBM paper   diazobenzyloxymethyl paper that binds all single-stranded DNA, RNA, and proteins by means of covalent linkages to the diazonium group; used in situations where nitrocellulose blotting is not technically feasible. *See* **nitrocellulose filter.**

DDT   an insecticide, dichlorodiphenyltrichlorethane (*q.v.*).

DEAE-cellulose   trademark for a family of anionic ion exchange resins.

deamination   the oxidative removal of NH_2 groups from amino acids to form ammonia.

decarboxylation   the removal or loss of a carboxyl group from an organic compound and the formation of CO_2.

decay constant   disintegration constant.

decay of variability   the reduction of heterozygosity because of the loss and fixation of alleles at various loci accompanying genetic drift.

deci-   a prefix meaning one-tenth, applied to metric units measurement.

deciduous   **1.** designating trees whose leaves fall off at the end of the growing season, as opposed to evergreen. **2.** designating teeth that are replaced by permanent teeth.

dedifferentiation   the loss of differentiation, as in the vertebrate limb stump during formation of a blastema.

defective virus   a virus that is unable to reproduce in its host without the presence of another "helper" virus.

deficiency   in cytogenetics, the loss of a microscopically visible segment of a chromosome. In a structural heterozygote (containing one normal and one deleted chromosome), the nondeleted chromosome forms an unpaired loop opposite the deleted segment when the chromosomes pair during meiosis.

deficiency loop   in polytene chromosomes, deficiency loops allow one to determine the size of the segment missing. The illustration below shows a portion of the X chromosome from the nucleus of a salivary gland cell of a *Drosophila* larva structurally heterozygous for a deficiency. Note that bands 3C2—3C11 are missing from the lower chromosome.

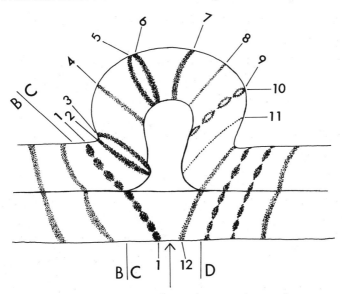

definitive host   the host in which a parasite attains sexual maturity.

deformylase an enzyme in prokaryotes that removes the formyl group from the NH_2-terminal amino acid; fMet is never retained as the NH_2-terminal amino acid in functional polypeptides. *See* **start codon.**

degenerate code one in which each different word is coded by a variety of symbols or groups of letters. The genetic code is said to be degenerate because more than one nucleotide triplet codes for the same amino acid (*q.v.*).

degrees of freedom the number of items of data that are free to vary independently. In a set of quantitative data, for a specified value of the mean, only (*n*-1) items are free to vary, since the value of the *nth* item is then determined by the values assumed by the others and by the mean.

dehiscent designating fruit that opens when ripe to release seeds.

deka- (also **deca-**) a prefix meaning ten, applied to metric units of measurement.

delayed dominance *See* **dominance.**

delayed hypersensitivity a cell-mediated immune response manifested by an inflammatory skin response 24–48 hours after exposure to antigen. *Compare with* **immediate hypersensitivity.**

delayed Mendelian segregation *See **Limnea peregra.***

deletion the loss of a segment of the genetic material from a chromosome. The size of the deletion can vary from a single nucleotide to sections containing a number of genes. If the lost part is at the end of a chromosome, it is called a terminal deletion. Otherwise it is called an intercalary deletion.

deletion mapping the use of overlapping deletions to localize the position of an unknown gene on a chromosome or linkage map.

deletion method a method of isolating specific messenger RNA molecules by hybridization with DNA molecules containing genetic deletions.

deletion-substitution particles a specialized transducing phage in which deleted phage genes are substituted by bacterial genes.

deliquesce to become liquified by absorbing moisture from the environment.

delta chain a component of hemoglobin A_2. *See* **hemoglobin.**

delta ray the track or path of an electron ejected from an atomic nucleus when an ionizing particle passes through a detection medium, especially through a photographic emulsion.

deme a geographically localized population within a species.

demographic transition a phenomenon in the history of industrialized nations wherein the death rate declines before the birth rate declines, producing an increase in population size.

denaturation the loss of the native configuration of a macromolecule resulting from heat treatment, extreme *p*H changes, chemical treatment, etc. Denaturation is usually accompanied by loss of biological activity. Denaturation of proteins often results in an unfolding of the polypeptide chains and renders the molecule less soluble. Denaturation of DNA leads to changes in many of its physical properties, including viscosity, light scattering, and optical density. This "melting" occurs over a narrow range of temperatures and represents the dissociation of the double helix into its complementary strands. The midpoint of this transition is called the melting temperature. *See* T_m.

denaturation map a map, obtained through electron microscopy, of a DNA segment indicating the location of denaturation loops induced by heating to the point where melting is just detected, then stabilized with formaldehyde; the loops represent regions of high AT content.

denatured DNA   *See* **denaturation.**

denatured protein   *See* **denaturation.**

dendrite   one of the many short, branching cytoplasmic projections of a neuron. Dendrites synapse with and receive impulses from the axons of other neurons. These impulses are then conducted toward the perikaryon.

dendritic evolution   the lineage relation which when diagrammed resembles a tree. The characteristic evolutionary relation seen for the interrelated species of the animal kingdom. *See* **reticulate evolution, cladogram.**

Denhardt's solution   a solution consisting of Fircoll, polyvinylpyrrolidone, and bovine serum albumin, each at a concentration of 0.02% (w/v). Preincubation of nucleic acid-containing filters in this solution prevents non-specific binding of single-stranded DNA probes.

de novo   **1.** arising from an unknown source. **2.** denoting synthesis of a specified molecule from very simple precursors, as opposed to the formation of the molecule by the addition or subtraction of a side chain to an already complex molecule.

de novo **pathway**   a process for synthesizing ribonucleoside monophosphates from phosphoribosylpyrophosphate, amino acids, CO_2, and NH_3, rather than from free bases, as in the salvage pathway.

densitometer   an instrument used for measuring the light transmitted through an area of interest. Densitometers are used for scanning chromatograms and electropherograms and for measuring the blackening of photographic films.

density-dependent factor   an ecological factor (e.g., food) that becomes increasingly important in limiting population growth as the population size increases.

density gradient equilibrium centrifugation   *See* **centrifugation separation.**

density gradient zonal centrifugation   *See* **centrifugation separation.**

density-independent factor   an ecological factor (e.g., temperature) that is uncorrelated with variations in size of a population.

dent corn   *See* **corn.**

deoxyadenylic, deoxycytidylic, deoxyguanylic acids   *See* **nucleotide.**

deoxyribonuclease   any enzyme that digests DNA to oligonucleotide fragments. DNAase I (also known as pancreatic DNAase) hydrolyzes DNA into four base fragments; DNAase II (also known as spleen DNAase) generates six base pair fragments. Both enzymes are endonucleases that introduce single strand breaks in the double helix.

deoxyribonucleic acid   DNA, the molecular basis of heredity. DNA consists of a polysugar-phosphate backbone from which the purines and pyrimidines project. The backbone is formed by bonds between the phosphate molecule and carbon 3 and carbon 5 of adjacent deoxyribose molecules. The nitrogenous base extends from carbon 1 of each sugar. According to the Watson-Crick model, DNA forms a double helix that is held together by hydrogen bonds between specific pairs of bases (thymine to adenine and cytosine to guanine). Each strand in the double helix is complementary to its partner strand in terms of its base sequence. The antiparallel strands form a right-handed helix which undergoes one com-

103

plete revolution with each ten nucleotide pairs. DNA molecules are the largest biologically active molecules known, having molecular weights greater than 1×10^8 daltons.

Thymine   Adenine

Guanine   Cytosine

At the top of the next page is shown a diagram of the ladder-like DNA molecule. "Uprights" of the ladder consist of alternating phosphate (P) and deoxyribose sugar (S) groups. The "cross rungs" consists of purine-pyrimidine base pairs which are held together by hydrogen bonds (represented here by dashed lines). Note that the AT pairs are held together less strongly than the GC pairs. In reality, the ladder is twisted into a right-handed double-helix. Each nucleotide pair is rotated 36° with respect to its neighbor. Thus ten successive nucleotide pairs complete one full turn of the helix. Only five nucleotide pairs are shown. A DNA molecule of molecular weight 2.5×10^7d would be made up of approximately 40,000 nucleotide pairs. A, T, G, and C represent adenine, thymine, guanine, and cytosine, respectively. The type of DNA described above is the B form that occurs under hydrated conditions and is thought to be the principal biological conformation. The A form occurs under less hydrated conditions. Like the B form, it too is a right-handed double helix; however, it is more compact with 11 base pairs per turn of the helix. The bases of the A form are tilted 20° away from perpendicular and displaced laterally in relation to the diad axis. The Z form of DNA is a left-handed double helix. It has 12 base pairs per turn of the helix, and presents a zigzag conformation (hence the symbolic designation). Z DNA has been localized in the interband regions of *Drosophila* salivary gland chromosomes and is hypothesized to exert regulatory controls over neighboring genes.

deoxyribonucleoside a molecule containing a purine or pyrimidine attached to deoxyribose.

deoxyribonucleotide a compound consisting of a purine or pyrimidine base bonded to deoxyribose, which in turn is bound to a phosphate group.

104

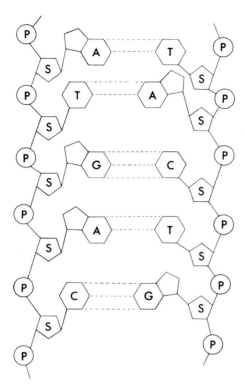

deoxyribonucleic acid

deoxyribose the sugar characterizing DNA.

$$
\begin{array}{c}
\overset{H \quad H}{\underset{5}{OH-C}} \quad \overset{O \quad H}{\underset{1}{C}} \\
\overset{4}{C} \quad \\
H \quad H \quad H \quad OH \\
\overset{3}{C} - \overset{2}{C} \\
OH \quad H
\end{array}
$$

dependent differentiation differentiation of an embryonic tissue caused by a stimulus coming from other tissue and dependent upon that stimulus.

depolymerization the breakdown of an organic compound into two or more molecules of less complex structure.

derepression an increased synthesis of gene product accomplished by preventing the interaction of a repressor with the operator portion of the operon in question. In the case of inducible enzyme systems, the inducer derepresses the operon. A mutation of the regulatory gene that blocks synthesis of the repressor or a mutation of the operator gene which renders it insensitive to a normal repressor will also result in derepression. *See* **regulator gene.**

derived the more recent stages or conditions in an evolutionary lineage; the opposite of primitive.

dermatoglyphics the study of the patterns of the ridged skin of the palms, fingers, soles, and toes.

105

desmids green algae that exist as pairs of cells with their cytoplasms joined at an isthmus that contains a single shared nucleus. *See* **classification,** Protoctisa: Gamophyta.

desmin a 51,000 d cytoskeletal protein. Desmin molecules fall into the intermediate filament class and are found in glial and muscle cells.

desmosome an intercellular attachment device. It is a discontinuous button-like structure consisting of two dense plaques on the opposing cell surfaces, separated by an intercellular space about 25 mμ wide. On each symmetrical half-desmosome a thin layer of dense material coats the inner leaf of the cell membrane, and bundles of fine cytoplasmic filaments converge upon and terminate in this dense substance.

desynapsis the failure of homologous chromosomes that have synapsed normally during pachynema to remain paired during diplonema. Desynapsis is usually the result of a failure of chiasma formation. *Contrast with* **asynapsis.**

detached X an X chromosome formed by the detaching of the arms of an attached X chromosome (*q.v.*), generally through crossing over with the Y chromosome.

determinant in immunology, the portion of the antigen which is responsible for the specificity of the response and which is recognized by the binding sites of immunoglobulins and antigen-recognizing lymphocytes.

determinant cleavages a successive series of cleavages that follow a specific three-dimensional pattern such that with each division, cells are produced each of which can be shown to serve as the progenitor of a specific type of tissue. In developing mollusc eggs, for example, cell 4d which is formed at the sixth cleavage is always the progenitor of all primary mesodermal structures.

determination the establishment of a single kind of histogenesis for a part of an embryo, which it will perform irrespective of its subsequent situations.

deutan *See* **color blindness.**

deuteranomaly *See* **color blindness.**

deuteranopia *See* **color blindness.**

deuterium *See* **hydrogen.**

deuteron the nucleus of a deuterium atom, containing one proton and one neutron.

Deuterostomia a subdivision of the Bilateria containing species in which the mouth does not arise from the blastopore. The coelom arises from the primitive gut. *See* **classification, Protostomia.**

deuterotoky parthenogenesis in which both males and females are produced.

developer a chemical that serves as a source of reducing agents which will distinguish between exposed and unexposed silver halide and convert the exposed halide to metallic silver, thus producing an image on a photographic film.

development an orderly sequence of progressive changes resulting in an increased complexity of a biological system. *See* **differentiation, morphogenesis.**

developmental control genes genes that have as their primary functions the control of developmental decisions. Such genes that regulate cell fates during development have been extensively studied in *Caenorhabditis elegans* (*q.v.*). For example in this nematode a specific set of cells can differentiate in two different ways to form the vulva or part of the uterus. Greenwald *et al.* have shown that the *lin-12* locus acts as a binary switch to affect the alternative cell fates. High activity of the locus specifies a ventral uterine precursor cell;

low activity specifies an anchor cell which organizes the development of the vulva. *See* **cell lineage mutants.**

developmental genetics the study of mutations that produce developmental abnormalities in order to gain understanding of how normal genes control growth, form, behavior, etc.

developmental homeostasis canalization (*q.v.*).

developmental homology anatomical similarity due to derivation from a common embryological source; e.g., the halteres of flies are developmentally homologous to the hind wings of moths.

deviation the departure of a quantity (derived from one or more observations) from its expected value (usually the mean of a series of quantities).

devolution regressive evolution.

Devonian a period in the Paleozoic era. *See* **geologic time divisions.**

Devoret test a test for potentially carcinogenic substances, based upon induction of lysis of phage lambda in lysogenic *E. coli*. The Devoret test is capable of identifying weak carcinogens whose detection is beyond the sensitivity of the Ames test (*q.v.*).

dex dextrorotatory. *See* **optical isomerism.**

dextran a polysaccharide composed of a single repeating simple sugar.

dextrose glucose (*q.v.*).

d.f., D/F degrees of freedom (*q.v.*).

diabetes insipidus excessive excretion of urine of normal analysis; brought about because of inadequate output of pituitary antidiuretic hormone.

diabetes mellitus a hereditary metabolic disorder characterized by glucosuria and caused by the inability to manufacture insulin, often detected by excessive sugar in the urine.

diakinesis *See* **meiosis.**

diallelic referring to a polyploid in which two different alleles exist at a given locus. In a tetraploid $A_1A_1A_2A_2$ and $A_1A_2A_2A_2$ would be examples.

dialysis the separation of molecules of differing size from a mixture by their differential diffusibility through a porous membrane. In the procedure knowns as *equilibrium dialysis,* soluble molecules of the same size are allowed to reach equivalent concentrations on either side of a semipermeable membrane. At equilibrium, if more molecules are detected on one side of the membrane, it indicates that they have become bound to some other larger molecules (e.g., repressor proteins, transport proteins, antibodies, etc.) present only on that side of the membrane, and thus are too large to pass through the pores of the membrane. This procedure is also used in immunology as a method of determining association constants for hapten-antibody reactions.

2,6-diaminopurine a mutagenically active purine analogue.

diapause   a period of inactivity and suspension of growth in insects accompanied by a greatly decreased metabolism. In a given species diapause usually takes place in a specific stage in the life cycle, and it often provides a means of surviving the winter.

diasteromer   epimer (*q.v.*).

diauxy   the adaptation of a microorganism to culture media containing two different sugars. The organism possesses constitutive enzymes for one of the sugars which it utilizes immediately. Induced enzyme synthesis is required before the second sugar can be metabolized.

dicentric   designating a chromosome or chromatid having two centromeres.

dichlorodiphenyltrichloroethane   DDT; an insecticide to which many insect species have developed resistant races.

```
                          Cl
      H  H    Cl | Cl    H  H
      |  |     \ | /     |  |
      C = C     C       C = C
     /     \    |      /     \
 Cl-C       C - C - C        C-Cl
     \\    //   |    \\     //
      C - C     H     C - C
      |  |            |  |
      H  H            H  H
```

2,4-dichlorophenoxyacetic acid   a phytohormone used as a weed killer.

```
        H  H
        |  |
        C - C
       /     \
   Cl-C       C-O-CH₂-COOH
       \     /
        C - C
       /     \
      H       Cl
```

2,6-dichlorophenoxyacetic acid   an antiauxin (*q.v.*).

dichogamous   referring to flowers or hermaphroditic animals characterized by male and female sex organs which become mature at different times.

dictyosome   the Golgi apparatus in plants.

Dictyostelium discoideum   *See* **Acrasiomycota.**

dictyotene stage   a prolonged diplotene stage of meiosis seen in oocytes during vitellogenesis. The chromosomes that have already undergone crossing over may remain in this stage for months or even years in long-lived species.

2′,3′-dideoxynucleoside triphosphates   analogues of normal 2′-deoxyribonucleoside triphosphates used in a modified "minus" technique for base sequencing of DNA molecules. Because these analogs have no oxygen at the 3′ position in the sugar, they act as specific chain-terminators (*q.v.*) for primed synthesis techniques (*see* **DNA sequencing techniques**). Nucleotides in which arabinose is substituted for deoxyribose also exhibit this chain-terminating effect.

differential affinity   the failure of two partially homologous chromosomes to pair during meiosis when a third chromosome is present that is more completely homologous to one of the two. In its absence, however, pairing if the partially homologous chromosomes can occur. *See* **autosyndesis, homoeologous chromosomes.**

differential segment   *See* **pairing segment.**

differentiation   the complex of changes involved in the progressive diversification of the structure and functioning of the cells of an organism. For a given line of cells, differentiation results in a continual restriction of the types of transcription that each cell can undertake. *See* **development, morphogenesis.**

differentiation antigen   a cell-surface antigen that is expressed only during a specific period of embryological differentiation.

diffuse centromere or kinetochore   *See* **centromere.**

diffusion   the tendency for molecules because of their random heat motion to move in the direction of a lesser concentration, and so make the concentration uniform throughout the system.

digital computer   *See* **computer.**

dihaploid   a diploid cell, tissue, or organism having arisen from a haploid cell by chromosome doubling.

dihybrid   a genotype characterized by heterozygosity at two loci. Mendel found that crosses between pure lines of peas that differed with respect to two unrelated traits produced genetically uniform F_1 dihybrid offspring. Intercrossing F_1 dihybrids produced parental and recombinant types in the F_2 population.

dihydrouridine   *See* **rare bases.**

2,5-dihydroxyphenylacetic acid   homogentisic acid (*q.v.*).

dihydroxyphenylalanine (DOPA)   a precursor of melanin. (*q.v.*).

dimer   a chemical entity consisting of an association of two monomeric subunits: e.g., the association of two polypeptide chains in a functional enzyme. If the two subunits are identical, they form a homodimer; if they are nonidentical, they form a heterodimer. Hexosaminidase (*q.v.*) is an example of a heterodimeric enzyme.

dimethylguanosine   *See* **rare bases.**

dimethyl sulfate protection   a method for identifying specific points of contact between a protein (such as RNA polymerase) and DNA based on the principle that, within an endonuclease-protected region (*see* **DNAase protection**), the adenines and guanines in the site of contact are not available to be methylated by exposure to dimethyl sulfate.

dimorphism   the phenomenon of morphological differences that split a species into two groups, as in the sexual dimorphic traits distinguishing males from females.

dinitrophenol (DNP)   a metabolic poison that prevents the uptake of inorganic phosphate and the production of energy-rich phosphorus compounds like ATP. DNP is a commonly used hapten in immunological experiments. See formula on p. 110.

dinitrophenol (DNP)

$$\begin{array}{c} OH \\ | \\ C \\ \diagup \diagdown \\ HC \quad\quad C - NO_2 \\ | \quad\quad\quad || \\ HC \quad\quad CH \\ \diagdown \diagup \\ C \\ | \\ NO_2 \end{array}$$

dioecious having male and female flowers or other reproductive organs on separate uni-sexual plants, or male and female reproductive organs in separate unisexual animals. *See* **flower.**

diploblastic having a body made of two cellular layers only (ectoderm and endoderm), as the coelenterates.

diplochromosome a chromosome arising from an abnormal duplication in which the centromere fails to divide and the daughter chromosomes fail to move apart. The resulting chromosome contains four chromatids.

Diplococcus pneumoniae the pneumonia bacterium. The classical studies on transformation were performed on this organism.

diplo-haplont an organism (such as an embryophyte) in which the products of meiosis form haploid gametophytes which produce gametes. Fertilization generates a diploid sporophyte in which meiosis takes place. Thus diploid and haploid generations alternate. *Contrast with* **diplont, haplont.**

diploid condition the chromosome state in which each type of chromosome (except the sex chromosomes) is represented twice (2N).

diploidy the diploid condition.

diplonema *See* **meiosis.**

diplont an organism (such as any multicellular animal) characterized by a life cycle in which the products of meiosis function as gametes. There is no haploid multicellular stage as in a diplo-haplont and haplont (*q.v.*).

diplophase the diploid phase of the life cycle between the formation of the zygote and the meiosis.

diplospory a type of apomixis in plants in which a diploid gametophyte is formed after mitotic divisions of the spore-forming cells.

diplotene *See* **meiosis.**

dipole a molecule carrying charges of opposite sign at opposite poles.

Diptera an insect order containing midges, mosquitoes, and flies. *See* **classification.**

directional selection selection resulting in a shift in the population mean in the direction desired by the breeder or in the direction of greater adaptation by nature. For example, the breeder might select for a number of generations seeds from only the longest ear of corn in the population. *See* **disruptive selection.**

direct repeats identical or closely related DNA sequences present in two or more copies in the same orientation in the same molecule, although not necessarily adjacent.

DIS *Drosophila Information Service. (q.v.)*

disassortative mating the nonrandom pairing of males and females that differ from each other phenotypically in characters other than sexually dimorphic ones.

discoidal cleavage cleavage occurring at the surface of an enormous yolk mass.

discontinuous distribution a collection of data recorded as whole numbers, and thus not yielding a continuous spectrum of values; e.g., the number of leaves per plant in a population of plants. *See* **continuous distribution.**

discontinous replication *See* **replication of DNA.**

discontinuous variations variations that fall into two or more non-overlapping classes.

discordant twins are said to be discordant with respect to a trait if one shows the trait and the other does not.

disintegration constant the fraction of the number of atoms of a radioactive element that decay in unit time; d in the equation $N = N_o e^{-dt}$, where N_o is the initial number of atoms present, and N is the number of atoms present after some time, t.

disjunction the moving apart of chromosomes during anaphase of mitotic or meiotic divisions.

dispersal mechanism any means by which a species is aided in extending its range. For example, sticky seeds can cling to animals and be transported by them to new regions.

dispersive replication an obsolete model of DNA replication in which parental and newly synthesized daughter molecules are interspersed in an essentially random fashion.

disruptive selection the selection of divergent phenotypic extremes in a population until, after several generations of selection, two discontinuous strains are obtained. For example, the breeder might select for a number of generations seeds from the longest and the shortest ears of corn in a population. *See* **directional selection.**

disseminule a plant part that gives rise to a new plant.

***Dissociation-Activator* system** *See **Activator-Dissociation* system.**

dissociation constant (pK) a relation between proton-donor and proton-acceptor molecules in solution. It is represented by: $\log \dfrac{[pa]}{[pd]} = pH - pK$; where [pa] is the concentration of the molecule functioning as a proton acceptor, [pd] is the concentration of the molecule functioning as a proton donor, and pH is the hydrogen ion concentration of the aqueous solution. A single molecular species may be present, if the molecule is amphiprotic. Thus, if the pK of a carboxyl group is 4 and the pH is 7, $\log \dfrac{[COO^-]}{[COOH]} = 7 - 4 = 3$. Thus there will be 1000 ionized groups for each un-ionized carboxyl group in a neutral aqueous solution.

distal situated away from the place of attachment. In the case of a chromosome, the part farthest from the centromere.

distributive pairing the pairing of chromosomes at metaphase I of meiosis which leads to their proper distribution to daughter cells. Synaptonemal complexes play no role in this type of chromosomal association.

distylic species a plant species composed of two types of individuals each characterized by a different flower morphology.

disulfide linkage the sulfur-to-sulfur S-S bonding of adjacent cysteine residues in or between molecules.

diurnal 1. pertaining to the daytime. 2. recurring in the period of a day; daily.

divergence in molecular biology, the percent difference between nucleotide sequences of two related DNA segments or between amino acid sequences of the two related polypeptide chains.

divergent transcription the transcriptional orientation of different DNA segments in opposite directions from a central region.

diversifying selection *See* **disruptive selection.**

diversity in ecology, the number of species or other taxa in a particular ecological unit.

dizygotic twins fraternal twins produced by two separate ova fertilized by separate sperms. *See* **monozygotic twins.**

D loop 1. a displacement loop formed early in the replication of duplex DNA (either circular or linear) consisting of a single, unreplicated, parental strand on one side, and a double-stranded branch (composed of one parental strand paired with the leading strand, *q.v.*) on the other side. Because the leading strand displaces the unreplicated parental strand, the replication "bubble" or "eye" is called a displacement or D loop. 2. A segment of mitochondrial DNA in which one strand is displaced by a short stretch of RNA base paired to the other complementary strand.

DNA deoxyribonucleic acid (*q.v.*).

DNA-agar technique a technique for testing the degree of homology between nucleic acid molecules from different sources by allowing fragments of radioactive nucleic acid from one source to react with nonlabeled nucleic acids from another source trapped in an agar gel. This procedure binds to the gel radioactive polynucleotide fragments that are complementary to those trapped in the agar. *See* **hybrid duplex molecule.**

DNAase symbol for deoxyribonuclease (*q.v.*).

DNAase protection a method for estimating the size of a DNA region that is interacting with a protein (e.g., the size of the region occupied by RNA polymerase during transcription). After the protein is bound to DNA, an endonuclease is added that degrades most of the DNA outside the region of interaction to mono- and dinucleotides. *See* **dimethyl sulfate protection.**

DNA clone a DNA segment that has been inserted via a viral or plasmid vector into a host cell with the following consequences: the segment has replicated along with the vector to form many copies per cell, the cells have mutiplied into a clone, and the insert has been magnified accordingly.

DNA complexity a measure of the amount of nonrepetitive DNA characteristic of a given DNA sample. In an experiment involving reassociation kinetics (*q.v.*), DNA complexity represents the combined length in nucleotide pairs of all unique DNA fragments. The DNA of evolutionarily advanced species is more complex than that of primitive species.

DNA-dependent RNA polymerase RNA polymerase (*q.v.*).

DNA-driven hybridization reaction a reaction involving the reassociation kinetics of complementary DNA strands when DNA is in great excess of a radioactive RNA tracer; employed in cot analysis to determine the repetition frequencies of the corresponding genome sequences.

DNA duplex a DNA double helix. *See* **deoxyribonucleic acid.**

DNA fiber autoradiography light microscopic autoradiography of tritiated thymidine-labeled DNA molecules attached to millipore filters. The technique was devised by Cairns

(1963) for studying DNA replication in *E. coli* and later adapted by Huberman and Riggs (1968) for visualizing the multiple replicons of mammalian chromosomes.

DNA grooves   the DNA double helix contains a narrow and a wide groove which run its length. During replication, DNA polymerase is believed to occupy the wide groove. During transcription, RNA polymerase may occupy the narrow groove.

DNA gyrase   *See* **gyrase.**

DNA hybridization   a technique for determining the degree of similarity of DNA base sequences in two species, involving radioisotope labeling of DNA from one species and annealing it to the DNA of the other species. *See in situ* **hybridization.**

DNA ligases   enzymes that catalyze the formation of a phosphodiester bond between adjacent 3′-OH and 5′-P termini in DNA. DNA ligases function in DNA repair to seal single-stranded nicks between adjacent nucleotides in a duplex DNA chain. *See* **blunt end ligation, cohesive end ligation, cut and patch repair.**

DNA modification   *See* **modification.**

dna *mutations*   mutations of *E. coli* that influence DNA replication. The *dna A, dna B* and *dna C* mutations are defective in proteins that interact with replication origins. The *dna E, dna X,* and *dna Z* genes encode subunits of DNA polymerase III, and *dna G* encodes primase (*q.v.*). *See* **DNA polymerase, replicon.**

DNA polymerase   an enzyme that catalizes the formation of DNA from deoxyribonucleoside triphosphates, using single-stranded DNA as a template. Three different DNA polymerases (pol I, pol II, and pol III) have been isolated from *E. coli*. Pol III is the major enzyme responsible for cellular DNA replication in this bacterium. The other two enzymes function primarily in DNA repair. Eukaryotes contain many different DNA polymerases. They reside in different parts of the cell (nucleus, cytosol, or mitochondria) and have different functions (replication, repair, and recombination).

DNA probe   *See* **probe.**

DNA puff   *See* **chromosomal puff.**

DNA relaxing enzyme   *See* **topoisomerase.**

DNA restriction enzyme   any of the specific endonucleases (*q.v.*) present in many strains of *Escherichia coli* that recognize and degrade DNA from foreign sources. These nucleases are formed under the directions of genes called *restriction alleles*. Other genes called *modification alleles* determine the methylation pattern of the DNA within a cell. It is this pattern that determines whether or not the DNA is attacked by a restriction enzyme. *See* **modification methylases, restriction endonucleases.**

DNA-RNA hybrid   a double helix consisting of one chain of DNA hydrogen bonded to a complementary chain of RNA.

DNase   also symbolized *DNAase*. *See* **deoxyribonuclease.**

DNase protection   *See* **DNAase protection.**

DNA sequencing techniques   **1.** the method developed by F. Sanger and A. R. Coulson (1975) is known as the "plus and minus" method or the "primed synthesis" method. DNA is synthesized *in vitro* in such a way that it is radioactively labeled and the reaction terminates specifically at the position corresponding to a given base. After denaturation, fragments of different lengths are separated by electrophoresis and identified by autoradiography. In the "plus" protocol, only one kind of deoxyribonucleoside triphosphate (dNTP) is available for elongation of the ^{32}P-labeled primer. In the "minus" protocol, one of the four dNTPs is missing; alternatively, specific terminator base analogues (2′,3′-dideoxyribonucleoside triphosphates, *q.v.*) can be used instead of the "minus" technique. **2.** in the 1977 procedure of A. M. Maxam and W. Gilbert (the "chemical" method), single-stranded DNA (derived from double-stranded DNA and labeled at the 5′ end with ^{32}P) is subjected to several chemical (dimethyl sulfate–hydrazine) cleavage protocols that selectively make breaks on one side of a particular base; fragments are separated according to size by electrophoresis on acrylamide gels and identified by autoradiography.

DNA topoisomerase *See* **topiosomerase.**

DNA unwinding protein a protein that binds to single-stranded DNA and facilitates the unwinding of the DNA duplex during replication and recombination. *See* **gene 32 protein.**

DNA vector a vehicle (such as a virus or a plasmid) for transferring DNA from one cell to another.

DNP **1.** 2:4 dinitrophenol. **2.** DNA-protein complex.

dog any of a large number of domesticated breeds of the species *Canis familiaris.* Popular breeds include: TERRIERS: Welsh, Bedlington, Dandie Dinmont, West Highland White, Skye, Cairn, Scottish, Sealyham, Fox (Smooth), Fox (Wire), Schnauzer, Airedale, Irish, Kerry Blue, Bull, Manchester. POINTERS: German Shorthaired Pointer, Irish Setter, English Setter, Gordon Setter, Weimaraner, Pointer, Brittany Spaniel. COURSING HOUNDS: Irish Wolfhound, Scottish Deerhound, Greyhound, Whippet, Borzoi, Saluki, Afghan. TRAILING HOUNDS: Basenji, Bloodhound, Dachshund, Bassett, Beagle, Black and Tan Coonhound. MISCELLANEOUS HOUNDS: Otter hound, Norwegian Elkhound. FLUSHING SPANIELS: English Springer, English Cocker, American Cocker, Welsh Springer. RETRIEVERS: Golden Retriever, Labrador Retriever, Chesapeake Bay Retriever, Irish Water Spaniel, Curly-coated Retriever. SHEEP DOGS: Briard, Kuvasz, Shetland Sheepdog, Collie, Belgian Sheepdog. SLED DOGS: Siberian Huskie, Eskimo, Samoyed, Alaskan Malamute. GUARD DOGS: Bouvier de Flandres, Mastiff, Rottweiller, Boxer, Great Dane, Bull Mastiff, Schnauzer, German Shepherd, Dobermann Pinscher. MISCELLANEOUS WORKING DOGS: St. Bernard, Welsh Corgi (Cardigan), Welsh Corgi (Pembroke), Newfoundland, Great Pyrenees. TOYS: Maltese, Pug, Japanese Spaniel, English Toy Spaniel (King Charles), Pekingese, Pomeranian, Yorkshire Terrier, Griffon, Chihuahua, Papillon, Poodle (Toy), Mexican hairless. NONSPORTING BREEDS: Lhasa Apso, Poodle (Standard), Poodle (Miniature), Dalmation, Chow Chow, Keeshond, Schipperke, English Bulldog, French Bulldog, Boston Terrier.

Dollo's law the proposition that evolution along any specific lineage is essentially irreversible. For example, no modern mammal can de-evolve back to a form identical in all respects to the mammal-like, reptilian ancestor from which it was derived.

domain **1.** a homology unit; i.e., any of the three or four homologous regions of an immunoglobulin heavy chain that apparently evolved by duplication and diverged by mutation. **2.** any discrete, continuous part of a polypeptide sequence that can be equated with a particular function. **3.** any region of a chromosome within which supercoiling is independent of other domains. **4.** an extensive region of DNA including an expressed gene that exhibits pronounced sensitivity to degradation by DNAase I.

domesticated species *See* **Appendix B.**

dominance referring to alleles that fully manifest their phenotype when present in the heterozygous, heterokaryotic, or heterogenotic state. The alleles whose phenotypic expressions are masked by dominant alleles are termed *recessive alleles.* Sometimes the dominant allele expresses itself late in development (e.g., Huntington's chorea, *q.v.*), in which case the allele is said to show *delayed dominance.*

dominance variance genetic variance for a polygenic trait in a given population attributed to the dominance effects of contributory genes.

dominant complementarity *See* **complementary genes.**

dominant gene *See* **recessive gene.**

donkey *Equus asinus,* a close relative of the horse. The female is referred to as a jennet, the male as a jack. *See* **horse-donkey hybrids.**

donor an individual from whom tissue is taken for transfer to another.

donor splicing site *See* **left splicing junction.**

DOPA dihydroxyphenylalanine (*q.v.*).

dosage compensation a mechanism that regulates the expression of sex-linked genes that differ in dose between females and males in species with an XX-XY method of sex determination. The hypothesis put forth by Mary Lyon proposes that dosage compensation in mammals is accomplished by inactivating at random one of the two X chromosomes in all somatic cells of the female. The inactivated X forms the Barr body or sex chromatin. In cases where multiple X chromosomes are present, all but one are inactivated. *See* **Fabry disease, glucose-6-phosphate dehydrogenase deficiency, Lesch-Nyhan syndrome, ocular albinism.**

dose 1. gene dose—the number of times a given gene is present in the nucleus of a cell. 2. radiation dose—the radiation delivered to a specific tissue area or to the whole body. Units for dose specifications are the roentgen, rad, and rep.

dose-action curve dose-response curve (*q.v.*).

dose fractionation the administration of radiation in small doses at regular intervals.

dose-response curve the curve showing the relation between some biological response and the administered dose of radiation.

dosimeter an instrument used to detect and measure an accumulated dosage of radiation.

Dotted a gene symbolized by *Dt* residing on chromosome 9 of maize which influences the rate at which *a* mutates to *A*. *A* is on chromosome 3.

double cross the technique used for producing hybrid seed for field corn. Four different inbred lines (A, B, C, and D) are used. A × B → AB hybrid and C × D → CD hybrid. The single cross hybrids (AB and CD) are then crossed and the ABCD seed is used for the commercial crop.

double crossover *See* **double exchange.**

double diffusion technique synonymous with Ouchterlony technique (*q.v.*).

double exchange breakage and interchange occurring twice within a tetrad involving two, three, or four of the chromatids.

double fertilization in flowering plants the union of one sperm nucleus with the egg nucleus to form the diploid zygote and of the other sperm nucleus with the polar nuclei to form a triploid endosperm nucleus. *See* **pollen grain.**

double helix the Watson-Crick model of DNA structure, involving plectonemic coiling (*q.v.*) of two hydrogen-bonded polynucleotide, antiparallel (*q.v.*) strands wound into a right-handed spiral configuration. *See* **deoxyribonucleic acid.**

double infection infection of a bacterium with two genetically different phages.

double-sieve mechanism a model that explains the rarity of misacylation of amino acids by proposing that an amino acid larger than the correct one is rarely activated because (1) it is too large to fit into the active site of the tRNA synthetase (first sieving), and (2) the hydrolytic site of the same synthetase is too small for the correct amino acid (second sieving). Thus an amino acid smaller than the correct one can be removed by hydrolysis.

doublet an aberrant race of *Paramecium aurelia* containing the cortical features of a single animal in duplicate.

double transformation *See* **cotransformation.**

double X in *Drosophila melanogaster* an acrocentric, double length X chromosome arising as a radiation-induced aberration. Such double X chromosomes are superior to the ordinary metacentric, attached X chromosomes (*q.v.*) for most stock-keeping operations, since they do not break up, by crossing over with the Y. *See also* **detached X.**

doubling dose that dose of ionizing radiation which doubles the spontaneous mutation rate of the species under study.

doubling time the average time taken for the cell number in a population to double. The doubling time will equal the generation time (*q.v.*) only if (1) every cell in the population is capable of forming two daughter cells, (2) every cell has the same average generation time, and (3) there is no lysis of cells. The doubling time is generally longer than the generation time.

Dowex trademark of a family of ion exchange resins.

down promoter mutations promoter alterations that decrease the frequency with which transcription is initiated relative to wild type; promoters with this property are called low-level or weak promoters.

downstream monomeric sequences of a linear polynucleotide proceeding in the direction of gene expression; **1.** during transcription, the direction of deoxyribonucleotide sequences on the sense strand of DNA from 3′ to 5′; **2.** during translation, the direction of ribonucleotide sequences on mRNA from 5′ to 3′; **3.** in a polynucleotide chain, the direction in which amino acids are linked togther from the free amino end toward the free carboxyl end. *Compare with* **upstream.**

Down syndrome a type of mental retardation due to trisomy of autosome 21. Since the eyelid openings of the patient are oblique and the inner corner of the eyelid may be covered by an epicanthic fold the condition is often called mongolism. *See* **translocation Down syndrome.**

DPN diphosphopyridine nucleotide. NAD is the preferred nomenclature.

drawing tube a device attached to a compound microscope which permits the image and a drawing surface to be viewed concurrently, so that the object under study may be drawn accurately.

drift genetic drift. (*q.v.*)

dRNA DNA-like RNA. RNA molecules that are not included in the rRNA and tRNA classes. Much of the dRNA is of high molecular weight, short half-life, and never leaves the nucleus.

Drosophila a genus of flies containing about 900 described species. The most extensively studied of all genera from the standpoint of genetics and cytology. The genus is subdivided into eight subgenera: (1) *Hirtodrosophila*, (2) *Pholadoris*, (3) *Dorsilopha*, (4) *Phloridosa*, (5) *Siphlodora*, (6) *Sordophila*, (7) *Sophophora*, and (8) *Drosophila. D. melanogaster*, the multicellular organism for which the most genetic information is available, belongs in the Sophophora.

Drosophila Information Service a yearly bulletin that lists all publications concerning *Drosophila* that year, the stock lists of major laboratories, the addresses of all *Drosophila* workers, descriptions of new mutants and genetic techniques, research notes, and new teaching exercises. *See* **Appendix D.**

***Drosophila* salivary gland chromosomes** the most extensively studied polytene chromosomes. During larval development, the cells of the salivary gland undergo 9 or 10 cycles of endomitotic DNA replications to produce chromosomes that contain 1,000–2,000 times the haploid amount of DNA. The cytological map of the chromosomes contains slightly over 5,000 bands. It is divided into 102 divisions, distributed as shown below.

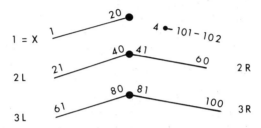

The solid circles represent the centromeres. Each division is subdivided into subdivisions lettered A-F, and the subdivisions contain varying numbers of bands. Genes have been localized within these bands by studying overlapping deficiencies and, more recently, by *in situ* hybridization with ^{3}H-labeled probes.

drosopterin one of the bright red pteridine (*q.v*) pigments of the *Drosophila* eye. The drosopterins are readily decomposed by light, and these reactions may play a major role in photoreception. *See* **sepiapterin.**

drumstick a small protrusion from the nucleus of the human polymorphonuclear leukocyte, found in 3 to 5% of these cells in females, but not in males.

drupe a simple, fleshy fruit, such as an olive, derived from a single carpel and usually single seeded.

Dryopithecus a genus of fossil primates from which the great apes and humans are thought to have diverged about 25 million years ago.

dsDNA double-standed DNA.

D_1 trisomy syndrome a well-defined set of congenital defects in man caused by the presence of an extra chromosome of the D group (13). Also called Patau's syndrome. *See* **human mitotic chromosomes.**

dual recognition an immunological model proposing that a T cell has two receptors, both of which must simultaneouly bind specific molecules in order to activate the cell; one receptor binds to the antigen, the other binds to a self-molecule of the major histocompatibility system (*q.v.*); a form of *associative recognition* (*q.v.*)

Duchenne muscular dystrophy　a form of muscular dystrophy (*q.v.*) due to a gene on the distal portion of the short arm of the human X chromosome.

Duffy blood group system　the first genetic locus in humans to be assigned to a specific autosome (i.e., No. 1). The recessive Duffy allele *fy* (governing absence of Duffy antigens a and b) is present at an extremely high frequency in the gene pools of African populations, presumably because it confers resistance to the malarial parasite, *Plasmodium vivax*. *See* **malaria.**

duplex　*See* **autotetraploid.**

duplex DNA　DNA molecules as described in the Watson-Crick model; that is, with the two polynucleotide chains of opposite 3'-5' polarity intertwined and annealed.

duplication　*See* **chromosomal aberration.**

durum wheat　*Triticum durum* (N = 14), an ancient wheat grown since Egyptian times. Used today in the manufacture of macaroni. *See* **wheat.**

dwarf　a short person with abnormal proportions, usually suffering from achondroplasia (*q.v.*). *See* **ateliosis.**

dyad　**1.** a pair of sister chromatids (*q.v.*). **2.** the products resulting from disjunction of tetrads at the first meiotic division; dyads are contained in the nuclei of secondary gametocytes. *See* **meiosis.**

Dynamic State of Body Constituents　a text published in 1942 by R. Schoenheimer which first put forward the concept of turnover (*q.v.*) as demonstrated through the use of isotopically tagged compounds in metabolic studies.

dynein　polypeptide components of the arm-like extensions of the A microtubules of the nine peripheral doublets of axonemes (*q.v.*). Dynein proteins are enzymes that function as ATPases. In *Drosophila melanogaster* the Y chromosomes appear to contain structural genes that encode the dyneins of sperm axonemes.

dysgenic　genetically deleterious.

E

e *See* **logarithm.**

E₁, E₂, E₃ the first, second and third generation following some experimental manipulation, such as irradiation with x rays.

Ea antigens alloantigens encoded by a group of genes in the mouse, originally defined as histocompatibility genes, but confined to erythrocytes.

early genes genes expressed during early development. In T_4 bacteriophage, those genetic elements that function in the period of phage infection before the start of DNA replication. *Compare with* **late genes.**

ecdysis molting, the periodic shedding of the cuticle of arthropods.

ecdysone a hormone produced by the prothoracic gland (*q.v.*) of insects. It is required for molting and puparium formation and is also called the *molting hormone.* The structural formulas for α and β ecdysone are shown on page 120. Beta ecdysone is sometimes called 20-hydroxy-ecdysone (20 HE). The precursor of α ecdysone is dietary cholesterol (*q.v.*). The α ecdysone synthesized by the prothoracic gland is converted in peripheral tissues, such as the fat body, to β ecdysone, which is the active molting hormone. The conversion involves the addition of a hydroxyl group at the position indicated by the arrow. The relative concentrations of ecdysone and allatum hormone (*q.v.*) determine whether juvenile or adult development will follow a given molt. Temperature sensitive ecdysone mutants are known for *Drosophila melanogaster.*

ECG electrocardiogram.

echinoderm an organism with an external skeleton of calcareous plates and internal water-vascular system (starfish, sea urchins, sea lilies, sea cucumbers, etc.). *See* **classification.**

eclipsed antigens antigenic determinants of parasite origin that resemble antigenic determinants of their hosts to such a degree that they do not elicit the formation of antibodies by the host. The formation of eclipsed antigens by parasites is termed *molecular mimicry.*

eclipse period the time interval between infection and the first intracellular reappearance of infective phage particles.

eclosion the emergence of the adult insect from its pupal case.

ecdysone

CH₃ OH
H₂ CH₃ C—CH-CH—CH₂—CH₂—C—OH
C   CH₃
H₂C   CH₂   CH₃
H₂   C
C CH₃ CH   C————CH₂
HO—CH   C   C OH
HO—CH   CH   CH
C   C
H₂   O

α Ecdysone

CH₃ OH
H₂ CH₃ C—C—CH—CH₂—CH₂—C—OH
C   OH   CH₃
H₂C   C   CH₂   CH₃
H₂   C
C CH₃ CH   C————CH₂
HO—CH   C   C OH
HO—CH   CH CH
C   C
H₂ O

β Ecdysone

ecodeme a deme (*q.v.*) associated with a specific habitat (a cypress swamp, for example).

ecogeographical divergence the evolution from a single ancestral species of two or more different species each in a different geographical area and each adapted to the local peculiarities of its habitat.

ecogeographic rules any of several generalizations concerning geographic variation within a species that correlate adaptations with climate or other environmental conditions: e.g., Allen's rule, Bergman's rule, Gloger's rule.

ecological isolation a premating (prezygotic) isolating mechanism in which members of different species seldom, if ever, meet because each species prefers to live (is adapted to) different habitats.

ecological niche the position occupied by a plant or animal in its community with reference both to its utilization of its environment and its required associations with other organisms.

ecology the study of the relationships of an organism and its external environment.

ecophenotype a nongenetic phenotypic modification in response to environmental conditions.

Eco RI *See* **restriction endonuclease.**

ecosystem the interrelated organisms in a particular environment.

ecotropic virus a virus that replicates only in cells of its host, not in cells from other species. *Compare with* **amphotropic virus.**

ecotype a race (within a species) genetically adapted to a certain environment.

ectoderm a germ layer forming the external covering of the embryo and neural tube (from which develop the brain, spinal cord, and nerves). Ectodermal derivatives include all nervous tissues, the epidermis (including cutaneous glands, hair, nails, horns, the lens of the eye, etc.), the epithelia of all sense organs, the nasal cavity sinuses, the anal canal, and the mouth (including the oral glands, and tooth enamel), and the hypophysis.

ectopic pairing nonspecific pairing of intercalary and proximal heterochromatic segments of *Drosophila* salivary chromosomes.

ectoplasm the superficial cytoplasm of ciliates.

edaphic race a race that is more affected by properties of the soil than by other environmental variables.

edema swelling due to the escape of fluids from the blood vessels into the intercellular spaces of the connective tissue.

editing *See* **proofreading.**

Edman degradation a technique for determining the amino acid sequence of short polypeptide fragments having at most 15 residues. The technique uses phenyl isothiocyanate to remove the amino acid at the N terminus of a polypeptide; after it has been identified, the next amino acid is reacted, removed, and identified; etc.

EDTA ethylene dinitrilotetra-acetic acid (*q.v.*).

Edward syndrome a well-defined set of congenital defects in humans caused by the presence of an extra chromosome 18. Also called trisomy 18 syndrome or E_1 trisomy syndrome. *See* **human mitotic chromosomes.**

EEG electroencephalogram (*q.v.*).

effective fertility *See* **fertility.**

effective lethal phase the stage in development at which a given lethal gene generally causes the death of the organism carrying it.

effective population size the average number of individuals in a population which contribute genes to succeeding generations. If the population size shows a cyclical variation as a function of season of the year, predation, parasitism, and other factors, the effective population size is closer to the number of individuals observed during the period of maximal contraction.

effector an organ or cell that reacts to a nervous stimulus by doing chemical or mechanical work; for example, muscles, glands, electric organs, etc.

effector cell in immunology, a cell (usually a T lymphocyte) which carries out cell mediated cytotoxicity.

effector molecules small molecules that combine with repressor molecules and activate or inactivate them with respect to their ability to combine with an operator gene. *See* **inducible system, repressible system, regulator genes.**

efferent leading away from the cell or organ involved. In immunology, the stages involved in the responses of the sensitized immune system.

egg a femal gamete; an ovum.

egg chamber the insect ovarian follicle. In *Drosophila* it consists of a cluster of 16 interconnected cystocytes surrounded by a monolayer of follicle cells. The oocyte is the most posterior cystocyte. The remaining 15 cystocytes function as nurse cells (*q.v.*).

Ehlers-Danlos syndrome a family of hereditary diseases characterized by overelasticity and brittleness of the skin and by excessive extensibility of the joints. The underlying defects involve blocks in the synthesis of collagen (*q.v.*). In type VI, for example, a hydroxylysine-deficient collagen is produced that is unable to form intermolecular cross-links. Both X-linked and autosomal genes are involved.

eidos a Platonic form; an ideal "type." *See* **typology.**

einkorn wheat *Triticum monococcum* (N = 7), "one-grained" wheat, so called because it has a single seed per spikelet. A wheat in cultivation since stone age times. *See* **wheat.**

ejaculate **1.** the process of semen release in higer vertebrates; also called *ejaculation.* **2.** the semen released in a given copulatory interaction or in a given artificially induced response.

elaioplast an oil-rich plastid.

electroblotting *See* **blotting.**

electrode either terminal of an electrical apparatus.

electroencephalogram the record of the rhythmical changes in the electrical potential of the brain.

electrolyte a substance that when dissolved in water conducts an electric current.

electron a negatively charged particle that is a constituent of every neutral atom. Its mass is 0.000549 atomic mass units.

electron carrier an enzyme, such as a flavoprotein or cytochrome, which can gain and lose electrons reversibly.

electron-dense said of a dense area seen on an electron micrograph, since the region has prevented electrons from passing through it. High electron density may mean that the area contains a high concentration of macromolecules or that it has bound the heavy metals (Os, Mn, Pb, U) used as fixatives and/or stains.

electron microscope a magnifying system that uses beams of electrons focused in a vacuum by a series of magnetic lenses and that has a resolving power hundreds of times that of the best optical microscopes. Two main types of electron microscopes are available. In the *transmission electron microscope* (TEM), the image is formed by electrons that pass through the specimen. In the *scanning electron microscope* (SEM), image formation is based on electrons that are reflected back from the specimen. Thus the TEM resembles a standard light microscope, which is generally used to look at tissue slices, and the SEM resembles a stereoscopic dissecting microscope, which is used to examine the surface properties of biological materials.

electron microscope techniques *See* **freeze-etching, freeze-fracture, Kleinschmidt spreading technique, negative staining.**

electron pair bond convalent bond.

electron transport chain a chain of molecules localized in mitochondria and acting as hydrogen and electron acceptors. The chain functions to funnel electrons from a given substrate to O_2. The energy released is used to phosphorylate ADP. *See* **cytochrome system.**

electron volt a unit of energy equivalent to the amount of energy gained by an electron passing through a potential difference of one volt. Larger multiple units of the electron volt (eV) are frequently used: keV = kilo (thousand) eV; MeV = mega (million) eV; and GeV = giga (billion) eV.

electropherogram a supporting medium containing a collection of molecules that have been separated by electrophoresis. The medium is generally in the form of a sheet, much longer and wider than it is thick, and it is often a gel such as agarose.

electrophoresis   the movement of the charged molecules in solution in an electrical field. The solution is generally held in a porous support medium such as filter paper, cellulose acetate (rayon), or a gel made of starch, agar, or polyacrylamide. Electrophoresis is generally used to separate molecules from a mixture, based upon differences in net electrical charge and also by size or geometry of the molecules, dependent upon the characteristics of the gel matrix. The SDS-PAGE technique is a method of separating proteins by exposing them to the anionic detergent sodium dodecyl sulfate (SDS) and polyacrylamide gel electrophoresis (PAGE). When SDS binds to proteins, it breaks all noncovalent interactions so that the molecules assume a random coil configuration, provided no disulfide bonds exist (the latter can be broken by treatment with mercaptoethanol). The distance moved per unit time by a random coil follows a mathematical formula involving the molecular weight of the molecule, from which the molecular weight can be calculated. *See* **zonal electrophoresis.**

electrostatic bond   the force holding ions together in a crystal such as common salt (NaCl)

element   a pure substance consisting of atoms of the same atomic number and which cannot be decomposed by ordinary chemical means. *See list at* **chemical elements.** *See also* **periodic table.**

Elodea canadensis   a common pond weed that behaves as a facultative apomict (*q.v.*).

elongation factors   proteins that complex with ribosomes to promote elongation of polypeptide chains; they dissociate from the ribosome when translation is terminated. Elongation factor G (EF-G), also called *translocase,* is associated with the movement of the peptidyl tRNA from the "A" site to the "P" site of the ribosome. Elongation factor T (EF-T) is responsible for alignment of the AA⁓tRNA complex in the "A" site of the ribosome.

elytra   the modified front wings of beetles.

emarginate   notched or indented.

emasculation   1. the removal of the anthers from a flower. 2. castration.

EMB agar   a complex bacterial growth medium containing two *p*H sensitive dyes, eosin and methylene blue, and a sugar such as lactose. Lac⁺ cells preferentially metabolize lactose oxidatively, resulting in excretion of hydrogen ions; under low *p*H, the dyes become deep purple. Lac⁻ cells that use amino acids as a source of energy release NH_3, thus raising *p*H; this decolorizes the dyes and produces a white colony. These color differences occur for all metabolizable sugars and also for many other carbon sources.

Embden-Meyerhof-Parnas pathway   glycolysis (*q.v.*).

embryo   an organism in early stages of development, especially before hatching from the egg when it is dependent upon its own yolk supply for nutrition. The developing human is designated as an embryo up to the beginning of the third month of pregnancy. Subsequently when the body shape is essentially formed, the term *fetus* is used.

embryo culture   a method of inducing the artificial growth of embryos by excising the young embryos under aseptic conditions and placing them on suitable nutrient media.

embryonic induction   *See* **induction.**

embryo sac   the female gametophyte of angiosperms. It contains several haploid nuclei formed by the division of the haploid megaspore nucleus.

embryo transfer (transplantation)   artificial introduction of an early embryo into the oviduct or uterus of the biological mother or of a surrogate mother. For example, a woman with blocked Fallopian tubes may choose to have an egg fertilized *in vitro,* and the resulting embryo may be introduced into her uterus for implantation. In the case of a superior dairy or beef cow, many eggs can be simultaneously artificially induced (superovulation), flushed from the oviducts, fertilized *in vitro,* and each embryo transferred to a different surrogate mother for subsequent development.

emergent properties   in a hierarchial system, properties that are manifest at higher levels because they are formed by measures or processes involving aggregations of lower level units. For example, in an ecological hierarchy, populations have properties not expressed by an individual; a community has properties not expressed in a population, etc.

emmer wheat   *Triticum dicoccum* (N = 14) (also called *starch wheat* or *two-grained spelt*), a wheat cultivated since Neolithic times. *See* **wheat.**

EMS   ethyl methane sulfonate (*q.v.*).

emulsion   a relatively stable dispersion of droplets of a liquid in another liquid with which it is immiscible.

enantiomers, enantiomorphs   compounds showing mirror-image isomerism.

endemic   confined to a given geographic region. Said of a disease continuously occurring in a particular area.

endergonic reaction   a reaction requiring energy from the outside before reactants will form products.

end labeling   the attachment of a radioactive chemical (usually ^{32}P) to the 5′ or 3′ end of a DNA strand.

endocrine system   a system of ductless glands which controls metabolism through the synthesis and release into the blood stream of hormones (*q.v.*).

endocytosis   the uptake by a cell of particles, fluids, or specific macromolecules by phagocytosis, pinocytosis, or receptor-mediated endocytosis, respectively.

endocytotic vesicles   intracellular vesicles ranging in diameter from 50 to 250 nm that are coated with bristle-like structures on their cytoplasmic surfaces. These coated vesicles arise by the invagination and pinching off of coated regions of the plasmalemma called *coated pits.* The cytoskeleton of each coated pit is a basket-like network of hexagons and pentagons formed by the assembly of three-legged protein complexes called *triskelions.* Each triskelion is composed of three molecules of *clathrin,* a 180 kd protein, and three smaller polypeptides.

endoderm   the layer of cells lining the primitive gut (archenteron) in an early embryo, beginning in the gastrula stage. The endoderm forms the epithelial lining of the intestine and all of the outgrowths of the intestine (gill pouches and gills, the larynx, wind pipe and lungs, the tonsils, thyroid, and thymus glands, the liver, the gall bladder and bile duct, the pancreas, and the urinary bladder and adjacent parts of the urogenital system).

endogenote   that portion of the original chromosome of a merozygote that is homologous to the exogenote (*q.v.*).

endogenous   originating within the organism.

endogenous virus   an inactive virus that is integrated into the chromosome of its host cell, and can therefore exhibit vertical transmission (*q.v.*).

endometrium   the glandular lining of the uterus of mammals which undergoes cyclical growth and regression during sexual maturity.

endomitosis   chromosomal replication within a cell nucleus that does not divide.

endomixis   a process of self-fertilization in which the sperm and egg nuclei from one individual unite.

endonuclease   an enzyme that breaks the internal phosphodiester bonds in a DNA molecule. Endonucleases of somatic tissue hydrolyze DNA by introducing double-strand breaks. Endonucleases isolated from cells in meiotic prophase produce single-strand breaks in the DNA with 5′ hydroxy termini. Single-strand breaks are essential first steps in replication and recombination. *See* **restriction endonuclease.**

endoplasmic reticulum (*shortened form,* **er**)   a system of membranes forming sheets and vesicles in the cytoplasm of many cells. In places the er is continuous with the plasma membrane or the outer membrane of the nuclear envelope. If the outer surfaces of the er membranes are coated with ribosomes the er is said to be rough-surfaced; otherwise it is said to be smooth-surfaced.

endopolyploidy   the occurrence in a diploid individual of cells containing 4C, 8C, 16C, 32C, etc., amounts of DNA in their nuclei. The nurse-cell nuclei in *Drosophila* egg chambers are good examples of endopolyploidy.

Endopterygota   Holometabola (*q.v.*).

endorphins   one of a group of mammalian peptide hormones liberated by proteolytic cleavage from a 29 kd prohormone called pro-opiocortin. Beta-endorphins are found in the intermediate lobe of the pituitary gland and are potent analgesics (pain suppressors). The name "endorphin" is a contraction of "endogenous morphine" because it is produced within the body (endogenously), attaches to the same neuron receptors that bind morphine, and produces similar physiological effects.

endoskeleton   an internal skeleton, as in the case of the bony skeleton of vertebrates.

endosperm   triploid nutritive cells surrounding and nourishing the embryo in seed plants.

endosymbiont theory   the proposal that self-replicating eukaryotic organelles originated by fusion of formerly free-living prokaryotes with primitive nucleated cells. *See* **serial symbiosis theory.**

endotherm   an animal that can metabolically regulate its body temperature.

endothermic reaction   a reaction during which heat is absorbed.

endotoxins   complex lipopolysaccharide molecules that form an integral part of the cell wall of many Gram-negative bacteria and are released only when the integrity of the cell is disturbed. Endotoxins are capable of non-specifically stimulating certain immune responses. *Compare with* **exotoxins.**

end product   a compound that is the final product in a chain of metabolic reactions.

end product inhibition   the situation in which the activity of an initial enzyme in a chain of enzymatic reactions is inhibited by the final product; feedback inhibition.

end product repression   the situation in which the final product in a chain of metabolic reactions functions as a corepressor and shuts down the operon turning out the enzymes that control the reaction chain.

energy   capacity for doing work.

energy-rich bond   a chemical bond that releases a large amount of energy upon its hydrolysis. The ATP molecule has an energy-rich phosphate bond, for example.

Engystomops pustulosus a tropic frog particularly suited to biochemical investigations of oogenesis because of the synchronous development of the egg masses.

enhancer a 72 bp sequence in the DNA of virus SV40 that enhances the transcriptional activity of genes present on the same molecule. For example, reconstruction experiments show that if a β-globin gene is placed on a DNA molecule containing the enhancer, the transcript of the β-globin gene increases 200 times, even if the enhancer is over a thousand bp upstream or downstream from the startpoint. The enhancer is not a promoter, and its mechanism of action is under study.

enkephalins pentapeptides with opiate-like activity (*compare with* **endorphins**), first isolated in 1975 from pig brain. Met-enkephalin has the amino acid sequence Tyr-Gly-Gly-Phe-Met; Leu-enkephalin has the sequence Tyr-Gly-Gly-Phe-Leu. *See* **polyprotein.**

enol forms of nucleotides *See* **tautomeric shifts.**

enrichment methods for auxotrophic mutants *See* **filtration enrichment, penicillin enrichment.**

enterovirus a member of a group of RNA-containing viruses (including poliomyelitis virus) inhabiting the human intestine.

entoderm endoderm (*q.v.*).

entomophilous designating flowers adapted for pollination by insects.

entropy **1.** in thermodynamics, the measure of the energy in a system which is unavailable to do work; **2.** the degree of randomness or disorder of a molecular system. Protein denaturation results in an increase in entropy, for example.

enucleate to remove the nucleus from a cell.

enucleated lacking a nucleus.

environment the complex of physical and biotic factors within which an organism exists.

environmental grain the heterogeneity of the environment as "perceived" by a population. A fine-grained environment offers little resistance to the population occupying it uniformly; a course-grained environment tends to restrict a population to a small portion of it.

environmental variance that portion of the phenotypic variance caused by differences in the environments to which the individuals in a population have been exposed.

enzyme any protein that catalyzes a specific chemical reaction without being consumed by the reaction. *See* **activation energy.**

enzyme induction *See* **inducible enzyme.**

eobiogenesis the first instance of the generation of living matter from inorganic material.

Eocene an epoch in the Tertiary period beginning approximately 55 million years ago, and lasting 20 million years. *See* **geologic time divisions.**

Epa antigens alloantigens in the mouse which are specific to epidermal cells.

Ephestia kühniella the Mediterranean meal moth. Numerous mutants affecting the color pattern of the forewing and eye pigmentation have been studied in this species. It is parasitized by *Habrobracon juglandis* (*q.v.*).

epicanthus a fold of skin extending over the inner corner of the eye characteristic of members of the Mongoloid race.

126

epicotyl the portion of the seedling above the cotyledons which develops into the shoot and its derivatives.

epidemiology the study of disease epidemics, with an effort to tracing down the cause.

epidermal growth factor (EFG) a serum protein that stimulates cell division of a wide variety of cells in tissue culture and presumably can also do so *in vivo*. EGF binds to the plasma membrane of sensitive cells and thereby stimulates the enzymatic activity of a protein kinase located within the cell membrane. This kinase phosphorylates tyrosine residues in polypeptides as does the product of gene *src* of the Rous sarcoma virus. Both kinases are antigenically related but are not structurally identical. *See* **vinculin.**

epigamic serving to attract or stimulate members of the opposite sex during courtship.

epigamic selection *See* **sexual selection.**

epigean *See* **epigeous.**

epigenesis the concept that an organism develops by the new appearance of structures and functions, as opposed to the hypothesis that an organism develops by the unfolding and growth of entities already present in the egg at the beginning of development (preformation).

epigenetics the study of the mechanisms by which genes bring about their phenotypic effects.

epigenotype the series of interrelated developmental pathways through which the adult form is realized.

epigeous developing above the soil.

Epilobium hirsutum hairy willow-herb, the subject of classic studies on cytoplasmic inheritance.

epimers organic compounds that are partial isomers in that they differ from each other only in the three-dimensional positioning of atoms about a single asymmetric carbon atom.

epinephrine a hormone from the adrenal medulla which elevates blood glucose by mobilizing glycogen reserves.

episome a class of genetic elements of which phage lambda and sex factor F are examples in *Escherichia coli*. Episomes may behave (1) as autonomous units replicating in the host independent of the bacterial chromosome, or (2) as integrated units attached to the bacterial chromosome and replicating with it. *Compare with* **plasmid.**

epistasis the nonreciprocal interaction of nonallelic genes. The situation in which one gene masks the expression of another. The recessive gene *apterous (ap)* in *Drosophila* produces wingless homozygotes. In such individuals any other recessive gene affecting wing morphology will have its action masked. The *apterous* gene is said to be epistatic to a gene like *curled wing* which is hypostatic to *ap*.

epistatic gene   *See* **epistasis.**

epithelium   a tissue that forms the surface of an organ or organism. For example, the outer skin is an epithelial tissue; the cells lining the gut and respiratory cavities are also epithelial cells.

epitope   *See* **antigenic determinant.**

epizoite   a nonparasitic sedentary animal living attached to another animal.

eponym   a term or phrase derived from the name of a person or place. Eponyms often honor the discoverer of a law or phenomenon, as in *Sewall Wright effect;* or the inventor of a technique, as in *Miller spread.* Genetic diseases are often named after the physicians who first described them (i.e., *Tay-Sachs disease*) or, more rarely, after the patient (i.e., *Christmas disease*). Geographic eponyms also occur in the genetic literature, as in *Bombay blood group.*

epoxide   a family of chromosome-breaking, alkylating agents. Di(2,–3 epoxy)propyl ether is an example.

$$
O^{\diagdown}_{\diagup}\!\!\!\!\!\quad CH_2-CHCH_2 \atop CH_2-CHCH_2 \quad ^{\diagup}_{\diagdown}O
$$

Epstein-Barr virus   a DNA virus of the herpes group discovered in 1964 by M. A. Epstein and Y. M. Barr in cultures of Burkitt's lymphoma cells. EBV is the cause of infectious mononucleosis, and it has an integration site on human chromosome 14. *See* **Burkitt's lymphoma.**

equational division   a division of each chromosome into equal longitudinal halves that are incorporated into two daughter nuclei. It is the type of division seen in mitosis.

equatorial plate   *See* **mitosis.**

equilibrium centrifugation   *See* **centrifugation separation.**

equilibrium dialysis   *See* **dialysis.**

equilibrium population   a population in which the allelic frequencies of its gene pool do not change through successive generations. An equilibrium can be established by counteracting evolutionary forces (e.g., a balance between selection and mutation pressures) or by the absence of evolutionary forces. *See* **Hardy-Weinberg law.**

equine   referring to members of the horse family, especially the domestic horse *Equus caballus.*

Equus   the genus that includes *E. asinus,* the donkey, and *E. caballus,* the horse.

er   endoplasmic reticulum (*q.v.*).

ergastoplasm   rough-surfaced endoplasmic reticulum.

ergosome   polysome (*q.v.*).

erythema an abnormal redness of the skin due to distension of the capillaries with blood. It can be caused by many different agents (heat, certain drugs, ultraviolet rays, ionizing radiation, etc.).

erythroblastosis the discharge of nucleated red blood corpuscles from the blood-forming centers into the peripheral blood.

erythroblastosis fetalis a hemolytic disease of infants due to Rh-incompatibilities between the fetus and its mother.

erythrocyte the hemoglobin-containing cell found in the blood of vertebrates.

erythrolabe *See* **color blindness.**

erythromycin an antibiotic produced by *Streptomyces erythieus.*

escaper *See* **breakthrough.**

Escherichia coli the colon bacillus, the organism about which the most molecular genetics is known.

essential amino acids amino acids required in the diet of a species because these molecules cannot be synthesized from other food materials; in contrast to nonessential amino acids that can be synthesized by normal members of the species. *See* **amino acid.**

established cell line a cell line (*q.v.*) that demonstrates the potential to be subcultured indefinitely *in vitro.* HeLa cells (*q.v.*) represent an established cell line.

estivate aestivate (*q.v.*).

estradiol a steroidal estrogen.

estrogen   an ovarian hormone which prepares the mammalian uterus for implantation of an embryo; also responsible for development of secondary sexual characteristics in females.

estrous cycle   a seasonal cycle of reproductive activity dependent upon endocrine factors. If the organism has one estrous period per year, it is called monestrous, if more than one, polyestrous.

estrus   **1.** the period of reproductive activity. **2.** the estrous cycle.

ethidium bromide   a compound used to separate covalent DNA circles from linear duplexes by density gradient centrifugation. Because more ethidium bromide is bound to a linear molecule than to a covalent circle, the linear molecules have a higher density at saturating concentrations of the chemical and can be separated by differential centrifugation.

Ethiopian   designating or pertaining to one of the six biogeographic realms (*q.v.*) of the world, comprising Africa, Iraq south of the Tropic of Cancer, Madagascar, and the adjacent islands.

ethological   behavioral.

ethological isolation   the failure of related species or semispecies to produce hybrid offspring because of differences in their mating behaviors.

ethology   the scientific study of animal behavior, particularly under natural conditions.

ethylene dinitrilotetra-acetic acid   a chelating agent (*q.v.*).

ethylenimine   an aziridine mutagen (*q.v.*).

ethyl methane sulfonate   one of the most commonly used mutagenic alkylating agents.

130

The most common reaction of EMS is with guanine (*q.v.*), where it causes an ethyl group to be added to the number 7 nitrogen. Alkylation of guanine allows it to pair with thymine. Then during replication the complementary strand receives thymine rather than cytosine. Thus EMS causes base substitutions of the transition type.

$$CH_3CH_2-O-\overset{\displaystyle O}{\underset{\displaystyle O}{\overset{\|}{\underset{\|}{S}}}}-CH_3$$

etiolation   a plant syndrome caused by suboptimal light, consisting of small, yellow leaves and abnormally long internodes.

etiology   the study of causes, especially of disease.

E$_1$ trisomy syndrome   a spectrum of congenital defects in man caused by the presence of an extra chromosome of the E group (18). *See* **Edward's syndrome, human mitotic chromosomes.**

Eubacteriales   spherical, ovoid, or rod-like unicellular microorganisms, which multiply by transverse, binary fission.

eucaryote   *See* **eukaryote.**

Eucaryotes   *See* **Eukaryotes.**

Euchlaena mexicana   Mexican fodder grass, a species hybridizing with maize. *See* **teosinte.**

euchromatic   containing euchromatin.

euchromatin   that chromatin which shows the staining behavior characteristic of the majority of the chromosomal complement. It is uncoiled during interphase and condenses during mitosis, reaching a maximum density at metaphase. In polytene chromosomes the banded segments contain euchromatin. *See* **heterochromatin.**

Euentoma   insects with genitalia, Malpighian tubes, and a tracheal system.

eugenics   the improvement of humanity by altering its genetic composition by encouraging breeding of those presumed to have desirable genes (positive eugenics), and discouraging breeding of those presumed to have undesirable genes (negative eugenics).

Euglena   a genus of unicellular flagellated protoctists often used in studes of molecular and developmental genetics. *See* Euglenophyta, **classification.**

eukaryon   the highly organized nucleus of an eukaryote.

eukaryote   a member of the superkingdom Eukaryotes (*q.v.*).

Eukaryotes   the superkingdom containing all organisms that are, or consist of, cells with true nuclei bounded by nuclear envelopes and that undergo meiosis. Cell division occurs by mitosis. Oxidative enzymes are packaged within mitochondria. The superkingdom contains four kingdoms: the Protoctista, the Fungi, the Animalia, and the Plantae. *See* **classification, Procaryotes.**

eumelanin   one of the pigment molecules found in the coat and pigmented retinal epithelium of mammals. It is derived from the metabolism of tyrosine and is normally black in color. Its precise coloration is affected by a large number of mutant genes. *See* **melanin.**

Eumetazoa   the subdivision of the animal kingdom containing organisms possessing organ systems, a mouth, and digestive cavity. *See* **classification.**

euphenics the amelioration of genotypic maladjustments brought about by efficacious treatment of the genetically defective individuals at some time in their life cycles.

euploid polyploid whose chromosome number is an exact multiple of the basic number of the species from which it originated.

eupyrene sperm *See* **sperm polymorphism.**

eusocial a social system in which certain individuals incur obligate sterility, but enhance their fitness by aiding their collateral kin to rear their offspring. For example, sterile female worker bees may rear the offspring of their fertile sister queens. *See* **inclusive fitness.**

euthenics the control of the physical, biological, and social environments for the improvement of humanity.

eV electron volt.

evagination an outpocketing.

eversible capable of being turned outward or inside out.

eversion projection by turning inside out. Thus, if a rubber glove is turned inside out by blowing into it, the fingers can be everted.

eversporting referring to a strain characterized by individuals that, instead of breeding true, produce variations of a specific sort in succeeding generations. Such strains generally contain mutable genes.

evocation the morphogenetic effect produced by an evocator.

evocator the morphogenically active chemical emitted by an organizer.

evolution **1.** potentially reversible gene frequency changes within a population gene pool; microevolution. **2.** irreversible (*see* **Dollo's law**) genetic changes within a genological lineage producing anagenic or phyletic speciation. **3.** cladogenesis or splitting of one species into two. **4.** production of novel adaptive forms worthy of recognition as new taxa; e.g., the appearance of feathers on a reptilian ancestor is considered to define a new taxon—Aves (birds). Items 2 through 4 are considered to be macroevolution. At the present time there are two conflicting hypotheses to explain the mechanism of evolution. The older theory, generally called the *gradualism model,* represents updating of the original ideas set forth by Charles Darwin. According to this model, those individuals with hereditary traits that best adapt them to their habitat are more likely to survive and to transmit these adaptive genes to their offspring. As a result with the passage of time, the frequencies of beneficial genes rise in the population, and when the composition of the gene pool of the evolving population becomes sufficiently different from that of the original population, a new species will have arisen. Since beneficial mutations must spread through an entire population to produce detectable evolutionary changes, speciation will be a gradual and continuous process. The alternative hypothesis was first proposed by N. Eldredge and S. J. Gould in 1972 and has been called the *punctuated equilibrium model.* According to this model, speciation occurs as a result of rapid genetic changes within small, geographically isolated groups of organisms. The genetic changes envisioned are those having profound effects (sometimes called macromutations). For example, mutations of a regulator gene *(q.v.)* might activate or inactivate a battery of operons and thereby drastically modify the development of the organism. If the change is an adaptive one, the isolated group may expand and quickly supplant the parent species. According to this view, the macromutations generate an incipient species and natural selection weeds out the misfits. Thus the evolutionary record will be one of periods of equilibrium punctuated by periods of rapid change.

evolutionarily derived character *See* **phylogenetic classification.**

132

evolutionarily primitive character *See* **phylogenetic classification.**

evolutionary clock the rate at which mutations accumulate in a given gene.

evolutionary rate rapid evolution is called *tachytelic,* slow evolution is called *bradytelic,* and evolution at an average rate is called *horotelic.*

Evolution in Mendelian Populations a text published in 1931 by S. Wright which constitutes a milestone in population genetics.

evulsion the act of plucking or pulling out.

ewe a female sheep. *See* **ram.**

exaptation a character currently subject to selection, but whose origin can be ascribed to processes other than selection or to selection for a different function. *See* **aptation.**

exchange pairing the type of pairing of homologous chromosomes that allows genetic crossing over to take place. Synaptonemal complexes play a critical role in exchange pairing.

excision the enzymatic removal of a polynucleotide segment from a nucleic acid molecule.

excision repair synonomous with cut-and-patch repair (*q.v.*).

excitation the reception of a quantum of energy by an atomic electron; an altered arrangement of planetary electrons in orbit about an atom resulting from absorption of electromagnetic energy.

exclusion principle the principle according to which two species cannot coexist in the same locality if they have identical ecological requirements.

exclusion reaction the healing reaction of a phage-infected bacterium that strengthens its envelope and prevents entry of additional phages.

exconjugant **1.** the separated eukaryotic protozoan cells (e.g., paramecia) that were partners in conjugation and therefore have exchanged genetic material. **2.** a female (F^-) recipient bacterial cell that has separated from a male (Hfr) donor partner after conjugation and therefore contains some of the donor's DNA.

exergonic reaction a reaction proceeding spontaneously and releasing energy to its surroundings.

exocytosis the discharge from a cell of materials by reverse endocytosis (*q.v.*).

exogamy the tendency of an individual to mate selectively with non-relatives.

exogenic heredity transmission from generation to generation of information in the form of knowledge and various products of the human mind (i.e., books, laws, inventions, etc.).

exogenote the new chromosomal fragment donated to a merozygote (*q.v.*).

exogenous virus a virus that replicates vegetatively (productively in lytic cycle) and is not vertically transmitted in a gametic genome.

exon a portion of a split gene (*q.v.*) that is included in the transcript of a gene and survives processing of the RNA in the cell nucleus to become part of a spliced messenger of a structural RNA in the cell cytoplasm. Exons generally occupy three distinct regions of genes that encode proteins. The first, which is not translated into protein, signals the beginning of RNA transcription and contains sequences that direct the mRNA to the ribosomes for protein syntheisis. The exons in the second region contain the information that is translated into the amino acid sequence of the protein. Exons in the third region are transcribed into the part of the mRNA that contains the signals for the termination of translation and for the addition of a polyadenylate tail. *See* **intron, polyadenylation.**

exon shuffling the creation of new genes by bringing together, as exons of a single gene, several coding sequences that had previously specified different proteins or different domains of the same protein, through intron-mediated recombination.

exonuclease an enzyme that digests DNA, beginning at the ends of the strands.

exonuclease III an enzyme from *E. coli* that attacks the DNA duplex at the 3′ end on each strand; used together with S1 nuclease (*q.v.*) to create deletions in cloned DNA molecules. *Compare with* **Bal 31 exonuclease.**

exonuclease IV an enzyme that specifically degrades single stranded DNA. It initiates hydrolysis at both the 3′ and 5′ ends to yield small oligonucleotides. This enzyme is active in the presence of EDTA.

exopterygota hemimetabola (*q.v.*).

exoskeleton a skeleton covering the outside of the body, characteristic of arthropods.

exothermic reaction a reaction in which heat is evolved.

exotoxin a poison excreted into the surrounding medium by an organism (e.g., certain Gram-positive bacteria such as those causing diphtheria, tetanus, and botulism). Exotoxins are generally more potent and specific in their action than endotoxins (*q.v.*).

explant an excised fragment of a tissue or an organ used to initiate an *in vitro* culture.

exponential growth phase that portion of the growth of a population characterized by an exponential increase in cell number with time. *See* **stationary phase.**

exponential survival curve a survival curve without a shoulder, or threshold region, and which plots as a straight line on semilog coordinates.

expression vector cloning vehicles designed to promote the expression of gene inserts. Typically, a restriction fragment carrying the regulatory sequences of a gene is ligated *in vitro* to a plasmid containing a restriction fragment possessing the gene but lacking its regulatory sequences. The plasmid with this new combination of DNA sequences is then cloned under circumstances that promote the expression of the gene under the control of the regulatory sequences.

expressivity the range of phenotypes expressed by a given genotype under any given set of environmental conditions or over a range of environmental conditions. For example, *Drosophila* homozygous for the recessive gene "eyeless" may have phenotypes varying from no eyes to completely normal eyes, but the usual condition is an eye noticeably smaller than normal.

expressor protein the product of a regulatory gene, necessary for the expression of one or more other genes under its positive transcriptional control.

extant living at the present time, as opposed to extinct.

extinction termination of an evolutionary lineage without descendents. *See* **pseudoextinction, taxonomic extinction.**

extrachromosomal inheritance *See* **extranuclear inheritance.**

extragenic reversion a mutational change in a second gene that eliminates or suppresses the mutant phenotype of the first gene. *See* **suppressor mutation.**

extranuclear inheritance   heredity attributed to DNA in organelles such as mitochondria or chloroplasts; also called extrachromosomal inheritance, cytoplasmic inheritance, maternal inheritance.

extrapolation number   in target theory (*q.v.*), the intercept of the extrapolated multitarget survival curve with the vertical logarithmic axis specifying the survival fraction. The extrapolation number gives the number of targets that each must be hit at least once to have a lethal effect on the biological system under study.

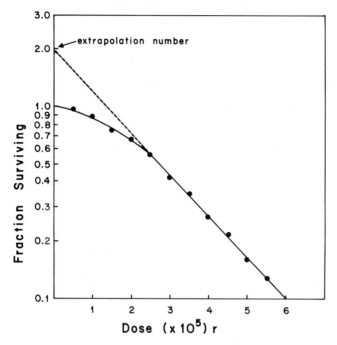

eyespot   a simple organ specialized for the detection of light, but which does not produce a true image. Many of the simpler invertebrates, including a number of protozoan species, possess eyespots.

F

F 1. Wright's inbreeding coefficient 2. fluorine. 3. Fahrenheit. 4. fertility factor.

F_1 first filial generation; the offspring resulting from first experimental crossing of the plants or animals. The parental generation with which the genetic experiment starts is referred to as P_1 (*q.v.*).

F_2 the progeny produced by intercrossing or self-fertilization of F_1 individuals.

Fab that fragment of a papain-digested immunoglobulin molecule which bears a single antigen-binding site and contains one intact light chain and a portion of one heavy chain. *See* **immunoglobulin.**

F(ab)$_2$ fragment that fragment of a pepsin-digested immunoglobulin molecule which contains portions of two heavy chains and two intact light chains and bears two antigen binding sites. *See* **immunoglobulin.**

Fabry disease a hereditary disease of glycosphingolipid metabolism in humans. The disease results from a mutation in the X-linked gene which codes for lysosomal *alpha galactosidase A* (*q.v.*). The enzyme defect causes the accumulation of glycosphingolipids in lysosomes of blood vessels and leads to vascular malfunctions. Fabry disease can be detected *in utero* by demonstration of a lack of alpha galactosidase A activity in amniocytes of XY fetuses. Heterozygous females can be documented by cloning cultured skin fibroblasts followed by the demonstration of two cell populations, one normal and one lacking enzyme activity (*see* **dosage compensation**). Hemizygotes survive to sexual maturity, so the mutation is transmitted by both heterozygotes and hemizygotes. Prevalence 1/40,000 males.

facet ommatidium (*q.v.*).

factorial a continuing product of factors, calling for the multiplication together of all the integers from one to the given number. Factorial four means $1 \times 2 \times 3 \times 4 = 24$. The usual factorial symbol is an exclamation point (!) following the largest number in the series. Thus factorial four is written 4!

facultative having the capacity to live under more than one specific set of environmental conditions. For example, a facultative parasite need not live as a parasite (if another food source is available). A facultative apomict may reproduce sexually or asexually depending upon environmental conditions.

136

facultative heterochromatin   *See* **heterochromatin.**

FAD   flavin adenine dinucleotide, a coenzyme *(q.v.).*

Fagus sylvatica   the European beech tree.

Fahrenheit   *See* **temperature.**

falcate   sickle-shaped.

falciparum malaria   *See* **malaria.**

fallout   the radioisotopes generated by nuclear weapons that are carried aloft and eventually fall to the earth's surface and contaminate it.

fallow   cropland left idle, usually for one growing season, while the soil is being cultivated to control weeds and conserve moisture.

familial   transmitted in families.

familial Down syndrome   *See* **translocation Down syndrome.**

familial hypercholesterolemia (FHC)   a human hereditary disease characterized by an elevation in the plasma concentration of low density lipoproteins (LDL). FHC is inherited as an autosomal dominant at a locus on chromosome 14. The prevalence of heterozygotes is about 1/500 among American, European, and Japanese populations, and this makes FHC among the most common hereditary diseases. Homozygotes are rare (1 per million in the U.S.A.). The mutant gene product is a LDL receptor. Receptor activity is absent in homozygotes and one half normal in heterozygotes. Deficient receptor-mediated endocytosis causes LDL to accumulate in the plasma. Cholesterol is deposited on arterial walls and atherosclerosis results. The disease is far more serious in homozygotes than in heterozygotes. *See* **plasma lipoproteins.**

family selection   artificial selection of an individual(s) to participate in mating(s) based upon the merits of other members of the same family exclusive of parents and ancestors (e.g., full sibs or half sibs).

Fanconi anemia   the first inherited disease in which hereditary chromosome fragility was established. The disease is characterized by a pronounced reduction in the number of erythrocytes, all types of white blood cells, and platelets in the circulating blood. Chromosome aberrations are common and usually involve nonhomologous chromosomes. In the Bloom syndrome *(q.v.),* most chromosome interchanges are between homologous chromosomes. The disease may result from a defect in the transport of enzymes functioning in DNA repair from cytoplasm to nucleus.

farrow   to give birth to pigs.

fast component   **1.** in reassociation kinetics, the first components to renature, containing highly repetitive DNA. **2.** in electrophoresis, the molecules that move furthest in a given time from the origin.

fast green   an acidic dye used in cytochemistry. See p. 138.

fat   a glycerol ester of fatty acids. Glycerol tripalmitate may be taken as example:

fat

$$H_2-C-O-CO-(CH_2)_{14}CH_3$$
$$H-C-O-CO-(CH_2)_{14}CH_3$$
$$H_2-C-O-CO-(CH_2)_{14}CH_3$$

fat body   the adipose tissue found in larval insects.

fate map   a map of an embryo in an early stage of development which indicates the various regions whose prospective significance has been established by marking methods.

fatty acid   an acid present in lipids, varying in carbon content from C_2 to C_{34}. Palmitic acid may be taken as an example: $CH_3(CH_2)_{14}COOH$.

fauna   the animal life in a given region or period of time.

favism   a hemolytic response to the consumption of beans produced by *Vicia faba* (*q.v.*). Various compounds in fava beans are enzymatically hydrolyzed to quinones, which generate oxygen radicals. Red blood cells deficient in glucose-6-phosphate dehydrogenase (*q.v.*) have a marked sensitivity to oxidating agents and lyse when oxygen radicals are abundant. The toxic properties of fava beans have been known for centuries in Mediterranean cultures where glucose-6-phosphate dehydrogenase deficiency (*q.v.*) is a common hereditary disease.

F$^+$ cell   a bacterial cell possessing a fertility (F) factor extrachromosomally in a plasmid. An F$^+$ cell can donate the F factor to an F$^-$ cell during conjugation. If the F factor integrates into the bacterial chromosome, the cell becomes an Hfr (*q.v.*), capable of transferring chromosomal genes.

F$^-$ cell   a bacterial cell devoid of a fertility (F) factor and that therefore acts only as a recipient ("female") in bacterial conjugation.

F$_c$ fragment   that *c*rystallizable *f*ragment (hence the name) of a papain-digested immunoglobulin molecule which contains only portions of two heavy chains and no antibody binding sites. This fragment does, however, bind complement and is responsible for the binding of immunoglobulin to various types of cells in a non-antigen-specific manner. *See* **immunoglobulin.**

F$_c$ receptor   a cell surface component of many cells of the immune system responsible for binding the F$_c$ portion of immunoglobulin molecules.

F-duction   *See* **sexduction.**

fecundity   potential fertility or the capability of repeated fertilization. Specifically the term refers to the quantity of gametes, generally eggs, produced per individual over a defined period of time. *See* **fertility.**

feedback the influence of the result of a process upon the functioning of the process.

feedback inhibition end product inhibition (*q.v.*).

feeder cells irradiated cells, capable of metabolizing but not of dividing, that are added to culture media to help support the growth of unirradiated cells.

feline leukemia virus an oncogenic RNA virus.

Felis catus the domesticated cat; for a listing of breeds *see* **cat.**

female carrier a woman who is heterozygous for a recessive, X-chromosomal gene.

female gonadal dysgenesis Turner syndrome (*q.v.*).

female pronucleus the haploid nucleus of a female gamete, which functions in syngamy.

female-sterile mutation one of a class of mutations that cause female sterility generally because of a developmental block during oogenesis. Recessive female-sterile mutations are common in *Drosophila melanogaster* and *Bombyx mori.* Dominant female-steriles are much rarer.

female symbol ♀. The zodiac sign for Venus, the goddess of love and beauty in Roman mythology. The sign represents a looking glass.

F-episome *See* **fertility factor.**

F′-episome an F-episome carrying a genetically recognizable fragment of bacterial chromosome.

feral pertaining to formerly domesticated animals now living in a wild state.

fermentation the energy-yielding enzymatic breakdown of sugar molecules that takes place in bacteria and yeasts under anaerobic conditions.

ferredoxin a protein of high iron content involved during photosynthesis in electron transfer and in the production and consumption of gaseous hydrogen.

ferritin an iron storage protein found in the liver and spleen, containing up to 20% of its weight in the form of iron. It consists of a protein component (apoferritin) and colloidal micelles of ferric hydroxide-ferric phosphate. Ferritin is often conjugated to proteins such as immunoglobulins, thus enabling their locations within tissues to be determined in electron micrographs due to the great electron scattering property of the iron atoms.

ferritin-labeled antibodies *See* **ferritin.**

fertility the productivity of an individual or population in terms of generating viable offspring. The term is often used to refer to the minimum number of offspring generated by a female during her reproductive period. In human genetics, the term *effective fertility* is used to refer to the mean number of offspring produced by individuals suffering from a hereditary disease as compared to the mean number of offspring produced by individuals free from the disease, but otherwise very similar. Effective fertility thus gives an indication of the selective disadvantage of the disease. *Compare with* **fecundity.**

fertility (F) factor *See* **circular linkage map, F-prime factor, Hfr strain.**

fertility restorer a dominant nuclear gene in corn that nullifies the effect of a cytoplasmic male-sterility factor.

fertilization the union of two gametes to produce a zygote. *See* **double fertilization.**

fertilization cone a conical projection protruded from the surface of certain eggs at the point of contact with the fertilizing sperm.

fertilization membrane   a membrane that grows outward from the point of contact of the egg and sperm and rapidly covers the surface of the egg.

fertilizin   a substance secreted by the ovum of some species, which attracts sperm of the same species.

fetus   *See* **embryo.**

Feulgen-positive   stained by the Feulgen procedure and therefore containing DNA.

Feulgen procedure   a cytochemical test that utilizes the Schiff reagent *(q.v.)* as a stain and is specific for DNA.

F factor (fertility factor)   an episome, symbolized by F, that determines the sex of the bacterium. The presence of this factor in the cell makes it function as a male. The F factor enables the bacterium to produce an F-pilus which serves as a hollow tube through which DNA is transferred during conjugation. The F factor is a circular DNA molecule, which in *E. coli* is about 2.5% of the length of the host chromosome. *See* **circular linkage map, F-prime factor, Hfr strain.**

F' factor   *See* **F-prime factor.**

fibrin   *See* **blood clotting.**

fibrinogen   *See* **blood clotting.**

fibroblasts   spindle-shaped cells responsible for the formation of extracellular fibers such as collagen *(q.v.)* in connective tissues.

fibrocystic disease of the pancreas   cystic fibrosis *(q.v.)*.

fibroin   the major protein component of silk *(q.v.)*.

Ficoll   a nonionic synthetic polymer of sucrose.

field   *See* **prepattern.**

figure eight   in molecular genetics, two DNA circles connected by a recombination bridge that has not yet been completed.

filaform   thread-shaped.

filial generations   any generation following the parental generation. Symbolized F_1, F_2, etc.

filly   a female horse less than four years old.

film badge   a pack of photographic film used for the approximate measurement of the radiation exposure of personnel.

filter hybridization   exposing DNA, denatured to single strands and immobilized on a nitrocellulose filter, to a solution of radioactively labeled RNA or DNA; only hybrid double-stranded molecules remain on the filter after washing. *Compare with* **liquid hybridization.**

filter route   a migration path along which only a few species can easily disperse. *Compare with* **corridor.**

filtration enrichment   a method for the isolation of nutritional mutants in fungal genetics. Mutagenized spores are placed upon a minimal medium. Normal spores germinate and send out an extensive mycelial network. These colonies are then filtered off, and the remaining germinated spores that show poor mycelial development are grown upon a supplemented medium, where each produces enough mycelia to allow further propagation and study.

finalism a philosophy that views evolution as being directed (by some rational force) toward an ultimate goal. *See* **teleology** (of which this is a special case); *see also* **orthogenesis.**

fine structure genetic mapping the high resolution analysis of intragenic recombination down to the nucleotide level.

fingerprint the pattern of the ridge skin of the distal surface of a finger.

fingerprint technique a method for combining electrophoresis and chromatography to separate the components of a protein hydrolysate.

first-arriver principle a theory proposing that the first individuals to colonize a new environment or to become adapted to a specific niche acquire thereby a selective advantage over later arrivals, merely because they got there first; also known as "king-of-the-mountain" principle.

first cousin *See* **cousin.**

first division segregation ascus pattern in ascomycetes, a 4-4 linear order of spore phenotypes within an ascus. This pattern indicates that a pair of alleles (e.g., those controlling spore pigmentation) separated at the first meiotic division, because no crossovers occurred between the locus and the centromere. *See* **ordered tetrad.**

first law of thermodynamics *See* **thermodynamics.**

first-order kinetics the progression of an enzymatic reaction in which the rate at which the product is formed is proportional to the prevailing substrate concentration, with the result that the rate slows gradually, and the reaction never goes to completion. *See* **zero-order kinetics.**

fission 1. binary fission (*q.v.*). 2. nuclear fission (*q.v.*).

fitness the relative ability of an organism to survive and transmit its genes to the next generation.

fixation the first step in making permanent preparations of tissues for microscopic study. The procedure aims at killing cells and preventing subsequent decay with the least distortion of structure. *See* **fixative.**

fixation, genetic the status of a locus in which all members of a population are homozygous or hemizygous for a given allele; the frequency of the fixed allele is 1.0; all other alleles at that locus have been lost and therefore their frequencies are zero. *See* **monomorphic populations.**

fixative a solution used for the preparation of tissues for cytological or histological study. It precipitates the proteinaceous enzymes of tissues and so prevents autolysis, destroys bacteria that might produce decay of the tissue, and causes many of the cellular constituents to become insoluble.

fixing in photography, the removal of the unchanged halide after the image is developed. An aqueous solution of sodium thiosulfate (hypo) is used.

fixity of species theory the theory held by Linné that organisms of one species could give rise only to others like themselves and therefore that all species were independently created.

flabellum   a fan- or a leaf-like process.

flagellate   a protoctist belonging to the Zoomastigina or Euglenophyta. *See* **classification.**

flagellin   the major protein of bacterial flagellae. Its molecular weight is about 20,000 d. Flagellin molecules appear to be wound in a tight spiral within each flagellum.

flagellum (*plural,* **flagella**)   **1.** in bacteria, a whip-like motility appendage present on the surface of some species. Bacterial flagellae are composed of the protein flagellin *(q.v.)* and range in length from 2 to 20 μm. Bacteria having a single flagellum are called *monotrichous;* those with a tuft of flagella at one pole are called *lophotrichous;* and those with flagella covering the entire surface are called *peritrichous.* Antigens associated with flagella are called *H antigens. Compare with* **pilus. 2.** in eukaryotes, flagellum refers to a thread-like protoplasmic extension used to propel flagellates and sperm. Flagella have the same basic structure as cilia *(q.v.)*, but are longer in proportion to the cell bearing them and are present in much smaller numbers (most sperm are monotrichous). *See* **axoneme, undulipodium.**

flanking DNA   nucleotide sequences on either side of the region under consideration. For example, the hallmarks of a transposon *(q.v.)* are (1) it is flanked by inverted repeats at each end, and (2) the inverted repeats are flanked by direct repeats.

flavin adenine dinucleotide (FAD)   a coenzyme composed of riboflavin phosphate and adenylic acid. FAD forms the prosthetic group of enzymes such as d-amino acid oxidase and xanthine oxidase.

flavin mononucleotide   riboflavin phosphate, a coenzyme for a number of enzymes including l-amino acid oxidase and cytochrome c reductase.

flavoprotein   a protein requiring FMN or FAD to function.

flint corn   *See* **corn.**

flora   the plant life in a given region or period of time.

flour corn   *See* **corn.**

flower   the specialized reproductive shoot of angiosperms (*q.v.*). Perfect flowers bear both pistils and stamens. Imperfect flowers bear either pistils or stamens. A plant bearing solely functional staminate or pistillate flowers is said to be *dioecious*. A plant bearing (1) imperfect flowers of both sexual types (corn is an example), or (2) perfect flowers (the garden pea), or (3) staminate, pistillate, and perfect flowers (the red maple, for example) is said to be *monoecious*.

fluctuation test   a statistical analysis first used by Luria and Delbrück to prove that selected variants (*q.v.*), such as bacteriophage-resistant bacteria, are spontaneous mutants which arose prior to the exposure to the selective agent. They reasoned that if bacterial mutation is an event which is rare, discontinuous, and random, then there should be a marked fluctuation in the number of resistant variants present, at a given time, in a large number of independent cultures, each of which was grown from a small inoculum. The number of variants per sample would fluctuate because some cultures would contain large numbers of variants arising from the division of an early mutant, while in other cultures mutants occuring later during the growth of the culture would produce clones of smaller size. Conversely, in separate samples taken from a single culture inoculated under identical conditions the variability in the number of mutants should be much less. However, if the agent used for selection induced the mutations, then the distribution of the number of mutants in any population of samples should be independent of the previous history of the culture. Since it was found that the variance was much larger when samples came from independent cultures than when they were taken from the same culture, it was concluded that spontaneous mutation was the source of the variants.

fluid mosaic concept   a model of the cell membrane put forth by S. J. Singer and G. L. Nicholson. In it, the cell membrane is considered to be a two-dimensional viscous solution consisting of a bilayer of highly oriented lipids. The layer is discontinuous, being interrupted by protein molecules that penetrate one or both layers. *See* **lipid bilayer model.**

fluke   a common name for flatworms belonging to the class Trematoda. Flukes of medical importance are members of the order Digenea. These parasites have molluscs as intermediate hosts. *See* **schistosomiasis.**

fluourescein   resorcinolphthalein, an orange-red compound that yields a bright green fluorescence when exposed to ultraviolet light. Commonly used derivitives of fluorescein are eosin, phloxine, and rose bengal.

fluorescence   *See* **luminescence.**

fluorescence microscopy   the usual methods of microscopical examination are based on observing the specimen in the light transmitted or reflected by it. Fluorescence microscopical preparations are self-luminous. In most biological preparations the tissue sections are stained with a *fluorochrome,* a dye that emits light of longer wavelength when exposed to blue or ultraviolet light. The fluorescing parts of the stained object then appear bright against a dark background. The staining techniques are extremely sensitive and can often be used in living materials.

fluorescent antibody technique   a method for localizing a specific protein or other antigen in a cell by staining a section of the tissue with an antibody specific for that antigen. The antibody is tagged directly or indirectly by a fluorochrome for detection under a fluorescence microscope. In the *direct method,* the tissue antigen binds directly to the tagged antibody. In the *indirect method,* the tissue antigen is first bound to an unlabeled primary antibody and then a secondary antibody (labeled with a flurochrome and specific for the primary antibody) is bound to the antigen-primary antibody complex. For example, the primary antibody could be produced in rabbits against a *Xenopus* vitellogenin and the secondary antibody could be a fluorescein-labeled, anti-rabbit immunoglobulin from goats.

fluorescent screen   a sheet of material coated with a substance such as calcium tungstate or zinc sulfide that will emit visible light when irradiated with ionizing radiation. Such screens are used in TV sets and as the viewing screens of electron microscopes.

fluorine   a biological trace element. Atomic number 9; atomic weight 18.9984; valence 1^-; most abundant isotope ^{19}F.

fluorochrome   *See* **fluorescence microscopy.**

flush end   synonomous with blunt end. *See* **restriction endonuclease.**

F-mediated transduction   sexduction.

f-met-tRNA   the complex between N-formyl methionine and its transfer RNA.

FMN   flavine mononucleotide, a coenzyme.

foci   **1.** regions of growth of tumor cells appearing as raised clusters above a confluent monolayer of cells in tissue culture. **2.** opaque pocks appearing on the chorioallantoic membrane of a developing chick embryo that has been infected with certain viruses such as herpesviruses.

focus map   a fate map (*q.v.*) for regions of the *Drosophila* blastoderm determined to become adult structures, inferred from the frequencies of specific kinds of mosaics.

foldback DNA   single-stranded regions of DNA that have renatured by intrastrand reassociation between inverted repeats; hairpin DNA.

folic acid   the anti-pernicious-anemia vitamin. Folic acid plays a vital role in nucleic acid metabolism. Synonymous with pteroylglutamic acid.

follicle-stimulating hormone   a glycoprotein hormone that stimulates the growth of ovarian follicles and estrogen secretion. It is produced by the adenohypophysis of vertebrates. Abbreviated FSH.

footprinting   a technique for identifying a segment of a DNA molecule that is bound to some protein of interest, on the principle that the phosphodiester bonds in the region covered by the protein are protected against attack by endonucleases. A control sample of pure DNA and one of protein-bound DNA are subjected to endonuclease attack. The resulting fragments are electrophoresed on a gel that separates them according to their lengths. For every bond position that is susceptible, a band is found on the control gel. The gel prepared from the protein-bound DNA will lack certain bands, and the missing bands identify the length of the site covered by the protein.

Forbes disease   a hereditary glycogen storage disease in humans arising from a deficiency

144

of the enzyme amylo-1,6-glucosidase. Inherited as an autosomal recessive. Prevalence 1/100,000.

forensic medicine medical science in its relation to the law.

formaldehyde $H_2C=0$, a colorless gas readily soluble in water.

formalin an aqueous solution of formaldehyde (*q.v.*) commonly used as a fixative, which functions through cross linking protein molecules.

formamide a small organic molecule that combines with the free NH_2 groups of adenine and prevents the formation of A-T base pairs, thereby causing denaturation of double-stranded DNA.

$$\underset{\displaystyle H-C-NH_2}{\overset{\displaystyle O}{\parallel}}$$

forme fruste an extremely mild expression of an abnormality, disease, or syndrome.

formylkynurenine the vermilion-plus substance (*q.v.*) in *Drosophila melanogaster.*

formyl methionine *See* **N-formyl methionine.**

Forssman antigen a carbohydrate antigen discovered in guinea pig tissues by J. Forssman. It is present in many other species including horse, sheep, and mice, but is absent in humans and rabbits.

forward mutation a change in a gene from wild type (normal) allele to a mutant (abnormal) allele.

fossil any remains or traces of now extinct life, including shells, bones, footprints, etc.

fossorial fitted for digging.

founder cells *See* **compartmentalization.**

founder effect the principle that when a small sample of a larger population establishes itself as a newly isolated entity, its gene pool carries only a fraction of the genetic diversity represented in the parental population. The evolutionary fates of the parental and derived populations are thus likely to be set along different pathways because the different evolutionary pressures in the different areas occupied by the two populations will be operating on different gene pools.

fox *See Vulpes vulpes.*

fowl *See* **poultry breeds.**

fowl achondroplasia hereditary chondrodystrophy affecting certain breeds of chickens (Scots, Dumpies, Japanese Bantams). The condition is inherited as an autosomal dominant. Homozygotes die as embryos.

fowl leukosis *See* **leukemia.**

F-pilus *See* **fertility factor.**

F-prime (F′) factor   a bacterial episomal fertility (F) factor containing an additional portion of the bacterial genome. F-prime factors have been most extensively studied in *E. coli.*

fractionated dose   the treatment of an organism by a series of short exposures to mutagenic radiations.

fraction collector   an automated instrument that collects consecutive samples of fluids percolating through a column packed with porous material.

Fragaria chiloensis   The majority of cultivated strawberries are derived from this species.

fragile chromosome site   a non-staining gap of variable width that usually involves both chromatids and is always at exactly the same point on a specific chromosome derived from an individual or kindred. Such fragile sites are inherited in a Mendelian codominant fashion and exhibit fragility as shown by the production of acentric fragments and chromosome deletions.

fragile X-associated mental retardation   a moderate degree of mental retardation (IQs around 50) found in males carrying an X chromosome that has a fragile site at the interface of bands q27 and q28. The frequency of such hemizygotes is about 1.8 per 1,000. X-linked mental retardation accounts for about 25% of all mentally retarded males.

frame shift mutation   *See* **reading frame shift.**

framework region   the highly conserved, relatively invariant portion of the variable (V) region of an immunoglobulin chain, as distinguished from the hypervariable segments of the V region.

fraternal twins   two individuals born at the same time from common parents, but originating from different zygotes; sometimes called *dizygotic twins.*

free energy   that component of the total energy of a system which can do work. *See* **thermodynamics, second law of.** *See also* **entropy.**

freemartin   a mammalian intersex arising due to the masculinization of a female twin by hormones from its male sibling when the fetal circulations are continuous.

free radical   an unstable molecule generated by the interaction of ionizing radiation with water. Hydroxyl and hydroperoxyl radicals are examples of free radicals that are potent oxidizing agents.

freeze-drying   a method of dehydrating a cell or solution by rapidly freezing its moisture content to ice. The solid material is then dried in the frozen state under vacuum, so that ice sublimes directly to water vapor with a minimization of shrinkage. *See* **lyophilize.**

freeze-etching   a technique for preparing biological material for electron microscopy. Live or fixed specimens are frozen in a liquid gas, such as freon or nitrogen, and then placed in a Balzers freeze fracture apparatus. This is an instrument that allows frozen tissues to be sectioned in a vacuum. The exposed surface is allowed to sublime slightly (to etch), so that surface irregularities which reflect the type and distribution of cell constituents are accentuated. The surface is then replicated, and the replica is stripped away and viewed under the electron microscope. Preparations made in this way provide useful information concerning three-dimensional organization of protein particles imbedded in the lipoidal membranes of cells.

freeze fracture   a method for preparing samples for electron microscopy; frozen samples are fractured with a knife and the complementary surfaces are cast in metal.

frequency-dependent fitness   a phenomenon in which the adaptive value of a genotype varies with changes in allelic frequencies. For example, in Batesian mimicry, mimics have greater fitness when they are rare relative to their models. *See* **mimicry.**

146

frequency-dependent selection selection involving frequency-dependent fitness.

Freund's adjuvant a widely used adjuvant containing killed, dried mycobacteria suspended in the oil phase of a water-in-oil emulsion.

Friend leukemia virus a virus-inducing leukemia in mice and rats.

Fritillaria a genus of lilies. Species of this genus are widely used in cytogenetic investigations.

frizzle a feather mutation in domestic fowl. *FF* individuals are "extreme frizzle" with bristle feathers which wear off easily; whereas *Ff* individuals are "mild frizzle" with more normal curly feathers. Frizzle feather keratin shows a poorly ordered crystalline structure, and its amino acid composition is abnormal.

frontal plane a longitudinal plane that separates a bilaterally symmetrical animal into dorsal and ventral halves. Synonymous with horizontal plane.

fructification 1. a reproductive organ or fruiting body 2. the generation of fruit or spore-producing structures by plants.

fructose a six carbon hexose sometimes called levulose. It is a component of sucrose.

$$HO-CH_2 \quad O \quad OH$$
$$| \qquad\qquad |$$
$$C \qquad\qquad C$$
$$| \ H \quad OH / |$$
$$H \ | \quad | / CH_2OH$$
$$C----C$$
$$| \qquad |$$
$$OH \quad H$$

fructose intolerance a disorder of carbohydrate metabolism inherited as an autosomal recessive. Patients lack fructose-1,6-diphosphatase. Symptoms disappear if dietary fructose is restricted.

fruit the ripened ovary of the flower which encloses the seeds.

fruit fly *See* **Drosophila.**

FSH follicle-stimulating hormone (*q.v.*).

F⁻ strain *Escherischia coli* behaving as recipients during unidirectional genetic transfer.

F⁺ strain *Escherischia coli* behaving as donors during unidirectional genetic transfer.

F test *See* **analysis of variance.**

fundamental theorem of natural selection *See* **natural selection.**

Fungi the kingdom of fungi, eukaryotic animals that form spores and lack flagellae at all stages of their life cycle. *See* **classification, Eukaryotes.**

funiculus the plant stalk bearing on ovule.

fused protein a hybrid protein molecule produced when a gene of interest is inserted by recombinant DNA techniques into a recipient plasmid and displaces the stop codon for a plasmid gene. The fused protein begins at the amino end with a portion of the plasmid protein sequence and ends with the protein of interest. *Compare with* **polyprotein.**

fusidic acid an antibiotic that prevents translation by interfering with elongation factor G.

fusion gene a hybrid gene composed of parts of two other genes, arising from deletion of a chromosomal segment between two linked genes or by unequal crossing-over. Hemoglobin Lepore (*q.v.*) is an example of a fused gene.

147

G

g   gravity; employed in describing centrifugal forces. Thus 2000 × g refers to a sedimenting force 2000 times that of gravity.

G   guanine or guanosine.

G1, G2 phases   *See* **cell cycle**.

galactose   a six carbon sugar which forms a component of the disaccharide lactose and of various cerebrosides and mucoproteins.

galactosemia   a hereditary disease in humans inherited as an autosomal recessive due to a gene on the short arm of chromosome 9. Homozygotes suffer from a congenital deficiency of the enzyme galactosyl-1-phosphate uridyl-transferase, and galactose-1-phosphate accumulates in their tissues. They exhibit enlargement of the liver and spleen, cataracts, and mental retardation. Symptoms regress if galactose is removed from the diet. Prevalance 1/62,000.

galactosidase   *See* **alpha galactosidase, beta galactosidase**.

Galapagos finches   *See* **Geospizinae**.

Galapagos Islands   a cluster of 14 islands located 600 miles west of the coast of Ecuador. Darwin landed there on September 15, 1833, and spent five weeks surveying the plant and animal life. The tremendous abundance of related species suggested to Darwin that the islands were colonized by a few species from the mainland and that these species later evolved various daughter species which became specialized to live in each of the many new environments provided.

gall   an abnormal growth of plant tissues.

gallinaceous   resembling domestic fowl.

148

Gallus domesticus the domesticated chicken, the genetically best known avian species. *See* **poultry breeds.**

Galton's apparatus an apparatus consisting of a glass-faced case containing an upper reservoir where balls are stored. Below the reservoir are arranged row after row of equally spaced pegs which stand out from the wall and below these are a series of vertical slots. The balls are allowed to fall one at a time through a central opening at the bottom of the reservoir. Since each ball after striking a peg has an equal probability of bouncing to the left or right, most will follow a zigzag course through the pegs and will eventually land in a central slot. The final distribution of balls in the slots will be a bell-shaped one. The apparatus demonstrates how the compounding of random events will generate a family of bell-shaped curves.

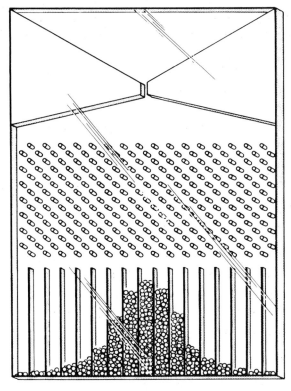

gametangium an enclosed compartment in which gametes are developed in ferns, mosses, and fungi.

gamete a haploid germ cell.

game theory a mathematical theory dealing with the determination of optimum strategies where the policies adopted depend on the most likely behaviors of two or more competitors. Game theory is employed in mathematical models of species competition.

gametic meiosis referring to meiosis which immediately precedes gamete formation, as in diplonts *(q.v)*.

gametic number the haploid number of chromosomes, symbolized by N.

gametocyte a cell that will form gametes through division; a spermatocyte or oocyte.

gametogenesis the formation of gametes.

gametophyte the haploid phase (of the life cycle of plants undergoing an alternation of generations) during which gametes are produced by mitosis. *See* **sporophyte**.

gamma chain one of the two polypeptides found in fetal hemoglobin (*q.v.*).

gamma field a field where growing plants may be exposed to chronic irradiation from a centrally placed multicurie ^{60}Co gamma-ray source.

gamma globulin an antibody-containing protein fraction of the blood.

gamma ray an electromagnetic radiation of short wavelength emitted from an atomic nucleus undergoing radioactive decay.

gamone a compound produced by a gamete to facilitate fertilization. Chemotactic sperm attractants produced by eggs are examples.

ganglion a small nervous tissue mass containing numerous cell bodies.

ganglioside a family of complex lipids containing sphingosine, fatty acids, carbohydrates, and neuraminic acid. A ganglioside that accumulates in the brain of patients with Tay-Sachs disease (*q.v.*) is shown below.

gargoylism   a term covering two genetically distinct hereditary diseases of connective tissue in man, Hunter syndrome (*q.v.*) and Hurler syndrome (*q.v.*).

Garrod's disease   *See* **alkaptonuria.**

gas chromatography   a chromatographic technique in which an inert gas is used to sweep through a column the vapors of the materials to be separated.

gas-flow radiation counter   a counter in which an appropriate atmosphere is maintained in the sensitive volume by allowing a suitable gas to flow slowly through it.

gastrin   a hormone secreted by the stomach which causes secretion of digestive enzymes by other stomach cells.

Gastropoda   the class of molluscs containing the snails. *See* **classification.**

gastrula   the stage of embryonic development when the gastrulation movements occur.

gastrulation   the complex cell movements that carry those cells whose descendants will form the future internal organs from their larger superficial position in the blastula to approximately their definitive positions inside the animal embryo. Prior to gastrulation the amphibian embryo relies on RNA molecules pre-loaded into the ooplasm during oogenesis. During gastrulation, however, newly synthesized nuclear gene products are required.

Gaucher cells   large, lipid-filled cells scattered throughout the reticuloendothelial system of patients with Gaucher disease (*q.v.*).

Gaucher disease   the most common hereditary disorder of glycolipid metabolism, inherited as an autosomal recessive. The disease is due to subnormal activity of lysosomal glucocerebrosidase, which cleaves glucose from cerebrosides. As a result, glucocerebrosides accumulate in lysosomes. Heterozygotes can be identified because their peripheral leukocytes show lower levels of lysosomal glucocerebrosidases. Homozygotes can be detected *in utero* by assays run on cells obtained by amniocentesis.

Gause's law   *See* **competitive exclusion principle.**

Gaussian curve   normal distribution.

G banding   *See* **chromosomal banding techniques.**

Geiger-Mueller (G-M) counter   a sensitive gas-filled radiation-measuring device.

gelding   a castrated horse.

gemmules   pangenes. *See* **pangenesis.**

gene   a hereditary unit that, in the classical sense, occupies a specific position (locus) within the genome or chromosome; a unit that has one or more specific effects upon the phenotype of the organism; a unit that can mutate to various allelic forms; a unit that recombines with other such units. Three classes of genes are now recognized: (1) *structural genes* that are transcribed into mRNAs that are then translated into polypeptide chains, (2) genes that are transcribed into rRNA or tRNA molecules that are used directly, and (3) *regulatory genes* that are not transcribed, but serve as recognition sites for enzymes and other proteins involved in DNA replication, restriction, modification, recombination, transcription, and repair processes. Examples of regulatory genes are promoters, terminators, operators, attenuators, and enhancers. Seymour Benzer coined the following names for the gene in terms of its operational definitions. A *cistron* is the amount of genetic material that codes for one complete, mature tRNA, rRNA, or polypeptide chain. A cistron includes (if present) regions preceding and following the coding region (leader and trailer), as well as intervening sequences (introns) between coding segments (exons). A *muton* is

151

the smallest unit that, when modified, changes the genetic code (corresponding to a nucleotide). A *recon* is the smallest unit that can experience recombination (corresponding to an adjacent pair of nucleotides in cis position).

gene activation *See* **genetic induction.**

genealogy a record of the descent of a family, group, or person from an ancestor or ancestors; lineage; pedigree.

gene amplification any process by which specific DNA sequences are replicated disproportionately greater than their representation in the parent molecules. During development, some genes become amplified in specific tissues; e.g., ribosomal genes amplify and become active during oogenesis, especially in some amphibian oocytes (*see* **rDNA amplification, Xenopus**); genes for *Drosophila* chorion and for chicken alpha-actin amplify during development.

gene cluster *See* **multigene family.**

gene conversion a situation in which the products of meiosis from an *AA'* individual are 3*A* and 1*A'* or 1*A* and 3*A'*, not 2*A* and 2*A'* as is usually the case. Thus one gets the impression that one *A* gene has been converted to an *A'* gene (or vice versa). A postulated cause for this departure from the 2:2 ratio is a switch in templates during chromosomal replication. As a result the *A* gene is copied twice and the *A'* gene not at all (or vice versa).

gene dosage the number of times a given gene is present in the nucleus of a cell.

gene duplication the production of a tandem repeat of a DNA sequence by unequal crossing-over or by an accident of replication.

gene expression the manifestation of the genetic material of an organism as a collection of specific traits.

gene family *See* **multigene family.**

gene flow the exchange of genes between different populations of the same species produced by migrants, and commonly resulting in simultaneous changes in gene frequencies at many loci in the recipient gene pool.

gene frequency the percentage of all alleles at a given locus in a population represented by a specific allele. Also referred to as *allelic frequency* (*q.v.*).

gene insertion any technique that inserts into a cell a specific gene or genes from an outside source, including cell fusion, gene splicing, transduction, and transformation.

gene interaction interaction between different genes residing within the same genome in the production of a particular phenotype. Such interactions often occur when the products of the non-allelic genes under study function at steps in a sequence of reactions that result in compounds that generate the phenotype in question. These interactions can produce variations from the classical genetic ratios. An example would be the inheritance of aleurone color in *Zea mays*. In order for the corn kernel to possess colored aleurone at least one *A* and one *C* gene must be present. Given *A* and *C* in the heterozygous or homozygous condition and, in addition, *R* in the heterozygous or homozygous condition, then a red pigment is produced. Purple pigment is synthesized if *P* is present in addition to *A, C,* and *R. A, C, P,* and *R* reside on different chromosomes. Thus if a plant of genotype *AaCCRRPp* is self-pollinated, the offspring will contain the following aleurone classes: purple, red, and white in a 9:3:4 ratio. Here the 9:3:3:1 ratio has been converted into a 9:3:4 ratio, because the *P* gene cannot be expressed in the absence of *A*.

gene library a random collection of fragments of the DNA of a given species inserted into plasmids and cloned in a suitable host. The collection must be large enough to include all of the unique nucleotide sequences of the genome.

gene locus (*plural* loci)   the specific position on a chromosome where a gene is located.

gene mapping   assignment of a locus to a specific chromosome and/or determining the sequence of genes and their relative distances from one another on a specific chromosome.

gene pair   in a diploid cell, the two representative genes (either identical or nonidentical alleles) at a given locus on homologous chromosomes.

gene pool   the total genetic information possessed by the reproductive members of a population of sexually reproducing organisms.

gene 32 protein   the first DNA unwinding protein (*q.v.*), discovered by B. M. Alberts and L. Frey in 1970. It is the product of gene 32 of phage T_4 and is essential for its replication. The protein has a molecular weight of 35,000 d and binds to a stretch of DNA about 10 nucleotides long.

generalized   in evolution theory, an unspecialized condition or trait, usually considered to have a greater potential for evolving into a variety of alternative conditions than that possessed by a highly specialized one. Primitive traits tend to be generalized; derived or advanced traits tend to be more specialized.

generalized transduction   *See* **transduction.**

generation time (T_g)   the time required for a cell to complete one growth cycle. *See* **doubling time.**

gene redundancy   the presence in a chromosome of many copies of a gene. For example, the nucleolus organizer of *Drosophila melanogaster* contains hundreds of duplicate copies of the cistrons which code for the 18S and 28S rRNA molecules.

gene splicing   *See* **recombinant DNA research.**

gene substitution   the replacement of one gene by its allele, all the other genes (or all other relevant genes) remaining unchanged.

gene therapy   addition of a functional gene(s) to a cell by gene insertion (*q.v.*) to correct a hereditary disease.

Genetical Theory of Natural Selection   a text, published in 1930, by R. A. Fisher which constitutes a milestone in population genetics.

genetic assimilation   the process by which a phenotypic character initially produced only in response to some environmental influence becomes, through a process of selection, taken over by the genotype, so that it is formed even in the absence of the environmental influence that at first had been necessary.

genetic block   a block in a biochemical reaction generally due to a mutation that prevents the synthesis of an essential enzyme or results in the formation of a defective enzyme. If the defective enzyme has limited activity, the block may be a partial one, and the mutant is referred to as "leaky."

genetic burden   *See* **genetic load.**

genetic coadaptation   *See* **coadaptation.**

genetic code   the consecutive nucleotide triplets (codons) of DNA and RNA that specify the sequence of amino acids for protein synthesis. The code appears to be universal for all organisms, but a few exceptions are known in some mitochondrial systems (*see* **universal**

code theory). The mRNA nucleotide sequences are written 5′ to 3′ right to left, respectively, because that is the direction in which translation occurs.

| FIRST BASE | SECOND BASE | | | | THIRD BASE |
	U	C	A	G	
U	phe	ser	tyr	cys	U
	phe	ser	tyr	cys	C
	leu	ser	Ter	Ter	A
	leu	ser	Ter	trp	G
C	leu	pro	his	arg	U
	leu	pro	his	arg	C
	leu	pro	gln	arg	A
	leu	pro	gln	arg	G
A	ile	thr	asn	ser	U
	ile	thr	asn	ser	C
	ile	thr	lys	arg	A
	met and f met	thr	lys	arg	G
G	val	ala	asp	gly	U
	val	ala	asp	gly	C
	val	ala	glu	gly	A
	val	ala	glu	gly	G

See **amino acids** for symbols. Ter, terminator codon; f met, N-formyl methione.

Thus, an mRNA segment specifying proline-tryptophan-methionine would be (5′) CCU-UGG-AUG (3′), whereas its complementary, antiparallel DNA template strand would be (3′) GGA-ACC-TAC (5′). The code is degenerate in that all amino acids, except methionine and tryptophan, are specified by more than one codon. Most of the degeneracy involves the third nucleotide at the 3′ end of the codon (*see* **wobble hypothesis**). the code is read from a fixed starting point, in one direction, in groups of three consecutive nucleotides. The start codon is AUG, and in bacteria it specifies the insertion of N-formyl methionine (*q.v.*). When AUG occupies an internal position in the mRNA it specifies methionine.

genetic code dictionary *See* **amino acids.**

genetic colonization introduction of genetic material from a parasite into a host, thereby inducing the host to synthesize products that only the parasite can use. *See* **opines.**

genetic counseling the analysis of risk of producing genetically defective offspring within a family, and the presentation to family members of available options to avoid or ameliorate possible risks.

genetic death death of an individual without reproducing.

genetic detasseling a breeding technique used in the commercial production of corn seed. The breeding scheme produces pollen abortion with the result that the plants are no longer hermaphroditic and can only be cross fertilized.

genetic disease *See* **hereditary disease.**

genetic dissection analysis of the genetic basis of a biological phenomenon through the study of mutations that effect that phenomenon. For example, spermatogenesis can be "genetically dissected" by inducing and then characterizing mutations that sterilize male *Drosophila.*

genetic distance a measure of the number of allelic substitutions per locus that have occurred during the separate evolution of two populations or species.

genetic drift   the random fluctuations of gene frequencies due to sampling errors. While drift occurs in all populations, its effects are most evident in very small populations.

genetic engineering   the experimental or industrial technologies used to alter the genome of a living cell so that it can produce more or different chemicals, or perform new functions. The manipulation of genes in ways that bypass normal sexual or asexual transmission.

genetic equilibrium   the situation reached in a population containing, as an example, the allelic genes *A* and *a,* where the frequencies of both alleles are maintained at the same values generation after generation.

genetic fine structure   *See* **fine structure genetic mapping.**

genetic fitness   the contribution to the next generation of a specified genotype in a population in relation to the contributions of all other genotypes of that same population.

genetic homeostasis   the tendency of a population to equilibrate its genetic composition and to resist sudden changes.

genetic identity   a measure of the proportion of genes that are identical in two populations.

genetic induction   the process of gene activation by an inducer molecule, resulting in transcription of one or more structural genes. *See* **inducible system.**

genetic information   the information contained in a sequence of nucleotide bases in a nucleic acid molecule. *See* **exon, intron.**

genetic load   **1.** the average number of lethal equivalents per individual in a population. **2.** the relative difference between the actual mean fitness of a population and the mean fitness that would exist if the fittest genotype presently in the population were to become ubiquitous.

genetic map   the linear arrangement of mutable sites on a chromosome as deduced from genetic recombination experiments.

genetic marker   a gene, whose phenotypic expression is usually easily discerned, used to identify an individual or a cell that carries it, or as a probe to mark a nucleus, chromosome, or locus.

genetic polymorphism   the long-term occurrence in a population of two or more genotypes in frequencies that cannot be accounted for by recurrent mutation. Such polymorphism may be due to mutations which are (a) advantageous at certain times and under certain conditions and (b) disadvantageous under other circumstances, and which exist in habitats where situations (a) and (b) are encountered frequently. Genetic polymorphism may also result if genotypes heterozygous at numerous loci are generally superior to any homozygous genotype.

genetics   the scientific study of heredity.

Genetics and the Origin of Species   a text, published in 1937 by T. Dobzhansky, which constitutes a milestone in evolutionary genetics.

genetic system   the organization and method of transmission of the genetic material for a given species.

genetic variance   the phenotypic variance of a trait in a population attributed to genetic heterogeneity.

genic balance   a mechanism of sex determination, originally discovered in *Drosophila,* that depends upon the ratio of X chromosomes to sets of autosomes (A). Males develop when the X/A ratio is 0.5 or less; females develop when the X/A ratio is 1.0 or greater; an intersex develops when the ratio is between 0.5 and 1.0. *See* **metafemales, metamales.**

genital disc   the imaginal disc from which the reproductive duct system and the external genitalia are devied in *Drosophila.*

genome   a term used to refer to all of the genes carried by a single gamete (i.e., by a single representative of each of all chromosome pairs).

genomic blotting   *See* **Southern blotting.**

genomic exclusion   an abnormal form of conjugation occurring in *Tetrahymena pyriformis* between cells with defective micronuclei and normal cells. The progeny are heterokaryons; each has an old macronucleus but a new diploid micronucleus derived from one meiotic product of the normal mate.

genomic formula   a mathematic representation of the number of genomes (sets of genetic instructions) in a cell or organism. Examples: n (haploid gamete or monoploid somatic cell), 2n (diploid), 3n (triploid), 4n (tetraploid), $2n - 1$ (monosomic), $2n + 1$ (trisomic), $2n - 2$ (nullisomic), etc.

genomic library (bank)   a collection of recombinant DNA molecules (cloned fragments) representative of the individual's genome.

genopathy   a disease resulting from a genetic defect.

genophore   the chromosome equivalent in viruses and prokaryotes. Such genophores contain nucleic acid, but lack associated histones.

genotype   the genetic consitution of an organism, as distinguished from its physical appearance (its phenotype).

genotype-environment interaction   an inference drawn from the observation that the phenotypic expression of a given genotype varies when measured under different environmental conditions.

genus (*plural* genera)   a taxon that includes one or more species presumably related by descent from a common ancestor, *See* **hierarchy.**

geochronology   a science that deals with the measurement of time in relation to the earth's evolution.

geographical isolate   a population separated from the main body of the species by some geographical barrier.

geographic speciation   the splitting of a parent species into two or more daughter species following geographic isolation of two or more parental populations; allopatric speciation.

geologic time divisions   *See* page 157.

geometric mean   the square root of the product of two numbers; more generally, the *n*th root of the product of a set of *n* positive numbers.

Geospizinae   a sub-family of finches to which the 14 species found on the Galapagos Islands belong. The most marked distinction between the different species is in the size and shape of the bill, which is directly related to the bird's food habits. Charles Darwin was the first to suggest that the modern populations of these birds are the end product of an adaptive radiation from a single ancestral species. The evolutionary divergences resulted from adaptations that allowed different populations to utilize different food resources and so to avoid competition.

geotropism   the response of plant parts to the stimulus of gravity.

Eon	Era	Period		Epoch	
Phanerozoic Eon *(Phaneros = "evident"; Zoon = "life")*	Cenozoic Era	Quaternary		Recent or Holocene	
				Pleistocene	
		Tertiary	Neogene	Pliocene	−2
					−5
				Miocene	
					−23
			Paleogene	Oligocene	
					−37
				Eocene	
					−55
				Paleocene	
	Mesozoic Era	Cretaceous			−65
		Jurassic			−140−
		Triassic			−195−
	Paleozoic Era	Permian			−230−
		Carboniferous	Pennsylvanian		−280−
			Mississippian		−310−
		Devonian			−345−
		Silurian			−400−
		Ordovician			−435−
		Cambrian			−570−
Precambrian or Cryptozoic Eon		Proterozoic			−600−
		Archean			−2600−
					−3800−

Geologic Time Divisions

Each number gives the date in millions of years before the present. In the figure, the relative widths of each time division are not proportional to the absolute time spent in them. Thus, the Holocene epoch occupies only the last 10,000 years while the Cryptozoic eon constitutes about 85% of the entire geological record. The time segment between the formation of the earliest rocks (3.8 billion years ago) and the origin of the earth (4.6 billion years ago) has been named the Hadean eon.

Adapted from *The Earth Through Time*, second edition, by Harold L. Levin, W.B. Saunders, Co. (1983).

157

germarium   the anterior, sausage-shaped portion of the insect ovariole. It is in the germarium that cystocyte divisions occur and the clusters of cystocytes become enveloped by follicle cells. *See* **insect ovary types.**

germ cell   a sex cell or gamete; egg (ovum) or spermatozoan; a reproductive cell that fuses with one from the opposite sex in fertilization to form a single-celled zygote.

germinal cells   cells that produce gametes by meiosis: e.g., oocytes in females and spermatocytes in males.

germinal choice   the concept advocated by H. J. Muller of progressive human evolution by the voluntary choice of germ cells. Germ cells donated by individuals possessing recognized superior qualities would be frozen and stored in germ banks. In subsequent generations these would be available for couples who wished to utilize these rather than their own germ cells to generate a family. Such couples are referred to as "pre-adoptive" parents.

germinal mutations   genetic alterations occurring in cells destined to develop into germ cells.

germinal selection   **1.** selection by man of the germ cells to be used in producing a subsequent generation of a domesticated species. Such selection has been suggested for human beings. **2.** selection during gametogenesis against induced mutations that retard the proliferation of the mutated cells. Such selection introduces errors in estimating the frequency of mutations induced in gonial cells.

germinal vesicle   the diploid nucleus of a primary oocyte during vitellogenesis. The nucleus is generally arrested in a postsynaptic stage of meiotic prophase.

germination inhibitor   any of the specific organic molecules present in seeds that block processes essential to germination and therefore are often the cause of dormancy.

germ line   pertaining to the cells from which gametes are derived. When referring to species, the cells of the germ line, unlike somatic cells, bridge the gaps between generations.

germ plasm   the hereditary material transmitted to offspring through the germ cells.

gerontology   the study of aging.

gestation period   in a viviparous animal, the time from conception to birth.

GH   growth hormone (*q.v.*).

gibberellin   one of a family of phytohormones of widespread distribution in plants. Many single-gene, dwarf mutants in *Pisum, Vicia,* and *Phaseolus* are cured by the application of gibberellins. Gibberellins also promote seed germination, the breaking of dormancy, and flowering, and can induce parthenocarpy.

Gibberellin A$_1$

158

giga- a prefix meaning one billion (10^9). Used with metric measurement units.

gilt a female swine less than a year old, which has not farrowed.

gland an organ that synthesizes specific chemical compounds (secretions) which are passed to the outside of the gland.

gln glutamine. *See* **amino acid.**

globins a widespread group of respiratory proteins including the tetrameric hemoglobins of protochordates and various invertebrates, the monomeric myoglobins, and the leg-hemoglobins. In the case of myoglobins and the vertebrate hemoglobins, phylogenetic trees have been constructed showing that the globin sequences have diverged during evolution at a constant rate. From these data, the timing of the gene duplications that gave rise to the family of hemoglobins has been deduced. *See* **hemoglobin gene, myoglobin gene.**

globulin any of certain proteins that are insoluble in distilled water, but soluble in dilute aqueous salt solution. *See* **albumin, immunoglobin.**

Gloger's rule an ecological rule stating that, in a geographically variable species, those populations living in the warmer, moister (more tropical) parts of the range tend to be of darker color than those living in the warmer, drying (more desert) parts of the range; those living in cold, typically moist (more alpine or arctic) parts of the range tend to be colored very pale or white.

glu glutamate; glutamic acid. *See* **amino acid.**

glucagon a polypeptide hormone from the alpha cells of the pancreas which promotes the breakdown of liver glycogen and the consequent elevation of blood glucose.

glucocerebroside any of certain compounds related to sphingomyelins. They differ in that the phosphorylcholine seen in sphingomyelin (*q.v.*) is replaced by glucose. *See* **cerebroside.**

glucocorticoids steroid hormones produced by the adrenal cortex having effects on intermediary metabolism, such as stimulating glycogen deposition by the liver. Some glucocorticoids, such as cortisone, produce anti-inflammatory effects.

glucose a six-carbon sugar widely distrubuted in plants, animals, and microorganisms. It is a component of the disaccharide lactose and is present in polysaccharides such as cellulose, starch, and glycogen.

glucose-6-phosphate dehydrogenase (G6PD) an enzyme catalyzing the conversion of glucose-6-phosphate to 6-phosphogluconate. G6PD from human erythrocytes is a dimer consisting of identical subunits, each with a molecular weight of 55,000 d. Over 80 genetic variants of this enzyme are known. The four most common forms of the enzyme are listed on p. 160.

ENZYME	ACTIVITY	SOURCE
B	100%	widespread
A	90%	African blacks
A⁻	8%–20%	African blacks
M	0–7%	Mediterranean whites

glucose-6-phosphate dehydrogenase deficiency the most common disease-producing, enzyme deficiency of humans. The gene encoding G6PD resides at the distal end of the long arm of the X chromosome. As a result of random X-chromosome inactivation (*see* **dosage compensation**), females heterozygous for the G6PD gene have some erythrocytes with the normal enzyme and some with the defective enzyme. The red cells of males with the A⁻ or M forms of the enzyme have a reduced life span, and exposure to antimalarial drugs such as primaquine results in life-threatening hemolysis. However, the possession of mutant proteins with reduced enzymatic activity appears to confer a resistance to malaria.

glucose-sensitive operons bacterial operons whose activity is inhibited by the presence of glucose. This lowers the level of cyclic AMP, thereby blocking a required positive control signal.

glucosylceramide lipidosis Gaucher disease (*q.v.*).

glume a chaffy bract, pairs of which enclose the base of grass spikelets.

glutamic acid *See* **amino acid.**

glutamine *See* **amino acid.**

glutathione a tripeptide containing glutamic acid, cysteine, and glycine and capable of being alternately oxidized and reduced. Glutathione plays an important role in cellular oxidations.

gly glycine. *See* **amino acid.**

glycerol a trihydric alcohol, which combines with fatty acids to form fats (*q.v.*).

glycogen (*also* "**animal starch**") a soluble polysaccharide built up of numerous glucose molecules. Carbohydrate is stored as glycogen by vertebrates, especially in liver and muscles.

glycogenesis glycogen synthesis.

glycogenolysis   the liberation of glucose from glycogen.

glycogenosis   glycogen storage disease (*q.v.*).

glycogen storage disease   any of a group of congenital and familial disorders characterized by the deposition of either abnormally large or abnormally small quantities of glycogen in the tissues. Anderson, Forbes, von Gierke, Hers, McArdles, and Pompe diseases are examples.

glycolipid   a lipid containing carbohydrate.

glycolysis   the sequential series of anaerobic reactions diagrammed on page 162 found in a wide variety of tissues which constitutes the principal route of carbohydrate breakdown and oxidation. The process starts with glycogen, glucose, or fructose and ends with pyruvic or lactic acids. The conversion of a molecule of glucose to two molecules of pyruvic acid generates two ATP molecules. Under aerobic conditions the yield is eight ATP molecules. The pyruvic acid formed is broken down by way of the citric acid cycle (*q.v.*). *See* **pentosephosphate pathway.**

glycolytic   participating in glycolysis.

glycoprotein   a protein containing small amounts of carbohydrate (usually less than 4%). *See* **mucoprotein.**

glycoside   a compound yielding a sugar upon enzymatic hydrolysis.

glycosidic bonds   the bonds coupling the monosaccharide subunits of a polysaccharide.

glyoxylate cycle   a series of metabolic reactions involved in photorespiration of plants, including oxidation of glycolate to glyoxylate by the enzyme glycolate oxidase within microbodies or peroxisomes (*q.v.*).

Glyptotendipes barbipes   a midge possessing exceptionally large polytene chromosomes in the salivary glands of larvae and therefore a favorite for cytological studies.

gnotobiosis   the rearing of laboratory animals in a germ-free state or containing only microorganisms known to the investigator.

gnotobiota   the known microfauna and microflora of a laboratory animal in gnotobiosis (*q.v.*).

goiter   a chronic enlargement of the thyroid gland which is due to hyperplasia, not neoplasia.

Golgi apparatus (or complex or material)   a cell organelle identified in electron micrographs as a complex made up of closely packed broad cisternae and small vesicles. The Golgi apparatus is distinguished from the endoplasmic reticulum by the arrangement of the membranous vesicles and by the lack of ribosomes. The Golgi apparatus is a segregating system capable of collecting and sequestering substances synthesized by the endoplasmic reticulum.

gonad   a gamete-producing organ of an animal; in the male, the testis; in the female, the ovary.

gonadotropic hormones   pituitary hormones (such as LH and FSH, *q.v.*) that stimulate the gonads; also called *gonadotropins.*

gonadotropin-releasing hormone (GnRH)   a neurohormone from the hypothalamus that stimulates release of luteinizing hormone (LH) and follicle-stimulating hormone (FSH) from the pituitary gland.

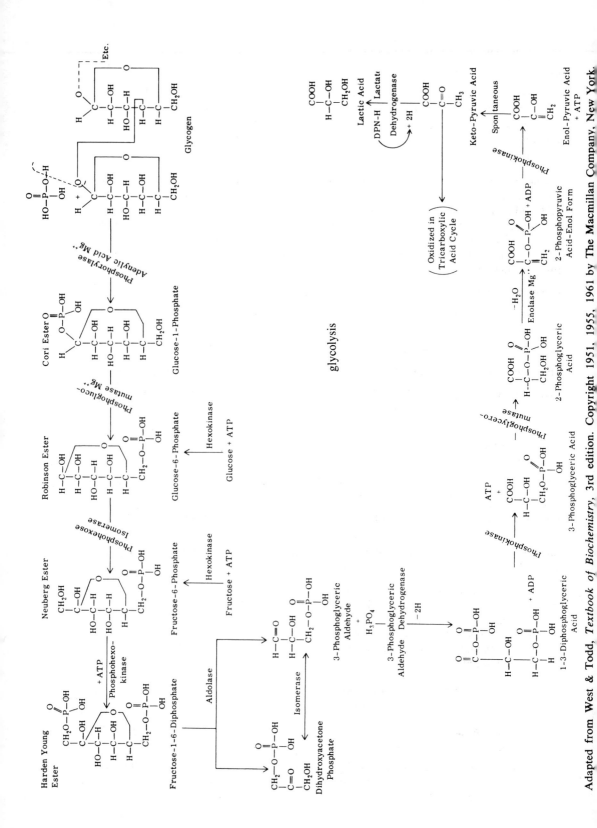

glycolysis

Adapted from West & Todd, *Textbook of Biochemistry*, 3rd edition. Copyright 1951, 1955, 1961 by The Macmillan Company, New York.

gonochorism   a sexual system in which each individual is either a male or a female. *Compare with* **hermaphrodite, monoecy.**

gonophore   **1.** in sessile coelenterates the bud producing the reproductive medusae. **2.** in higher animals any accessory sexual organ, such as an oviduct, or a sperm duct. **3.** in plants, a stalk that bears stamens and pistils.

gonomery   the separate grouping of maternal and paternal chromosomes during the first few mitoses following fertilization as occurs in some insect embryos.

gonotocont   an auxocyte (*q.v.*).

Gossypium   a genus composed of about 35 diploid and tetraploid species. Two tetraploid species *G. hirsutum* and *G. barbadense* are the source of most commercial cotton. The comparative genetics of these species has been intensively studied.

gout   a hereditary disorder of purine metabolism characterized by increased amounts of uric acid in the blood and recurrent attacks of acute arthritis.

Gowen's crossover suppressor   a recessive gene on the third chromosome of *Drosphila melanogaster* symbolized by *c(3)G*. Females homozygous for this gene are unable to form synaptonemal complexes (*q.v.*) in their oocyte nuclei and crossing over is abolished.

G6PD   glucose-6-phosphate dehydrogenase (*q.v.*).

G6PD deficiency   *See* **glucose-6-phosphate dehydrogenase deficiency.**

G_1 period, G_2 period   *See* **cell cycle.**

Graafian follicle   a fluid-filled spherical vesicle in the mammalian ovary, containing an oocyte attached to its wall.

grade   a stage of evolutionary advance. A level reached by one or more species in the development of a structure, physiological process, or behavioral character. Different species may reach the same grade because they share genes that respond in the same way to an environmental change. When two species that do not have a common ancestry reach the same grade, the term *convergence* is used to describe this evolutionary parallelism.

gradient   a gradual change in some quantitative property of a system over a specific distance (e.g., a clinal gradient; density gradient).

gradualism   the theory that evolution always proceeds by imperceptibly small, cumulative steps rather than by abrupt, major changes; usually associated with Darwinism. *See* **evolution.**

graduational model of evolution   *See* **gradualism, evolution.**

Graffi leukemia virus   a virus that induces myeloid leukemia in mice and rats.

graft   **1.** a relatively small piece of plant or animal tissue implanted into an intact organism. **2.** the act of transferring a part of an organism from its normal position to another position in the same or another organism. *See* **autograft, heterograft, homograft, scion, stock.**

graft hybrid   a plant made up of two genetically distinct tissues due to fusion of host and donor tissues after grafting.

graft rejection   a cell-mediated immune response to transplanted tissue that causes destruction of a graft. Rejection is evoked by the histocompatibility antigens (*q.v.*) of the foreign cells.

graft-versus-host reaction   a syndrome arising when an allograft, containing immunocompetent cells, mounts an immune response against a host that is unable to reject it because the host is immunologically immature or immunologially compromised or suppressed (e.g., by radiation or drugs); synonymous with allogeneic disease, runt disease (*q.v.*).

gram atomic weight   the quantity of an element that has a mass in grams numerically equal it its atomic weight.

gram equivalent weight   the mass of an acid or a base which will release or neutralize one gram molecule (mole) of hydrogen ion. A 1 M solution of H_2SO_4 contains 2 gram equivalents. A thousandth of a gram equivalent weight is a millequivalent.

gramicidin S   a cyclic antibiotic synthesized by *Bacillus brevis.* The molecule has the structure

$$\begin{bmatrix} \text{d-phe–pro–val–orn–leu} \\ \text{leu–orn–val–pro–d–phe} \end{bmatrix}$$

and it contains amino acids not usually found in proteins—namely, ornithine (orn) and d-phenylalanine (rather than the usual l-isomer). The synthesis of gramicidin S constitutes one of the best understood examples of a polypeptide that is not synthesized on a ribosome. Two enzymes are required (E_1 and E_2), which are bound together, forming a unit called gramicidin synthetase. One molecule each of proline, valine, ornithine, and leucine bind in that sequence to sulfhydryl groups of E_1. E_2 functions to isomerize l- to d-phenylalanine and to transfer it to the proline attached to E_1. The two identical polypeptides are then joined head to tail to form a decapeptide. This rare type of polypeptide synthesis is very uneconomical and cannot generate molecules greater than 20 amino acids long.

gram molecular weight   the quantity of a compound that has a mass in grams numerically equal to its molecular weight. Gram molecular weight is often shortened to *gram-mole* or *mole.*

Gram staining procedure   a staining technique perfected by the Danish physician Hans Christian Gram (1853–1938). It allows bacteria to be divided into two groups, Gram-positive (which stain deep purple) and Gram-negative (which stain light pink). The staining differences lie in the permeability properties of the cell walls of the two groups of bacteria.

grana (*singular* granum)   long columns in chloroplasts made up of stacks of dense discs. Grana are sites where chlorophyll is located. Each disc contains a double layer of quantosomes. *See* **chloroplast.**

granulocytes   white blood cells possessing distinct cytoplasmic granules and a multilobed nucleus; includes basophils, eosinophils, and neutrophils; also called *polymorphonuclear leukocytes.*

grass   any species of monocotyledon belonging to the family Gramineae. Such species are characterized by leaves with narrow, spear-shaped blades, and flowers borne in spikelets of bracts.

gratuitous inducer   a compound not found in nature that acts as an inducer although it cannot be metabolized. *See* **IPTG, ONPG.**

gravid   swollen from accumulated eggs or embryos.

Greek alphabet   *See* list on page 165.

grid   **1.** a network of uniformity spaced horizontal and vertical lines; **2.** a specimen screen used in electron microscopy.

griseofulvin   an antibiotic synthesized by certain *Penicillium* species that is used as a fungicide.

164

Greek alphabet

CAPS	LOWER CASE	GREEK NAME	CAPS	LOWER CASE	GREEK NAME	CAPS	LOWER CASE	GREEK NAME
A	α	Alpha	I	ι	Iota	P	ρ	Rho
B	β	Beta	K	κ	Kappa	Σ	$\sigma\ s$	Sigma
Γ	γ	Gamma	Λ	λ	Lambda	T	τ	Tau
Δ	$\delta\ \partial$	Delta	M	μ	Mu	Υ	υ	Upsilon
E	ϵ	Epsilon	N	ν	Nu	Φ	ϕ	Phi
Z	ζ	Zeta	Ξ	ξ	Xi	X	χ	Chi
H	η	Eta	O	o	Omicron	Ψ	ψ	Psi
Θ	$\theta\ \vartheta$	Theta	Π	π	Pi	Ω	ω	Omega

Gross leukemia virus   a virus isolated by Gross which induces lymphoid leukemia in mice.

group selection   natural selection acting upon a group of two or more individuals by which traits are selected that benefit the group rather than the individual. *See* **Hamilton's genetical theory of social behavior.**

group-transfer reactions   chemical reactions involving the exchange of functional groups between molecules (excluding oxidations or reductions and excluding water as a participant). The enzymes that catalyze group-transfer reactions are called transferases or synthetases. For example, activation of an amino acid involves transfer of an adenosine monophosphate group from ATP to the COO^- group of the amino acid.

growth curve   a curve showing the change in the number of cells in a growing culture as a function of time.

growth factor   a specific substance that must be present in the growth medium to permit cell multiplication.

growth hormone   a protein, growth-stimulating hormone secreted by the adenohypophysis of vertebrates. It is a single polpypeptide chain containing 191 amino acids. Human growth hormone is encoded by a gene on the long arm of chromosome 17.

GSH   reduced glutathione.

GT-AG rule   intron junctions start with the dinucleotide GT and end with the dinucleotide AG, corresponding to the left and right ("donor and acceptor") splicing sites, respectively.

GTP   guanosine triphosphate (*q.v.*).

guanine   a purine base found in nucleic acids.

guanine deoxyriboside   *See* **nucleoside.**

guanine-7-methyl transferase   *See* **methylated cap.**

guanosine   *See* **nucleoside.**

guanosine triphosphate   an energy-rich molecule (analogous to ATP) that is required for the synthesis of all peptide bonds during translation.

guanylic acid   *See* **nucleotide.**

guanylyl transferase   *See* **methylated cap.**

guinea pig   *See **Cavia porcellus.***

guppy   *See **Lebistes reticularis.***

gymnosperm   a primitive plant having naked seeds (conifers, cycads, ginkgos, etc.).

gynander   synonomous with **gynandromorph** *(q.v.).*

gynandromorph   an individual made up of a mosaic of tissues of male and female geno-types. The fruit fly illustrated below is a bilateral gynandromorph with the right side female, the left side male. The zygote was $+ +/w\ m$. Loss of the X chromosome containing the dominant (+) genes occurred at the first nuclear division. The cell with the single X chromosome containing the recessive marker genes gave rise to the male tissues. Therefore, the left eye is white and the left wing is miniature. Note the male abdominal pigmentation and the sex comb.

gynodioecy   a sexual dimorphism in plants having both bisexual and separate female individuals.

gynoecium   a collective term for all the carpels of a flower.

gynogenesis   reproduction by parthenogenesis requiring stimulation by a spermatozoan for the activation of the egg.

gyrase   the colloquial name for a type II topoisomerase *(q.v.)* of *E. coli* that converts relaxed closed-circular, duplex DNA to a negatively superhelical form both *in vitro* and *in vivo*. This enzyme prepares the DNA for proteins that require unwinding of the duplex or single-stranded regions in order to participate in such processes as replication, transcription, repair, and recombination. Several drugs are known to inhibit gyrase, including cou-mermycin, naladixic acid, novobiocin, and oxilinic acid.

166

H

H   hydrogen.

H1, H2A, H2B, H3, H4   *See* **histones.**

³H   *See* **tritium.**

habitat   the natural abode of an organism.

Habrobracon juglandis   *See* **Microbracon hebetor.**

Haemanthus katherinae   the African blood lily. A favorite species used for the time-lapse photographic study of mitotic endosperm cells.

haemoglobin   *See* **hemoglobin.**

hairpin loops   any double-helical regions of DNA or RNA formed by complementary base pairing between adjacent inverted sequences on the same strand.

Haldane's law   the generalization that if first generation hybrids are produced between two species, but one sex is absent, rare, or sterile, that sex is the heterogametic sex.

half-chromatid conversion   *See* **chromatid conversion.**

half-life, biological   the time required for the body to eliminate one-half of the dose of a given substance. This time is approximately the same for both stable and radioactive isotopes of any element.

half-life, radioactive   the time required for half the mass of a radioactive substance to decay into another substance. Each radionuclide has a unique half-life.

half-sib mating   mating between half brother and half sister. Such individuals have one parent in common.

half tetrad analysis   recombinational analysis where two of the four chromatids of a given tetrad can be recovered, as in the case of attached X chromosomes in *Drosophila*.

half value layer   the thickness of a specified material that reduces the flux of radiation by one-half.

halide   a fluoride, chloride, bromide, or iodide.

halogen   fluorine (F), chlorine (Cl), bromine (Br), or iodine (I).

haltere   the modified hind-wing of a dipteran.

Hamilton's genetical theory of social behavior   a theory put forth by W. D. Hamilton to explain how altruism can evolve when it increases the fitness of relatives. The theory proposes that a social act is favored by natural selection if it increases the *inclusive fitness* of the performer. Inclusive fitness consists of the individual's own fitness as well as his effects on the fitness of any genetically related neighbors. The idea is that genetic alleles change in frequency in a population owing to effects on the reproduction of relatives of the individual in which the character is expressed, rather than on the personal reproduction of the individual. Therefore, Hamilton's theory is often referred to as *kin selection.* For example, a mutation that affected the behavior of a sterile worker bee so that she fed her fertile queen but starved herself would increase the inclusive fitness of that worker because, while her own fitness decreased, her actions increased the fitness of a close relative.

hamster   common laboratory rodent. *See **Cricetulus griseus, Mesocricetus auratus.***

hanging drop technique   a method for microscopic examination of organisms suspended in a drop on a special concave microscope slide.

Hansenula wingei   a yeast which has provided information concerning the genetic control of mating type (*q.v.*).

H antigens   1. histocompatibility antigens governed by histocompatibility genes (*q.v.*). **2.** flagellar protein antigens of motile Gram-negative enterobacteria.

haplo-   the prefix haplo- when followed by a symbol designating a particular chromosome, indicates an individual whose somatic cells lack one member of the designated chromosome pair. Thus, in *Drosophila,* haplo-IV means a fly that is monosomic for chromosome 4.

haplodiploidy   a genetic system found in some animals (such as the honey bee) in which males develop from unfertilized eggs and are haploid; whereas the females develop from fertilized eggs and are diploid.

haploidization   a phenomenon taking place during the parasexual cycle in certain fungi during which a diploid cell is transformed into a haploid cell by the progressive loss of one chromosome after another by nondisjunction.

haploid number   the gametic chromosome number, symbolized by N.

haploid parthenogenesis   the situation in which a haploid egg develops without fertilization, as in the honey bee.

haplont   an organism in which only the zygote is diploid (as in the algae, protozoa, and fungi). It immediately undergoes meiosis to give rise to the haplophase. *See **diplo-haplont, diplont.***

Haplopappus gracilis   a species of flowering plant showing the lowest number of chromosomes (N = 2), and therefore studied by cytologists.

haplophase   the haploid phase of the life cycle of an organism, lasting from meiosis to fertilization.

haplosis   the establishment of the gametic chromosome number by meiosis.

haplotype   1. a set of alloantigens coded by a single allele or by closely linked genes, as occurs with antigens of the major histocompatibility system; formerly known as a *phenogroup.* **2.** a set of closely linked genes that tends to be inherited together, as occurs with the A, B, and C loci of the human leukocyte antigen (HLA) gene complex.

hapten   an incomplete antigen; a substance that cannot induce antibody formation by itself, but can be made to do so by coupling it to a larger carrier molecule (e.g., a protein).

Complex haptens can react with specific antibodies and yield a precipitate; *simple haptens* behave as monovalent substances that cannot form serological precipitates.

haptoglobin   a plasma glycoprotein that forms a strong and stable complex with hemoglobin to aid the recycling of heme iron. In man this protein is encoded by a gene on the short arm of chromosome 16.

Hardy-Weinberg law   the law stating that both gene frequencies and genotype frequencies will remain constant from generation to generation in an infinitely large, interbreeding population in which mating is at random and there is no selection, migration, or mutation. In a situation where a single pair of alleles (*A* and *a*) is considered, the frequencies of germ cells carrying *A* and *a* are defined as p and q, respectively. At equilibrium the frequencies of the genotypic classes are p^2 (*AA*), 2 pq (*Aa*), and q^2 (*aa*).

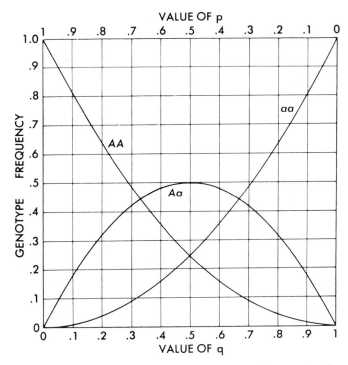

The relationships between the frequencies of genes *A* or *a* and the frequencies of genotypes *AA*, *Aa*, and *aa* as predicted by the Hardy-Weinberg law.

harlequin chromosomes   *See* **5-bromodeoxyuridine.**

Harvey murine sarcoma virus   a virus carrying the oncogene *v-ras^H* that is homologous to the cellular proto-oncogene *c-ras^H*, which resides on chromosome 11.

HAT medium   a tissue culture medium containing *h*ypoxanthine, *a*minopterin, and *t*hymidine. Mutant cells deficient in or lacking the enzymes thymidine kinase (TK$^-$) and hypoxanthine-guanine-phosphoribosyltransferase (HGPRT$^-$) cannot grow in HAT medium because aminopterin blocks endogenous (*de novo*) synthesis of both purines and pyrimidines. Normal TK$^+$ HGPRT$^+$ cells can survive by utilizing the exogenous hypoxanthine and thymidine via the *salvage pathway* (*q.v.*) of nucleotide synthesis. HAT medium has been used to screen for hybridomas (*q.v.*) by mixing TK$^+$ HGPRT$^-$ myeloma cells with antigen-stimulated TK$^-$ HGPRT$^+$ spleen cells. The hybrid TK$^+$ HGPRT$^+$ clones that survive in HAT medium are then assayed for monoclonal antibodies specific to the immunizing antigen.

Hayflick limit   an experimental limit to the number of times a normal animal cell seems capable of dividing; mouse and human cells divide 30 to 50 times before they enter the "crisis period" (*q.v.*); named after its discoverer, Leonard Hayflick.

Hb   hemoglobin. Hb^A symbolizes normal hemoglobin; Hb^F, fetal hemoglobin; Hb^S, sickle hemoglobin, etc.

HbO$_2$ oxyhemoglobin (*q.v.*).

H chain   *See* **immunoglobulin.**

H-2 complex   the major histocompatibility complex of the mouse lying in a segment of chromosome 17, which carries a number of polymorphic loci associated with various aspects of the immune system. It consists of four regions (K, I, S, D) that contain genes coding for classical transplantation antigens, Ia antigens and complement components, as well as immune response genes. The I region is further subdivided (*see* **I region**).

headful mechanism   a mechanism of packaging DNA in a phage head (e.g., T4) in which concatemeric DNA is cut, not at a specific position, but rather when the head is filled. This mechanism accounts for the observations of terminal redundancy and cyclic permutation in T4.

heat   in reproductive biology, that period of the sexual cycle when female mammals will permit coitus, estrus.

heat shock proteins   proteins synthesized in *Drosophila* cells within 15 minutes after a heat shock. These proteins are named in accordance with their molecular weights (in kd). After their synthesis in the cytoplasm, most heat shock proteins move to the nucleus and bind to chromatin. Here they protect the chromosomes from thermal damage in some unknown way.

heat shock puffs   a unique set of chromosomal puffs induced in *Drosophila* larvae exposed to elevated temperatures (for example, 40 minutes at 37°C). In *D. melanogaster* there are nine heat-inducible puffs; in *D. hydei* there are six. Heat shock results in the transcription of a specific set of mRNAs by the genes that form the puffs in polytene chromosomes. In culture, heat-shocked *Drosophila* cells also transcribe these same mRNAs into specific heat shock proteins.

heat shock response   the transcriptional activity that is induced at a small number of chromosomal loci following exposure of cells to a brief period of elevated temperature. At the same time, other loci that were active prior to the heat shock are switched off. This phenomenon appears to be universal, since it has been observed in *Drosophila, Tetrahymena,* sea urchin embryos, soy beans, chick fibroblasts, etc. Furthermore, heat shock proteins and the genes that encode them show a high degree of conservation during evolution.

heavy chain   in a heteromultimeric protein, the polypeptide chain with the higher molecular weight (e.g., in an immunoglobulin molecule, the heavy chains are about twice the length and molecular weight of the light chains); abbreviated *H chains;* the smaller molecules are called *light (L) chains.* The heavy chain determines the class to which the immunoglobulin belongs.

heavy chain class switching   the switching of a B lymphocyte from synthesis of one class of antibodies to another. For example, a B lymphocyte first synthesizes IgM, but it may later switch to the synthesis and secretion of IgG, and both antibodies will have the same antigen specificity. Thus only the constant regions of the heavy chains differ between these two classes of antibodies. Class switching of this type involves both somatic recombination prior to transcription and processing of transcripts by eliminating some segments and splicing the fragments.

heavy isotope   an atom, such as ^{15}N, which contains more neutrons than the more frequently occurring isotope, and thus is heavier.

heavy-metal stain   one of the elements of high atomic weight often used as stains in electron microscopy (U, Pb, Os, Mn).

heavy water   nontechnical term for deuterium oxide.

hebephrenia   a variety of schizophrenia marked by regression, silliness, mannerisms, introversion, hallucinations, and delusions.

hecto-   a prefix meaning one hundred. Applied to metric measurement units.

Heidelberg man   an extinct sub-species of primitive man known from fossil remains found near Heidelberg. Now classified as *Homo erectus heidelbergensis.*

HeLa cells   An established cell line (*q.v.*) consisting of an aneuploid strain of human epithelial-like cells, maintained in tissue culture since 1951; originating from a specimen of tissue from a carcinoma of the cervix in a patient names *Henrietta Lacks.*

Helianthus annuus   the sunflower.

helicase   an enzyme that unwinds a DNA double helix molecule ahead of DNA polymerase III in *E. coli;* e.g., the rep protein.

heliotropism   phototropism.

helix   a curve on the surface of a cylinder or cone that cuts all the elements of the solid at a constant angle. Applied especially to the *circular helix* on a right circular cylinder. It resembles the thread of a bolt.

Helix   a genus of gastropod containing common garden snails. Species from this genus are favorite experimental animals for the experimental analysis of molluscan development.

helix-destablizing proteins   any protein that binds to single-stranded regions of duplex DNA created by "breathing" (*q.v.*), and thereby causes unwinding of the helix; e.g., helicase (*q.v.*); also called *relaxation proteins* or *melting proteins.*

Hellin's rule of multiple births   a rule that allows calculation of the expected frequency of triplets, quadruplets, etc., from the observed frequency of twins. If the observed frequency of twins is *n,* then the expected frequency of triplets is n^2, of quadruplets is n^3, etc. This rule applies only to spontaneously occurring multiple births, not to cases of artificially induced twinning by the use of clomiphenes (*q.v.*) or other hormones resulting in multiple ovulations.

Helminthosporium maydis   the corn leaf blight. A major effort is being made to produce corn strains resistant to this fungus, which produced a loss to the 1970 U.S. corn crop of approximately a billion dollars.

helper T lymphocytes   a subpopulation of T lymphocytes that are programmed to amplify the function of B lymphocytes, other T lymphocytes, and macrophages. Helper T (T_H) lymphocytes constitute about one-third of the population of mature T lymphocytes.

helper virus   a virus which, by its infection of a cell already infected by a defective virus, is able to supply something the defective virus lacks, thus enabling the latter to multiply.

hemagglutinins   1. antibodies involved in specific aggregation of red blood cells. **2.** glycoproteins, formed on the surfaces of cells infected with certain viruses (e.g., influenza virus) or on the surfaces of enveloped viruses released from such cells, that can aggregate red blood cells of certain species. **3.** lectins (*q.v.*).

hematopoiesis   the formation of red blood cells.

hematopoietic   pertaining to hematopoiesis.

heme   an iron-containing porphyrin molecule that forms the oxygen-binding portion of hemoglobin. *See* **bilirubin, hemoglobin.**

hemel   an aziridine mutagen (*q.v.*).

Hemimetabola   a superorder of primitive insects having incomplete metamorphosis (mayflies, dragon flies, stoneflies, roaches, etc.). Synonomous with Exopterygota. *See* **classification.**

hemimetabolous   belonging to the Hemimetabola.

hemipteran   a true bug; a member of the order Hemiptera. *See* **classification.**

hemizygous gene   a gene present in single dose. It may be a gene in a haploid organism, or a sex-linked gene in the heterogametic sex, or a gene in the appropriate chromosomal segment of a deficiency heterozygote.

172

hemocoel a body cavity of arthropods and mollusca that is an expanded part of the blood-system. The hemocoel never communicates with the exterior, and never contains germ cells.

hemocyte an amoeboid blood cell of an insect. It is analogous to a mammalian leukocyte.

hemoglobin the oxygen-carrying pigment (molecular weight 64,500 d) of red blood cells. Hemoglobin is a conjugated protein composed of four separate chains of amino acids and four iron-containing ring compounds (heme groups). The protein chains of adult hemoglobin (HbA) occur as two pairs, a pair of alpha chains and a pair of beta chains. Each alpha chain contains 141 amino acids, and each beta chain contains 146 amino acids. Normal human adults also have a minor hemoglobin component (2%) called A_2. This hemoglobin has two alpha chains and two delta peptide chains. The delta chains have the same number of amino acids as the beta chains, and 95% of their amino acids are in sequences identical to those of the beta chains. The hemoglobin of the fetus (HbF) is made up of two alpha chains and two gamma chains. Each gamma chain also contains 146 amino acids. There are two types, G gamma and A gamma, which differ only in the presence of glycine or alanine, respectively, at position 136. The earliest embryonic hemoglobin tetramer consists of two zeta (alpha-like) and two epsilon (beta-like) chains. Beginning about the eighth week of gestation, the zeta and epsilon chains are replaced by alpha and gamma chains, and just before birth the gamma chains begin to be replaced by beta chains. *See* **leghemoglobin.**

hemoglobin Bart's *See* **hemoglobin fusion genes.**

hemoglobin C a hemoglobin with an abnormal beta chain. Lysine is substituted for glutamic acid at position 6. *See* **hemoglobin S.**

hemoglobin Constant Spring a human hemoglobin possessing abnormal alpha chains which contain 172 amino acids, rather than the normal number of 141. Hb Constant Spring seems to be the result of a nonsense mutation that converted a stop codon into a codon specifying an amino acid (glutamine, in this case). The thirty additional amino acids must be coded by adjacent base sequences that normally are not transcribed, or, if transcribed, not translated.

hemoglobin fusion genes abnormal hemoglobin genes arising as a result of unequal crossing-over between genes sharing homologous nucleotide sequences. Examples are the fused gamma-beta gene which codes for the non-alpha chain of Hb Kenya and the variety of fused delta-beta genes that code for the non-alpha chains of the Hb Lepore (*q.v.*).

hemoglobin genes the genes coding for human hemoglobins are located on chromosomes 11 and 16. Chromosome 11 contains the epsilon chain gene, two gamma chain genes (symbolized by G_γ and A_γ), a delta chain gene, and a beta chain gene. Chromosome 16 contains two zeta chain genes (symbolized ζ_2 and ζ_1) and two alpha chains (symbolized α_2 and α_1).

Between ζ_1 and α_2 lies a DNA segment, with no coding function, that resembles the α_2 gene in about 75% of its nucleotide sequences. This is called the pseudogene $\psi_{\alpha2}$. It is assumed that certain neighboring genes (G_γ and A_γ, δ and β, ζ_2 and ζ_1, and α_2 and α_1) arose during evolution as the results of duplications, and the same, of course, applies to the $\psi_{\alpha2}$. *See* **myoglobin gene.**

hemoglobin H *See* **hemoglobin homotetramers, thalassemias.**

hemoglobin homotetramers abnormal hemoglobins made up of four identical polypeptides. Examples are Hb Bart's (only gamma chains) and Hb H (only beta chains).

hemoglobin Kenya *See* **hemoglobin fusion genes.**

hemoglobin Lepore an abnormal hemoglobin named after the Italian family in which it was first found. The protein contains a pair of normal alpha chains and a pair of abnormal chains, each chain made up of 146 amino acids. Each abnormal chain appears to be a hybrid molecule consisting of an N-terminal end containing amino acids in a sequence characteristic of the left end of the delta chain and a C-terminal end containing amino acids in a sequence characteristic of the right end of a beta chain. Hemoglobin Lepore is presumed to have arisen by unequal crossing-over between a mispaired delta and beta cistron.

hemoglobin S a hemoglobin with an abnormal beta chain. Valine is substituted for glutamic acid at position 6. Thus the mutations resulting in the hemoglobin S and C beta chains appear to involve the same codon. Unlike all other substitutions in the beta chain, this mutation affects the solubility of hemoglobin under conditions of low oxygen tension. As the oxygen concentration declines, HbS polymerizes to form crystals which distort erythrocytes. These misshapen red blood cells are a diagnostic feature of sickle cell anemia (*q.v.*). HbC also causes sickling of erythrocytes.

hemoglobinuria the excretion of hemoglobin in solution in the urine.

hemolysis the rupturing of blood cells.

hemolytic anemia *See* **anemia.**

hemophilia a sex-linked disease in man in which blood-clotting is defective.

hemophilia A classical hemophilia, a sex-linked blood clotting disease due to a deficiency in factor VIII (anti-hemophilic globulin). The gene HEMA is at the end of the long arm of the X chromosome. Hemophilia A afflicts about 1 out of every 10,000 males.

hemophilia B also called *Christmas disease.* A sex-linked blood clotting disease due to deficiency of factor IX (plasma thromboplastin component, PTC). The genes for hemophilia A and B (HEMA and HEMB) are distant from one another. Hemophilia B is encountered about one-fifth as often as hemophilia A.

Hemophilis pertussis the bacterium responsible for whooping cough.

hemopoiesis *See* **hematopoiesis.**

hemopoietic hematopoietic.

hemopoietic histocompatibility (Hh) used in reference to transplantation of bone marrow in inbred strains of mice and their hybrids. As measured by the growth of the transplanted marrow cells in the spleen, a unique form of transplantation genetics seems to apply whereby hybrid recipients resist parental bone marrow grafts, but parental recipients accept hybrid bone marrow.

hempa an aziridine mutagen (*q.v.*).

$$H_3C \diagdown N \diagup CH_3$$
$$O = P - N \diagup CH_3$$
$$H_3C \diagup N \diagdown CH_3$$

Hensen-Weinberg rule a formula for calculating the expected frequency of monozygotic

174

twins (M) from the observed frequencies of all twins (T) and dizygotic twins (D): M = T − 2D.

hepat- a prefix referring to the liver, as in hepatoma, a cancer of the parenchymal cells of the liver.

herb a plant with no persistent parts above ground, as distinguished from a shrub or tree.

herbaceous herb-like.

herbage herbs collectively, especially the aerial portion.

herbarium a collection of dried and pressed plant specimens.

herbicide a chemical used to kill herbaceous plants.

herbivore a plant-eater.

hereditary disease a pathological condition caused by a mutant gene.

Hereditary Genius the book published by F. Galton in 1869 which founded the study of human heredity.

heredity a familial phenomenon wherein biological traits appear to be transmitted from one generation to another. The science of genetics has shown that heredity results from the transmission of genes from parents to offspring. The genes interact with one another and with their environment to produce distinctive characteristics or phenotypes. Offspring therefore tend to resemble their parents or other close relatives rather than unrelated individuals who do not share as many of the same kinds of genes.

heritability an attribute of a quantitative trait in a population that expresses how much of the total phenotypic variation is due to genetic variation. In the broad sense, heritability is the degree to which a trait is genetically determined, and it is expressed as the ratio of the total genetic variance to the phenotypic variance (V_G/V_P). In the narrow sense, heritability is the degree to which a trait is transmitted from parents to offspring (i.e., breeding value), and it is expressed as the ratio of the additive genetic variance to the total phenotypic variance (V_A/V_P). The concept of additive genetic variance makes no assumption concerning the mode of gene action involved. Heritability estimates are most commonly made by (1) regression-correlation analyses of close relatives (e.g., parent-offspring, full sibs, half-sibs), (2) experiments involving response to selection, and (3) analysis of variance components. Traits with high heritabilities respond readily to selection.

hermaphrodite an individual having both male and female reproductive organs. A *simultaneous hermaphrodite* has both types of sex organs throughout life. A *sequential hermaphrodite* may have the ovary first (protogyny), to be replaced by a testis later, or may develop the testis first (protandry) to be replaced later by an ovary. *See* **consecutive sexuality.**

herpes virus one of a group of animal viruses having a duplex DNA molecule within an icosahedral capsid. They range in size from 180 to 250 mμ and produce internuclear inclusions in host cells. Type 1 herpes simplex causes cold sores. Type 2 herpes simplex is associated with genital lesions and is sexually transmitted; both types 1 and 2 have been implicated in certain human cancers. The vericella-zoster virus is the cause of chicken pox and shingles. The Epstein-Barr virus (EBV) and the cytomegalovirus (CMV) also belong to the herpes group.

Hers disease a hereditary glycogen storage disease in humans arising from a deficiency of the enzyme hepatic phosphorylase. It is inherited as an autosomal recessive trait with an incidence of 1/200,000.

het   a partially heterozygous phage.

heterauxesis   the relation of the growth rate of a part of a developing organism to the growth rate of the whole or of another part. If the organ in question grows more rapidly than the organism as a whole it shows tachyauxesis; if less rapidly, bradyauxesis; if at the same rate, isauxesis.

heteroalleles   alternative forms of a gene that differ at nonidentical muton sites. Intragenic recombination between mutant heteroalleles can yield a functional cistron. *Compare with* **homoalleles.**

heteroallelic   referring to allelic mutant genes that have mutations at different sites (*q.v.*). Intragenic recombination can therefore yield a functional cistron. *Contrast with* **homoallelic.**

heterobrachial inversion   pericentric inversion.

heterocapsidic virus   *See* **segmented genome.**

heterocaryon   heterokaryon.

heterochromatin   chromosomal material that, unlike euchromatin (*q.v.*), shows maximal condensation in nuclei during interphase. Chromosomal regions behaving in this way are said to show *positive heteropycnosis.* When entire chromosomes (like some Y chromosomes) behave this way they are called *heterochromosomes.* In polytene chromosomes, heterochromatic regions adjacent to the centromeres of all chromosomes tend to adhere, forming a *chromocenter.* Heterochromatin is composed of repititious DNA, is late to replicate, and is transcriptionally inactive. Such heterochromatic segments are often called *constitutive* to distinguish them from chromosomal segments or whole chromosomes that become dense and compact at different developmental stages. In such cases, one homologous chromosome may behave differently from the other. An example is the condensed, inactivated X chromosome characteristic of the diploid somatic cells of mammalian females. Such chromosomes are sometimes said to contain *facultative heterochromatin,* although there is no evidence that they contain a type of DNA different from euchromatin.

heterochromosome   *See* **heterochromatin, B-chromosomes.**

heterochronic mutations   mutations that perturb the relative timing of developmental events. Certain of the cell lineage mutants (*q.v.*) of *Caenorhabditis elegans* are examples. In *lin-14* mutants certain aspects of development occur precociously with respect to the molting cycle, while in *lin-4* certain stages of development are repeated at abnormally late times.

heterochrony   a change during evolution in the onset of a developmental process so that the appearance or growth rate of a specific organ or other feature is altered in a descendant, relative to an ancestor. In the case of *acceleration,* the morphological feature appears earlier during the ontogeny of the descendant than it did in the ancestor. Conversely, *retardation* refers to the situation where the morphological feature appears later in the ontogeny of the descendant than it did during the ontogeny of the ancestor. *See* **heterauxesis.**

heterocyclic amino acid   any of the amino acids proline, hydroxyproline, tryptophan, and histidine. *See* **amino acid.**

heterocyclic compound   an organic compound forming a ring like an aromatic compound except that at least one atom in the ring is an element other than carbon. An example would be pyrimidine (*q.v.*).

heterodimer   a protein made up of paired polypeptides that differ in their amino-acid sequences.

176

heteroduplex 1. a DNA generated during genetic recombination by base pairing between complementary single strands from different parental duplex molecules. 2. a double-stranded nucleic acid in which each chain has a different origin and therefore is not perfectly complementary; e.g., the hybrid molecule generated by *in vitro* annealing of eukaryotic mRNA with its DNA.

heteroecious referring to a parasite requiring two or more hosts to complete its life cycle, as with some rust fungi and insects.

heterogametic sex the sex that produces gametes containing unlike sex chromosomes (e.g., male mammals produce X- and Y-bearing sperm, usually in equal proportions). *Compare with* **homogametic sex.**

heterogamy the alternation of bisexual reproduction with parthenogenetic reproduction.

heterogeneous nuclear RNA (hnRNA) the pool of extrachromosomal RNA molecules found in the nucleus, consisting of a heterogeneous mixture of primary transcripts, partly processed transcripts, discarded intron RNA, ubiquitous RNA, or small nuclear RNA. The term is often used to refer to the primary transcripts or to their modified products alone.

heterogenetic antigens the same or similar (cross-reacting) antigens shared by several species (e.g., Forssman antigen, *q.v.*). Antibodies produced against one of these antigens will also react with the other antigens of the system even though these are dervied from a different species; such antibodies are also called *heterophile antibodies. See* **heterophile antigen.**

heterogenotic merozygote a partially heterozygous bacterium carrying an exogenote containing differing alleles from those on the endogenote.

heterograft heterologous graft (*q.v.*).

heterokaryon a somatic cell that contains nuclei derived from genetically different sources. The nuclei do not fuse, but divide individually and simultaneously to form new cells, as commonly occurs in fungal hyphae.

heterokaryon test a test for organelle mutations based upon the appearance of unexpected phenotypes in uninucleate cells derived from specially marked heterokaryons. For example, suppose that heterokaryons form between colonies thought to carry a mitochondrial mutant (A) and colonies carrying a known nuclear mutation (B). If uninucleate progeny cells or spores exhibiting both the A and the B phenotypes can be derived from the heterokaryons, then the A mutation is probably in an extranuclear gene, because recombination of nuclear genes does not occur in heterokaryons.

heterokaryosis the condition in which fungus hyphae contain haploid nuclei of different genotypes, as a result of non-sexual fusion of the different types of hyphae.

heterokaryotypic referring to an individual carrying a chromosomal aberration in the heterozygous condition.

heterologous designating an antigen and antibody that do not correspond. Either may be said to be heterologous to the other.

heterologous graft a graft originating from a donor of a different species from the host.

Heterometabola Hemimetabola.

heteromixis in fungi, referring to the mating system where sexual reproduction involves the fusion of genetically different nuclei each from a different thallus. *Contrast with* **homomixis.**

heteromorphic bivalent a bivalent made up of chromosomes that are structurally different and consequently are only partly homologous (the XY bivalent is an example). *Contrast with* **homomorphic bivalent.**

heteromorphic chromosomes   homologous chromosomes that differ morphologically.

heteromorphosis   (*also* **homeosis**) the formation, whether in embryonic development or in regeneration, of an organ or appendage inappropriate to its site (for example, an antenna instead of a leg).

heteromultimeric protein   *See* **heteropolymeric protein.**

heterophile antigen   a substance that stimulates production in a vertebrate of antibodies capable of reacting with tissue components from other vertebrates or even from plants.

heteroplasmic   referring to a cell or organism that is heterozygous for one or more cytoplasmic genes.

heteroplastic transplantation   a transplantation between individuals of different species within the same genus.

heteroploid   referring in a given species to the chromosome number differing from the characteristic diploid number (or haploid number, if the species has a predominating haplophase).

heteropolymeric protein   a protein made up of more than one kind of polypeptide (e.g., hemoglobin).

heteropycnosis   referring to the appearance of chromosomes or chromosomal regions that have a coiling cycle out of phase with the rest of the genome. Positively heteropycnotic segments are more tightly coiled and negatively heteropycnotic segments are more loosely coiled than the rest of the chromosomal complement. *See* **isopycnotic, allocycly.**

heterosis   the greater vigor in terms of growth, survival, and fertility of hybrids, usually from crosses between highly inbred lines. Heterosis is always associated with increased heterozygosity.

heterosomal aberration   *See* **chromosomal aberration.**

heterospory   in higher plants, the existence (within the same species or within an individual organism) of two kinds of meiocytes (megasporocytes and microsporocytes) that produce two kinds of meiospores. *Compare with* **homospory.**

heterostyly   a polymorphism of flowers which ensures cross-fertilization by producing flowers having stamens and styles of unequal lengths.

heterothallic fungus   a fungal species producing a sexual spore that results from the fusion of genetically different nuclei that arose in different thalli. *Compare with* **homothallic fungus.**

heterotopic transplantation   transplantation of tissue from one site to another on the same organism.

heterotroph   an organism that requires complex nutrient molecules such as glucose, amino acids, etc., from which to obtain energy and to build macromolecules.

heterozygosis   heterozygosity.

heterozygosity   the condition of having one or more pairs of dissimilar alleles.

heterozygote   a diploid or polyploid individual that has inherited different alleles at one or more loci and therefore does not breed true. *See* **homozygote.**

hexaploid   a polyploid possessing six sets (6N) of chromosomes or genomes; e.g., bread wheat is thought to have originated by hybridizations involving three different species, each of which contributed two genomes to the allohexaploid. *See* **wheat.**

hexosaminidase   an enzyme functioning in the catabolism of gangliosides (*q.v.*). Hexosaminidase A is composed of alpha and beta subunits coded for by genes on human autosomes 15 and 5, respectively. Mutations at these loci result in Tay-Sachs and Sandhoff diseases (*q.v.*).

hexose monophosphate shunt   *See* **pentose phosphate pathway.**

Hfr strain   a strain of *Escherichia coli* that shows high frequencies of recombination, hence the abbreviation. In a cell from such a strain the *F*-factor is integrated into the bacterial chromosome. *See* **circular linkage map.**

HGPRT   hypoxanthine-guanine-phosphoribosyltransferase (*q.v.*); an enzyme involved in the *salvage pathway* (*q.v.*) of nucleotide synthesis. *See* **HAT medium.**

hibernate   to be dormant during winter. Many mammals, reptiles, amphibians, and certain invertebrates hibernate. *See also* **aestivate.**

hierarchy   an organization pattern involving groups within groups, as exemplified by the taxonomic heirarchy of organisms. *See* **classification.**

high-energy bond   a covalent chemical bond (e.g., the terminal phosphodiester bond of adenosine triphosphate) that liberates at least 5 kcal/mole free energy upon hydrolysis.

high-energy phosphate compound   a phosphorylated molecule which upon hydrolysis yields a large amount of free energy. *See* **phosphate bond energy, ATP.**

high-frequency of recombination cell   *See* **Hfr strain.**

highly repetitive DNA   the fast component in reassociation kinetics, usually equated with satellite DNA. *See* **repetitive DNA.**

hilum   the scar on a seed coat marking the former point of attachment of the seed to the ovule stalk.

Himalayan   a mutant allele at the albino locus (with known examples in the mouse, rat, guinea pig, hamster, and cat) associated with a very lightly pigmented body and somewhat darker extremities. The form of tyrosinase encoded by this gene is temperature senstive and normally functions well only in the extremities where the body temperature is lower. Such animals raised in cold environments show darker pigmentation.

hind-brain   the hindmost of the three divisions marked out by constrictions in the embryonic vertebrate brain. During development the hind-brain forms the cerebellum and the medulla.

Hin dII   *See* **restriction endonuclease.**

hinge region   *See* **immunoglobulin.**

hinny   *See* **horse-donkey hybrids.**

his   histidine. *See* **amino acid.**

histidine   *See* **amino acid.**

histidinemia   a hereditary disease in man arising from a deficiency of the enzyme histidase.

histidine operon   a polycistronic operon of *Salmonella typhimurium* containing nine genes involved in the synthesis of histidine.

histochemistry the study by specific staining methods of the distribution of particular molecules within sections of tissues.

histocompatibility antigen genetically encoded cell surface alloantigens that can cause the rejection of grafted tissues, cells, and tumors bearing them.

histocompatibility gene a gene belonging to the major histocompatibility (MHC) system or to any of numerous minor histocompatibility systems responsible for the production of histocompatibility antigens (*q.v.*)

histogenesis the development of histologically detectable differentiation.

histogenetic antigens or responses defined by means of cell mediated immunity.

histogram a bar graph.

histo-incompatibility intolerance to transplanted tissue.

histology the study of tissues.

histolysis tissue destruction.

histone genes in both the sea urchin and *Drosophila* these genes are repetitive and clustered. In *Strongylocentrotus purpuratus* the genes for H1, H4, H2B, H3, and H2A lie in a linear sequence, each spearated by a spacer. There are several hundred serial repeats of this five-membered unit transcribed as a polycistronic message. In *Drosophila* there are about 110 copies of the histone genes, and these are localized in a four band region in the left arm of chromosome 2. In man the histone gene family resides on chromosome 7.

histones small DNA-binding proteins. They are rich in basic amino acids and are classified according to the relative amounts of lysine and arginine they contain.

HISTONE SYNONYMS	MOLECULAR WEIGHT (d)	TOTAL AMINO ACIDS	% LYSINE	% ARGINE	RELATIVE AMOUNT PER 100 bp DNA
H1 = I = F1	21,000	207	27	2	0.5
H2A = IIbl = F2A2	14,500	129	11	9	1
H2B = IIb2 = F2B	13,700	125	16	6	1
H3 = III = F3	15,300	135	10	15	1
H4 = IV = F2A1	11,300	102	10	14	1

Histones are conserved during evolution. For example, H4 of calf thymus and pea differ at only two sites. The nucleosome (*q.v.*) contains two molecules each of H2A, H2B, H3, and H4, arranged along the DNA in the order H2A, H2B, H4, H3, H3, H4, H2B, H2A. *See* **nucleoprotein.**

historical science *See* **idiographic science.**

hitchhiking the spread of a neutral allele through a population because it is closely linked to a beneficial allele and therefore is carried along as the gene that is selected for increases in frequency.

HLA human leukocyte antigens concerned with the acceptance or rejection of tissue or organ grafts and transplants. These antigens are on the surface of most somatic cells except red blood cells, but are most easily studied on white blood cells (hence the name).

HLA complex the major histocompatibility complex of humans: it is located on chromosome 6 and consists of a cluster of genes coding for histocompatibility antigens (HLA-A,B,C,), for antigens similar to those of the mouse I region (HLA-D/DR), and for components of the complement system.

The HLA Complex

H locus   in humans, a genetic locus that encodes a fucosyl transferase enzyme that is required during an early step in the biosynthesis of the antigens of the ABO blood group system. *See* **A, B antigens, Bombay blood group.**

hnRNA   heterogeneous nuclear RNA (*q.v.*).

hog   *See* **swine.**

Hogness box   a segment 19-27 bp upstream from the startpoint of eukaryotic structural genes to which RNA polymerase II binds. The segment is 7 bp long, and the nucleotides most commonly found are TATAAAA; named in honor of David Hogness. *See* **canonical sequence, Pribnow box, promoter.**

holandric   appearing only in males. Said of a character determined by a gene on the Y chromosome. *See* **hologynic.**

holism   a philosophy maintaining that the entirety is greater than the sum of its parts. In biology, a holist is one who believes that an organism cannot be explained by studying its component parts in isolation. *See* **compositionism, reductionism.**

Holliday intermediate   *See* **Holliday model.**

Holliday model   a model proposed by R. Holliday that describes a series of breakage and reunion events occurring during crossing-over between two homologous chromosomes. The diagram on p. 182 illustrates his model. In (a) the duplex DNA molecule from two non-sister strands of a tetrad have aligned themselves in register so that the subsequent exchange does not delete or duplicate any genetic information. In (b) strands of the same polarity are nicked at equivalent positions. In (c) each broken chain has detached from its partner and paired with the unbroken chain in the opposite duplex. In (d) ligases have joined the broken strands to form an internal branch point. The branch is free to swivel to the right or left, and so it can change its position. This movement is called *branch migration,* and in (e) the branch is shown as having moved to the right. In (f) the molecule is drawn with its arms pulled apart, and in (g) the a b segment has been rotated 180° relative to the A B segment. The result is an x-shaped configuration, and in it one can readily see the four sectors of single-stranded DNA in the branch region. To separate the structure into two duplexes, cuts must be made across the single strands at the branch region. The cuts can occur horizontally (h) or vertically (i). In the horizontal case, the separated duplexes will appear as shown in (j), and when the nicks are ligased, the duplexes will each contain a patch in one strand, as shown in (k). In the vertical case, the separated duplexes will appear as shown in (l), and when the nicks are ligased, the duplexes will contain splices in every strand, as shown in (m). It is only in this case that single crossover chromatids (Ab and aB) will be detected. Structures similar to the molecule drawn in (g) have been observed under the electron microscope and have been named *Holliday intermediates. See* **chi structure.**

holoblastic cleavage   cleavage producing cells of approximately equivalent size.

Holocene   the geological epoch consisting of times since the end of the last ice age. A subdivision of the Quaternary period. *See* **geologic time divisions.**

Holliday model

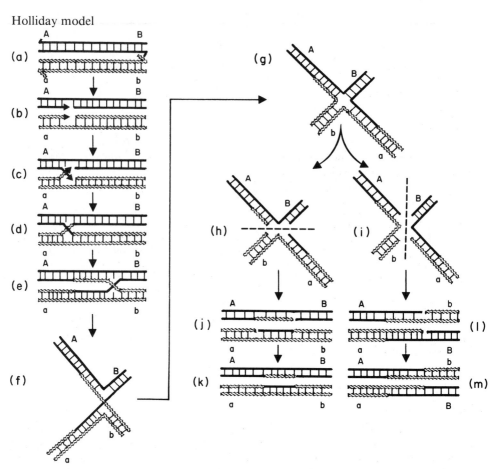

holocentric   referring to chromosomes with diffuse centromeres. *See* **centromere.**

holoenzyme   the functional complex formed by an apoenzyme (*q.v.*) and its appropriate coenzyme (*q.v.*).

hologynic   appearing only in females. Said of a trait passed from a P₁ female to all daughters occurring, for example, in the case of a gene linked to the W chromosome (*q.v.*). *See* **matrocliny, holandric.**

Holometabola   the superorder of insects containing species which pass through a complete metamorphosis with a larval and pupal stage interposed between the embryo and the adult. Synonomous with Endopterygota. *See* **classification.**

holometabolous   belonging to the insect subclass Holometabola.

holophyletic   an evolutionary lineage consisting of a species and all of its descendants.

holophytic nutrition   nutrition requiring only inorganic chemicals, as that of photosynthetic plants and protozoans.

holotype   the single specimen selected for the description of a species.

holozoic nutrition   nutrition requiring complex organic foodstuffs, as that of organisms other than photosynthetic plants and protoctists.

homeoplastic graft (*also* **homoeoplastic**)   a graft of tissue from one individual to another of the same species.

homeosis (*also* **homoeosis**)   heteromorphosis (*q.v.*).

homeostasis (*also* **homoeostasis**)   a fluctuation-free state. *See* **developmental homeostasis.**

homeotic   referring to mutations in which one developmental pattern is replaced by a different, but homologous one. The homeotic mutations of *Drosophila* cause an organ to differentiate abnormally and to form a homologous organ that is characteristic of an adjacent segment. Three such mutations are illustrated below: (A) A frontal view of the normal head. (B) The leg-like antenna of the mutation called *aristapedia.* (C) Leg-like structures extending from the proboscis of the mutation called *proboscipedia.* (D) A *bithorax* male in which halteres are changed into wing-like appendages. *See* **compartmentalization.**

hominid   a member of the human family Hominidae including *Australopithecus afarensis, Homo erectus,* and *Homo sapiens.*

hominoid   a member of the Primate superfamily Hominoidea, including the gibbon family (Hylobatidae), the great ape family (Pongidae), and the human family (Hominidae).

Homo   a genus containing two species, *H. erectus* and *H. sapiens. H. erectus* was the immediate predecessor of modern man, *H. sapiens. H. erectus* lived in the middle Pleistocene epoch between half a million and a million years ago. Various subspecies existed in different parts of the world: *H. erectus pekinensis* in east Asia, *H. erectus heidelbergensis* in Europe, *H. erectus erectus* and *H. erectus soloensis* in southeast Asia, and *H. erectus habilis* in Africa. The cranial capacity of this species was 700-1100 cc. *H. sapiens,* the extant species of man, arose about 500,000 years ago and has a cranial capacity ranging between 1300 and 1600 cc.

homoalleles   alternative forms of a gene that differ at the same muton site. Intragenic recombination between homoalleles is not possible. *Compare with* **heteroalleles.**

homoallelic   referring to allelic mutant genes that have mutations at the same site (*q.v.*). A functional cistron cannot be generated by intragenic recombination between homoalleles. *Contrast with* **heteroallelic.**

homobrachial inversion   paracentric inversion (*q.v.*).

homocaryon   homokaryon (*q.v.*).

homocystinuria   a hereditary disease in man arising from a deficiency of the enzyme serine dehydratase.

homodimer   a protein made up of paired identical polypeptides.

homoeologous chromosomes   chromosomes that are only partially homologous. Such chromosomes are derived from ancestral chromosomes which are believed to have been homologous. Evolutionary divergence has reduced the synaptic attraction of homoeologs. *See* **differential affinity, iso-anisosyndetic alloploid.**

homoeosis   homeosis. *See also* **homeoplastic graft, homeostasis, homeotic mutation.**

homoeotic   variant spelling of homeotic (*q.v.*).

homogametic sex the sex that produces gametes all of which carry only one kind of sex chromosome; e.g., the eggs of female mammals carry only an X chromosome. *Compare with* **heterogametic sex.**

homogamy the situation in which the male and female parts of a flower mature simultaneously.

homogenote a partial diploid (merozygote) bacterium in which the donor (exogenote) chromosomal segment carries the same alleles as the chromosome of the recipient (endogenote) cell.

homogentisic acid a compound derived from the metabolic breakdown of the amino acid tyrosine. *See* **alkaptonuria.**

homograft homeoplastic graft (*q.v.*).

homoimmunity the resistance of a lysogenic bacterium (harboring a prophage, *q.v.*) to superinfection by phage of the same kind as that carried in the prophage state. The excess repressor molecules generated by the prophage bind to operators on the infecting DNA molecules and prevent their transcription.

homokaryon a multinucleate mycelium containing nuclei of only one genotype.

homokaryotypic referring to an individual carrying a chromosomal aberration in the homozygous condition.

homolog variant spelling of homologue (*q.v.*).

homologous referring to structures or processes in different organisms that show a fundamental similarity because of their having descended from a common ancestor. Homologous structures have the same evolutionary origin although their functions may differ widely: e.g., the flipper of a seal and the wing of a bat. *See* **analogous.**

homologous chromosomes chromosomes that pair during meiosis. Each homologue is a duplicate of one of the chromosomes contributed at syngamy by the mother or father. Homologous chromosomes contain the same linear sequence of genes and as a consequence each gene is present in duplicate.

homologue *See* **homologous chromosome.**

homology the state of being homologous.

homomixis referring in fungi to the mating system in which sexual reproduction involves the fusion of genetically similar nuclei derived from one thallus.

homomorphic bivalent a bivalent made up of homologues of similar morphology. *See* **heteromorphic bivalent.**

homomultimer *See* **homopolymer.**

homonym   one name for two or more different things.

homoplasmic   referring to a cell or an organism that is homozygous for one or more cytoplastic genes.

homoplasy   parallel or convergent evolution; structural similarity in organisms not due directly to inheritance from a common ancestor or development from a common anlage.

homopolar bond   covalent bond (*q.v.*).

homopolymer   a polymer composed of identical monomeric units (poly U, for example).

homopolymer tails   a segment containing several of the same kind of deoxyribonucleotides arranged in tandem at the 3′ end of a DNA strand. *See* **terminal transferase.**

homosequential species   species with identical karyotypes, as found in many species of *Drosophila* endemic to the Hawaiian Islands.

homosomal aberration   *See* **chromosomal aberration.**

homospory   in plants of both mating types or sexes, the production of meiospores of equivalent size. *Compare with* **isogamy.**

homothallic fungus   a fungal species producing a sexual spore that results from the fusion of genetically different nuclei derived from the same thallus. *Compare with* **heterothallic fungus.**

homozygosity   the condition of having identical alleles at one or more loci in homologous chromosome segments.

homozygote   an individual or cell characterized by homozygosity.

homozygous   having identical rather than different alleles in the corresponding loci of homologous chromosomes and therefore breeding true. *See* **heterozygous.**

homunculus   a miniature individual imagined by early biologists to be present in a sperm.

hopeful monster   *See* **saltation.**

Hordeum vulgare   cultivated barley, a leading experimental organism in genetic studies of flowering plants.

horizontal classification   a system of evolutionary classification that tends to unite transitional forms with their ancestors; the opposite of *vertical classification* (*q.v.*).

horizontal evolution   the process of speciation whereby an ancestral population splits into two or more subgroups that evolve into separate species. *See* **speciation.**

horizontal section   any section through the body in a plane at right angles to the median plane and parallel to the dorsal and ventral surfaces.

horizontal transmission   *See* **vertical transmission.**

hormone   an organic compound produced in one part of an organism and transported to other parts where it exerts a profound effect. Mammalian hormones include ACTH, epinephrine, FSH, glucagon, GH, LH, insulin, intermedin, oxytocin, progesterone, prolactin, secretin, thyroxin, and vasopressin. *See also* **endocrine system.**

horotelic evolution   *See* **evolutionary rate.**

horse   any of a number of domesticated breeds of the species *Equus caballus*. Popular breeds include: DRAFT HORSES: Belgian, Clydesdale, Percheron, Shire, Suffolk. COACH HORSES: Cleveland Bay, French Coach, German Coach, Hackney. LIGHT HARNESS HORSES:

American Trotter. SADDLE HORSES: American Saddle Horse, American Quarter Horse, Appaloosa, Arabian, Morgan, Palomino, Tennessee Walking Horse, Thoroughbred, Lippizzaner. PONIES: Hackney Pony, Shetland Pony, Welsh Pony.

horse bean   broad bean (*q.v.*).

horse-donkey hybrids   since the horse has a diploid chromosome number of 64 and the donkey of 62, the hybrids have 63 chromosomes and are genetically sterile. The horse female × donkey male cross produces a mule; the reciprocal cross produces a hinny.

host   **1.** an organism infected by a parasite. **2.** the recipient of a graft.

host-cell reactivation   cut and patch repair (*q.v.*) of UV-induced lesions in the DNAs of bacteriophages once they infect a host cell. Host-cell reactivation does not occur in the case of ssDNA or RNA viruses.

host controlled restriction and modification   *See* **restriction enzyme.**

host range   the spectrum of strains of a bacterial species that can be infected by a given strain of phage.

host-range mutation   a mutation of a phage which enables it to infect and lyse a previously resistant bacterium.

hot spot   **1.** a site at which the frequencies of spontaneous mutation or recombination are greatly increased with respect to other sites in the same cistron. Examples are in the rII gene of phage T4 and in the lacZ and trpE genes of *E. coli.* **2.** a chromosomal site at which the frequencies of mutations are differentially increased in response to treatment with a specific mutagen. **3.** any DNA nucleotide or sequence exhibitng hyperactivity, such as for initiation of recombination.

housekeeping genes   constitutive loci that are theoretically expressed in all cells in order to provide the maintenance activities required by all cells: e.g., genes coding for enzymes of glycolysis and the Kreb's cycle.

HPRT   hypoxanthine-guanine-phosphoribosyltransferase (*q.v.*).

H substance   a precursor polysaccharide for production of A and B antigens of the ABO blood group system. It is usually unmodified on group O cells, but is modified by the addition of different sugars to produce the A or B antigens. The molecule is specified by a gene on human chromosome 19. *See* **Bombay blood group.**

hull   the capsule enclosing a seed or fruit.

human chromosome band designations   quinacrine and Giemsa-stained human metaphase chromosomes show characteristic banding patterns, and standard methods have been adopted to designate the specific patterns displayed by each chromosome. The X chromosome shown on p. 187 illustrates the terminology. In the diagram the dark bands represent those regions that flouresce with quinacrine or are darkened by Giemsa. The short (p) arm and the longer (q) arm are each divided into two regions. In the case of longer autosomes the q arm may be divided into three or four regions and the p arm into three regions. Within the major regions the dark and light bands are numbered consecutively. To give an example of the methods used for assigning loci, the G6PD gene is placed at q28, meaning it is in band 8 of region 2 of the q arm. The color-blindness genes are both assigned to q27-qter. This means they reside somewhere between the beginning of q27 and the terminus of the long arm. *See* **human mitotic chromosomes.**

human chromosome band

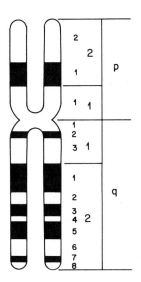

human cytogenetics *See* **symbols used in human cytogenetics, human mitotic chromosomes.**

human mitotic chromosomes the mitotic chromosomes of man are generally grouped into seven classes (A-G) according to the following cytological criteria: Group A (chromosomes 1-3)—large chromosomes with approximately median centromeres. Group B (chromosomes 4-5)—large chromosomes with submedian centromeres. Group C (chromosomes 6-12 and the X-chromosome)—medium-sized chromosomes with submedian centromeres. Group D (chromosomes 13-15)—medium-sized acrocentric chromosomes. Chromosome 13 has a prominent satellite on the short arm. Chromosome 14 has a small satellite on the short arm. Group E (chromosomes 16-18)—rather short chormosomes with approximately median (in chromosome 16) or submedian centromeres. Group F (chromosomes 19 and 20)—short chromosomes with approximately median centromeres. Group G (chromosomes 21, 22, and the Y-chromosome)—very short acrocentric chromosomes. Chromosome 21 has a satellite on its short arm. *See* **human chromosome band designations, symbols used in human cytogenetics.**

humoral immunity immunoglobulin (antibody)-mediated immune responses.

Hunter syndrome an X-linked disorder of connective tissue in man associated with the storage of mucopolysaccharides.

Huntington disease a human disease characterized by irregular, spasmodic, involuntary movements of the limbs and facial muscles, mental deterioration, and death, usually within 20 years of the onset of the symptoms. These generally do not appear until the heterozygote is 35–40 years of age. It is due to a dominant gene on chromosome 4; also called *Huntington's chorea.*

Hurler syndrome an autosomally linked disorder of connective tissue in man associated with the storage of mucopolysaccharides.

husk the coarse outer envelope of a fruit, such as the glumes of an ear of corn.

Hutchinson-Gilford syndrome *See* **progeria.**

HVL half value layer (*q.v.*).

H-X antigen   a histocompatibility antigen encoded by a locus on the X chromosome.

Hyalophora cecropia   the giant cecropia moth; because of its large size a favorite experimental insect.

hyaloplasm   cytosol.

hyaluronic acid   a mucopolysaccharide that is abundant in the jelly coats of eggs and in the ground substance of connective tissue. Hyaluronic acid is a polymer composed of glucosamine and glucuronic acid subunits.

D-glucosamine   D-glucuronic acid

hyaluronidase   an enzyme that digests hyaluronic acid.

H-Y antigen   an antigen detected by cell-mediated and humoral responses of homogametic individuals against heterogametic individuals of the same species, which are otherwise genetically identical. Antigenic responses of this sort have been demonstrated in mammals, birds, and amphibians. In mammals the antigen is called H-Y because it acts as a *H*istocombatibility factor determined by the *Y*-chromosome. The H-Y antigen is the major male-determining factor in the developing mammalian male fetus. The location of the gene encoding the H-Y antigen is not known. However, the gene which induces synthesis of the H-Y antigen in humans is located on the short arm of the Y-chromosome near the centromere. A homologous locus, which suppresses H-Y production, lies on the distal end of the short arm of the X. Evidently two doses of this gene are necessary for the complete suppression, since Turner syndrome (45, XO) females produce small amounts of the H-Y antigen. The H-Y locus is one of the areas that escapes X-chromosome inactivation (*q.v.*).

hybrid   1. a heterozygote (e.g., a monohybrid is heterozygous at a single locus; a dihybrid is heterozygous at two loci; etc.). 2. an offspring from genetically dissimilar parents, perhaps even different species.

hybrid arrested translation   a method for identifying the cDNA corresponding to an mRNA that depends upon the ability of cDNA to hybridize with its mRNA and thereby to inhibit its translation in an *in vitro* system; the disappearance of the translation product from the system indicates the presence of the cDNA.

hybrid breakdown   the reduction in fitness of F_2 and/or backcross populations from fertile hybrids produced by intercrossing genetically disparate populations or species; a postzygotic reproductive isolating mechanism.

hybrid corn   commercial corn grown from seed produced by the "double cross" (*q.v.*) procedure. Such corn is characterized by its vigor and uniformity.

hybrid DNA model   a model used to explain both crossing-over and gene conversion by postulating that a short segment of heteroduplex (hybrid) DNA is produced from both parental DNAs in the neighborhood of a chiasma. *See* **Holliday model.**

hybrid duplex molecule   an experimentally reconstituted molecule containing a segment of single-stranded DNA hydrogen bonded to a second RNA or DNA molecule of complementary base sequence.

hybrid dysgenesis a syndrome of correlated genetic abnormalities that occurs spontaneously when hybrids are formed between certain strains of *Drosophila*. The hybrids show germ line defects including chromosomal aberrations, high frequencies of lethal and visible mutations, and in most extreme cases, sterility. The cause of hybrid dysgenesis is a transposable element named P. Long established laboratory strains lack P elements. Those strains susceptible to P elements are called M strains. P elements are often present in strains established from newly captured wild flies. The cross P male × M female generates dysgenic F_1 individuals; the reciprocal cross does not. The P elements do not produce dysgenesis within P strains. P elements are dispersed over all the chromosome arms, and in P strains their transposition is repressed. When chromosomes carrying P elements are placed in M cytoplasm by suitable crosses, the P elements become derepressed and transpose at high rates, disrupting genetic loci and causing the dysgenic syndrome. *See* **P element.**

hybrid inviability a postmating (postzygotic) reproductive isolating mechanism in which hybrids between disparate populations fail to survive to reproductive age.

hybridization **1.** the mating of individuals belonging to genetically disparate populations or to different species. **2.** in Mendelian terms, the mating of any two unlike genotypes or phenotypes. **3.** the pairing of complementary RNA and DNA strands to produce an RNA-DNA hybrid, or the partial pairing of DNA single strands from genetically different sources.

hybridization competition a technique for distinguishing different mRNA molecules, using a variation of the basic filter hybridization (*q.v.*) technique. A specific DNA sequence trapped on a nitrocellulose filter is exposed to a tritiated RNA known to be complementary to that DNA. An unlabeled RNA of unknown specificity is added and, if complementary to the DNA, it will compete with the labeled molecules for hybridization to the DNA. Any diminution of labeling in the hybrid after equilibrium is reached is attributed to displacement by the unlabeled molecules.

hybridoma a cell resulting from the fusion of an antibody-producing plasma cell (a B lymphocyte) and a myeloma (a bone marrow cancer) cell. Such a hybrid cell produces a clone that can be maintained in tissue culture or as an animal tumor, and the clone may secrete only a single kind of antibody (monoclonal antibody).

hybrid resistance the phenomenon whereby tumors may grow more readily in homozygous recipients than in heterozygous recipients even though the tumor may be genetically histocompatible with both types of recipients.

hybrid sterility the failure of hybrids between different species to produce viable offspring.

hybrid swarm a continuous series of morphologically distinct hybrids resulting from hybridization of two species followed by crossing and backcrossing of subsequent generations.

hybrid tobacco mosaic virus an experimentally reconstituted virus containing nucleic acid and protein components arising from different sources.

hybrid vigor heterosis (*q.v.*).

hybrid zone a geographical zone where hybrids between two geographical races are observed.

hydrocarbon an organic compound composed only of carbon and hydrogen atoms.

hydrogen   the most abundant of the biologically important elements. Atomic number 1; atomic weight 1.00797; valence 1+, most abundant isotope ^1H, heavy isotope ^2H (deuterium); radioisotope ^3H (tritium, *q.v.*).

hydrogen bond   the weak electrostatic attraction that exists between a hydrogen atom that is covalently bonded to an O or N atom and an atom containing an unshared electron pair. Hydrogen bonds are weaker than covalent bonds, but stronger than the van der Waals attractive forces between molecules.

hydrogen ion concentration   expressed as the logarithm of the reciprocal of the concentration of hydrogen ions in grams per liter of solution; abbreviated as *p*H. the scale runs from 0 to 14, with values above 7 basic; those below, acidic.

hydrogen peroxide H_2O_2. *See* **catalase, peroxisomes.**

hydrolases   enzymes that catalyze the transfer of water between donor and receptor molecules. Proteolytic enzymes are a special class of hydrolases.

hydrolysis   the splitting of a molecule into two or more smaller molecules with the addition of the elements of water.

hydroperoxyl radical HO_2, an oxidizing agent formed during the interaction of ionizing radiation with oxygenated water.

hydrophilic   water attracting; referring to molecules or functional groups in molecules which readily associate with water. The carboxyl, hydroxyl, and amino groups are hydrophilic.

hydrophobic   water repelling; referring to molecules or functional groups in molecules (such as alkyl groups) that are poorly soluble in water. Populations of hydrophobic groups form the surface of water repellent membranes.

hydrophobic bonding   the tendency of nonpolar groups to associate with each other in aqueous solution, thereby excluding water molecules.

hydroponics   growing plants in aqueous chemical solutions.

hydrops   excessive accumulation of fluid in any tissue or body cavity.

hydrops fetalis   *See* **thalassemias.**

hydroquinone   *See* **quinone.**

hydroxyapatite   a form of calcium phosphate that binds to double-stranded DNA.

hydroxykynurenine   a precursor of xanthommatin (*q.v.*).

hydroxylamine NH_2OH, a mutagen that converts the NH_2 of cytosine to NHOH, which pairs only with adenine.

5-hydroxymethyl cytosine   a pyrimidine found instead of cytosine in the DNA of T-even coliphages. 5-Hydroxymethyl cytosine pairs with guanine. It is postulated that the phage-specific DNase that breaks down the host DNA is specific for DNA molecules containing cytosine.

$$
\begin{array}{c}
NH_2 \\
| \\
C \\
\diagup \diagdown \\
N \qquad C-CH_2OH \\
| \qquad\quad || \\
O=C \qquad CH \\
\diagdown \diagup \\
N \\
| \\
H
\end{array}
$$

hydroxyurea a compound which inhibits semiconservative DNA replication, but not repair synthesis.

hymenopteran an insect belonging to the order Hymenoptera (which includes bees, ants, wasps, etc.). *See* **classification.**

hyparchic genes *See* **autarchic genes.**

hyperammonemia a hereditary disease in man arising from a deficiency of the enzyme ornithine carbamoyl transferase.

hypercholesterolemia *See* **familial hypercholesterolemia.**

hyperchromic effect an increased absorbance of ultraviolet light by a solution containing DNA or RNA as these molecules are subjected to heat, alkaline conditions, etc. The hyperchromic effect implies a disturbance of the three-dimensional structure of the molecules. A pronounced hyperchromic effect accompanies the uncoiling of the DNA double helix.

hyperchromic shift an increase in the absorption of ultraviolet light by a solution of DNA caused by the disruption of hydrogen bonds of a DNA duplex to yield single-stranded structures.

hyperdontia the hereditary presence of one or more additional teeth.

hyperglycemia an increased glucose content in the blood.

hypergyny a mating that improves social standing, especially marriage of a woman to a man of higher social rank.

hyperlipemia an increased concentration of neutral fat in the blood serum.

hypermorph a mutant gene whose effect is similar to, but greater than, that of the standard or wild-type gene.

hyperplasia an increase in amount of tissue produced by an increase in the number of cells. Hyperplasia often accompanies the regeneration of a damaged organ. *See* **hypertrophy.**

hyperploid referring to cells or individuals containing one or more chromosomes or chromosome segments in addition to the characteristic euploid number.

hyperprolinemia a hereditary disease in man arising from a deficiency of the enzyme proline oxidase.

hypersensitivity the characteristic of responding with clinical symptoms to allergens in amounts which are innocuous to most individuals. *See* **allergy.**

hypertension an increased blood pressure.

hypertonic solution a solution having a solute concentration such that it gains water by osmosis across a semipermeable membrane from another aqueous solution having a lower solute concentration. *See* **hypotonic solution.**

hypertrophy   an increase in the size of a tissue or organ because of the increased volume of the component cells. *See* **hyperplasia.**

hypervariable (hv) sites   amino acid positions within the variable region of an immunoglobulin light chain or heavy chain, exhibiting great variation among antibodies of different specificity; these noncontiguous sites are brought together in the active site where antigen is bound (a paratope) by complex folding of the polypeptide chain. *See* **immunoglobulin.**

hypha   a filament of a fungus thallus.

hypo   *See* **fixing.**

hypochromic anemia   *See* **anemia.**

hypocotyl   that portion of the stem of a plant embryo or seedling below the cotyledons.

hypodermis   a layer of cells immediately internal to the epidermis.

hypodontia   the congenital absence of teeth.

hypogean, hypogeous   developing below the soil surface.

hypoglycemia   a decrease in sugar content of the blood serum.

hypomorph   a mutant gene acting in the same direction as the normal allele, but with less effect.

hypophosphatasia   a hereditary disease in man arising from a deficiency of the enzyme alkaline phosphatase.

hypophosphatemia   a decreased concentration of inorganic phosphate in the blood serum.

hypophysis   the pituitary gland.

hypoplasia   an arrested development of an organ or part. The opposite of hyperplasia (*q.v.*).

hypoploid   referring to cells or individuals containing one or more fewer chromosomes or chromosome segments than the characteristic euploid number.

hypostatic gene   *See* **epistasis.**

hypothalamus   the floor and sides of the vertebrate brain just behind the attachment of the cerebral hemispheres. The hypothalamus controls the secretion of anti-diuretic hormone and oxytocin, and probably supplies them to the pituitary.

hypothyroidism   a diminished production of thyroid hormone.

hypotonic solution   one having a solute concentration such that it loses water by osmosis across a semipermeable membrane separating it from another aqueous solution having a higher solute concentration. *See* **hypertonic solution.**

hypoxanthine   6-hydroxypurine. *See* **purine.**

hypoxanthine-guanine-phosphoribosyltransferase   the enzyme that catalyzes the transfer of the phosophoribosyl moiety of 5-phosphoribosyl-1-pyrophosphate to the 9 position of hypoxanthine and guanine to form inosine monophosphate and guanosine monophosphate. Abbreviated HPRT or HGPRT. The Lesch-Nyhan syndrome (*q.v.*) is caused by deficiency of HPRT. *See* **HAT medium.**

I

i   the regulator gene of the lactose operon in *Escherichia coli. See* **regulator genes.**

I   iodine.

I₁,I₂,I₃,etc.   the first, second, third, etc., generations obtained by inbreeding.

I^A,I^B,I^0   the allelic genes responsible for the ABO blood group system. *See* **A, B antigens.**

IAA   indole acetic acid (*q.v.*).

Ia antigens   alloantigens encoded by the Ia region of the mouse major histocompatibility complex (H-2). They are defined by serological methods and are found predominantly (but not exclusively) on B lymphocytes and macrophages.

icosahedron   a regular geometric polyhedron composed of 20 equilateral triangular faces with 12 corners. The capsids of many spherical eukaryotic viruses and bacteriophages are icosahedral. *See* **Shope papilloma virus.**

ICSH   interstitial cell-stimulating hormone. Identical to LH (*q.v.*).

identical twins   monozygotic twins (*q.v.*).

identifier sequences   nucleotide sequences about 80 base pairs long located in the introns of genes expressed in neural tissues. Such ID sequences are specifically transcribed in the brain. Sutcliffe *et al.* have shown that there are about 1×10^5 copies of ID sequences in the rat genome, and they suggest that ID sequences control tissue-specific gene expression, either by directing the splicing of mRNA precursors in the brain or by providing recognition sequences for brain-specific proteins that stimulate the activity of RNA polymerases.

idiocy   the most severe degree of mental retardation. An idiot reaches an intelligence level below that of a two-year-old child.

idiogram   a diagrammatic representation of the karyotype (*q.v.*) of an organism.

idiotypes   antigenic determinants that are specific for an individual. *Compare with* **allotypes, isotypes.**

idling reaction   production of ppGpp and pppGpp by ribosomes when an uncharged tRNA is present in the A site. *See* **translation.**

IF   initiation factor (*q.v.*).

IFN   interfron (*q.v.*).

Ig   immunoglobulin (*q.v.*).

IgA human immunoglobulin A, found as a 160 kd monomer or as a 320 kd dimer in mucous and secretory fluids and on the surface of cell membranes.

IgD human immunoglobulin D, found as a 185 kd monomer on the surface of lymphocytes.

IgE human immunoglobulin E, found as a 200 kd monomer and involved in allergic reactions. It forms a complex with antigen and then binds to the surface of mast cells, triggering the release of histamine.

IgG human immunoglobulin G, found as a 150 kd monomer, which is the predominant molecule involved in secondary immune responses. It fixes complement and is the only immunoglobulin that crosses the placenta.

IgM human immunoglobulin M, found as a 900 kd pentamer which is the predominant molecule involved in the primary immune response. It fixes serum complement and agglutinates effectively.

ile isoleucine. *See* **amino acid.**

imaginal discs inverted thickenings of epidermis containing mesodermal cells found in a holometabolous insect. During the pupal stage the imaginal discs give rise to the adult organs, and most larval structures are destroyed.

imago an adult insect.

imbecility mental retardation of medium severity. An imbecile reaches an intelligence level corresponding to that of the third to seventh year of life.

imbibition the absorption of liquids or vapors into the submicroscopic spaces and pores of materials such as cellulose.

imino forms of nucleotides *See* **tautomeric shifts.**

2-iminothiolane the most commonly used protein cross-linking reagent that causes a covalently linked bridge to form between two ε amino groups of lysine residues.

immediate hypersensitivity a type of hypersensitivity reaction which is mediated by antibodies and which occurs within minutes after exposure to the allergen or antigen in a previously sensitized individual. *Compare with* **delayed hypersensitivity.**

immigration the flow of genes into a population caused by immigrating individuals that interbreed with the residents.

immune competent cell a cell capable of producing antibody in response to an antigenic stimulus.

immune globulins *See* **antibody.**

immune reaction the reaction between a specific antigen and antibody.

immune response the physiological response(s) stemming from activation of the immune system by antigens, including beneficial immunity to pathogenic microorganisms, as well as detrimental autoimmunity to self-antigens, allergies, and graft rejection. The cells mainly involved in an immune response are T and B lymphocytes and macrophages. T cells produce lymphokines (*q.v.*) that influence the activities of other host cells, whereas B cells mature to produce immunoglobulins (*q.v.*) or antibodies that react with antigens. Macrophages "process" the antigen into immunogenic units that stimulate B lymphocytes to differentiate into antibody-secreting plasma cells, and stimulate T cells to release lymphokines. Complement (*q.v.*) is a group of normal serum proteins that can aid immunity by becoming activated as a consequence of antigen-antibody interactions. The first contact with an antigen "sensitizes" the animal and results in a *primary immune response.* Subsequent contact of the sensitized animal with that same antigen results in a more rapid and elevated reaction called the *secondary immune response* (also known as the "booster response" or the "anamnestic reaction") that is most easily demonstrated by monitoring the level of circulating antibodies in the serum. The immune response can be transferred from a sensitized to an unsensitized animal via serum or cells. It is highly specific for the inciting antigen, and is normally only directed against foreign substances.

immune response (Ir) gene   any gene that determines the ability of lymphocytes to mount an immune response to specific antigens. In the major histocompatibility complex of the mouse (the H-2 complex), the I region contains Ir genes and also codes for Ia (immune associated) antigens found on B cells and on some T cells and macrophages. In humans, the HLA D (DR) region is the homologue of the mouse H-2 I region.

immunity   1. the state of being refractive to a specific disease, mediated by the immune system (T and B lymphocytes and their products—lymphokines and immunoglobulins, respectively). *Active immunity* develops when an individual makes an immune response to an antigen; *passive immunity* is acquired by receiving antibodies or immune cells from another individual. 2. the ability of a prophage to inhibit another phage of the same type from infecting a lysogenized cell (phage immunity). 3. the ability of a plasmid to inhibit the establishment of another plasmid of the same type in that cell. 4. the ability of some transposons to prevent others of the same type from transposing to the same DNA molecule (transposon immunity). 5. phage resistant bacteria are usually "immune" to specific phages because they lack the cell surface receptors that define the host range of that phage.

immunity substance   a cytoplasmic factor produced in lysogenic bacteria that prevents them from being infected by bacteriophages of the same type as their prophages and also prevents the vegetative replication of said prophages.

immunization   administration of an antigen for the purpose of stimulating an immune response to it.

immunochemistry   the study of the chemistry of immune responses.

immunocompetent (immune competent) cell   a cell capable of carrying out its immune function when given the proper stimulus.

immunodominance   within a complex immunogenic molecule, the ability of a specific component 1. to elicit the highest titer of antibodies during an immune response, or 2. to bind more antibodies from a given polyvalent antiserum than any other component of that same molecule. For example, in a glycoprotein antigen, a specific monosaccharide may be the most highly antigenic component of the entire molecule and therefore exhibits immunodominance over other components of the same molecule.

immuno-electrophoresis   a technique that first separates a collection of different proteins by electrophoresis through a gel and then reacts them with a specific antiserum to generate a pattern of precipitin arcs. The proteins can thus be identified by their electrophoretic mobilities and their antigenic properties.

immunofluorescence   a visual examination of the presence and the distribution of particular antigens on or in cells and tissues using antibodies which have been coupled with fluorescent molecules such as rhodamine and fluorescein.

immunogen   a substance which causes an immune repsonse. Foreign proteins and glycoproteins generally make the most potent immunogens. *See* **antigen.**

immunogene   any genic locus affecting an immunological characteristic; examples: immune response genes, immunoglobulin genes, genes of the major histocompatibility complex *(all of which see).*

immunogenetics   studies using a combination of immunologic and genetic techniques, as in the investigation of genetic characters detectable only by immune reactions.

immunogenic   capable of stimulating an immune response.

immunoglobulin   an antibody secreted by mature lymphoid cells called plasma cells. Immunoglobulins are Y-shaped, tetrameric molecules consisting of two relatively long polypeptide chains called heavy (H) chains and two shorter polypeptide chains called light (L) chains. Each arm of the Y-shaped structure has specific antigen-binding properties and is referred to as an antigen-binding fragment (Fab). The tail of the Y-structure is a crystallizable fragment (Fc). Five H chain classes of immunoglobulin are based upon their antigenic structures. Immunoglobulin class G (IgG) is the most common in serum and is associated with immunological "memory"; class IgM is the earliest to appear upon initial exposure to an antigen. Class IgA can be secreted across epithelial tissues and seems to be associated with resistance to infectious diseases of the respiratory and digestive tracts. The antibodies associated with immunological allergies belong to class IgE. Not much is known about the functions of IgD. Antibodies of classes IgG, IgD, and IgE have molecular weights ranging from between 150,000 − 200,000d (7S); serum IgA is a 7S monomer, but secretory IgA is a dimer (11.4S); IgM is a pentamer (19S; 900,000d) of five 7S-like monomers.

Diagram of a typical IgG molecule. Within each immunoglobulin molecule, the two L chains are identical and the two H chains are identical. Numbers represent approximate amino acid residues from the N terminus of the respective chain. From W. D. Stansfield, *Serology and Immunology*, Macmillan Publishing Co., Inc., 1981.

In the case of IgG each heavy chain consists of four "domains" of roughly equal size. The variable (V_H) domain at the amino (N-terminus) end contains different amino acid sequences from one immunoglobulin to another, even within the same H chain class. The other three domains have many regions of homology that suggest a common origin by gene duplication and diversification by mutation. These "constant" domains (C_H1, C_H2, C_H3) are essentially invariate within a given H chain class. An L chain is about half as long as an H chain. Its amino end has a variable region (V_L); its carboxyl end has a constant region (C_L). An Fab fragment consists of an L chain and an Fd segment of an H chain ($V_H + C_H1$).

Within a tetrameric immunoglobulin molecule, the two L chains are identical and the two H chains are identical. The Fc fragment consists of carboxy-terminal halves of two H chains ($C_H2 + C_H3$). The region between C_H1 and C_H2 is linear rather than globular, and is called the "hinge region." Each mature antibody-producing plasma cell produces a single species of immunoglobulin, all of which contain identical L and H chains. *See* **hybridoma, IgA — IgM, monoclonal antibody.**

immunoglobulin chains   the components of the heteropolymeric immunoglobulin molecules. There are five groups of heavy chains, each characteristic of a specific class of immunoglobulin: gamma (IgG), mu (IgM), epsilon (IgE), alpha (IgA), and delta (IgD). The genes encoding all the heavy immunoglobulin chains are located on human chromosome 14. The constant region of each heavy chain makes up about ¾ of the molecule, and the gene segments encoding the constant regions are arranged in the sequence mu, delta, gamma, epsilon, and alpha in both man and mouse. There are two groups of light chains: kappa chains, encoded by gene segments on human chromosome 2, and lambda chains, encoded by gene segments on chromosome 22. *See* **immunoglobulin genes.**

immunoglobulin genes   genes encoding the light and heavy chains of the immunoglobulins. These genes are remarkable in that they are made up of segments that are shuffled as the B lymphocytes mature. The light chains contain segments that can be symbolized L-V, J, and C. The V or variable segment codes for the first 95 amino acids of the chain, whereas the C or constant segment codes for amino acids 108 to 214. The joining segment, J, codes for amino acids 96 to 107. L codes for a leader sequence 17–20 amino acids long. It functions in the transport of the molecule through the plasmalemma and is cleaved off the molecule in the process. There are about 300 L-V segments per light chain gene, and each of the V segments has a different base sequence. In the kappa gene there are six J segments, each with a different base sequence, and one C segment. During differentiation of a given B lymphocyte stem cell, an immunoglobulin gene is assembled containing one L-V, one J, and one C segment, and this gene is transcribed by the lymphocyte and all of its progeny. The lambda gene also contains about 300 L-V segments, but each of the six J segments has its own adjacent C segment. The heavy chain gene is over 100,000 nucleotides long and contains a series of segments that can be symbolized L-V, D, J, C_μ, C_δ, $C_{\gamma3}$, $C_{\gamma1}$, $C_{\gamma2b}$, $C_{\gamma2a}$, C_ϵ, and C_α. There are about 300 L-V segments, 10–50 D segments, 4 J segments, and one each of the C segments. Each D segment codes for about 10 amino acids. During differentiation the segments are shuffled so that the variable region of a heavy chain is encoded by a segment that contains one L-V, one D, and one J segment. The gene also contains mu, delta, gamma, epsilon, and alpha subsegments, and which one of these is trnascribed determines the class to which the antibody will belong. *See* **heavy chain class switching.**

immunological memory   the capacity of the immune system to respond more rapidly and more vigorously to the second contact with a specific antigen than observed in the primary response to the first contact; the booster or anamnestic response.

immunological suppression   a genetic or induced condition in which the ability of an individual's immune system to respond to most or all antigens is impaired. *See* **specific immune suppression.**

immunological surveillance theory   the theory that the cell-mediated immune system evolved primarily to continuously monitor the body for spontaneously arising cancerous cells or those containing foreign pathogens and to destroy them.

immunological tolerance   a state of nonreactivity towards a substance that would normally be expected to elicit an immune response. Tolerance to specific foreign antigens can be induced by the exposure of a bird or mammal to the foreign antigens during embryonic or neonatal (depending upon species) life. In adults, tolerance (usually of shorter duration) can be induced by using particular routes of administration for the antigens or administration of agents which are particularly effective against cells proliferating in response to antigen. Mechanisms may include actual deletion of potentially reactive lymphocytes or their "inactivation" by immunological suppression.

immunology the science dealing with immunity, serology, immunochemistry, immunogenetics, hypersensitivity, and immunopathology.

immunoselection a method for isolating cell-line variants lacking certain antigens, such as those of the major immunogene complex. By treating cells with a specific antiserum and complement, all cells die, except a few spontaneously arising variants. These do not express the corresponding antigen, and therefore they live and can be isolated. Many of these variants appear to be due to deletion mutations rather than to epigenetic changes or mitotic crossing-over. *Compare with* **antigenic conversion.**

immunosuppressive drug a compound that blocks the normal response of antibody-producing cells to antigen molecules.

impact theory the proposal that the mass extinctions of groups of organisms occurring late in the Devonian and Cretaceous periods of geological time resulted from the collision of the earth with an asteroid or comet.

impaternate offspring an offspring from parthenogenetic reproduction in which no male parent took part.

imperfect excision the release of a genetic element (e.g., an insertion sequence or prophage) from a DNA molecule in a way that either includes more than or less than the element itself.

imperfect flower *See* **flower.**

implant material artifically placed in an organism, such as a tissue graft, an electronic sensor, etc.

implantation **1.** attachment of a mammalian embryo to the uterine wall. **2.** the addition of tissue grafts to an organism without the removal of anything from it.

imprinting the imposition of a stable behavior pattern in a young animal by exposure, during a particular period in its development, to one of a restricted set of stimuli.

inactivation center a region of the mouse X chromosome that governs the degree to which translocated autosomal genes are inactivated when the associated X-linked genes are inactivated as the result of random X-inactivation. *See* **Cattanach's translocation, Lyon hypothesis.**

inactive X hypothesis Lyon hypothesis (*q.v.*).

Inarticulata a superphylum of invertebrates containing the unsegmented, coelomate protostomes. *See* **classification.**

inborn error a genetically determined biochemical disorder resulting in a metabolic defect that produces a metabolic block having pathological consequences.

Inborn Errors of Metabolism a treatise comprising the earliest study of the biochemical genetics of man (or any other species) published by A. E. Garrod in 1909.

inbred strain a group of organisms so highly inbred as to be identical, except for sexual differences.

inbreeding the crossing of closely related plants or animals.

inbreeding coefficient *See* **Wright's inbreeding coefficient.**

inbreeding depression decreased vigor in terms of growth, survival, or fertility following one or more generations of inbreeding.

incapsidation   the construction of a capsid around the genetic material of a virus.

inclusive fitness   *See* **Hamilton's genetical theory of social behavior.**

incompatibility   in immunology, genetic or antigenic differences between donor and recipient tissues that evoke an immunological rejection response.

incomplete dominance   failure of a dominant phenotype to be fully expressed in an organism carrying a dominant and a recessive allele. The result is usually a phenotype that is intermediate between the homozygous dominant and the recessive forms. *See* **semidominance.**

incompletely linked genes   genes on the same chromosome that can be recombined by crossing over.

incomplete sex linkage   the rare phenomenon of a gene being located on the homologous segments of both X and Y chromosomes.

incross   mating between individuals from the same inbred line or variety, often of the same genotype.

indehiscent   designating a fruit that remains closed at maturity.

independent assortment   the random distribution to the gametes of genes located on different chromosomes. Thus an individual of genotype *Aa Bb* will produce equal numbers of four types of gametes: *AB, Ab, aB,* and *ab. See* **Mendel's laws.**

index case   propositus (*q.v.*).

index fossil   a fossil that appears only in rocks of a relatively limited geological age span.

indole   a tryptophan precursor in microorganisms.

indoleacetic acid   auxin, a phytohormone.

induced mutation   a genetic alteration produced by exposure to a mutagen. *Compare with* **spontaneous mutation.**

inducer   any of the small organic molecules that cause the cell to produce larger amounts of the enzymes involved in their metabolism. Inducers are a class of effector molecules (*q.v.*). *See* **gratuitous inducer, regulator genes.**

inducible enzyme   an enzyme synthesized only in response to an inducer. *See* **regulator gene.**

inducible system   a regulatory system in which the product of a regulator gene (the repressor) is active and blocks transcription of the operon. The effector (called an inducer) inactivates the repressor and therefore allows mRNA synthesis to occur. Thus transcription occurs only in the presence of effector molecules. *See* **regulator gene.**

induction   **1.** the determination of the developmental fate of one cell mass by another. The morphogenic effect is brought about by an evocator acting upon competent tissue. **2.** the stimulation of a lysogenized bacterium to produce infective phage. **3.** the stimulation of synthesis of a given enzyme in response to a specific inducer.

inductor   any substance that carries out an induction similar to that performed by an organizer (*q.v.*).

industrial melanism   the prevalence in industrial regions of black or dark brown forms of various animal species. The pigments involved are most often melanins. The most famous example involves the moth *Biston betularia* (*q.v.*). Studies on this species illustrated the importance of predation determining the direction of evolution.

inelastic collision   *See* **collision.**

infectious nucleic acid   purified viral nucleic acid capable of infecting a host cell and causing the subsequent production of viral progeny.

infectious transfer   the rapid spread of extrachromosomal episomes (plus any integrated chromosomal genes) from donor to recipient cells in a bacterial population.

inflorescence   the arrangement of flowers on an axis.

informosome   a mRNA molecule which is protected against nuclease digestion by a protein coat. Such hypothetical mRNA protein particles are postulated to continually enter the cytoplasm from the nucleus.

inheritance of acquired characteristics   *See* **acquired characteristics, Lamarckism.**

initiation codon   *See* **start codon.**

initiation factors   proteins required for the initiation of protein synthesis. One (protein IF3) is required for the binding of the 30S particle to mRNA. A second (protein IF1) binds to f-met-tRNA and helps it attach to the 30S mRNA initiation complex. A third protein (IF2) is required although its precise function is unclear. *See* **N-formyl methionine, translation.**

initiator   a molecule which initiates replication once it binds to a replicator. *See* **replicon.**

innervation   the nerve supply to a particular organ.

inoculum   a suspension of cells used to start a new culture.

inosine   hypoxanthine riboside. *See* **rare bases.**

inquiline   an animal that lives in the abode of another species.

insect ovary types   three types of ovaries are found among insects. The *panoistic ovary* appears to be the ancestral type. Here all oogonia (except stem line oogonia) are eventually transformed to oocytes. In *meroistic* ovaries both oocytes and nurse cells (*q.v.*) are generated. These may be organized within the ovariole in two ways. In the *polytrophic meroistic* ovary, the nurse cells and oocytes alternate along the length of the ovariole. In the *telotrophic meroistic* ovary, the nurse cells are restricted to the germarium and are connected to oocytes in early stages of their development by cytoplasmic processes called nutritive chords. Panoistic ovaries are found in insects belonging to the more primitive orders

(Archeognatha, Zygentoma, Ephemeroptera, Odonata, Plecoptera, Phasmida, Orthoptera, and Dictyoptera). Polytrophic meroistic ovaries occurs in the Psocoptera, Phthiraptera, Hymenoptera, Trichoptera, Lepidoptera, and Diptera. Telotrophic ovaries occur in the Hemiptera, Coleoptera, Raphidioptera, and Megaloptera.

insertion the addition of one or more base pairs into a DNA molecule; a type of mutation commonly induced by acridine dyes or by mobile insertion sequences (*q.v.*).

insertional translocation *See* **translocation.**

insertion sequences small, transposable elements (*q.v.*) which usually encode proteins for their own transposition and catalyze the transposition of sequences they flank. They occasionally produce lethal mutations by inserting into essential genes or by causing deletions due to imperfect excision (*q.v.*), but they have no known effects beyond transposition and inactivation of the gene into which they insert. Insertion sequences are abbreviated IS. *See* **transposable elements.**

insertion vector *See* **lambda cloning vehicle.**

in situ **hybridization** a technique utilized to localize, within intact chromosomes, DNA segments complementary to specific radioactively-labeled nucleic acid molecules. In the case of chromosomal squash preparations on glass microscope slides, the specimens are treated so as to denature the DNAs and to remove adhering RNAs and proteins. Then they are incubated with tritium-labeled nucleic acid molecules of interest. The radioactive molecules hybridize with the DNA segments of specific chromosomal areas (*see* **hybrid duplex molecule**), and these are visualized in autoradiographs. The technique was first used to demonstrate that rDNA resided in the nucleolus organizer. The *in situ* hybridization technique can also be used to identify DNA sequences of interest in DNAs liberated from lysed colonies of bacteria on nitrocellulose filters. *See* **colony hybridization.**

instar the period between insect molts.

instinct an unlearned pattern of behavior.

instructive theory an early immunological theory in which it was believed that the specificity of antibody for antigen was conferred upon it by its initial contact with the antigen. This theory has been discarded in favor of the clonal selection theory (*q.v.*), in which specificity is pre-existing prior to contact with antigen.

insulin a polypeptide hormone produced by the beta cells in the islets of Langerhans. Insulin causes a fall in the sugar concentration of the blood, and its deficiency produces the symptoms of diabetes mellitus. Beef insulin was the first protein to have its amino acid sequence determined. This molecule (shown below) is made up of an A polypeptide (21 amino acids long) and a B peptide (containing 30 amino acids) joined by two disulfide bridges. In humans, insulin is encoded by a gene on the short arm of chromosome 11. *See* **proinsulin.**

```
         S ——————————— S
         |             |
gly—ile—val—glu—gln—cys—cys—ala—ser—val—cys—ser—leu—tyr—gln—leu—glu—asn—tyr—cys—asn
                         |                                                        |
                         S              A peptide                                 S
                         |                                                        |
                         S                                                        S
                         |                                                        |
phe—val—asn—gln—his—leu—cys—gly—ser—his—leu—val—glu — ala — leu — tyr — leu — val — cys—gly
                                                                                      |
                                                                                     glu
                     B peptide                                                        |
                                                                                     arg
                                                                                      |
                                    ala—lys—pro—thr—tyr—phe—phe—gly
```

integration   the insertion of a viral or another DNA sequence into a host genome.

integration efficiency   the frequency with which a foreign DNA segment is incorporated into the genotype of a recipient bacterium, particularly with reference to transformation.

intelligence quotient (IQ)   an individual can be assigned to a "mental age" group on the basis of performance on standardized intelligence tests. This mental age divided by the individual's chronological age and multiplied by 100 is the IQ.

intelligence quotient classification   according to the Binet-Simon classification, intelligence quotients can be grouped as follows: genius, 140 and over; very superior, 120–139; superior, 110–119; average, 90–109; dull, 80–89; borderline, 70–79; mild retardation (moron), 50–69; moderate retardation (imbecile), 25–49; and severe retardation (idiot), 0–24.

interallelic complementation   referring to the change in the properties of a multimeric protein as a consequence of the interaction of subunits coded by two different mutant alleles (in contrast to the protein consisting of subunits derived from a single mutant allele). The mixed protein (heteromultimer) may exhibit more activity (positive complementation) or less activity (negative complementation, *q.v.*) than the homomultimer. Also known as *intragenic complementation.*

interbands   the regions between bands in a polytene chromosome (*q.v.*). The DNA concentration in interbands is only a fraction of that in bands.

intercalary deletion   *See* **deletion.**

intercalating agent   a substance (e.g., acridine dyes) that inserts between base pairs in a DNA molecule, often disrupting the alignment and pairing of bases in the complementary strands. By causing addition or deletion of one or more base pairs during replication, a reading frame shift (*q.v.*) is often induced.

intercellular   between cells.

interchange   an exchange of segments between nonhomologous chromosomes resulting in translocations.

interchromosomal translocation   *See* **translocation.**

intercistronic region   the segment between the termination codon of one gene and the initiation codon of the next gene in a polycistronic transcription unit.

intercross   mating of heterozygotes (*a/+* × *a/+*).

interference   *See* **positive interference.**

interference filter   a filter used to produce a monochromatic light source.

interference microscope   like the phase microscope, the interference microscope is used for observing transparent structures. However, with the interference microscope *quantitative* measurements of the relative retardation of light by various objects can be made. Such measurements can be used to determine the dry mass per unit area of specimen or the section thickness.

interferons   a group of proteins with antiviral (and possibly anticancer) properties. Three types of interferon have been identified in humans. For historical reasons they were originally named leukocyte interferon, fibroblast interferon, and immune interferon, but now are designated alpha, beta, and gamma interferons, respectively. There are at least 14 different human genes that code for IFN-alpha, and at least 5 for IFN-beta. IFN-alpha and IFN-beta appear to have a common cell receptor that differs from that of IFN-gamma. IFN-alpha and IFN-beta both induce synthesis of a dozen polypeptides in human tissue culture cells; IFN-gamma induces four additional polypeptides; viral replication is inhibited by the concerted action of these polypeptides.

intergenic suppression *See* **suppression.**

interkinesis the abbreviated interphase between the first and second meiotic division. No DNA replciation occurs during interkinesis, unlike a premitotic interphase.

intermediary metabolism the chemical reactions in a cell which transform ingested nutrient molecules into the molecules needed for the growth of the cell.

intermediate filaments cytoplasmic filaments with diameters between 8 and 12 nm. They comprise a heterogeneous class of cytoskeletal proteins. In general, a given class of intermediate filaments is characteristic of a specific cell type. For example, keratin filaments are characteristic of epithelial cells, neurofilaments of neurons, vimentin filaments of fibroblasts, and desmin filaments of glial cells.

intermediate host a host essential to the completion of the life cycle of a parasite, but in which it does not become sexually mature.

intermedin a polypeptide hormone from the intermediate lobe of the pituitary gland which causes dispersion of melanin in melanophores. Also called melanocyte-stimulating hormone or MSH.

internal radiation the exposure to ionizing radiation from radioelements deposited in the body tissues.

internode the region of a stem between two successive nodes (*q.v.*).

interphase the period between succeeding mitoses.

interrupted mating experiment a genetic experiment in which the manner of gene transfer between conjugating bacteria is studied by withdrawing samples at various times and subjecting them to a strong shearing force in an electric blender.

intersex a class of individuals of a bisexual species which have sexual characteristics intermediate between the male and female.

interstitial cells cells that lie between the testis tubules of vertebrates and secrete testosterone.

intervening sequence *See* **intron.**

intrachange *See* **intrachromosomal aberration.**

intrachromosomal recombination sister chromatid exchange (*q.v.*).

intrachromosomal translocation *See* **translocation.**

intragenic complementation *See* **interallelic complementation.**

intragenic recombination recombination between mutons of a cistron. Such recombination is characterized by negative interference and by non-reciprocality (recovery of either wild-type or double-mutant recombinants, but not both from the same tetrad).

intragenic suppression *See* **suppression.**

intrasexual selection   *See* **sexual selection.**

introgression   *See* **introgressive hybridization.**

introgressive hybridization   the incorporation of genes of one species into the gene pool of another. If the ranges of two species overlap and fertile hybrids are produced, they tend to backcross with the more abundant species. This process results in a population of individuals most of whom resemble the more abundant parents but which possess also some characters of the other parent species.

intromittent organ   any male copulatory organ that implants sperm within the female.

intron   in split genes (*q.v.*) a segment that is transcribed into nuclear RNA, but is subsequently removed from within the transcript and rapidly degraded. Most genes in the nuclei of eukaryotes contain introns and so do mitochondrial genes and some chloroplast genes. The number of introns per gene varies greatly, from one in the case of rRNA genes to more than 30 in the case of the yolk protein genes of *Xenopus*. Introns range in size from less than 100 to more than 10,000 nucleotides. There is little sequence homology among introns, but there are a few nucleotides at each end that are nearly the same in all introns. These boundary sequences participate in excision and splicing reactions. *See* **exon, splice junctions, U1 RNA.**

intron intrusion   the disruption of a preexisting gene by the insertion of an intron into a functional gene. Intron intrusion and the exon shuffling (*q.v.*) have been proposed, along with junctional sliding (*q.v.*), as mechanisms for evolutionary diversification of genes.

intron-mediated recombination   *See* **exon shuffling.**

intussusception   **1.** the growth of an organism by the conversion of nutrients into protoplasm. **2.** the deposition of material between the microfibrils of a plant cell wall. **3.** the increase in surface area of the plasmalemma by intercalation of new molecules between the existing molecules of the extending membrane.

in utero   within the uterus.

inv   *See* **symbols used in human cytogenetics.**

in vacuo   in a vacuum.

invagination   an inpocketing or folding in of a sheet of cells or a membrane.

inversion   chromosome segments that have been turned through 180° with the result that the gene sequence for the segment is reversed with respect to that of the rest of the chromosome.

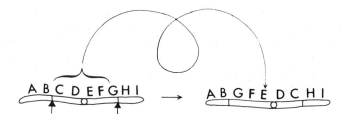

Inversions may include or exclude the centromere. An inversion that includes the centromere is called *pericentric* or *heterobrachial,* whereas an inversion that excludes the centromere is called *paracentric* or *homobrachial.* Paracentric inversions are found more often in nature than pericentric inversions. A paracentric inversion heterozygote forms a reverse loop pairing configuration during pachynema. The various types of single and double exchanges that can occur within the loop are illustrated below.

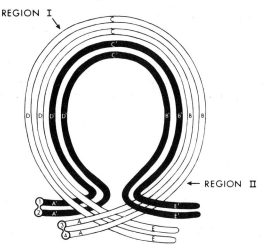

REGION I

REGION II

SINGLE EXCHANGE (REGION I, STRANDS 2 & 3)

A' D' C' B' E'

A' D' C B A

E D C B A

E' B' C' D E

TWO STRAND DOUBLE EXCHANGE (REGION I, STRANDS 2&3; REGION II, STRANDS 2&3)

A' D' C' B' E'   A' D' C B E'   E D C' B' A   E D C B A

THREE STRAND DOUBLE EXCHANGE (REGION I, STRANDS 2&3; REGION II, STRANDS 2&4)

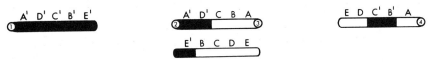

A' D' C' B' E'

A' D' C B A

E D C' B' A

E' B C D E

THREE STRAND DOUBLE EXCHANGE (REGION I, STRANDS 1&3; REGION II, STRANDS 2&3)

A' D' C B E'

A' D' C' B' A

E D C B A

E' B' C' D E

FOUR STRAND DOUBLE EXCHANGE (REGION I, STRANDS 1&4; REGION II, STRANDS 2&3)

A' D' C B A

A' D' C' B' A

E' B' C' D E

E' B C D E

Note that no monocentric, single-crossover chromatids are produced. For this reason, inversions give the impression of being *crossover suppressors,* and it was their action on crossing over that led to their discovery.

inversion heterozygote   an organism in which one of the homologues has an inverted segment while the other has the normal gene sequence.

invertebrate   an animal without a dorsal column of vertebrae; non-chordate metazoans.

inverted repeats (IR)   two copies of the same DNA sequence orientated in opposite directions on the same molecule. IR sequences are found at opposite ends of a transposon (*q.v.*). *See* **palindrome.**

inverted terminal repeats   short, related or identical sequences oriented in opposite directions at the ends of some transposons (*q.v.*).

in vitro   designating biological processes made to occur experimentally in isolation from the whole organism; literally "in glass", i.e., in the test tube. Examples: tissue cultures, enzyme-substrate reactions.

in vitro **complementation**   *See* **allelic complementation.**

in vitro **fertilization**   experimental fertilization of an egg outside the female body. In humans, this is usually done because the woman's Fallopian tubes are blocked. The resulting embryo can then be inserted into the uterus for implantation. *See* **embryo transfer.**

in vitro **marker**   a mutation induced in a tissue culture that allows subsequent phenotypic detection. Human *in vitro* markers include genes conferring resistance to various viruses, aminopterin, and purine analogues.

in vitro **mutagenesis**   experiments in which segments of genomic DNA are treated with reagents that produce localized chemical changes in the molecule. The subsequent ability of the mutated molecules to function during replication, transcription, etc., is assayed either by using cell-free systems or *in vivo,* after splicing the fragment into an appropriate plasmid.

in vitro **packaging**   the production of infectious particles from naked DNA by incapsidation of the DNA in question after supplying lambda phage packaging proteins and preheads.

in vitro **protein synthesis**   the incorporation in a cell-free system of amino acids into polypeptide chains.

in vivo   within the living organism.

in vivo **culturing of imaginal discs**   the technique developed by Hadorn in which an imaginal disc is removed from a mature *Drosophila* larva, cut in half, and the half organ implanted into a young larva. Here regenerative growth occurs, and once the host larva has reached maturity the implant is removed once again, bisected, and one of the halves transplanted to a new host. By multiple repetition of this procedure the cells are subjected to an abnormally long period of division and growth in a larval environment. If the regenerated disc is finally allowed to undergo metamorphosis, it shows an abnormally high probability of producing structures characteristic of different discs. A regenerated genital disc may produce antennae, for example. Hadorn terms such differentiation as *allotypic.* Since the allotypic organs appear in the offspring of cells which were previously determined to form genital structures, a change in determination must be postulated. This event is called *transdetermination.*

in vivo **marker**   a naturally occurring mutant mammalian gene that allows phenotypic detection of the tissue cultured cells bearing it. Examples are the genes causing galactosemia and glucose-6-phosphate dehydrogenase deficiency in man, and the genes producing certain cell surface antigens in the mouse.

iodine   a biological trace element. Atomic number 53; atomic weight 129.9044; valence 1⁻; most abundant isotope ^{127}I; radioisotope ^{131}I; half-life, 8.0 days; radiations emitted—beta particles and gamma rays.

iojap   a mutant nuclear gene in maize which induces changes in chloroplast characters. The mutant plastids behave autonomously thereafter.

ion exchange column   a column packed with an ion exchange resin. *See* **chromatography.**

ion exchange resin   a polymeric resin that has a higher affinity for some charged groups than it has for others. For example, resins with fixed cation groups will bind anions and thus can be used in column separation procedures.

ionic bond   electrostatic bond (*q.v.*).

ionization   any process by which a neutral atom or molecule acquires a positive or negative charge.

ionization chamber   any instrument designed to measure the quantity of ionizing radiation in terms of the charge of electricity associated with ions produced within a defined volume.

ionization track   the trail of ion pairs produced by an ionizing radiation during its passage through matter.

ionizing energy   the average energy lost by an ionizing radiation in producing an ion pair in a given gas. The average ionizing energy for air is about 33 eV.

ionizing event   the occurrence of any process in which an ion or group of ions is produced.

ionizing radiation   electromagnetic or corpuscular radiation that produces ion pairs as it dissipates its energy in matter.

ionophores   a class of antibiotics of bacterial origin that facilitate the movement of monovalent and divalent cations across biological membranes. Some of the major ionophores and the ions they transort are valinomycin (K^+, Rb^+), A 23187 (Ca^{++}, $2H^+$), nigericin (K^+, H^+), and gramicidin (H^+, Na^+, K^+, Rb^+).

ion pair   the electron and positive atomic or molecular residue resulting from the interaction of ionizing radiation with the orbital electrons of atoms.

2ip   6-(γ,γ-dimethylallylamino) purine (*q.v.*).

IPTG   isopropylthiogalactoside; a gratuitous inducer for the *E. coli lac* operon (*q.v.*). *See* **ONPG.**

IQ   intelligence quotient (*q.v.*).

I region   one of the central regions of the major histocompatibility complex (H-2) of the mouse. It contains genes coding for Ia antigens and controlling various immune responses. It has five subregions (A, B, J, E, C) and may be the equivalent of the D/DR region of the human major histocompatibility complex. *See* **HLA complex.**

Ir gene   *See* **immune response gene.**

iron   a biological trace element. Atomic number 26; atomic weight 55.847; valence $2,3^+$; most abundant isotope ^{56}Fe; principal radioisotope ^{59}Fe, half-life 46 days, radiation emitted—beta particle.

isauxesis   *See* **heterauxesis.**

islets of Langerhans clusters of hormone-secreting cells located in the pancreas of vertebrates. Two types of cells are found: alpha cells, which secrete glucagon (*q.v.*), and beta cells, which secrete insulin (*q.v.*).

isoacceptor transfer RNA one of a group of different tRNAs that accept the same amino acid but possess different anticodons. Higher organisms contain 2-4 isoacceptor tRNAs for each amino acid. *See* **amino acid.**

isoagglutinin an antibody directed against antigenic sites on the red blood corpuscles of the same species and which causes agglutination.

isoagglutinogen an antigenic factor on the surface of cells which is capable of inducing the formation of homologous antibodies (isoagglutinins) in some members of the same species.

iso-allele an allele whose effect can only be distinguished from that of the normal allele by special tests. For example, two + alleles $+^1$ and $+^2$ may be indistinguishable (i.e. $+^1/$ $+^1$, $+^2/+^2$, and $+^2/+^1$ individuals are phenotypically wild type). However, when compounded with a mutant allele *a,* $+^1$ and $+^2$ prove to be distinguishable (i.e. $a/+^1$ and $a/$ $+^2$ individuals are observably different).

isoanisosyndetic alloploid an allopolyploid in which some chromosomes derived from both species are homoeologous and undergo a limited synapsis. *See* **isosyndetic alloploid.**

isoantibody an antibody formed in response to immunization with tissue constituents derived from an individual of the same species as the recipient.

isocapsidic viruses *See* **segmented genome.**

isochromatid break an aberration involving breaks in both sister chromatids at the same locus, followed by lateral fusion to produce a dicentric chromatid and an acentric fragment.

isochromosome a metacentric chromosome produced during mitosis or meiosis when the centromere splits transversely instead of longitudinally. The arms of such a chromosome are equal in length and genetically identical. However, the loci are positioned in reverse sequence in the two arms.

isocoding mutation a point mutation which alters the nucleotide sequence of a codon but, because of the degeneracy of the genetic code, does not change the amino acid that the codon specifies.

isoelectric point the *p*H at which the net positive and negative charge on a protein is zero.

isoenzymes isozymes (*q.v.*).

isoforms families of functionally related proteins that differ slightly in their amino acid sequences. An example would be the isoforms of contractile porteins, such as tropomyosin (*q.v.*). The tropomyosins of *Drosophila* are encoded by a cluster of three genes. Two of these encode muscle isoforms of tropomyosin, and one encodes a cytoplasmic isoform. Moreover, one of the muscle genes can generate distinct isoforms in different tissues. The isoforms differ with respect to a 27 amino acid segment at their C termini. These modifications result from differential splicing of exons at the 3′ end of the gene.

isogamy that mode of sexual reproduction involving sex cells of similar size and morphology but opposite mating types. *See* **anisogamy.**

isogeneic referring to a graft involving genetically identical donor and host.

isogenic genetically identical (except possibly for sex); coming from the same individual or from a member of the same inbred strain.

isograft a tissue graft between two individuals of identical genotype.

isohemagglutinin isoagglutinin.

isoimmunization antibody formation in reaction to antigens of the same species.

isoionic point isoelectric point.

isolabeling labeling of both, or parts of both, daughter chromatids at the second metaphase following administration of tritiated thymidine during the S period. Normally both daughter chromatids are labeled at metaphase 1, but only one is labeled at metaphase 2.

isolate a segment of a population within which assortative mating occurs.

isolating mechanism a cytological, anatomical, physiological, behavioral, or ecological difference, or a geographical barrier which prevents successful mating between two or more related groups of organisms.

isolecithal egg one in which the yolk spheres are evenly distributed throughout the ooplasm. *See* **centrolecithal, telolecithal egg.**

isoleucine *See* **amino acid.**

isologous synonomous with isogeneic (*q.v.*).

isologous cell line cell lines derived from identical twins or from highly inbred animals.

isomers compounds with the same molecular formula but with different three-dimensional molecular shapes or orientations in space.

isometry isauxesis.

isonymous marriage marriage between persons with the same surname. Isonymous marriages are used as indications of consanguinity in population genetics.

isophene a line on a map which connects points of equal expression of a character that varies clinally.

isopropylthiogalactoside a gratuitous inducer of the *lac* operon (*q.v.*).

isopycnic having the same density; used to refer to cell constituents having similar buoyant densities. *See* **centrifugation separation.**

isopycnotic referring to chromosomal regions or entire chromosomes which are not heteropycnotic, that is, which are the same in appearance as the majority of the chromosomes. *See* **heteropycnosis.**

isoschizomers two or more restriction endonulceases (*q.v.*) isolated from different sources which cleave DNA within the same target sequences.

isosyndetic alloploid an allopolyploid where synapsis is restricted to the homologues derived from one species. *See* **isoanisosyndetic alloploid.**

isothermal process a reaction occurring at constant temperature.

isotonic solution a solution having the same osmotic pressure as a material within which it is compared (usually blood or protoplasm).

isotope one of the several forms of a chemical element. Different isotopes have the same number of protons and electrons, but differ in the number of neutrons contained in the atomic nucleus. Hence they have identical chemical properties, but differ in atomic weights. *See* **radioactive isotope.**

209

isotopic dilution analysis   a method of chemical analysis for a component of a mixture. The method is based on the addition to the mixture of a known amount of labeled component of known specific activity, followed by isolation of a quantity of the component and measurement of the specific activity of that sample.

isotopically enriched material   material in which the relative amount of one or more isotopes of a constituent has been increased.

isotropic   *See* **anisotropy.**

isotypes   antigenic determinants shared by all individuals of a given species, but absent in individuals of other species. *Compare with* **allotypes, idiotypes.**

isozymes   multiple forms of a single enzyme. While isozymes of a given enzyme catalyze the same reaction, they differ in properties such as the *p*H or substrate concentration at which they function best. Isozymes are complex proteins made up of paired polypeptide subunits. The lactic dehydrogenases, for example, are tetramers made up of two polypeptide units, A and B. Five isozymes exist and can be symbolized as follows: AAAA, AAAB, AABB, ABBB, and BBBB. Isozymes often have different isoelectric points and therefore can be separated by electrophoresis. The different monomers of which isozymes like lactic dehydrogenase are composed are specified by different gene loci. The term *allozyme* is used to refer to variant proteins produced by allelic forms of the same locus. *See* **allozymes.**

iteroparity   repeated periods of reproduction during the life of an individual. *Compare with* **semelparity.**

IVS   intervening sequence.

J

J   Jurassic.

Japanese quail   *See Coturnix coturnix japonica.*

jarovization   synonym for vernalization (*q.v.*).

Java man   an extinct sub-species of primitive man known from fossils obtained in central Java. Now classified as *Homo erectus erectus,* but formerly referred to as *Pithecanthropus erectus.*

J chain   a small protein of about 15,000 daltons that holds the monomeric units of a multimeric (polymeric) immunoglobulin together, as occurs in the classes IgM and IgA.

J genes   a tandem series of four or five homologous nucleotide sequences coding for part of the hypervariable regions of light and heavy chains of mouse or human immunoglobulins; so named because they help *join* one of the genes for the variable region upstream to one of the genes for the constant region downstream and therefore are an important part of the mechanism generating antibody diversity.

JH   juvenile hormone. *See* **allatum hormone.**

Jordan's rule   an ecological principle stating that the ranges of closely related species or subspecies are generally adjacent and separated by a barrier of some sort.

jumping genes   mobile or "nomadic" genetic entities such as insertion elements and transposons.

junctional complex   a term used in electron microscopy to refer to any specialized region of intercellular adhesion, such as a desmosome (*q.v.*).

junctional sliding   a term descriptive of the fact that the location of intron-exon junctions is not constant within members of a gene family, such as the serine proteases. Some variation in length of such gene products can be attributed to extension or contraction of exons at the intron junctions.

junk DNA   *See* **selfish gene theory.**

Jurassic   the middle period in the Mesozoic era. *See* **geological time divisions.**

juvenile hormone   *See* **allatum hormone.**

K

K **1.** degrees Kelvin. *See* **temperature. 2.** Cretaceous. **3.** potassium. **4.** the gene in *Paramecium aurelia* required for the maintenance of kappa. **5.** carrying capacity (*q.v.*).

kairomone a trans-specific chemical messenger the adaptive benefit of which falls on the recipient rather than the emitter. Kairomones are commonly nonadaptive to the transmitter. For example, a secretion that attracts a male to the female of the same species may also attract a predator. *See* **allomone.**

Kalanchoe a genus of succulent plants studied in terms of the genetic control of photoperiodic flowering response. *See* **phytochrome.**

K and r selection theory *See* **r and K selection.**

kanomycin an antibiotic that binds to the 70S ribosomes of bacteria and causes misreading of the mRNA.

K antigens *See* **O antigens.**

kappa symbiont *See* **killer paramecia.**

karyogamy the fusion of two gametic nuclei.

karyokinesis nuclear division as opposed to cytokinesis (*q.v.*).

karyolymph nucleoplasm (*q.v.*).

karyon nucleus.

karyoplasm nucleoplasm.

karyosome a Feulgen-positive body seen in the nucleus of the *Drosophila* oocyte during stages 3-13. During stages 3-5 it contains synaptonemal complexes.

karyosphere the condensed Feulgen-positive mass seen in the anterior, dorsal portion of the mature primary oocyte of *Drosophila melanogaster.* This mass of DNA is not surrounded by a nuclear envelope. The tetrads subsequently emerge from the karyosphere and enter metaphase of the first meiotic division. The karyosphere stage of oogenesis is the most radiation-sensitive one.

karyotheca nuclear envelope.

karyotype the somatic chromosomal complement of an individual or species. The term is often used for photomicrographs of the metaphase chromosomes arranged in a standard sequence, an idiogram (*q.v.*).

kb, kbp   *See* **kilobase.**

KB cells   a strain of cultured cells derived in 1954 by H. Eagle from a human epidermoid carcinoma of the nasopharynx.

kcal   *See* **calorie.**

K cells   killer cells that mediate antibody-dependent cellular cytotoxicity. These cells and natural killer (NK) cells have many similar properties, are under common genetic control, and may belong to the same cell lineage (lymphocyte or monocyte). Both K and NK cells do not have surface markers characteristic of either T- or B-cell lineages, but do possess Fc receptors for immunoglobulin molecules, and thus may acquire membrane-bound antibodies and react with target cells bearing the corresponding antigens.

kd   kilodalton.

Kell-Cellano antibodies   antibodies against the red-cell antigens specified by the *K* gene, named for the first patient known to produce them. *See* **blood groups.**

kelp   the largest of the marine brown algae.

Kelvin   *See* **temperature.**

keratin   the major component of hair, nails, horns, etc. There are many different types of keratins, encoded by a large family of genes. Epidermal cells synthesize a sequence of different keratins as they mature.

keto forms of nucleotides   *See* **tautomeric shifts.**

keV   *See* **electron volt.**

Kidd blood group   a blood group defined by a human red cell antigen encoded by the *Jk* gene on the short arm of chromosome 2.

killer cells   **1.** nonphagocytic cells of the immune system that destroy incompatible tissue grafts or neoplasms via lymphotoxins or other secreted products; *see* **K cells. 2.** protozoan cells that harbor killer symbionts (*q.v.*). *See* **kappa.**

killer paramecia   paramecia which secrete into the medium particles that kill other paramecia. The killer trait is due to kappa particles, which reside in the cytoplasm of those strains of *Paramecium aurelia* syngen 2 that carry the dominant *K* gene. Later it was found that kappa particles were symbiotic bacteria and that the particles with killing activity were defective DNA phages. The lysogenic, symbiotic bacterium has been named *Caenobacter taenospiralis,* and it is but one of many kinds of bacterial endosymbionts that occur in over 50% of the *P. aurelia* collected in nature. *See* ***Paramecium aurelia.***

killer particle   in yeast, a double-stranded RNA plasmid containing ten genes for replication and several others for synthesis of a killer substance similar to bacterial colicin (*q.v.*). It is the only known plasmid that does not contain DNA.

kilo   a prefix meaning one thousand, used with measurement units of the metric system.

kilobase   a unit of length for nucleic acids consisting of 1,000 nucleotides; abbreviated kb, or kbp for kilobase pairs (DNA).

kilocalorie   *See* **calorie.**

kilovolt   a unit of electrical potential equal to 1000 volts, symbolized by **kV.**

kinase   an enzyme that adds a phosphate group to its substrate.

kindred   a group of human beings each of which is related, genetically or by marriage, to every other member of the group.

kinetic complexity   *See* **DNA complexity.**

kinetin   6-furfurylaminopurine, a cytokinin.

kinetochore   *See* **centromere.**

kinetoplast   a highly specialized mitochondrion associated with the kinetosome of trypanosomes.

kinetosome   a self-duplicating organelle homologous to the centriole. Kinetosomes reside at the base of cilia and flagella and are responsible for their formation.

kinety   a row of interconnected kinetosomes on the surface of a ciliate.

king-of-the-mountain principle   *See* **first-arriver principle.**

kin selection   *See* **Hamilton's genetical theory of social behavior.**

Kjeldahl method   a technique often used for the quantitative estimation of the nitrogen content of biological material.

Kleinschmidt spreading technique   A procedure developed by A. K. Kleinschmidt that allows DNA molecules to be viewed under the electron microscope. The DNA is mounted in a postitively charged protein film formed on the surface of an aqueous solution. The film of protein serves to hold the DNA in a relaxed but extended configuration and allows the sample to be transferred to a hydrophobic electron microscope grid when it is touched to the surface film.

Klenow fragment   a portion of bacterial DNA polymerase I derived by proteolytic cleavage; it lacks the 5′ to 3′ exonuclease activity of the intact enzyme.

Klinefelter syndrome   a genetic disease that produces sterile males with small testes lacking sperm. The condition, which is sometimes associated with mental retardation, is due to the XXY AA karyotype.

Km   the Michaelis constant (*q.v.*).

knife breaker   a mechanical apparatus that provides a method for breaking strips of plate glass first into squares and then into triangles. These are used as knives against which plastic-embedded tissues are cut into ultrathin sections for observation under the electron microscope. *See* **ultramicrotome.**

knob   in cytogenetics, a heavily staining enlarged chromomere that may serve as a landmark, allowing certain chromosomes to be identified readily in the nucleus. In maize, knobbed chromatids preferentially enter the outer cells of a linear set of four megaspores during megasporogenesis and are therefore more likely to be included in the egg nucleus (*see* **meiotic drive**); genetic markers close to a knob tend to appear more frequently in gametes than those far from a knob.

214

Kornberg enzyme the DNA polymerase isolated from *E. coli* in 1959 by a group led by A. Kornberg; now called *DNA polymerase I;* it functions mainly in repair synthesis (*q.v.*).

Krebs' cycle critic-acid cycle. (*q.v.*).

Krebs'-Henseleit cycle ornithine cycle (*q.v.*).

K selection *See* **r and K selection theory.**

Kupffer cells liver macrophages.

kurtosis the property of a statistical distribution that produces a steeper or shallower curve than a normal distribution (*q.v.*) with the same parameters.

kuru a chronic, progressive, degenerative disorder of the central nervous system found in the Fore natives living in a restricted area of New Guinea. The disease was at one time thought to be genetically determined, but it is now believed to be caused by a prion (*q.v.*).

kV kilovolt.

kwashiorkor a severe nutritional disorder due to a deficiency of certain amino acids (especially lysine). Kwashiorkor occurs in humans that subsist on a diet of cereal proteins deficient in lysine.

L

l **1.** line **2.** levorotatory. **3.** liter.

label, electron dense *See* **ferritin.**

label, heavy a heavy isotopic element introduced into a molecule to facilitate its separation from otherwise identical molecules containing the more common isotope.

label, radioactive a radioactive atom introduced into a molecule to facilitate the study of its metabolic transformations.

labeled compound a compound containing a radioactive label. The compound or its breakdown products may be followed through a series of biological reactions by detecting its radioactivity.

***lac* operon** in *E. coli* a DNA segment about 6000 bp long containing an operator gene and the structural genes *lac Z, lac Y,* and *lac A.* The structural genes code for β-galactosidase, β-galactoside permease, and β-galactoside transacetylase, respectively. The three structural genes are transcribed into a single mRNA from a promotor lying to the left of the operator. Whether or not this mRNA is transcribed depends upon whether or not a repressor protein is bound to the operator. The repressor protein is encoded by *lac I,* a gene lying to the left of the *lac* promotor. β-Galactosidase (*q.v.*) catalyzes the hydrolysis of lactose (*q.v.*) into glucose and galactose. After glucose and galactose are produced, a side reaction occurs forming allolactose. This is the inducer that switches on the *lac* operon. It does so by binding to the repressor and inactivating it. *See* **IPTG, ONPG.**

lactic dehydrogenase *See* **isozyme.**

lactogenic hormone a protein hormone secreted by the anterior lobe of the pituitary which stimulates milk production in mammals and broodiness in birds.

lactose 4-(β-D-galactoside)-D-glucose. A disaccharide made up of two hexoses joined by a β-galactoside linkage. It is split into galatose and glucose by the enzyme β-galactosidase. Lactose differs from allolactose in that in lactose the galactose and glucose moieties are joined by a 1-4 linkage, whereas in allolactose the linkage is 1-6. As its name implies, lactose is abundant in the milk of mammals. *See* ***lac*** operon.

lacuna a space within a cell or tissue.

lagging delayed movement from the equator to the poles at anaphase of a chromosome so that it becomes excluded from the daughter nuclei.

216

lactose

$6\ CH_2OH$ ... $6\ CH_2OH$

(structural formula of lactose)

lagging strand the discontinuously synthesized strand of DNA containing ligated Okazaki fragments (*q.v.*). *See* **leading strand, replication of DNA.**

lag growth phase a period of time in the growth of a population during which little or no increase in the number of organisms occurs. The lag period precedes the exponential growth phase (*q.v.*).

Lamarckism a historically important, but no longer credited, theory (advanced in 1809 by J. B. Lamarck) that species can change gradually into new species by the willful striving of organisms to meet their own needs, together with the cumulative effects of use and disuse of body parts. All such acquired characteristics were thought to become part of the individual's heredity and as such could be transmitted to their offspring; otherwise known as the inheritance of acquired characteristics.

lambda (λ) microliter (μl). The volume contained in a cube of side 1 mm. One μl of water weighs 1 mg.

lambda cloning vector a lambda phage that is genetically engineered to serve as a receptor for foreign DNA fragments in recombinant DNA experiments. Vectors that have a single target site at which foreign DNA is inserted are called *insertion vectors,* those having a pair of sites that span a DNA segment that can be exchanged with a foreign DNA fragment are called *replacement* or *substitution vectors.*

lambda d *gal* (λd*gal*) a lambda (λ) phage carrying a gene for galactose fermentation (*gal*) and also defective (d) for some phage function (usually lacking genes for making tails).

lambda phage a double-stranded DNA virus that infects *E. coli.* Once inside the host cell, the lambda genome can enter a lysogenic cycle or a lytic cycle of replication. Which pathway is chosen depends upon an intricate balance of host and viral factors.

lamellipodia extensive, lamellar cellular projections involved in attachment of eukaryotic cells such as fibroblasts to solid surfaces. Lamellipodia mark the forward edges of moving cells such as macrophages and are also called *ruffled edges.*

lampbrush chromosome a chromosome characteristic of the primary oocytes of vertebrates. Those found in the oocytes of salamanders are the largest known chromosomes, and they are generally studied in diplonema. They have a rather fuzzy appearance when viewed at low magnification with a light microscope, and this is due to hundreds of paired loops that extend laterally from the main axis of each chromosome. The loops are active in RNA transcription.

Lampyris a genus of beetles commonly called fireflies. The light signals sent by male fireflies serve as a visual isolating mechanism (*q.v.*).

Laplacian curve. *See* **normal distribution.**

large angle x-ray diffraction a technique for the analysis of the small distances between individual atoms. *See* **x-ray crystallography, small angle x-ray diffraction.**

Laron dwarfism *See* **pituitary dwarfism.**

lampbrush
chromosome

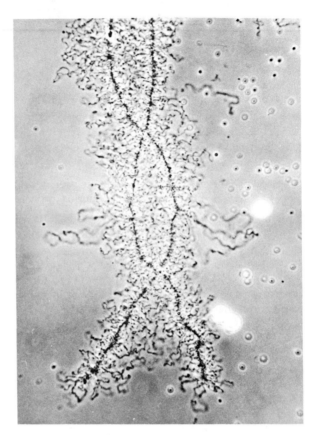

Phase contrast photomicrograph of a lampbrush chromosome of *Notophthalmus viridescens*.

larva   the pre-adult form in which some animals hatch from the egg. A larva is capable of feeding, though usually on a diet different from the adult, and is usually incapable of sexual reproduction.

larviparous   depositing larvae rather than eggs. Fertilized eggs develop internally up to the larval stage. The female then lays these larvae. Some blowflies, for example, are larviparous.

laser   an electronic device that generates and amplifies light waves coherent in frequency and phase in a narrow and extremely intense beam of light. The word is an acronym for *l*ight *a*mplification by *s*timulated *e*mission of *r*adiation. Microlaser beams are sometimes used to perform microcautery in experimental studies of cell division or morphogeneses.

laser microprobe   a technique that uses a laser beam focused by a microscope to vaporize a minute tissue area. The vapor is then analyzed spectrographically.

late genes   genes expressed late in the life cycle. In T_4 bacteriophage, those genes responsible for making capsid proteins, lysozyme, and other proteins after the initiation of phage DNA replication. *Compare with* **early genes.**

latent image   the pattern of changes occurring in a surface of silver halide crystals when they absorb photons (from light) or ionizing particles. The pattern may be developed into a photographic image by chemical treatment.

lateral   designating, located on, or extending toward or from the left or right side.

218

lateral element *See* **synaptonemal complex.**

late replicating X chromosome. in the mammalian somatic cell nucleus all X chromosomes but one coil up into a condensed mass (the Barr body or sex chromatin body) and do not function in transcription. Such X chromosomes complete their replication later than the functional X and the autosomes. *See* **Barr body, Lyon hypothesis, sex chromatin.**

Latimeria chalumnae a present-day coelacanth, popularly called a "living fossil."

Latin square (*also* **Roman square**) a set of symbols arranged in a checkerboard in such a fashion that no symbol appears twice in any row or column. The Latin square, long a mathematical curiosity, was discovered to be useful for subdividing plots of land for agricultural experiments so that treatments could be tested even though the field had soil conditions which might vary in an unknown fashion in different areas. The technique required that the field be subdivided by a grid into subplots and the differing treatments be performed at consecutive intervals to plants from different subplots. Thus, if the subplots are A, B, C, and D the experiments could be run as follows:

treatment	day 1	day 2	day 3	day 4
1	A	B	C	D
2	B	C	D	A
3	C	D	A	B
4	D	A	B	C

latitudinal parallel to the equator.

lattice a structure composed of elements arranged in a geometrical pattern with spaces between them.

lawn a continuous layer of bacteria on the surface of an agar plate.

law of parsimony *See* **Occam's razor.**

lazy maize a maize mutant characterized by a stalk that grows flat on the ground like a vine.

L cell *See* **mouse L cells.**

L chain *See* **immunoglobulin.**

LD50 the radiation dose required to kill half of a population of organisms within a specified time. Synonomous with median lethal dose.

LDH lactate dehydrogenase. *See* **isozyme.**

leader protein *See* **leader sequence peptide.**

leader sequence an untranslated segment of mRNA from its 5' end to the start codon. The leader may contain regulatory signals such as an attenuator region (*q.v.*) or ribosome binding sites.

leader sequence peptide a segment containing 16-20 amino acids that is located at the extreme amino terminus of protein molecules that are destined to enter the cysternal lumen of the endoplasmic reticulum. This leader sequence always contains hydrophobic amino acids that become embedded in the lipid bilayer membrane, and it functions to guide the nascent protein to a receptor protein which marks the position of a pore in the ER membrane. Once the protein passes into the cysternal lumen through the pore, the leader segment is cleaved from the protein. For example, the leader sequence peptide of an interferon protein allows the cell to secrete the interferon, but is removed from the mature molecule during the secretion process.

leading strand the DNA strand synthesized with few or no interruptions; as opposed to the *lagging strand,* which is produced by ligation of Okazaki fragments (*q.v.*). The leading strand is synthesized 5′ to 3′ toward the replication fork, whereas the lagging strand is synthesized 5′ to 3′ away from the replication fork. *See* **replication of DNA.**

leaky gene hypomorphic gene

leaky protein a mutant protein that has a subnormal degree of biological activity.

least squares, method of a method of estimation based on the minimization of sums of squares. *See* **line of best fit.**

Lebistes reticularis the guppy, a well-known tropical aquarium fish. The genetic control of sexuality has been extensively studied in this species.

lectins proteins capable of agglutinating certain cells, especially erythrocytes, by binding to specific carbohydrate receptors on the surfaces of these cells. This agglutination behavior resembles antibody-antigen reaction but is not a true immunological reaction. Those lectins extracted from plant seeds have been called phytohemagglutinins. However, lectins have also been isolated from sources as varied as bacteria, snails, horseshoe crabs, and eels. *See* **concanavalin A, phytohemagglutinin, pokeweed mitogen.**

left splicing junction the boundary between the left (5′) end of an intron and the right (3′) end of an adjacent exon in nRNA; also termed the donor splicing site.

leghemoglobin an oxygen-binding protein found in the root nodules of legumes. This molecule is very similar to vertebrate hemoglobin, and the genes that code for these proteins also show close homologies in legumes and vertebrates. How legumes got their globin-like genes remains a mystery.

lek a specific location where males of a species (especially birds) aggregate for courtship displays to females.

leptonema *See* **meiosis.**

leptotene stage *See* **meiosis.**

Lesch-Nyhan syndrome the most common human hereditary defect in purine metabolism, due to a recessive gene on the long arm of the X chromosome. Hemizygotes lack hypoxanthine-guanine-phosphoribosyltransferase (*q.v.*). The disease is characterized by excessive production of uric acid, developmental and mental retardation, and death before sexual maturity. As a result of random X-chromosome inactivation (*see* **dosage compensation**), female heterozygotes are mosaics. About 60% of their cloned fibroblasts show HGPRT activity, whereas 40% have no detectable activity. The disease is transmitted via heterozygous mothers. Prevalence is $\frac{1}{10,000}$ males.

LET linear energy transfer.

lethal equivalent value the average number of recessive deleterious genes in the heterozygous condition carried by each member of a population of diploid organisms times the mean probability that each gene will cause premature death when homozygous. Thus a genetic burden of eight recessive semilethals each of which, when homozygous, produced only a 50% probability of premature death would be classified as a burden of four "lethal equivalents."

lethal mutation a mutation which results in the premature death of the organism carrying it. Dominant lethals kill heterozygotes, whereas recessive lethals kill only homozygotes. *See* **aphasic lethal, monophasic lethal, polyphasic lethal.**

leu leucine. *See* **amino acid.**

leucine *See* **amino acid.**

220

leucocyte   a variant spelling of leukocyte (*q.v.*).

leukemia   a generally fatal disease characterized by an overproduction of white blood cells, or a relative overproduction of immature white cells. Leukemia is a common disease in the cow, dog, cat, mouse, guinea pig, and chicken, as well as man. Many virus-induced leukemias are known in the mouse and chicken. *See* **retrovirus.**

leukocyte   a white blood cell.

leukopenia   a decrease in the number of white blood corpuscles.

leukoplasts   colorless plastids of tubers, endosperm, and cotyledons.

leukosis   proliferation of leukocyte-forming tissue; leukemia (*q.v.*).

leukoviruses   *See* **retroviruses.**

lev   levorotatory. *See* **optical isomerism.**

levulose   fructose.

Lewis blood group   a blood group determined by an antigen specified by the *Le* gene on human chromosome 19. The antigens are present in body fluids and are absorbed to red blood cells only secondarily. *See* **blood groups, Lutheran blood group.**

Lex A repressor   *See* **SOS response.**

Leydig epidermal cell   a cell found in the epidermis and having the characteristics of a macrophage.

L forms   bacteria that have lost their cell walls. *See* **protoplast.**

LH   luteinizing hormone (*q.v.*).

library   *See* **genomic library.**

lichen   a composite plant consisting of a fungus and an alga living in symbiosis.

life cycle   the series of developmental changes undergone by an organism from fertilization to reproduction and death.

life history strategies   evolutionary adaptations in a biological lineage involving the timing of reproduction, fecundity, longevity, etc.

ligand   a molecule that will bind to a complementary site on a given structure. For example, oxygen is a ligand for hemoglobin and a substrate of an enzyme molecule is a specific ligand of that molecule.

ligase   *See* **DNA ligase.**

ligation   formation of a phosphodiester bond to join adjacent nulceotides in the same nucleic acid chain (DNA or RNA).

ligature   a thread for tying off or constricting a vessel or body region. Generally used to interrupt blood flow.

light chain   *See* **heavy chain.**

Lilium   the genus containing *L. longiflorum,* the Easter lily and *L. tigrinum,* the tiger lily, favorite species for cytological and biochemical studies of meiosis.

limited chromosome   a chromosome that occurs only in nuclei of the cells of the germ line and never in somatic nuclei. *See* **chromosome diminution.**

Limnea peregra   a fresh-water snail upon which were first performed classical studies on the inheritance of the direction of the coiling of the shell. The trait showed delayed Mendelian segregation, since the phenotype of the snail is determined by the genotype of the maternal parent.

line a homozygous, pure-breeding group of individuals that are phenotypically distinctive from other members of the same species.

lineage a linear evolutionary sequence from an ancestral species through all intermediate species to a particular descendant species.

linear accelerator *See* **accelerator.**

linear energy transfer the energy, in electron volts, dissipated per micron of tissue traversed by a particular type of ionizing particle.

linear regression regression line (*q.v.*).

linear tetrad a group of four meiotic products aligned linearly in such a way that sister products remain adjacent to one another. Ascospores show this order because the confines of the fungal ascus prevent nuclei from sliding past one another. *See* **ascus.**

line of best fit a straight line that constitutes the best moving average for a linear group of observed points. This requires that the sum of the squares of the deviations of the observed points from the moving average be a minimum.

linkage the greater association in inheritance of two or more non-allelic genes than is to be expected from independent assortment. Genes are linked because they reside on the same chromosome.

linkage disequilibrium nonrandom association of linked genes in the gametes of a population; the tendency for certain alleles of one locus to occur with certain alleles of another locus on the same chromosome with frequencies greater than that attributed to chance alone.

linkage group the group of genes having their loci on the same chromosome.

linkage map a chromosome map showing the relative positions of the known genes on the chromosomes of a given species.

linked genes *See* **linkage.**

linker DNA **1.** a short, synthetic DNA duplex containing the recognition site for a specific restriction endonuclease. Such a linker may be connected to ends of a DNA fragment prepared by cleavage with some other enzyme. **2.** a segment of DNA to which histone H1 is bound. Such linkers connect the adjacent nucleosomes of a chromosome.

linking number the number of times that the two strands of a closed-circular, double-helical molecule cross each other. The *twisting number* (T) of a relaxed closed-circular DNA is the total number of base pairs in the molecule divided by the number of base pairs per turn of the helix. For relaxed DNA in the normal B form, L is the number of base pairs in the molecule divided by ten. The *writhing number* (W) is the number of times the axis of a DNA molecule crosses itself by supercoiling. The linking number (L) is determined by the formula: $L = W + T$. For a relaxed molecule, $W = 0$, and $L = T$. The linking number of a closed DNA molecule cannot be changed except by breaking and rejoining of strands. The utility of the linking number is that it is related to the actual enzymatic breakage and rejoining events by which changes are made in the topology of DNA. Any changes in the linking number must be by whole integers. Molecules of DNA that are identical except for their linking numbers are called *topological isomers.*

lipase an enzyme that breaks down fats to glycerol and fatty acids.

lipid any of a group of biochemicals which are variably soluble in organic solvents like alcohol and barely soluble in water (fats, oils, waxes, phospholipids, sterols, carotenoids, etc.)

lipid bilayer model a model for the structure of cell membranes based upon the hydrophobic properties of interacting phospholiplids. The polar head groups face outward to the solvent, whereas the hydrophobic tails face inward. Proteins are embedded in the bilayer, sometimes being exposed on the outer surface, sometimes on the inner surface, and sometimes penetrating both surfaces. The proteins exposed on the outer surfaces of cells commonly serve as distinctive antigenic markers. Membrane proteins may serve a variety of functions such as communication, energy transduction, and transport of specific molecules across the membrane. *See* **fluid mosaic concept.**

lipopolysaccharide the active component of bacterial endotoxins associated with the cell walls of many Gram-negative species; a B-cell mitogen in some animal species.

liposomes single- or multilaminar vesicles (made from lecithins and other lipids) that can function as artificial membrane systems for delivery of polynucleotides and genetic materials to cells, with a resultant stabilized expression of new genes in recipient cells.

lipovitellin lipoprotein of molecular weight 400,000 found in amphibian yolk platelets. *See* **phosphovitin.**

liquid-holding recovery a special form of dark reactivation (*q.v.*) in which repair of ultraviolet damage to DNA is enhanced by delaying bacterial growth and DNA replication through post-irradiation incubation of cells in a warm, nutrient-free buffer for several hours before plating them on nutrient agar.

liquid hybridization formation of double helical nucleic acid chains (DNA with DNA, DNA with RNA, or RNA with RNA) from complementary single strands in solution. *Compare with* **filter hybridization.**

liquid scintillation counter an electronic instrument for measuring radioisotopes dissolved in a solvent containing a fluorescent chemical that emits a flash of light (a scintillation) when struck by an ionizing particle or photon of electromagnetic radiation. The flash is captured by a photomultiplier tube, transformed into an electric pulse, amplified, routed through a scaler, and counted.

liter standard unit of capacity in metric system, equal to 1 cubic decimeter.

litter animals of one multiple birth.

littoral pertaining to the shore.

Liturgosa a genus of mantid. The evolution of karyotype has been extensively studied in this genus.

locus (*plural,* **loci**) the position that a gene occupies in a chromosome.

Locusta migratoria the Old World "plague" locust.

logarithm the exponent of the power to which a fixed number must be raised to produce a given number. The fixed number is called the *base*. In Briggsian logarithms, the base is 10; while in Naperian, or natural, logarithms, the base is 2.7183818 . . ., a number designated by the symbol *e.*

logarithmic phase   the growth stage during which organisms are doubling their number each time a specified period of time elapses.

long-day plant   a plant in which the flowering period is initiated and accelerated by a daily exposure to light exceeding 12 hours.

longitudinal   meridional.

long period interspersion   a genomic pattern in which long segments of moderately repetitive and nonrepetitive DNA sequences alternate.

long terminal repeats (LTRs)   domains of several hundred base pairs at the ends of retroviral DNAs. LTRs may provide functions fundamental to the expression of most eukaryotic genes (e.g., promotion, initiation, and polyadenylation of transcripts).

lophophore   a circular crescentic, or double spirally coiled ridge bearing ciliated tentacles found in species belonging to the Tentaculata (*q.v.*).

low-energy phosphate compound   a phosphorylated compound yielding relatively little energy upon hydrolysis.

LSD   lysergic acid diethylamide (*q.v.*).

LTH   lactogenic hormone (*q.v.*).

Lucilia cuprina   the Australian blowfly, a major pest of the sheep industry in Australia. Good polytene chromosome preparations can be obtained from pupal trichogen cells (*q.v.*), and extensive cytogenetic studies have been made of chromosome rearrangements. The concept of pest control through the introduction into the field of chromosomally altered strains whose descendants manifest sterility has been tested in this species.

Lucké virus   a virus causing renal cancer in frogs.

luminescence   light emission that cannot be attributed to the temperature of the emitting body. The light energy emitted may result from a chemical reaction going on within the emitter or it may be initiated by the flow of some form of energy into the body from the outside. The slow oxidation of phosphorus at room temperature is an example of the former, while luminescence resulting from electron bombardment of gaseous atoms in a mercury vapor lamp is an example of the latter. *Fluorescence* is defined as a luminescence emission which continues after the source of exciting energy is shut off. This afterglow is temperature independent. In the case of *phosphorescence* the afterglow duration becomes shorter with increasing temperature.

lupus erythematosis   a connective tissue disorder characterized by autoantibody production against cellular components that are abundant and highly conserved. Among the antibodies produced by patients with lupus are those that recognize U1 RNA (*q.v.*).

luteinizing hormone   a glycoprotein hormone stimulating ovulation, growth of the corpus luteum and secretion of estrogen. LH is secreted by the adenohypophysis of vertebrates. Identical to ICSH.

luteotropin   lactogenic hormone (*q.v.*).

Lutheran blood group   a blood group determined by a red cell antigen specified by the *Lu* gene on human chromosome 19. Autosomal linkage in man was first demonstrated between *Lu* and *Le. See* **blood groups, Lewis blood group.**

224

luxury genes genes coding for specialized (rather than "household") functions. Their products are usually synthesized in large amounts only in particular cell types (e.g., hemoglobin in erythrocytes; immunoglobulins in plasma cells).

Luzula a genus of plants containing the wood rushes. Many species in this genus have diffuse centromeres. *See* **centromere.**

Lycopersicon esculentum the cultivated tomato. A vast wealth of genetic information exists for this species. Over 1000 mutant genes are known, and both cytological and genetic mapping are far advanced.

Ly **loci** a group of genes that determine antigens found on various subclasses and developmental stages of T and B lymphocytes.

Lymantria dispar the Gypsy moth. Classical studies on sex determination were performed on this species.

lymphatic tissue tissues in which lymphocytes are produced and/or matured, including the thymus, spleen, lymph nodes and vessels, and the bursa of Fabricius (*q.v.*).

lymphoblast the larger, cytoplasm-rich cell that differentiates from antigenically stimulated T lymphocytes.

lymphocyte a mononuclear, white blood cell with a round nucleus, present in lymph nodes, spleen, thymus, bone marrow, and blood. Lymphocytes that mature within a microenvironment of the bone marrow in mammals (or within the bursa of Fabricius in birds) are called B cells, and after contact with antigen they proliferate and differentiate into antibody-secreting plasma cells. Lymphocytes that mature within a microenvironment of the thymus gland are designated T cells. Upon antigenic stimulation, these cells secrete lymphokines (*q.v.*) that affect the activity of other host cells rather than interacting directly with antigens. Several subclasses of T cells have been recognized, including T_H (helper) cells that aid B cells to mature into antibody-secreting plasma cells, T_S (suppressor) cells that diminish the T or B cell responses, T_C (cytotoxic) cells that attack virally infected or cancerous cells, and T_{DH} cells that participate in delayed hypersensitivity responses.

lymphocytic choriomeningitis virus a natural infectious virus of mice often used in studies of immune responses to viral infection.

lymphokines a heterogeneous group of glycoproteins (molecular weights 10,000–200,000) released from T lymphocytes after contact with a cognate antigen. Lymphokines affect other cells of the host rather than reacting directly with antigens. Various lymphokines serve several major functions: (1) recruitment of uncommited T cells; (2) retention of T cells and macrophages at the site of reaction with antigen; (3) amplification of "recruited" T cells; (4) activation of the retained cells to release lymphokines; and (5) cytotoxic effects against cells bearing foreign antigens (including foreign tissue grafts and cancer cells). Examples of lymphokines produced by T cells following *in vitro* antigenic stimulation are as follows: migration inhibition factor (MIF) prevents migration of macrophages; lymphotoxins (LT) kill target cells; mitogenic factor (MF) stimulates lymphocyte division; interleukin 2 (IL2) is required for T helper activity; interferons (IFN) promote antiviral immunity; chemotactic factors attract neutrophils, eosinophils, and basophils; macrophage activation factor (MAF) activates macrophages.

lymphoma cancer of lymphatic tissue.

Lyon hypothesis *See* **dosage compensation.**

Lyonization a term used to characterize heterozygous females that behave phenotypically as if they were carrying an X-linked recessive in the hemizygous condition. An example would be a hemophilic mother that has produced non-hemophilic sons. It is assumed, according to the Lyon hypothesis, that occasionally a particular tissue comes to be made up entirely of cells that contain inactivated X chromosomes carrying the normal allele. The phenotype of such a female heterozygote would then resemble that of mutant males.

lyophilize   to render soluble by freeze drying.

lys   lysine. *See* **amino acid.**

lysate   a population of phage particles released from host cells via the lytic cycle.

Lysenkoism   a school of pseudoscience that fluorished in the Soviet Union between 1932 and 1965. Its doctrines were advanced by T. D. Lysenko (1898–1976), who did not accept the gene concept and believed in the inheritance of acquired characteristics.

lysergic acid diethylamide (LSD)   a serotonin antagonist that induces a schizophrenic-like state in humans.

lysine   *See* **amino acid.**

lysis   *See* **cell lysis.**

lysochrome   a compound that colors lipids by dissolving in them. *See* **Nile blue, Sudan black B.**

lysogen   a lysogenic bacterium.

lysogenic bacterium   one carrying a temperate virus in the prophage state.

lysogenic conversion   the change in phenotype of a bacterium (in terms of its morphology or synthetic properties) accompanying lysogeny (*q.v.*).

lysogenic immunity   a phenomenon in which a prophage prevents another phage of the same type from becoming established in the same cell.

lysogenic repressor   a phage protein responsible for maintenance of a prophage and lysogenic immunity.

lysogenic response   the response following infection of a nonlysogenic bacterium with a temperate phage. The infecting phage does not multiply but rather behaves as a prophage. *See* **lytic response.**

lysogenic virus   a virus that can become a prophage.

lysogenization the experimental production of a lysogenic strain of bacteria by exposing sensitive bacteria to a temperate phage.

lysogenized bacterium a bacterium harboring an experimentally introduced, temperate phage.

lysogeny the phenomenon in which genetic material of a virus and its bacterial host are integrated.

lysosomal diseases hereditary diseases in man arising from a deficiency of enzymes located in lysosomes. Fabry, Gaucher, Niemann-Pick, Pompe, and Wolman diseases arise from lysosomal enzyme deficiencies.

lysosome a membrane-enclosed intracellular vesicle that acts as the primary component for intracellular digestion in all eukaryotes. Lysosomes are known to contain at least 50 acid hydrolases, including phosphatases, glycosidases, proteases, sulfatases, lipases, and nucleases. Collectively they can hydrolyze all classes of macromolecules.

lysozyme an enzyme digesting mucopolysaccharides. Lysozymes having a bacteriolytic action have been isolated from diverse sources (tears, egg white, etc.). An important lysozyme is the enzyme synthesized under the direction of a phage which digests the cell wall of the host from within and thus allows the escape of the phage progeny.

lytic response lysis following infection of a bacterium by a virulent phage, as opposed to lysogenic response (*q.v.*).

lytic virus a virus whose intracellular multiplication leads to lysis of the host cell.

M

M   molar solution.

M13   a single-stranded bacteriophage cloning vehicle, with a closed circular DNA genome of approximately 6.5 kb. The major advantage of using M13 for cloning is that the phage particles released from infected cells contain single-stranded DNA that is homologous to only one of the two complementary strands of the cloned DNA, and therefore it can be used as a template for DNA sequencing analysis.

Macaca mulatta   the rhesus monkey. The original source of the Rh antigen.

macaroni wheat   *Triticum durum* (N = 14).

macha wheat   *Triticum macha* (N = 21).

macroconidia   *See* **conidia**.

macroevolution   a large evolutionary pattern usually viewed through the perspective of geologic time, such as the evolution of the horse from *Eohippus* to *Equus*. The evolutionary changes in taxonomic categories above the species level, and the evolutionary changes that bring about the origin of a new higher taxon.

macromolecule   a molecule of molecular weight ranging from a few thousand to hundreds of millions (proteins, nucleic acids, polysaccharides, etc.).

macromutation   *See* **evolution**.

macronucleus   the larger of the two types of nuclei in the ciliates; the "vegetative" nucleus. Macronuclei contain many copies of each gene and are transcriptionally active. *See* **micronucleus**.

macrophage   a large, phagocytic, mononuclear leukocyte found in tissues, but derived from blood monocytes. Macrophages are called histiocytes in connective tissues, Kupffer's cells in the liver, Leydig cells in the skin, microglial cells in the nervous system, and alveolar macrophages in the lung. To stimulate an immune response, most antigens must be "processed" by macrophages and presented on their surfaces to lymphocytes in association with self-Ia molecules. *See* **major immunogene complex**.

macrophage activation factor   a lymphokine (*q.v.*). that actives macrophages.

macroscopic   visible to the unaided eye.

macula adherans   desmosome (*q.v.*).

maggot   a worm-like larva of certain flies, with no appendages or eyes.

magnesium   an element universally found in small amounts in tissues. Atomic number 12; atomic weight 24.312; valence 2^+; most abundant isotope ^{24}Mg, radioisotope ^{28}Mg, half life 21 hrs, radiation emitted—beta particle.

major gene   a gene with pronounced phenotypic effects, in contrast to its modifiers.

major histocompatibility complex (MHC)   a large cluster of genes on human chromosome 6 and on mouse chromosome 17. The MHC controls many activities of immune cells, including the transplantation rejection process and the killing of virus-infected cells by specific killer T lymphocytes. The MHC is part of a larger *major immunogene complex* (*q.v.*) with more diverse functions. In different mammals different symbols have been assigned to the MHC, for example: chicken (B), dog (DLA), guinea pig (GPLA), human (HLA) mouse (H-2), and rat (Rt-1).

major immunogene complex (MIC)   a genetic region containing loci coding for lymphocyte surface antigens (e.g., Ia), histocompatibility (H) antigens, immune response (Ir) gene products, and proteins of the complement system. The genes specifying immunoglobulins assort independently of the MIC, but the plasma cells responsible for their production are under the control of the MIC.

MAK column   *See* **methylated albumen-Kieselguhr column.**

malaria   the single most critical infectious disease of mankind. It affects about 200 million people annually and causes death in about 2 million. The disease is caused by protozoans of the genus *Plasmodium* transmitted to man by female mosquitoes of the genus *Anopheles.* The most dangerous form of malaria is subtertian malaria caused by *P. falciparum.* Benign tertian malaria and quartan malaria are caused by *P. vivax* and *P. malariae,* respectively. *See* **Duffy blood group system, gluclose-6-phosphate dehydrogenase deficiency, sickle cell trait.**

male gametophyte   *See* **pollen grain.**

male pronucleus   the generative nucleus of a male gamete.

male symbol ♂   The zodiac sign for Mars, the Roman god of war. The sign represents a shield and spear.

maleuric acid $C_5H_6N_2O_4$, a mitotic poison.

malignancy   a cancerous growth.

Maloney leukemia virus   a retrovirus of mice producing lymphocytic leukemia. It can be transmitted from an infected mother to newborn progeny through her milk.

Malpighian tubule   the excretory tubule of insects which opens into the anterior part of the hind gut.

Malthusian   having to do with the theory advanced by the English social economist, T. R. Malthus, in his *An Essay on the Principle of Population,* published in 1798. According to this theory the world's population tends to increase faster than the food supply, and poverty and misery are inevitable unless this trend is checked by war, famine, etc.

Malthusian parameter   the rate at which a population with a given age distribution and birth and death rate will increase.

mammary tumor agent   a milk-borne retrovirus that induces mammary cancer in mice of the appropriate genotype.

man   *See Homo sapiens,* **human mitotic chromosomes.**

Mandibulata   a subphylum of arthropods containing those species possessing antennae and a pair of mandibles. *See* **classification.**

manganese   a biological trace element. Atomic number 25; atomic weight 54.9380; valences 2^+, 3^+, 4^+, 6^+, 7^+; most abundant isotope ^{55}Mn; radioisotope ^{57}Mn, half life 7 days, radiation emitted—beta particles.

manic depressive psychosis   a mental disorder characterized by periods of overelated hyperactivity alternating with depression, anxiety, and even stupor. The condition is associated with a dominant gene on the short arm of the X chromosome.

manometer   a device used to measure the pressure of a gas. The Warburg manometer is an instrument used in physiological measurements of the gas evolved or consumed by cells.

mantid   an insect belonging to the orthopteran family Mantidae of the Dictyoptera. *See* **classification.**

mantissa   the decimal part of a Briggsian logarithm. For example, in the logarithm 3.728539, the mantissa is .728539.

Manx syndrome   taillessness in domestic cats. The condition is inherited as an autosomal dominant. Homozygotes die *in utero.*

map   *See* **genetic map.**

map distance   the distance between genes expressed as map units or centiMorgans (cM).

mapping function   a mathematical formula developed by J. B. S. Haldane that relates map distances to recombination frequencies. The function is represented graphically below. It demonstrates that no matter how far apart two genes are on a chromosome, one never observes a recombination value greater than 50%. It also shows that the relation between recombination frequencies and map distances is linear for genes that recombine with frequencies less than 10%.

map unit   a measure of genetic distance between two linked genes corresponding to a recombination frequency of 1% or 1 centimorgan (cM). *See* **Morgan unit.**

mare   a female horse.

Marfan syndrome   a human hereditary disorder of connective tissue inherited as an autosomal dominant. The disease is characterized by excessively long fingers and toes and is often called arachnodactyly ("spider fingers") for this reason. Prevalence ½₀,₀₀₀.

marker   **1.** a gene with a known location on a chromsome and a clear-cut phenotype, used as a point of reference when mapping a new mutant. **2.** antigenic markers serve to distinguish cell types (e.g., the Thy-1 molecules are expressed on T lymphocytes, but not on B lymphocytes). **3.** marker DNAs, RNAs, and proteins are fragments of known sizes and/or properties that are used to calibrate an electrophoretic gel.

marker rescue   the phenomenon found when bacteria are mixedly infected with two genetically marked phages, of which only one type is irridiated. The progeny phages are predominately of the unirradiated type. However, some recombinants are found which contain genes from the irradiated parent. Such genes have been "rescued" by recombination.

marsupial   any of a group of primitive mammals, females of which have an abdominal pouch in which the young are carried and nourished (kangaroos, oppossums, etc.). *See* **classification.**

masked mRNA   messenger RNA that can be stored in large quantities because it is in some way inactivated and protected against digestion by nucleases (perhaps by being associated with proteins). Sea urchin eggs store mRNA in this masked form.

mass action, law of   the principle that the rate of a chemical reaction is proportional to the concentrations of the reacting substances, with each concentration raised to a power equal to the relative number of molecules of each species participating in the reaction.

mass number   the number of protons and neutrons in the nucleus of an atom (symbol, A).

mass spectrograph   an instrument for analyzing a substance in terms of the ratios of mass to charge of its components. It produces a focused mass spectrum of lines on a photographic plate.

mass unit   *See* **atomic mass unit.**

mast cells   tissue cells, thought to be the counterparts of blood basophils, that are abundant in lymph nodes, spleen, and bone marrow, in connective tissues, and in the skin. However, blood and lymph do not contain mast cells. The cytoplasm of mast cells contains granules rich in heparin, serotonin, and histamine. Mast cells possess receptors for IgE molecules. When antigen becomes attached to mast-cell-bound antibodies, these vasoactive amines are released and initiate an allergic reaction.

master-slave concept   the hypothesis advanced by H. G. Callan that each cistron is represented by numerous copies joined end to end. Within each serial gene there is one "master" gene copy (probably at the end), and the other gene copies in the same series are "slaves" which once per life cycle are corrected as necessary to conform to the master copy.

mate killers   *Paramecium aurelia* carrying mu particles. Such animals kill or injure sensitives with which they conjugate. The presence of mu in a paramecium protects it from the action of other mate killers.

maternal effect   a nonlasting influence of the genotype or phenotype of the mother upon the phenotype of the immediate offspring (i.e., phenotypic effects of stored material in the egg or maternally derived mRNAs transcribed by the F_1).

maternal inheritance   phenotypic differences controlled by cytoplasmic genetic factors (e.g., in mitochondria, long-lived mRNAs, viruses, or chloroplasts) derived solely from the maternal parent; also known as uniparental heredity, cytoplasmic inheritance, or extranuclear heredity.

mating   the union in pairs of individuals of opposite sexual type to accomplish sexual reproduction.

mating type many species of microorganisms can be subdivided into groups (or mating types) on the basis of their mating behavior; only individuals of different mating types will undergo conjugation. Individuals from a given mating type possess on their surfaces proteins that will bind to complementary proteins or polysaccharides found only on the coats of individuals of opposite mating type.

matrilinear inheritance the transmission of cytoplasmic particles only in the female line.

matroclinous inheritance *See* **matrocliny.**

matrocliny inheritance in which the offspring resembles the female parent more closely than the male. In *Drosophila* the daughters produced by attached-X females are matroclinous in terms of their sex-linked genes. *See* **hologynic inheritance.**

maturase a protein encoded by an exon-intron combination that helps catalyze the excision of the intron from its own primary transcript. A maturase is probably not an enzyme that catalyzes intron removal and exon splicing, but rather a factor that modifies the specificity of a preexisting splicing enzyme.

maturation divisions a series of nuclear divisions in which the chromosome complement of the nuclei is reduced from diploid to haploid number. *See* **meiosis.**

Maxam-Gilbert method *See* **DNA sequencing techniques.**

maximum permissible dose the greatest amount of ionizing radiation safety standards permit to be received by a person.

maze an experimental device consisting of a network of paths through which a test animal must find its way.

maze-learning ability the speed at which an animal learns to find its way through a maze without entering blind alleys.

mc millicurie. *See* **curie.**

McArdle disease a hereditary glycogen storage disease in humans with a frequency of 1/500,000. It arises from a deficiency of the enzyme, skeletal muscle glycogen phosphorylase, which is encoded by a gene on chromosome 11.

mean the sum of an array of quantities divided by the number of quantities in the group.

mean free path the average distance a subatomic particle travels between collisions.

mean square variance.

measurement units *See* Tables 1–3 below and on pp. 233 & 234.

Table 1. Metric measurement units with U.S. common equivalents

	Length				
Myriameter	10,000	meters	6.213	miles	(statute)
Kilometer	1,000	meters	0.621	mile	(statute)
Hectometer	100	meters	109.3	yards	
Dekameter	10	meters	10.93	yards	
Meter	1	meter	3.28	feet	
Decimeter	0.1	meter	3.937	inches	
Centimeter	0.01	meter	0.393	inch	
Millimeter	0.001	meter	0.039	inch	
Micron	0.000 001	meter	0.000 039	inch	

Table 1. Metric measurement units with U.S. common equivalents (continued)

Area

Hectare	10,000 square meters		2.471 acres
Are	100 square meters	119.6 square yards	0.025 acre

Volume

Kiloliter	1,000 liters	1 cubic meter	1.308 cubic yards	264.18 gallons
Hectoliter	100 liters	0.1 cubic meter	2.838 bushels	26.41 gallons
Dekaliter	10 liters	0.01 cubic meter	1.135 pecks	2.64 gallons
Liter	1 liter	0.001 cubic meter	61.02 cubic inches	1.05 quarts
Deciliter	0.1 liter	0.000 1 cubic meter	6.10 cubic inches	0.10 quart
Centiliter	0.01 liter	0.000 01 cubic meter	0.61 cubic inch	0.33 ounce
Milliliter	0.001 liter	0.000 001 cubic meter (1.0 cubic centimeter)	0.06 cubic inch	0.27 dram

Weight

Metric ton	1,000 kilograms	1 cubic meter water, 4°C	2,204.6 pounds
Myriagram	10,000 grams	1 dekaliter water, 4°C	22.05 pounds
Kilogram	1,000 grams	1 liter water, 4°C	2.20 pounds
Hectogram	100 grams	1 deciliter water, 4°C	3.52 ounces
Dekagram	10 grams	10 cubic centimeters water, 4°C	0.35 ounce
Gram	1 gram	1 cubic centimeter water, 4°C	15.43 grains
Decigram	0.1 gram	0.1 cubic centimeter water, 4°C	1.54 grains
Centigram	0.01 gram	10 cubic millimeters water, 4°C	0.15 grain
Milligram	0.001 gram	1 cubic millimeter water, 4°C	0.01 grain
Microgram	0.000 001 gram		0.000 015 grain

Table 2. U.S. common measurement units with metric equivalents

Inch	2.54 cm	Cord (128 cubic feet)	3.625 m³
Foot	30.48 cm	Pint, liquid, (16 oz)	0.473 liter
Yard	0.914 m	Quart, liquid, (32 oz)	0.946 liter
Fathom (2 yards)	1.829 m	Quart, dry, (2 pints)	1.101 liters
Rod (5½ yards)	5.029 m	Gallon, (4 quarts)	3.785 liters
Furlong (220-yards)	201.16 m	Peck, dry, (8 quarts)	8.809 liters
Mile, statute (1,760 yards)	1.609 km	Bushel, dry, (4 pecks)	35.24 liters
Mile, nautical (2,026 yards)	1.854 km	Pound, avdp (16 oz)	453.592 g
Square inch	6.452 cm²	Pound, troy (12 oz)	373.24 g
Square foot	0.092 m²	Ounce, avdp (16 drams)	28.35 g
Square yard	0.836 m²	Ounce, troy (480 grains)	31.103 g
Square rod	25.29 m²	Dram, avdp	1.177 g
Square mile (640 acres)	2.59 km²	Grain, troy	0.065 g
Acre	40.468 ares	Carat (precious stones)	0.200 g
Cubic inch	16.387 cm³	Ton, metric (2,204.6	
Cubic foot (1,728 cubic inches)	0.028 m³	pounds)	10⁶ g
Cubic yard (27 cubic feet)		Ton, long (2,240 pounds)	1,016 metric ton
Board foot (144 cubic inches)	0.0025 m³ 0.7646 m³	Ton, short (2,000 pounds)	0.907 metric ton

Table 3. Metric prefixes

Prefix	Symbol	Multiple
deka-	da	10
hecto-	h	10^2
kilo-	k	10^3
mega-	M	10^6
giga-	G	10^9
tera-	T	10^{12}
deci-	d	10^{-1}
centi-	c	10^{-2}
milli-	m	10^{-3}
micro-	μ	10^{-6}
nano-	n	10^{-9}
pico-	p	10^{-12}
femto-	f	10^{-15}
atto-	a	10^{-18}

mechanical isolation   reproductive isolation because of the incompatibility of male and female genitalia.

mechanistic philosophy   the point of view that holds life to be mechanically determined and explicable by the laws of physics and chemistry. Opposed to vitalism (*q.v.*).

meconium   the intestinal contents formed before birth, hatching, or eclosion.

medaka   *See Oryzias latipes.*

medial complex   *See* **synaptonemal complex.**

medial lethal dose   the dose of radiation required to kill 50% of the individuals in a large group of organisms within a specified period. Synonomous with LD50.

median   the middle value in a group of numbers arranged in order of size.

mediolateral axis   *See* **transverse axis.**

medium   the nutritive substance provided for the growth of a given organism in the laboratory.

medusoid stage   the free-swimming sexual (jelly-fish) stage in the life cycle of many coelenterates.

mega-   a prefix meaning one million. Used with measurement units of the metric system.

megakaryocyte   a large cell with a multilobed nucleus present in the bone marrow, but not circulating in the blood. Megakaryocytes bud off platelets (*q.v.*).

megasporangium   a spore sac containing megaspores.

megaspore   in angiosperms, one of four haploid cells formed from a megasporocyte during meiosis. Three of the four megaspores degenerate. The remaining megaspore divides to produce the female gametophyte or embryo sac (*q.v.*).

megaspore mother cell   the diploid megasporocyte in an ovule that forms haploid megaspores by meiotic division.

megasporocyte   megaspore mother cell.

megasporogenesis   the production of megaspores.

meiocyte   an auxocyte (*q.v.*).

meiosis   in most sexually reproducing organisms the doubling of the gametic chromosome number which accompanies syngamy is compensated for by a halving of the resulting zygotic chromosome number at some other point during the life cycle. These changes are brought about by a single chromosomal duplication followed by two successive nuclear divisions. The entire process is called meiosis, and it occurs during animal gametogenesis or sporogenesis in plants. The prophase stage is much longer in meiosis than in mitosis. It is generally divided into five consecutive stages: *leptonema, zygonema, pachynema, diplonema,* and *diakinesis.*

During the leptotene stage the chromosomes appear as thin threads with clearly defined chromomeres. The chromosomes are often all oriented with one or both of their ends in contact with one region of the nuclear membrane, forming the so-called *bouquet configuration.* Although each chromosome appears single, it is actually made up of two chromatids. However, the doubleness of the chromosome does not become obvious until pachynema. The DNA replication which doubles the diploid value occurs before leptonema.

In diploid somatic cells the 2N chromosomes are present as N pairs, and each chromosome is a replica of one contributed by the male or female parent at fertilization. In the somatic nuclei of most organisms the homologous chromosomes do not pair. However, during the zygotene stage of meiotic prophase *synapsis* of homologous chromosomes takes place. This pairing begins at a number of points and extends "zipper-like" until complete. Pairing is accompanied by the formation of synaptonemal complexes (*q.v.*). When synapsis is finished the apparent number of chromosome threads is half what it was before, and the visible bodies in the nucleus are now *bivalents* rather than single chromosomes.

During the pachytene stage each paired chromosome separates into its two component sister chromatids (except at the region of the centromere). As a result of the longitudinal division of each homologous chromosome into two chromatids there exist in the nucleus N groups of four chromatids lying parallel to each other called *tetrads*. A type of localized breakage followed by exchange between non-sister chromatids occurs. This process, called crossing over, is accompanied by the synthesis of an amount of DNA amounting to less than 1% of the total in the nucleus. The exchange between homologous chromatids results in the production of crossover chromatids containing genetic material of both maternal and paternal origin.

During the diplotene stage one pair of sister chromatids in each of the tetrads begins to separate from the other pair. However, the chromatids are prevented from separating at places where exchanges have taken place. In such regions the overlapping chromatids form a cross-shaped structure called a chiasma (pl. chiasmata). The chiasmata slip along laterally toward the ends of the chromatids with the result that the position of a chiasma no longer coincides with that of the original crossover. This *terminalization* proceeds until during diakinesis all the chiasmata reach the ends of the tetrads and the homologues can separate during anaphase.

The chromosomes coil tightly during diakinesis and so shorten and thicken to produce a group of compact tetrads lying well spaced out in the nucleus, often near its membrane. Terminalization is completed, and the nucleolus disappears.

The nuclear envelope disappears during division I and the tetrads are arranged at the equator of the spindle. The chromatids of a tetrad disjoin in such a way that there is separation of maternal from paternal chromosomal material with the exceptions of regions distal to where crossing over has occurred. Division I produces two *secondary gametocytes* which contain *dyads* surrounded by a nuclear envelope.

Division II commences after a short interphase during which the chromosomes do not uncoil. The nuclear membrane disappears, and the dyads arrange themselves upon the metaphase plate. The chromatids of each dyad are equivalent (again with the exception of regions distal to points of crossing over); the centromere divides and thus allows each chromosome to pass to a separate cell. In animals division II produces four *spermatids* (*q.v.*) or *ootids* (*q.v.*) which contain *monads* surrounded by a nuclear membrane. Meiosis therefore provides a mechanism whereby (1) an exchange of genetic material may take place between homologous chromosomes and (2) each gamete receives but one member of each chromosome pair. *See* **interkinesis, oogenesis, spermatogenesis.**

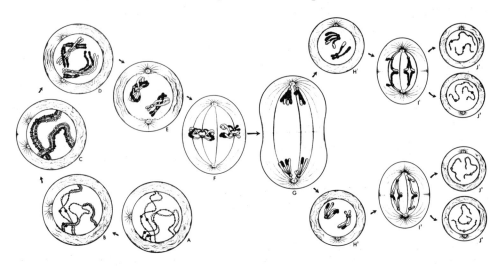

(A) A primary spermatocyte in leptonema. The organism has two pairs of chromosomes (one metacentric and one submetacentric). The maternal and paternal homologous chromosomes are drawn in different shades. The centromeres are represented by open circles. Note that the chromosomes are oriented with one of their ends in contact with one region of the nuclear membrane, forming a bouquet configuration. The chromosomes are uncoiled and maximally extended. In this state the chromomeres and chromonemata can be seen readily. A centrosome encompassed by astral rays is present. It contains a mother and daughter centriole, oriented at right angles to each other.
(B) During zygonema, synapsis of homologous chromosomes takes place. This pairing begins at one or more points and extends "zipper-like" until complete. Subsequently the cell contains two bivalents. The centrosome has divided into two daughter centrosomes, each containing a single centriole.
(C) During pachynema each paired chromosome splits into two sister chromatids, except at the region of the centromere. As a result the bivalents are converted into tetrads. A synaptonemal complex extends the length of the bivalent. The centrosomes move apart.
(D) During diplonema the medial complex of the synaptonemal complex disappears, and in each tetrad one pair of sister chromatids begins to separate from the other pair. However, the chromatids are prevented from separating at places where interchanges have taken place. The points where the chromatids form cross-shaped configurations are called chiasmata.
(E) At diakinesis, the chromatids are shorter and thicker, and terminalization is occurring. The centrosomes have reached the poles, and the nuclear envelope is beginning to disappear.
(F) At metaphase I, the tetrads are arranged at the equator of the spindle.
(G) At late anaphase I, the chromsomes have separated and have moved to each pole. However, the centromeres have not divided. As a consequence maternal and paternal chromosomal material has been separated (except in regions distal to points of crossing over).
(H^1, H^2) Secondary spermatocytes containing dyads.
(I^1, I^2) During early anaphase II, centromeres divide and allow separation of sister chromotids.
(J^1, J^2, J^3, J^4) The four spermatids containing monads. Chromosomes once again uncoil and elongate. (J^1) and (J^4) each contain one single and one double crossover chromatid.

meiospore   a spore produced by meiosis.

meiotic cycle   a first (reductional) division, followed by a second (equational) division. *See* **meiosis.**

236

meiotic drive   any meiotic mechanism which results in the unequal recovery of the two types of gametes produced by a heterozygote.

Melandrium album   a dioecious plant with a method of sex determination similar to man (that is, maleness is determined by the Y chromosome).

melanic   of darker color.

melanin   a dark brown to black pigment responsible for the coloration of skin, hair, and the pigmented coat of the retina. The macromolecule consists of polymers of indole 5,6-quinone and 5,6-dihydroxyindole-2-carboxylic acid and is formed by the enzymatic oxidation of tyrosine or tryptophan. A segment of the structure is shown below. *See* **eumelanin, phaeomelanin.**

```
                    H2      |
                    C       C
                  /    \\   / \
          H—C      C      C=0
              ||    |      |
              C  —  C      C=0
                \  /    \  /
                 C        C
               /    \\   / |
          0=C      C  —  C
              |    |      ||
          0=C      C      CH
              \   / \\   /
               C        C
               |        H2
              H2        C
               C      /    \\
             /   \\  C      C=0
          H—C      |  |      |
              ||    C— C      C=0
              C        \  /  \  /
                \  /    C        C
                 C    /    \\   / |
               /    C      C  —  C
          0=C      |  |      ||
              |    0=C      C      CH
          0=C      \   / \\   /
               \   / C        C
                C        · H2
                |        C
               H2      /    \\
                C      C      C=0
              /   \\  |  |      |
          H—C      C— C      C=0
              ||      \  /  \  /
              C        C        C
               \      /    \\   / |
                C    C      C  —  C
              /   \\ |  |      ||
          0=C      0=C      C      CH
              |        \   / \\   /
          0=C           C        C
               \\      /  |
                C    C   H2
                |   H2
```

melanism   the hereditary production of increased melanin resulting in darker coloring. *See* **albinism, industrial melanism.**

melanocyte   a pigment cell containing melanin granules.

melanocyte-stimulating hormone   intermedin (*q.v.*).

melanoma   a cancer composed of melanocytes (*q.v.*).

237

Melanophus femur-rubrum   a species of grasshopper widely used for the cytological study of meiosis.

melanosome   an intracellular organelle containing aggregations of tyrosinase, found in melanocytes.

melanotic tumor of *Drosophila*   *See* **pseudotumor.**

Meleagris gallopavo   the domesticated turkey.

melphalan   a mutagenic, alkylating agent.

$$\text{HOOC} - \underset{\underset{\text{NH}_2}{|}}{\text{CH}} - \text{CH}_2 - \underset{\underset{\underset{\text{H}}{|}}{\text{C}}}{\overset{\overset{\text{H}}{|}}{\text{C}}} \quad \underset{\underset{\text{H}}{|}}{\overset{\overset{\text{H}}{|}}{\text{C}}} \quad \text{C} - \text{N} \underset{\text{CH}_2\text{CH}_2\text{Cl}}{\overset{\text{CH}_2\text{CH}_2\text{Cl}}{}}$$

melting   denaturation of double-stranded DNA to single strands.

melting-out temperature   the temperature at which the bonds joining the constituent strands of a DNA/DNA or DNA/RNA duplex are broken and the molecules dissociate.

melting profile   a curve describing the degree of dissociation of the strands in a DNA/DNA or DNA/RNA duplex in a given time as a function of temperature. The stabilities of duplexes are a function of their molecular weights, so that the melting profiles shift to the right as the chain lengths increase.

melting proteins   *See* **helix-destabilizing proteins.**

melting temperature   the temperature at which fifty percent of the double helices have denatured; the midpoint of the temperature range over which DNA is denatured. *See* **denaturation.**

memory   in immunology, the ability to mount a specific secondary response to an antigen which had been encountered at some previous time. *See* **immune response.**

menarche   the beginning of the first menstrual cycle during puberty in human females. *Compare with* **menopause.**

Mendeleev's table   the periodic table of chemical elements (page 291).

Mendelian character   a character that in inheritance follows Mendel's laws.

Mendelian genetics   referring to the inheritance of chromosomal genes following the laws governing the transmission of chromosomes to subsequent generations; also called *Mendelism.*

Mendelian population   an interbreeding group of organisms sharing a common gene pool.

mendelize   to segregate according to Mendel's laws.

Mendel's laws   1. *The Law of Segregation.* The factors of a pair of characters are segregated. In modern terms this law refers to the separation into different gametes and thence into different offspring of the two members of each pair of alleles possessed by the diploid parental organism. **2.** *The Law of Independent Assortment.* The members of different pairs of factors assort independently. A restatement of the law in modern terms is that the members of different pairs of alleles are assorted independently into gametes during gametogenesis (provided they reside on different chromosomes), and that the subsequent pairing of male and female gametes is at random.

meniscus the curved upper surface of a column of liquid.

menopause the cessation of menstrual cycles in human females, usually occurring between age 50 and 60.

meq milliequivalent.

mercaptoethanol $SHCH_2CH_2OH$, a mitotic poison.

mercaptopurine a purine analogue.

Mercenaria mercenaria *See* **Pelecypoda.**

mericlinal chimera an organism or organ composed of two genetically different tissues one of which partly surrounds the other.

meridional at right angles to the equator.

Merino a breed of sheep of Spanish origin noted for its fine wool.

meristem the undifferentiated, mitotically active tissues of plants.

meristematic having or pertaining to meristems.

meristic variation variation in characters that can be counted, like the number of bristles, leaves, scales, etc.

meroblastic cleavage cleavage producing cells some of which are larger than others because of a polarized distribution of yolk.

merogony the development of an experimentally produced egg fragment (containing a diploid nucleus or a haploid nucleus of male or female origin) into a small-sized embryo, called a merogone.

meroistic pertaining to an insect ovariole containing nutritive cells. If apical nurse cells are joined to the oocytes by nutritive cords, the ovariole is called acrotrophic or telotrophic. If nurse cells are included with each oocyte in a follicle, the ovariole is called polytrophic. *See* **insect ovary types.**

See **insect ovary types.**meromixis genetic exchange in bacteria which involves a unidirectional transfer. *See* **F factor.**

merospermy the situation in which the nucleus of the fertilizing sperm does not fuse with the egg nucleus and later degenerates. Subsequent development is by gynogenesis.

merozygote a partially diploid bacterial zygote containing an exogenotic chromosomal fragment donated by the F^+ mate. The exogenote may also be introduced during transduction or sexduction. *See* **endogenote, exogenote.**

Mertensian mimicry *See* **mimicry.**

mesenchyme an embryonic type of connective tissue, consisting of amoeboid cells with many processes. Populations of such cells form a loose network. Most mesenchyme is derived from mesoderm. The mesenchyme produces the connective tissue and the circulatory system during the development of vertebrates.

Mesocricetus auratus the golden hamster, a rapidly breeding rodent used in the laboratory.

mesoderm the middle layer of embryonic cells between the ectoderm and endoderm in triploblastic animals. Mesoderm forms muscle, connective tissue, blood, lymphoid tissue, the linings of all the body cavities, the serosa of the viscera, the mesenteries, and the epithelia of the blood vessels, lymphatics, kidney, ureter, gonads, genital ducts, and suprarenal cortex.

mesophyte a plant growing in conditions of moderate moisture.

mesosome one of the invaginated segments of the plasma membranes in certain bacteria to which DNA molecules are attached.

mesothorax the middle of the three thoracic segments of an insect. It bears a pair of legs and (in winged insects) a pair of wings.

Mesozoic the era extending from 63 to 230 million years before present; the age of reptiles. *See* **geologic time divisions.**

messenger RNA an RNA molecule that functions during translation (*q.v.*) to specify the sequence of amino acids in a nascent polypeptide. In eukaryotes, mRNA is formed in the nucleus from pre-messenger RNA molecules (*q.v.*). *See* **exon, polysome.**

met methionine. *See* **amino acid.**

metabolic pathway a series of stepwise biochemical changes in the conversion of some precursor substance to an end product, each step usually catalyzed by a specific enzyme.

metabolic poison a compound poisoning a metabolic process. *See* **dinitrophenol.**

metabolism the sum of all the physical and chemical processes by which living protoplasm is produced and maintained and by which energy is made available for the uses of the organism.

metabolite a product of metabolism.

metacentric designating a chromosome with a centrally placed centromere.

metachromasy the phenomenon where one dye stains more than one color. A substance that stains metachromatically is called a chromatrope. Chromatropes are high molecular weight structures with serially arranged charged groups. Acidic mucopolysaccharides and nucleic acids are prime examples. Azure B (*q.v.*) is an example of a metachromatic dye. The color it produces depends on the way dye molecules stack on the chromatrope. As the amount of stacking increases, the color changes from green, to blue, to red. In an azure B-stained tissue section, chromsomes generally stain green, nucleoli and cytoplasmic ribosomes blue, and mucopolysaccharide-containing deposits red.

metachromatic dye a dye that stains tissues two or more colors. *See* **azure B, metachromasy.**

metafemale in *Drosophila,* a female phenotype of relatively low viability in which the ratio of X chromosomes to sets of autosomes exceeds 1.0; previously called a superfemale. *See* **intersex, metamale.**

metagenesis alternation of generations (*q.v.*) among animals. Metagenesis is commonly seen in invertebrates, especially coelenterates. Unlike the situation among plants, both generations are diploid.

metalloenzyme   a protein combined with one or more metal atoms which functions as an enzyme.

metallothioneins   small proteins that bind heavy metals and thereby protect cells against their toxic effects. Genes coding for metallothioneins are activated by the same metal ions that these proteins bind.

metamale   in *Drosophila,* a poorly viable male characterized by cells containing one X and three sets of autosomes; previously called a supermale. *See* **intersex, metafemale.**

metamerism   a phenomenon, common in annelids and arthropods, involving the repetition of a pattern of elements belonging to each of the main organ systems of the body. It is exhibited along the antero-posterior axis of the body. The term is also applied to a comparable repetition along the axis of an appendage.

metamorphosis   the transformation from larval to adult form.

metaphase   *See* **mitosis.**

metaphase arrest   referring to the accumulation of metaphase figures in a population of cells poisoned with colchicine, Colcemid, or some other spindle poison.

metaphase plate   the grouping of the chromosomes in a plane at the equator of the spindle during the metaphase stage of mitosis (*q.v.*).

metastable state   an excited state of an atomic nucleus, which returns to its ground state by the emission of radiation.

metastasis   the spread of malignant neoplastic cells from the original site to another part of the body.

metatarsus   the basal tarsal segment of the insect leg. In the male of *Drosophila melanogaster* the metatarsus of each foreleg bears a sex comb.

metathorax   the hindmost of the three thoracic segments of an insect. It bears a pair of legs and (in many winged insects) a pair of wings. In flies it bears the halteres.

methionine   *See* **amino acid.**

method of least squares   *See* **least squares.**

methotrexate   a folic acid antagonist.

methylated albumen-Kieselguhr column   an ion exchange column used in the chromatographic separation of nucleic acid molecules.

methylated cap   a modified guanine nucleotide terminating eukaryotic mRNA molecules. The cap is introduced after transcription by linking the 5′ end of a guanine nucleotide to the 5′ terminal base of the mRNA and adding a methyl group to position 7 of this terminal guanine. The addition of the terminal guanine is catalyzed by the enzyme *guanylyl transferase.* Another enzyme, *guanine-7-methyltransferase* adds a methyl group to the 7 position of the terminal guanine. Unicellular eukaryotes have a cap with this single methyl group (cap 0). The predominant form of the cap in muticellular eukaryotes (cap 1) has another methyl group added to the next base at the 2'-o position by the enzyme 2'-o-*methyl transferase*. More rarely, a methyl group is also added to the 2'-o position of the third base, creating cap 2 type. Capping occurs shortly after the initiation of transcription and precedes all excision and splicing events. The function of the cap is not known, but it may protect the mRNA from degradation by nucleases or provide a ribosome binding site.

methylcholanthrene   a carcinogenic hydrocarbon.

methyl green   a basic dye often used in cytochemistry to detect DNA. In 2 molar solution of magnesium chloride at pH 5.7 methyl green stains only undegraded DNA. *See* **pyronin Y.**

methyl guanosine   *See* **rare bases.**

methyl inosine   *See* **rare bases.**

2'-o-methyl transferase   *See* **methylated cap.**

metric equivalents and prefixes   *See* **measurement units** (pp. 232–234).

metric traits   *See* quantitative character, continuous variation.

Mg   magnesium.

MHC   major histocompatibility complex (*q.v.*).

MHC molecules   molecules that are anchored to cell membranes and determine the specificity of antigen recognition by T lymphocytes.

MIC   major immunogene complex (*q.v.*).

242

micelle   a spherical array of amphipathic molecules in which the nonpolar tails form a hydrocarbon microdroplet enclosed in a shell composed of the polar heads.

Michaelis constant (K_m)   the substrate concentration at which the reaction rate of an enzyme is half maximal; the concentration at which half of the enzyme molecules in the solution have their active sites occupied by a substrate molecule.

Michurinism   the genetic theories expounded by the Russian horticulturist I.V. Michurin, particularly those dealing with the modification of the genetical constitution of a scion by grafting. Michurin's theories were subsequently incorporated into Lysenkoism (*q.v.*).

Micrasterias thomasiana   a disc-shaped unicellular desmid, whose cortical differentiation has been extensively studied.

micro-   **1.** a prefix meaning one millionth; used with units of measurements in the metric system. **2.** a prefix meaning microscopic or minute.

microbe   microorganism (a nontechnical term).

microbeam irradiation   the use of a beam of microscopic diameter to selectively irradiate portions of a cell with ionizing radiation or ultraviolet light.

microbial genetics   the genetic study of microorganisms.

microbiology   the scientific study of microorganisms.

microbody   a cytoplasmic organelle bounded by a fragile membrane and containing a variety of oxidases (urate oxidase, catalase, etc.).

Microbracon hebetor   a wasp (called *Habrobracon juglandis* or *Bracon hebetor* in the early genetic literature). It is the arrhenotokous species in which the genetic control of sex determination was elucidated.

microcinematography   *See* **time-lapse microcinematography.**

micrococcal nuclease   an endonuclease used to cleave eukaryotic chromatin preferentially between nucleosomes.

microconidia   *See* **conidia.**

microevolution   an evolutionary pattern usually viewed over a short period of time, such as changes in gene frequency within a population over a relatively few generations (industrial melanism, for example). *See* **macroevolution.**

microfilaments   elongated intracellular fibers 5–7 nm in diameter, containing polymerized actin, and thought to function in maintenance of cell structure and movement.

Micrographia   a book published in 1665 by R. Hooke in which the first description of cells is given.

micromanipulator   an instrument allowing surgery on and injection into microscopic specimens. It is also useful in isolating single cells.

micron (μ)   a unit of length (1×10^{-6} meters) convenient for describing cellular dimensions.

micronucleus   the smaller reproductive nucleus as distinguished from the larger, vegetative macronucleus (*q.v.*) in the cells of ciliates. Micronuclei are diploid and are transcriptionally inactive. They participate in meiosis and autogamy.

microorganism   an organism too small to be observed with the unaided eye.

micropyle   **1.** a canal through the coverings of the nucellus through which the pollen tube passes during fertilization. In a mature seed the micropyle serves as a minute pore in the seed coat through which water enters when the seed begins to germinate. **2.** the pore in the egg membranes of an insect oocyte which allows entry of the sperm.

microsome fraction a cytoplasmic component, obtained upon centrifugation of homogenized cells, consisting of ribosomes and torn portions of the endoplasmic reticulum.

microspectrophotometer an optical system combining a microscope with a spectrophotometer. With this apparatus one can determine the amount of light of specified wave length passing through a given cytoplasmic region relative to some standard area. From such measurements estimates can be made of the concentrations of dye-binding or ultra-violet-light-absorbing materials present in the cytoplasmic area in question.

microspikes thin ($\sim$0.2 μ diameter) cellular projections that can reach 20 μ in length. They extend and contract rapidly and are speculated to have sensory functions.

microspore the first cell of the male gametophyte generation of seed plants. Each becomes a pollen grain.

microsporidian a protoctist belonging to the Cnidosporidia (*see* **classification**). Many microsporidian species parasitize insects (including *Drosophila*).

microsporocyte the pollen mother-cell, which undergoes two meiotic divisions to produce four microspores.

microsporogenesis the production of microspores.

microsporophyll the stamen. *See* **flower.**

microsurgery surgery done while viewing the object through a microscope and often with the aid of a micromanipulator.

microtome a machine for cutting thin (1 to 10 mμ thick) slices of paraffin-embedded tissue. These sections are later stained and examined with the light microscope. *See* **ultramicrotome.**

microtomy the technique of using the microtome in the preparation of sections for study under the microscope.

microtrabecular lattice a network of thin filaments interconnecting the three major cytoskeletal elements (microtubules, microfilaments, and intermediate filaments). This three-dimensional lattice can be visualized in freeze-etched preparations viewed with the electron microscope.

microtubule organizing centers (MCs) the structures or loci that give rise to microtubular arrays. Centrioles and kinetosomes function as MCs in some organisms. In others, the MC consists of a deposit of amorphous granulofibrillar material. MCs contain RNA, and RNA replication accompanies their multiplication.

microtubules long, non-branching, thin cylinders with an outside diameter about 24 nm and a central lumen about 15 nm diameter. The lengths are at least several μm. The tubules are composed of strands called protofilaments, and there are usually 13 of these. Each protofilament in turn is composed of a linear array of subunits, and each subunit is a dimer containing an alpha and a beta tubulin molecule.

Microtus the genus of meadow mice and voles subjected to extensive cytogenetic study. *M. agrestis,* the field vole, has giant sex chromosomes.

microvillus a finger-like projection of the plasmalemma of a cell.

middle lamella the outermost layer of a plant cell wall which connects it to its neighbor.

middle piece midpiece.

midparent value the mean of two parental values for a quantitative trait in a specific cross. *See* **quantitative inheritance.**

midpiece the portion of the sperm behind the head containing the nebenkern.

mid-spindle elongation the elongation during anaphase of the mid-region of the mitotic spindle. This elongation serves to draw the chromosomes poleward once the movement of chromsomes along traction fibers has ceased.

migrant selection selection based on the different migratory abilities of individuals of different genotypes. If, for example, individuals carrying gene *M* found new colonies more often than those bearing gene *m,* then gene *M* is said to be favored by migrant selection.

migration in population genetics, the movement of individuals between different populations of a species, resulting in gene flow (*q.v.*).

migration coefficient the proportion of a gene pool represented by migrant genes per generation.

migration inhibition factor a lymphokine which inhibits the movement of macrophages under *in vitro* culture conditions.

Miller spreads chromosomal whole mounts spread for electron microscopy by a method developed by O. L. Miller. In this method, chromosomes from ruptured nuclei are centrifuged through a solution of 10% formalin in 0.1 M sucrose and onto a membrane-coated grid. The grid is then treated with an agent that reduces the surface tension as the grid dries. After being stained with phosphotungstic acid, the preparation is ready for viewing under the electron microscope.

Miller trees transcribing rRNA genes. O. L. Miller and B. R. Beatty developed a technique of spreading chromatin for the electron micrographic observation of transcriptionally active genes. In their 1969 paper, they showed nucleolar genes isolated from salamander oocytes. Each rRNA transcription unit (rTU) was about 2.5 μm long, corresponding to a molecule of 8,000 bp. The rRNA molecules attached to the chromatin fiber give rTUs a Christmas tree morphology (hence the term Miller tree), and such rTUs are arranged in tandem, separated by fiber-free spacers.

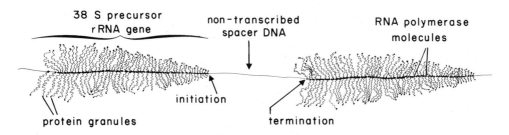

milli- a prefix meaning one-thousandth. Used with metric units of measurement.

millicurie one-thousandth of a curie; the amount of radioactive material in which 3.7×10^7 disintegrations occur per second.

milliequivalent *See* **gram equivalent weight.**

milliliter (ml) one-thousandth of a liter; the volume contained in a cube of side 1 cm. One ml of water weighs one gram at 4°C.

245

Millipore filter   a disc-shaped synthetic filter having holes of specified diameter through its surface. The available pore sizes range between 0.005 and 8 μ. The discs are used to filter microorganisms out of nutrient fluids that will not stand sterilization by autoclaving.

mimicry   the similarity in appearance of one species of animal to another which affords one or both protection. In *Batesian mimicry* one of two species is poisonous, distasteful, or otherwise protected from predators, and often conspicuously marked. The mimic is innocuous, gaining protection from predators by its similarity to the model. In *Mullerian mimicry* both species, are distasteful to predators, and they gain mutually from having the same warning coloration, since predators learn to avoid both species after tasting one. *Peckhammian mimicry* is an aggressive mimicry in which the predator is the pretender (mimic); e.g., a female firefly of one species mimics the flashing sexual signals of another species, thus attracting a male of that other species which becomes a meal for the pretender. In *Mertensian mimicry,* one species is mildly poisonous (e.g., false coral snakes) and thus serves as a model for a fatally poisonous species (e.g., true coral snakes). Obviously a predator can learn only if it survives the encounter. *See* **automimicry.**

minichromosomes   beaded DNA structures of certain viruses, similar to the beaded nucleosomal structures characteristic of eukaryotic chromatin. The circular DNA duplexes of papovaviruses (*q.v.*) are bound throughout the replication cycle to host histones, except H1.

minigenes   segments on chromosomes that code for the variable regions of immunoglobulin heavy and light chains. There are several hundred of these segments, but during the differentiation of B lymphocytes a single minigene is attached to the gene segment coding for the constant region. *See* **immunoglobulin genes.**

minimal medium   in microbiology a medium providing only those compounds essential for the growth and reproduction of wild-type organisms.

mink   *See* **Mustela vison.**

Minute   a mutation in *Drosophila melanogaster* which is dominant and lethal when homozygous. Heterozygotes have a prolonged developmental period, and the adults are small, semi-sterile, with small bristles, and abnormal wing veins. Many *Minutes* are small deficiencies. *Minute* loci occur on all the chromosomes, and they may represent the cistrons coding for tRNAs.

Miocene   an epoch in the Tertiary period lasting from 5 to 23 million years before present. *See* **geologic time divisions.**

Mirabilis jalapa   the four-o'clock, a variegated dicotyledonous plant extensively studied in terms of plastid inheritance.

mismatch repair   correction of nucleotide mispairings by "cut and patch repair" (*q.v.*).

mispairing   the presence in one chain of a DNA double helix of a nucleotide not complementary to the nucleotide occupying the corresponding position in the other chain.

missense mutant   a mutant in which a codon is mutated to one directing the incorporation of a different amino acid, making an inactive or perhaps unstable enzyme.

Mississippian   an epoch in the Carboniferous period. *See* **geologic time divisions.**

mistranslation   the insertion of an incorrect amino acid into a site on a growing polypeptide chain that is the result of environmental factors or mutations that effect either the tRNA, or the enzymes that attach specific amino acids to specific tRNAs, or the ribosome itself.

mitochondrial DNA   the mitochondrial genome consists of a circular DNA duplex, and there are generally 5–10 copies per organelle. The mammalian mtDNA has a molecular weight of about 11×10^6 d and therefore is less than 10^{-5} times that the size of the nuclear genome. The genetic code of mitochondria differs slightly from the "universal" genetic code. Since, of the two gametes, only the egg cell contributes significant numbers of mitochondria to the zygote, mtDNA is maternally inherited. *See* **universal code theory.**

mitochondrial sex locus (ω)   A gene on the mitochondrial DNA of yeast cells that can produce markedly different recombination frequencies when homozygous or heterozygous in mitochondrial "crosses."

mitochondrion   a self-reproducing organelle that occurs in the cytoplasm of all eukaryotes. Each mitochondrion is surrounded by a double limiting membrane. The inner membrane is highly invaginated. Its projections, called *cristae*, may be tubular or lamellar. Mitochondria are the sites of the reactions of oxydative phosphorylation (*q.v.*), which result in the formation of ATP. Mitochondria contain distinctive ribosomes, tRNAs, aminoacyl-tRNA synthetases, and elongation and termination factors. Mitochondria depend upon genes within the nucleus of the cells in which they reside for many essential mRNAs. The proteins translated from these mRNAs in the cell cytosol are imported into the organelle. Mitochondria are believed to have arisen from aerobic bacteria which established a symbiotic relationship with primitive protoeukaryotes. *See* **citric acid cycle, cytochrome system,** *Pelomyxa palustris,* **petites.**

mitogen   a compound that stimulates cells to undergo mitosis.

mitomycin   a family of antibiotics produced by *Streptomyces caespitosus*. Mitomycin C prevents DNA replication by crosslinking the complementary strands of the DNA double helix.

mitosis   Mitosis or nuclear division is generally divided into four phases: *prophase, metaphase, anaphase,* and *telophase.*
Prophase. During mitotic prophase the centriole divides and the two daughter centrioles move apart. The chromosomes become visible within the nucleus because they coil up to produce a series of compact gyres. Each chromosome is longitudinally double except in the region of the centromere, and each chromosome is called a chromatid. The nucleolus and the nuclear envelope break down.
Metaphase. During this phase the chromosomes move about within the spindle and eventually arrange themselves in the equatorial region of the spindle. The two chromatids are now ready to be separated and to move under the action of the traction fibers to the poles of the spindle.
Anaphase. During this phase the centromere becomes functionally double, and the chromatids are converted to independent chromosomes which separate and move to opposite poles.
Telophase. During telophase the spindle disappears and reconstruction of nuclear envelopes about the two groups of daughter chromosomes begins. As nuclear envelopes form around each group the chromosomes return to their extended state, and nucleoli reappear.

Next, cytokinesis occurs, and the cytoplasm is divided into two parts by a cleavage furrow in the case of animal cells or by a cell plate in the case of plant cells. The result of mitosis and cytokinesis is the production of two daughter cells with precisely the same nuclear content and approximately equal amounts of cytoplasm.

mitospore a spore produced by mitosis and therefore having the same chromosome number as the mother cell.

mitotic apparatus an organelle consisting of three components: (1) the asters, which form about each centrosome, (2) the gelatinous spindle, and (3) the traction fibers, which connect the centromeres of the various chromosomes to either centrosome. *See* **mitosis, multipolar spindle.**

mitotic center the agent that defines the poles toward which the chromosomes move during anaphase. The centrioles function as mitotic centers in most animal cells. In plants with anastral mitosis the nature of the mitotic center is unclear.

mitotic chromosome *See* **human mitotic chromosome.**

mitotic crossover somatic crossing-over.

mitotic index the fraction of cells undergoing mitosis in a given sample.

mitotic poison a compound that prevents mitosis. *See* **chloral hydrate, colchicine, maleuric acid, mercaptoethanol, podophyllin, vincaleukoblastine.**

mitotic recombination somatic crossing-over.

mixoploidy the presence of more than one chromosome number in a cellular population.

MLD medial lethal dose (*q.v.*).

mM millimolar concentration.

Mn manganese.

MN blood group a human blood group system defined by red cell antigens specified by a gene on the short arm of chromosome 4.

Mo molybdenum.

mobile genetic element *See* **transposable genetic elements.**

Möbius strip a topological figure, named after the German astronomer. A. F. Möbius, made by putting a 180-degree twist in a long, rectangular strip, then pasting the ends together. The strip has only one surface and one edge. If a Möbius strip is cut longitudinally, it forms a double sized ring. If a ring with a double twist is cut longitudinally, it forms interlocked rings. A Möbius strip demonstrates the behavior of a twisted ring chromosome when it attempts replication.

modal class the class that contains more individuals than any other in a statistical distribution.

mode modal class.

modification any changes made to DNA or RNA nucleotides after their original incorporation into a polynucleotide chain: e.g., methylation, deamination, formylation, etc. *See* **modification methylase.**

modification allele *See* **DNA restriction enzyme.**

modification methylase bacterial enzymes that bind to the DNA of the cell at specific sites determined by specific base sequences. Here they attach methyl groups to certain bases.

This methylation pattern is unique to and protects the species from its own restriction endonucleases. Modification methylases are coded for by modification alleles. *See* **restriction and modification model.**

modified bases   postsynthetically altered nucleotides of the four usual bases (A, T, G, and C) of DNA. *See* **modification methylase.**

modifier   a gene that modifies the phenotypic expression of a nonallelic gene.

modulating codon   specific triplets which code for rare tRNAs. The translation of a mRNA molecule is slowed down when a modulating codon is encountered.

moiety   one of two roughly equal parts.

mol   variant spelling of mole (*q.v.*).

molal solution   one containing a gram molecular weight of solute dissolved in one liter of water. *See* **molar solution.**

molar solution   on containing Avogadro's number of molecules of dissolved solute in sufficient water to make a liter of solution. *See* **molal solution.**

mole   gram molecular weight (sometimes spelled mol).

molecular biology   a modern branch of biology concerned with explaining biological phenomena in molecular terms. Molecular biologists often use the techniques of physical chemistry to investigate genetic problems.

molecular clock   *See* **evolutionary clock.**

molecular cloning   *See* **recombinant DNA research.**

molecular drive   a hypothesis proposing that cohesive (jointly occurring) genetic changes can occur in a population through a constant turnover of repeated nucleotide sequences (homogenization) produced by one or more of three mechanisms (unequal crossing-over, transposition, and gene conversion). Molecular drive propels the new variant through a population in a cohesive manner, even if it confers no reproductive or other advantage. Accidental speciation might result from such a cohesive population change if the repeat family involved in some way affects development or if it affects the chromosomes so that viable hybrids with members of other populations cannot form.

molecular genetics   that subdivision of genetics which studies the structure and functioning of genes at the molecular level.

molecular mimicry   *See* **eclipsed antigens.**

molecular sieve   a crystalline alumino-silicate pellet used to absorb water, carbon dioxide, hydrogen sulfide, and similar gases from gas mixtures and organic solvents. Molecular sieves are also sometimes used as ion exchange media.

molecular weight   the sum of the atomic weights of all of the atoms in a given molecule.

molecule   that ultimate unit quantity of a compound which can exist by itself and retain all the chemical properties of the compound.

molting hormone   *See* **ecdysone.**

moltinism   a polymorphism in terms of the number of larval molts undergone by a given species. For example, in *Bombyx mori* there are strains which are known to molt three, four, of five times.

molybdenum   a biological trace element. Atomic number 24; Atomic weight 95.94; most abundant isotopes ^{92}Mo, ^{94}Mo, ^{95}Mo, ^{96}Mo, ^{97}Mo, ^{98}Mo; radioisotope ^{99}Mo, half life 67 hours, radiation emitted—beta particles.

monad   *See* **meiosis.**

Monarch butterfly   *Danaus plexippus.* Monarch larvae generally feed upon plants, such as milkweeds, which contain molecules toxic to vertebrates. These molecules are sequestered by the insect, and they serve to render the animals unpalatable to avian predators.

Monera   one of the five kingdoms of living organisms. It contains the bacteria and blue-green algae and therefore all prokaryotic species. *See* **classification.**

monestrous mammal   *See* **estrous cycle.**

mongolism   Down syndrome (*q.v.*).

Mongoloid   pertaining to a race of mankind, characterized by a faintly yellowish skin, an epicanthic fold, sparse body hair, and black straight head hair.

moniliform   bead-like; having rounded segments.

monitoring   the periodic or continuous determination of the amounts of ionizing radiation or radioactive contamination present in an occupied region as a safety measure.

monoallelic   referring to a polyploid in which all alleles at a given locus are the same. In a tetraploid, for example, $A_1A_1A_1A_1$.

monocentric   designating a chromosome having a single centromere.

monochromatic light   light of a single wave length.

monochromatic radiation   electromagnetic radiation of a single wave length, or in which all of the photons have the same energy.

monocistronic mRNA   messenger ribonucleic acid coding for a single polypeptide chain; the typical mRNA of eukaryotic cells.

monoclonal antibodies   immunoglobulins derived from a single clone of plasma cells. Since all immunoglobulins produced by a given plasma cell (or clone thereof) are chemically and structurally identical, these antibodies constitute a pure population with highly specific antigen-binding properties. *See* **hybridoma.**

monocyte   the largest leukocyte found in the blood; they are phagocytic, amoeboid cells. *See* **agranulocytes, macrophages.**

monoecy   having both male and female reproductive organs on the same plant. *See* **flower.**

monoenergetic radiation   radiation of a given type (alpha, beta, neutron, gamma) in which all particles or photons have the same energy.

monogamy   an animal reproductive strategy in which a specific male and female form an exclusive mating pair during a reproductive cycle, a season, or a lifetime. *Compare with* **polygamy.**

monogenic character   a character determined by a single gene.

monohybrid   an individual that is heterozygous (e.g., *Aa*) for a pair of alleles at the locus under study.

monohybrid cross   a mating between two individuals, both of which are heterozygotes genetically identical at a given locus: e.g., *Aa* × *Aa.*

monolayer   a single layer of cells growing on a surface.

monolepsis   transmission to the offspring of characteristics of but one parent.

monomer   a simple compound from which, by repetition of a single reaction, a polymer is made. For example, uridylic acid (U) can be polymerized to form polyuridylic acid (UUU. . . .)

monomorphic population   a population showing only one trait (of potentially variable expression) due to fixation of one allelic form of the gene responsible for that trait.

mononuclear leukocyte   *See* **agranuloctye.**

monophasic lethal   a mutation having one effective lethal phase (*q.v.*).

monophyletic group   a natural taxon composed of two or more species. It includes the ancestral species (known or hypothesized) and all of its descendants. Members of a monophyletic group are sister taxa. *See* **polyphyletic group.**

monoploid   **1.** the basic chromosome number in a polyploid series. **2.** a somatic cell or individual having only one set of chromosomes.

monosome   **1.** a single mRNA-ribosome complex. *See* **polysome. 2.** a chromosome lacking a homologue.

monosomy   the condition in which one chromosome of one pair is missing. A monosomic diploid has one chromosome less than the normal diploid number (2N-1). An XO *Drosophila* is monosomic for the sex chromosome.

monospermy   fertilization by only on spermatozoon.

monothetic group   in systematics, organisms that are grouped together because they possess a unique set of features that is both sufficient and necessary for membership in the group thus defined. For example, all mammals and only mammals have hair and nurse their young with milk.

monotreme   an egg-laying mammal such as the duckbilled platypus or the echidna.

monotrichous   having a single flagellum.

monotypic   a taxonomic group containing only one taxon of the next lower rank; for example, a monotypic genus contains only one species; a monotypic species has no recognized subgroups (races, subspecies); the opposite of polytypic.

monozygotic twins   twins formed by the fission into two, at some stage of its early development, of the embryo derived from a single fertilized egg. Such twins are genetically identical. *See* **dizygotic twins.**

Morgan unit   a unit for expressing the relative distance between genes on a chromosome. One morgan (M) equals a crossover value of 100%. A crossover value of 10% is a decimorgan (dM); 1% is a centimorgan (cM); named in honor of Thomas Hunt Morgan.

Mormoniella vitripennis   a calcidoid wasp also referred to in the scientific literature as *Nasonia brevicornis* and *Nasonia vitripennis*. The wasp is often used for classroom demonstrations of parthenogenesis (*q.v.*). In the laboratory it is provided with pupae of the fly *Sarcophaga bullata* to parasitize.

morph   any of the individuals of a polymorphic population; any phenotypic or genotypic variant.

morphogenes   genes involved directly or indirectly in the control of growth and morphogenesis: e.g., genes for hormones, inducers, mitogens, inhibitors, cell cycle control factors, etc.

morphogenesis   the developmental processes leading to the characteristic mature form of a cell, an organism, or part of an organism. *See* **development, differentiation.**

morphogenetic movements   the cell movements that change the shape of differentiating cells of tissues, *e.g.,* in an embryo (invagination, expansion, longitudinal extension, dorsal convergence, etc.).

morphogenetic stimulus   a stimulus exerted by one part of the developing embryo upon another, leading to morphogenesis in the reacting part.

morphology   the science dealing with the visible structures of organisms and the developmental and evolutionary history of these structures.

morphometric cytology   the determination of quantitative parameters of cytological structures in tissue sections.

morula   an embryo that consists of a cluster of cleaving blastomeres. A stage prior to the blastula.

mosaic   an individual composed of two or more cell lines of different genetic or chromosomal constitution, both cell lines being derived from the same zygote; in contrast with a *chimera* (*q.v.*).

mosaic development   in mosaic development the fates of all the parts of the embryo are already fixed at or before fertilization, so that any localized ablation will later be manifested by the absence of the part in question. *See* **regulative development.**

mosaic evolution   evolutionary change in one or more body parts without simultaneous changes in other parts. For example, in the evolution of birds from dinosaurs, the origin of feathers occurred long before specialized bones (e.g., keel) and powerful flight muscles developed.

motility symbiosis   the situation where motility is conferred upon an organism by its symbiont. For example, many protists contain cortical populations of symbiotic spirochaetes whose coordinated movements propel the host through its aqueous medium.

mottled   marked with irregular patches of different colors.

mouse   *See **Mus musculus.***

mouse inbred lines   laboratory strains of mice propagated by brother-sister matings for many generations and hence highly homozygous and genetically uniform. In some strains, the inbreeding program has been carried out for 40 years. The strains most commonly mentioned in the literature are: albino (A, Ak, BALB, R_{III}), black (C_{57} black, C_{58}), black agouti (CBA, C3H), brown (C_{57} brown), dilute brown (DBA/2), dilute brown piebald (I).

mouse L cells   a strain of fibroblast-like cells carried in tissue culture. The cells originated from subcutaneous areolar and connective tissue derived from a male c3H mouse in 1940.

mouse satellite DNA   a DNA making up about 10% of the DNA isolated from a wide variety of mouse tissues. It forms a band slightly separated from the main peak when mouse DNA is spun to equilibrium in a CsCl density gradient. *See* **centrifugation separation.** Mouse satellite DNA consists of about one million copies per genome of a sequence some 400 nucleotide pairs in length. *In situ* hybridization experiments show that most of this DNA is located in the pericentric heterochromatin.

MPD   maximum permissible dose (*q.v.*).

M phase   *See* **cell cycle.**

mRNA   messenger RNA.

mRNA coding triplets   *See* **amino acid, start codon, stop codon.**

MS2   an RNA phage from which an RNA-replicase has been isolated.

MSH   melanocyte-stimulating hormone. *See* **intermedin.**

MTA   mammary tumor agent (*q.v.*).

mtDNA   mitochondrial DNA (*q.v.*).

M5 technique   a technique used to detect sex-linked lethal and viable mutations in *Drosophila melanogaster*. The sc^{S1} *B InS* w^a sc^8 chromosome is employed.

mt mRNA, mt rRNA, mt tRNA   symbols for mitochondrial messenger, ribosomal, and transfer RNAs.

mu   **1.** map unit. **2.** mate killers.

mucopolysaccharide   a polysaccharide composed of sugars and sugar derivatives, such as amino sugars and uronic acids. *See* **chondroitin sulfuric acid.**

mucoprotein   a protein containing more than 4% carbohydrate. *See* **glycoprotein.**

mulatto   the hybrid from a Negro-white cross.

mule   *See* **horse-donkey hybrids.**

Mullerian mimicry   *See* **mimicry.**

multifactorial   polygenic.

multiforked chromosome   a bacterial chromosome containing more than one replication fork, due to a second initiation having begun before completion of the first replication cycle.

multigene family   a set of genes descended by duplication and variation from some ancestral gene. Such genes may be clustered together on the same chromosome or dispersed on different chromosomes. Examples of multigene families include those which encode the histones, hemoglobins, immunoglobulins, histocompatibility antigens, actins, tubulins, keratins, collagens, heat shock proteins, salivary glue proteins, chorion proteins, cuticle proteins, yolk proteins, and phaseolins. *See* **reiterated genes.**

multimer   a protein molecule made up of two or more polypeptide chains, each referred to as a monomer. The terms dimer, trimer, tetramer, pentamer, etc., are used if the number of monomers per multimer is known.

multiparous   bearing or producing more than one offspring at a birth.

multiple allelism   *See* **allele.**

multiple choice mating   referring to an experimental design in studies of behavior genetics where a test organism is allowed to choose between two (or more) genetically different mates.

multiple codon recognition   *See* **wobble hypothesis.**

multiple-event curve   a curve (relating relative survival to radiation dose) which contains an initial flat portion. This finding indicates that there is little biological effect until a certain dose has accumulated, and suggests that the sensitive target must be hit more than once (or that there must be multiple targets, each of which must be destroyed) to produce a biologically measurable effect. *See* **single-event curve, target theory.**

multiple factor hypothesis   *See* **quantative inheritance.**

multiple genes   *See* **polygenes, multiple factor hypothesis, quantitative inheritance.**

multiple infection   simultaneous invasion of a bacterial cell by more than one phage, often of different genotypes in experiments designed to promote phage recombination; superinfection.

multiple myeloma   cancer of plasma cells, presumably due to clonal proliferation of a single plasma cell that escapes the normal control of division. Such cells reproduce and secrete a specific homogeneous protein related to γ globulins (Bence-Jones proteins).

multiplicity of infection   the average number of phages that infect a bacterium in a specific experiment. The fraction of bacteria infected with 0, 1, 2, 3, . . ., *n* phage follows a Poisson distribution.

multiplicity reactivation   the production of recombinant virus progeny following the simultaneous infection of each host cell by two or more virus particles, all of which are incapable of multiplying because they carry lethal mutations induced by exposure to a mutagen.

multipolar spindle   a spindle with several poles found in cells with multiple centrioles. Such cells are seen infrequently, but they can be produced in large numbers by irradiation. *See* **mitotic apparatus.**

multivalent   designating an association of more than two chromosomes whose homologous regions are synapsed by pairs (as in autotetraploids and translocation heterozygotes).

multivoltine   producing more than one brood in a year, as in certain birds and moths.

mu **phage**   a phage "species" whose genetic material behaves like insertion sequences, being capable of transposition, insertion, inactivation of host genes, and causing rearrangements of host chromosomes.

murine   belonging to the family of rodents that contains the mice and rats.

murine mammary tumor virus   an oncogenic RNA virus. *See* **mammary tumor agent.**

Musa paradisaica sapientum   the banana. *See* **parthenocarpy.**

Musca domestica   the house fly. DDT-resistance in this species has been extensively studied by geneticists.

muscular dystrophy   a group of genetically determined diseases of man characterized by a progressive degeneration of the proximal musculature of the girdles and limbs. Duchenne (pseudohypertrophic) muscular dystrophy is inherited as a sex-linked recessive. The disease is also known in the mouse as the result of single gene mutations.

Mus musculus   the laboratory mouse. It is the laboratory mammal that has been studied most extensively from the genetic standpoint. *See* **mouse inbred lines.**

mustard gas   sulfur mustard (*q.v.*).

Mustela   the genus that includes *M. erminea,* the ermine; *M. lutreola,* the European mink; *M. vison,* the North American mink.

mutable gene   in multicellular organisms a gene that spontaneously mutates at a sufficiently high rate to produce mosaicism.

mutable site   a site on a chromosome at which mutations can occur.

mutagen   a physical or chemical agent that raises the frequency of mutation above the spontaneous rate.

mutagenic   causing mutation.

mutagenize   to expose to a mutagenic agent.

mutant   an organism bearing a mutant gene that expresses itself in the phenotype of the organism.

mutant hunt   the isolation and accumulation of a large number of mutations affecting a given process, in preparation for mutational dissection of the gene(s) governing that process. For example, one might select for mutations that confer phage resistance in *E. coli.*

mutation   **1.** process by which a gene undergoes a structural change. **2.** modified gene resulting from mutation. **3.** by extension, the individual manifesting the mutation. *See* **isocoding mutation, point mutation.**

mutational dissection   *See* **genetic dissection.**

mutational load   the genetic disability sustained by a population due to the accumulation of deleterious genes generated by recurrent mutation.

mutation breeding   induction of mutations by mutagens to develop new crop varieties that can increase agricultural productivity.

mutation distance   the smallest number of mutations required to derive one DNA sequence from another.

mutation event   the actual origin of a mutation in time and space, as opposed to the phenotypic manifestation of such an event, which may be generations later.

mutation frequency   the proportion of mutants in a population.

mutation pressure   the continued production of an allele by mutation.

mutation rate   the number of mutation events per gene per unit time (e.g. per cell generation).

mutator gene   a mutant gene that increases the spontaneous mutation rate of one or more other genes. The Dotted (*q.v.*) gene of corn is an example. Mutator genes may be mutant alleles of genes that normally function in the replication or repair of DNA. For example, a mutation in a gene coding for a DNA polymerase could generate errors in the replication of many nucleotide sequences. These would manifest themselves as mutations at sites distinct from the cistron coding for the polymerase.

mutein   a mutant protein, such as a CRM (*q.v.*).

muton   the smallest unit of DNA in which a change can result in a mutation (a single nucleotide).

mutual exclusion   a phenomenon observed among ciliary antigens of certain protozoans in which only one genetic locus for a serotype is active at a given time. For example, in *Paramecium primaurelia* and *Tetrahymena pryriformis* mutual exclusion of serotypes in heterozygotes occurs with allelic genes as well as with nonallelic genes.

mutualism   a symbiosis in which both species benefit.

myc   a gene originally described in the avian MC29 myelocytomatosis virus, an oncovirus of the chicken. A homologous gene is located on the long arm of human chromosome 8. The viral gene is often symbolized *v-myc* and the cellular gene *c-myc* (pronounced "see-mick"). *See* **Burkitt's lymphoma.**

mycelium   the vegetative portion of a fungus composed of a network of filaments called hyphae.

Mycoplasma   a genus of primitive bacteria containing species that are the simplest forms capable of independent life. *See* **classification** (Aphragmabacteria), **pleuropneumonia-like organisms.**

Mycostatin   a trade name for nystatin (*q.v.*).

myelin sheath   the insulating covering of an axon formed by the plasma membrane of a Schwann cell.

myeloblasts   cells that differentiate by aggregation to form multinucleated, striated muscle cells.

myeloma   a tumor derived from an immunoglobulin-secreting cell that usually produces a homogeneous product. *See* **Bence-Jones proteins.**

myeloma protein   a partial or complete immunoglobulin molecule secreted by a myeloma (*q.v.*).

Myleran   a trade name for busulfan (*q.v.*).

myoglobin   the monomeric heme protein of vertebrate muscle. Human myoglobin contains 152 amino acids. The myoglobin cistron is thought to be derived directly from the ancestral cistron which by duplication produced the gene that served as the ancestor of the alpha chain cistron. *See* **hemoglobin.**

myoglobin gene   the gene which encodes myoglobin is remarkable in that less than 5% of its structure codes for message. All the genes of the α- and β-hemoglobin families are made up of three coding regions interrupted by two introns. The myoglobin gene contains four exons and three introns, and each of these introns is much longer than any of those found in hemoglobin genes.

myria   a rarely used prefix meaning ten thousand. Used with metric units of measurement.

myriapod   a millipede or centipede.

Mytilus edulis   *See* **Pelecypoda.**

myxomatosis   a fatal virus disease affecting rabbits. The virus was introduced into wild populations of rabbits in Australia as a means of controlling them.

Myxomycota   the phylum containing the plasmodial slime molds. These protoctists generate multinucleate plasmodia that feed by phagocytosis and subsequently form stalked, fungus-like fruiting structures. From the standpoint of genetics, *Physarum polycephalum* is the best known species.

N

n neutron.

N **1.** the haploid chromosome number. **2.** normal solution. **3.** nitrogen.

Na sodium.

N-acetyl serine an acetylated serine thought to function in mammalian systems as N-formylmethionine does in bacterial translation.

$$CH_3 - \overset{\overset{\displaystyle O}{\|}}{C} - \overset{\overset{\displaystyle H}{|}}{N} - \underset{\underset{\displaystyle OH}{\underset{|}{CH_2}}}{\overset{|}{CH}} - COOH$$

NAD nicotinamide-adenine dinucleotide (*q.v.*).

NADP nicotinamide-adenine dinucleotide phosphate (*q.v.*).

Naegleria a genus of soil amoebae capable of transforming into flagellates. Species from this genus are often studied in terms of the morphogenesis of flagella.

nail patella syndrome a herediatry disease in man. Individuals afflicted with this disorder have misshapen fingernails and small kneecaps or lack them. The disease is due to a dominant gene residing on chromosome 9.

nalidixic acid a drug that inhibits *E. coli* DNA gyrase.

nano a prefix meaning one billionth (10^{-9}). Used with metric units of measurement.

Naperian logarithm *See* **logarithm.**

narrow heritability *See* **heritability.**

nascent polypeptide chain the forming polypeptide chain which is attached to the 50 S subunit of a ribosome through a molecule of tRNA. The free end of the nascent polypeptide contains the N-terminal amino acid. *See* **translation.**

nascent RNA an RNA molecule in the proscess of being synthesized (hence incomplete) or a complete, newly synthesized RNA molecule before any alterations have been made (e.g., prior to nuclear processing, *q.v.*).

Nasonia brevicornis   another name for *Mormoniella vitripennis* (*q.v.*).

Nasonia vitripennis   another name for *Mormoniella vitripennis* (*q.v.*).

native   indigenous. A native species is not introduced into an area by man, either intentionally or accidentally.

natural acquired immunity   a resistance acquired by a person's having had the disease in question.

natural immunity   an outmoded concept that some immunities are inherited in the apparent absence of prior contact with an antigen. The prevailing paradigm is that all immunity ultimately requires contact with a sensitizing antigen and therefore is acquired.

natural killer cell   large leukocytes derived from the bone marrow and related to T lymphocytes. NK cells are found in the spleen, lymph nodes, and peripheral blood. They are activated by interferon, and they attack tumor cells without prior immunization. NK cells thus differ from K cells (*q.v.*), which kill tumor cells only after previous immunization.

natural selection   the differential fecundity (*q.v.*) in nature between members of a species possessing adaptive characters and those without such advantages. The *fundamental theorem of natural selection* developed by R. A. Fisher proposes that the increase in fitness of a population at any given time is directly proportional to the genetic variance in fitness of its members. *Compare with* **artificial selection.**

n_D refractive index.

Neanderthal man   a sub-species of *Homo sapiens* that lived in Europe, North Africa, the Near East, Iraq, and Central Asia in the middle and upper Pleistocene.

Nearctic   one of the six biogeographic realms (*q.v.*) of the earth, comprising North America, Greenland, and extending to the Mexican plateau.

nearest neighbor sequence analysis   a biochemical technique used to estimate the relative frequencies with which pairs of each of the four bases lie next to one another. The results of such analyses reveal that DNA molecules from different sources have specific patterns of nearest neighbor sequences even though they may have similar base sequences.

nebenkern   a two-stranded helical structure surrounding the proximal region of the tail filament of a spermatozoon. The nebenkern is derived from clumped mitochondria.

nectary   a plant organ where nectar is stored.

negative complementation   suppression of the wild-type activity of one subunit of a multimeric protein by a mutant allelic subunit.

negative contrast technique   *See* **negative staining.**

negative eugenics   *See* **eugenics.**

negative gene control   prevention of gene expression by the binding of a specific controlling factor to DNA. For example, in bacterial operons (either inducible or repressible), the binding of a repressor protein to the operator prevents transcription of structural genes in that operon. *See* **regulator gene.** *Compare with* **positive gene control.**

negative interference   a situation in which the coefficient of coincidence is greater than 1. In such cases the occurrence of one exchange between homologous chromosomes appears to increase the liklihood of another in its vicinity.

negative staining   a staining technique developed by S. Brenner and R. W. Horne in 1959 for high resolution electron microscopy of viruses. A virus suspension is mixed with a phosphotungstic acid solution and poured into an atomizer sprayer. The mixture is then sprayed upon electron microscope grids previously coated with a film of carbon. The phosphotungstic acid enters the contours of the specimen, which is viewed as a light object against a dark background.

negative supercoiling *See* **supercoiling.**

Neisseria a genus of bacteria to which the species causing meningitis and gonorrhea belong.

Nembutal a trademark for an anesthetic commonly used on mammals; sodium pentobarbital.

neobiogenesis the concept that life has been generated from inorganic material repeatedly in nature.

neo-Darwinism the post-Darwinian concept that species evolve by the natural selection of adaptive phenotypes caused by mutant genes.

Neolithic pertaining to the later stone age during which agriculture and animal husbandry originated and flourished.

neo-Malthusianism the doctrine that the numbers of human beings can be controlled and the standards of living can be raised through birth control. *See* **Malthusian.**

neomorph a mutant gene producing a qualitatively new effect that is not produced by the normal allele.

neomycin an antibiotic produced by *Streptomyces fradiae.*

neontology the study of living (extant) species, as opposed to paleontology (the study of extinct species).

neoplasm a localized population of proliferating cells in an animal which are not governened by the usual limitations of normal growth. The neoplasm is said to be *benign* if it does not undergo metastasis and *malignant* if it undergoes metastasis.

neotenin synonym for allatum hormone (*q.v.*).

neoteny the retention of larval characteristics throughout life with reproduction during the larval period, as in the axolotl.

Neotropical one of the six biogeographic realms (*q.v.*) of the earth, comprising Central and South America (south of the Mexican plateau) and the West Indies.

neuraminic acid a component of many phospholipid membranes (formula on p. 260).

neurohormone a hormone synthesized and secreted by specialized nerve cells; e.g., gonadotropin-releasing hormone produced by neurosecretory cells located in the hypothalamus.

neuraminic acid

$$
\begin{array}{c}
\text{COOH} \\
| \\
\text{C}-\text{OH} \\
| \\
\text{CH}_2 \\
| \\
\text{CHOH} \\
| \\
\text{H}_2\text{N}-\text{C}-\text{H} \\
| \\
\text{C}-\text{H} \\
| \\
\text{H}-\text{C}-\text{OH} \\
| \\
\text{H}-\text{C}-\text{OH} \\
| \\
\text{CH}_2\text{OH}
\end{array}
$$

neurohypophysis the portion of the hypophysis that develops from the floor of the diencephalon.

neurological mutant a mutant producing gross malformations of the central nervous system or striking abnormalities in locomotion. Nearly one hundred such mutants are known for the laboratory mouse.

neuron a nerve cell.

neuropathy a collective term for a great variety of behavioral disorders that may have hereditary components.

neurosecretory spheres electron-dense spheres $0.1–0.2\ \mu$ in diameter, synthesized by and transported in the axoplasm of specialized neurons.

Neurospora crassa an ascomycete fungus upon which many of the classical studies of biochemical genetics were performed.

neurula the stage of development of a vertebrate embryo at which the neural axis is fully formed and histogenesis is proceeding rapidly.

neutral equilibrium *See* **passive equilibrium.**

neutral gene theory of molecular evolution a theory first advanced by M. Kimura according to which the majority of the nucleotide substitutions in the course of evolution are the result of the random fixation of neutral or nearly neutral mutations, rather than the result of positive Darwinian selection. Many protein polymorphisms are selectively neutral and are maintained in a population by the balance between mutational input and random extinction. Neutral mutations are not functionless; they are simply equally effective to the ancestral alleles in promoting the survival and reproduction of the organisms that carry them. However, such neutral mutations can spread in a population purely by chance because only a relatively small number of gemetes are "sampled" from the vast supply produced in each generation and therefore are represented in the individuals of the next generation.

neutralism a nonsymbiotic relationship between members of different species that neither directly compete for any resources of the habitat nor depend directly upon one another for food.

neutral mutation **1.** a genetic alteration whose phenotypic expression results in no change in the organism's adaptive value or fitness for present environmental conditons. **2.** a mutation that has no measurable phenotypic effect as far as the study in question is concerned.

260

neutron   an elementary nuclear particle with a mass approximately the same as that of a hydrogen atom and electrically neutral; its mass is 1.0087 mass units.

neutron contrast matching technique   a technique that involves determining the neutron scattering densities of particles irradiated in solutions containing various concentratons of light and heavy water. Pardon *et al.* used this technique on nucleosomes (*q.v.*) and found that under conditions where neutron scattering from DNA dominated the reaction, the radius of gyration was 50Å. When scattering from the histone proteins was dominant the radius was 30Å. The larger radius for DNA proved that it was located on the surface of the nucleosome.

N-formylmethionine   a modified methionine molecule that has a formyl group attached to its terminal amino group. Such an amino acid is "blocked" in the sense that the absence of a free amino group prevents the amino acid from being inserted into a growing polypeptide chain. N-formylmethionine is the starting amino acid in the synthesis of all bacterial polypeptides. *See* **start codon.**

$$\begin{array}{c} \overset{O}{\overset{\|}{H-C}}-\overset{H}{\overset{|}{N}}-CH-COOH \\ | \\ CH_2-CH_2-CH_2-S-CH_3 \end{array}$$

niacin   an early name for nicotinic acid.

niche   *See* **ecological niche.**

niche preclusion   *See* **first-arriver principle.**

nick   the absence of a phosphodiester bond between adjacent nucleotides in one strand of duplex DNA. *Compare with* **cut.**

nickase   an enzyme that causes single-stranded breaks in duplex DNA, allowing it to unwind.

nick-closing enzyme   *See* **topoisomerase.**

nick translation   an *in vitro* procedure used to radioactively label a DNA of interest to a high specific activity. First, nicks are introduced into the unlabeled DNA by an endonuclease, generating 3' hydroxyl termini. *E. coli* DNA polymerase I is then used to add radioactive residues to the 3' hydroxy terminus of the nick, with concomitant removal of the nucleotides from the 5' side. The result is an identical DNA molecule with the nick displaced further along the duplex.

Nicotiana   a genus containing about 60 species, many of which have been intensively studied genetically. Much interest has been generated from the finding that tumors arise spontaneously at high frequency in certain interspecific hybrids, such as those plants produced by the cross *N. langsdorffii* × *N. glauca*. The species of greatest commercial importance is *N. tabacum,* the source of tobacco.

nicotinamide-adenine dinucleotide   a coenzyme (formerly called DPN or coenzyme 1) functioning as an electron carrier in many enzymatic oxidation-reduction reactions.

nicotinamide-adenine dinucleotide phosphate   an electron carrier formerly called TPN or coenzyme 2. *See* **citric acid cycle, cytochrome system.**

nicotinamide-adenine dinucleotide
(NAD)   R = H

nicotinamide-adenine dinucleotide
phosphate
(NADP)   R = PO(OH)$_2$

nicotine   a poisonous, volatile alkaloid present in the leaves of *Nicotiana tabacum* and responsible for many of the effects of tobacco smoking. It functions in the plant as a potent insecticide.

nicotinic acid   one of the B vitamins. Also called niacin in the older literature.

nidicolous   reared in a nest, as in the case of birds.

Niemann-Pick disease   a group of human disorders characterized by enlargement of the spleen and liver and by the accumulation of sphingomyelin (*q.v.*) and other lipids throughout the body. The disease is due to a defective lysosomal sphingomyelinase coded for by an autosomal recessive gene. Amniocentesis and testing of fetal cells for sphingomyelinase activity permits monitoring of pregnancies at risk. Heterozygotes can be identified, since their leukocytes contain about 60% the normal activity of sphingomyelinase.

nif* genes** genes for nitrogen fixation present in certain bacteria, especially those that live in root nodules of leguminous plants. *See **Rhizobium.

nigericin *See* **ionophore.**

Nile blue a mixture of two dyes: Nile blue A, a water soluble basic dye; and Nile red, a lysochrome formed by spontaneous oxidation of Nile blue A (an example of allochromacy, *q.v.*).

Nile blue A   Nile red

ninhydrin an organic reagent that reacts with and colors amino acids. Ninhydrin solutions are sprayed on chromatographs, and the separated amio acids and polypeptides are then rendered visible as ninhydrin-positive spots.

nitrocellulose filter a filter made of nitrocellulose fibers that selectively traps double-stranded DNA or DNA-RNA hybrids while allowing single-stranded molecules to pass through.

nitrogen the fourth most abundant of the biologically important elements. Atomic number 7; atomic weight 14.0067; valence 3^-, 5^+; most abundant isotope ^{14}N; heavy isotope ^{15}N.

nitrogen fixation the enzymatic incorporation of nitrogen from the atmosphere into organic compounds.

nitrogen mustard di(2-chloroethyl) methylamine; an alkylating agent (*q.v.*) that is a potent mutagen and chromosome-breaking agent. *See* **sulfur mustard.**

263

nitrogenous base a purine or pyrimidine; more generally an aromatic, nitrogen-containing molecule that has basic properties (is a proton acceptor).

nitrous acid HNO_2, a mutagen that converts the NH_2 groups of the purines and pyrimidines to OH groups.

NK cells *See* **K cells.**

nocturnal pertaining to the night time; of animals, feeding at night.

node **1.** in vascular plants, a slightly enlarged portion of a stem where leaves and buds arise and where branches originate. **2.** in a circular DNA superhelix, the point of contact in a figure-of-eight; if the left strand in the upper part of 8 is closest to the viewer at the node, it is called a positive node; if the left strand in the upper part of 8 is in back of the other strand at the node, it is called a negative node.

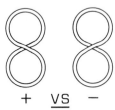

noise in colloquial usage, variation in an experiment attributed to uncontrolled effects; usually assoicated with a variance component called experimental error.

Nomarski differential interference microscope an optical system which, like the phase contrast microscope, permits the visualization of transparent structures in a living cell. However, in the Nomarski system the field is quite shallow so that there is freedom from phase disturbances from structures above and below the plane of focus. The observation method is comparable to that with extreme oblique illumination, and the specimen therefore appears in relief.

nomenclature the naming of taxa in accordance with certain international rules.

nomogram a graph (also called a nomograph) consisting of two or more coplanar curves (usually parallel straight lines) each graduated for a different variable so that a numerical value on one scale can be converted to a numerical value on another scale by drawing a straight line between points on any two scales. For example, the relative centrifugal force of a centrifuge can be determined by drawing a straight line between points on the speed (rpms) and radius scales, and then observing where that line intersects the rcf scale. For a nomogram that converts Fahrenheit to Celsius temperatures, *see* **temperature.**

nonautogenous anautogenous (*q.v.*).

nonbasic chromosomal proteins acidic or neutral proteins (therefore not histones) associated with chromosomes: e.g., certain enzymes such as DNA polymerases.

noncrossover gamete a sex cell containing no recombination of the linked genes under consideration; a parental-type gamete.

non-Darwinian evolution genetic changes in populations produced by forces other than natural selection; a term usually associated with the neutralist view of evolution.

nondisjunction the failure of homologous chromosomes (in meiosis I, primary nondisjunction) or sister chromatids (in meiosis II, secondary nondisjunction; or mitosis) to separate properly and to move to opposite poles. Nondisjunction results in one daughter cell receiving both and the other daughter cell none of the chromosomes in question.

nonessential amino acids   *See* **essential amino acids.**

nonhomologous chromosomes   chromosomes that do not synapse during meiosis.

noninducible enzyme   constitutive enzyme (*q.v.*).

nonlinear tetrad   a group of four meiotic products that are randomly arranged in the ascus. *See* **linear tetrad.**

non-Mendelian ratio   in the progeny of a cross, unusual phenotypic ratios that fail to follow Mendel's laws, suggesting that gene conversion or another aberrant mechanism is responsible.

nonparametric statistics   *See* **statistics.**

nonparental ditype   *See* **tetrad segregation types.**

nonpermissive cells   *See* **permissive cells.**

nonpermissive conditions   environmental settings in which conditional lethal mutants fail to survive.

nonpolar   referring to water-insoluble chemical groups, such as the hydrophobic side chains of amino acids.

nonreciprocal recombination   *See* **unequal crossing-over.**

nonreciprocal translocation   *See* **translocation.**

nonrecurrent parent   that parent of a hybrid which is not again used as a parent in backcrossing.

nonrepetitive DNA   segments of DNA exhibiting the reassociation kinetics expected of unique sequences; single sequence DNA.

nonsense codon   synonomous with stop codon *(q.v.)*.

nonsense mutation   a mutation that converts a sense codon to a chain terminating codon or *vice versa*. The results following translation are abnormally short or long polypeptides, generally with altered functional properties.

nonsense suppressor   a gene coding for a tRNA that is mutant in its anticodon and therefore able to recognize a nonsense (stop) codon; nonsense suppressors cause extension of polypeptide chain synthesis through stop codons. *See* **amber suppressor, ochre suppressor, readthrough.**

nonspherocytic hemolytic anemia   *See* **glucose-6-phosphate dehydrogenase deficiency.**

nopaline   *See* **opine.**

NOR   nucleolar organizer region.

noradrenaline   norepinephrine.

norepinephrine   a hormone of the adrenal medualla which causes vasoconstriction and raises the blood pressure.

***n* orientation**   one of two possible orientations for inserting a target DNA fragment into a vector; in the *n* orientation, the genetic map of both target and vector have the same orientation; in the *u* orientation, the target and the vector are in different orientations.

Norite   trademark for a widely used group of activated charcoals.

normal distribution   the most commonly used probability distribution in statistics. The formula for the normal curve is $Y = \dfrac{1}{\sqrt{2\pi}\,\sigma}\, e^{-(X-m)2/2\,\sigma 2}$ where m = the mean, σ = the standard deviation, e = the base of natural logarithms, π = 3.1416, and Y = the height of the ordinate for a given value of X. The graph of this formula, the normal curve, also called Laplacian or Gaussian, is bell shaped. The value of m locates the curve along the abscissa and that of σ determines its shape. The larger the standard deviation the broader the curve. In nature a vast number of continuous distributions are normally distributed.

normalizing selection   the removal of those alleles that produce deviations from the average population phenotype by selection against all deviant individuals. Such selection will reduce the variance in subsequent generations. Also called *stabilizing selection, centripetal selection.*

normal solution   one containing a gram equivalent weight of solute dissolved in sufficient water to make a liter of solution.

norm of reaction   the phenotypic variability produced by a given genotype under the range of environmental conditions common to the natural habitat of the species or under the standard culture or experimental conditons. *See* **adaptive norm.**

northern blotting   *See* **Southern blotting.**

Notch   a deficiency of band 3C7 in the X chromosome of *Drosophila melanogaster.* Females carrying but one dose of the plus allele of *Notch* (N^+) show notches in trailing edges of their wings. Male hemizygotes die as embryos.

Notophthalmus viridescens   the common spotted newt of the eastern United States. The amplification of rDNA has been extensively studied using the oocytes of this species. *See* **lampbrush chromosomes, ribosomal DNA,** *Triturus.*

novobiocin   an antibiotic produced by *Streptomyces niveus.*

np   nucleotide pair.

nRNA   nuclear RNA.

N-terminal end   proteins are conventionally writeen with the amino (NH_2) end to the left. The assembly of amino acids into a polypeptide starts at the N-terminal end. *See* **translation.**

266

N-terminus N-terminal end.

nu (ν) body particles arranged like beads on a string along interphase chromosomes. These are most clearly seen in electron micrographs of negatively stained Miller spreads (*q.v.*). Nu bodies correspond to the nucleosomes (*q.v.*) of the biochemist.

nucellus the central tissue of the ovule. It contains the embryo sac and is surrounded by integuments.

nuclear duplication mitosis (*q.v.*).

nuclear emulsion a photographic emulsion especially compounded to make visible the individual tracks of ionizing particles.

nuclear envelope an envelope surrounding the nucleus, composed of two membranes enclosing a perinuclear cisterna. The outermost membrane is studded with ribosomes. The perinuclear cisterna is traversed by ring-shaped pores.

nuclear family a pair of parents and their children.

nuclear fission a transformation of atomic nuclei characterized by the splitting of a nucleus into at least two other nuclei and the release of amounts of energy far greater than those generated by conventional chemcial reactions.

nuclear fusion the coalescence of two or more atomic nuclei with the release of relatively vast amounts of energy.

nuclear processing of RNA a primary transcript (*q.v.*) experiences excision of introns and splicing of exons. If it is to serve as an mRNA, a methylated cap (*q.v.*) is added to the 5′ end and a poly-A tail is usually added to the 3′ end prior to transport from the nucleus to the cytoplasm. *See* **pre-ribosomal RNA.**

nuclear reactor the apparatus in which nuclear fission may be sustained in a self-supporting chain reaction. A source of energy and radioisotopes.

nuclear RNA RNA molecules found in the nucleus either associated with chromosomes or in the nucleoplasm. *See* **chromosomal RNA, heterogeneous nuclear RNA.**

nuclear transfer the injection of a diploid somatic nucleus into an enucleated egg. The nature of the ensuing development reveals the developmental potentialities of the implanted nucleus. Various amphibian species are generally used *(Rana pipiens, Xenopus laevis).*

nuclease any enzyme which breaks down nucleic acids.

nucleic acid a nucleotide polymer. *See* **deoxyribonucleic acid, ribonucleic acid.**

nucleocapsid a virus nucleic acid and its surrounding capsid. *See* **capsomere.**

nucleo-cytoplasmic ratio the ratio of the volume of nucleus to the volume of cytoplasm.

nucleoid **1.** a DNA-containing region within a prokaryote, mitochondrion, or chloroplast. **2.** in an RNA tumor virus, the core of genetic RNA surrounded by an icosahedral protein capsid.

nucleolonema *See* **nucleolus.**

nucleolus an RNA-rich, spherical body associated with a specific chromosomal segment, the nucleolus organizer. In maize, the nucleolus organizer resides on chromosome 6. The illustration shows this chromosome and its nucleolus as they appear in meiotic prophase.

nucleolus

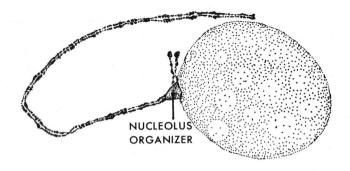

The nucleolus organizer contains the ribosomal RNA genes (*q.v.*), and the nucleolus is composed of the primary products of these genes, their associated proteins, and a variety of enzymes (e.g., RNA polymerase, RNA methylases, and RNA endonucleases). Under the electron microscope, the nucleolus is seen to consist of a fibrous central core (pars amorpha) and a granular cortex (pars granulosa). *See* **Miller trees, pre-ribosomal RNA, rDNA amplification.**

nucleolus organizer   *See* **nucleolus.**

nucleolus organizer region (NOR)   *See* **nucleolus.**

nucleon   a constituent particle of an atomic nucleus.

nucleoplasm   the protoplasmic fluid contained in the nucleus.

nucleoprotein   a compound of nucleic acid and protein. Either one of two main classes of basic proteins are found combined with DNA: one of low molecular weight (protamine) and one of high molecular weight (histone). The basic amino acids of these proteins neutralize the phosphoric acid residues of the DNA.

nucleosidase   any enzyme that catalyzes the splitting of nucleosides into bases and pentoses.

nucleoside   a purine or pyrimidine base attached to ribose or deoxyribose. The nucleosides commonly found in DNA or RNA are cytidine, cytosine deoxyriboside, thymidine, uridine, adenosine, adenine deoxyriboside, guanosine, and guanine deoxyriboside. Note that thymidine is a deoxyriboside and cytidine, uridine, adenosine, and guanosine are ribosides. *See* **rare bases, inosine.**

nucleosome   a bead-like structure of eukaryotic chromosomes, consisting of a core of eight histone molecules (two each of proteins H2A, H2B, H3, and H4) wrapped by a DNA segment about 146 base pairs in length and separated from adjacent nucleosomes by a "linker" DNA sequence of variable length (8–114 bp). *See* **histones.**

nucleotide   one of the monomeric units from which DNA or RNA polymers are constructed, consisting of a purine or pyrimidine base, a pentose, and a phosphoric acid group. The nucleotides of DNA are deoxyadenylic acid, thymidylic acid, deoxyguanilic acid, and deoxycytidylic acid. The corresponding nucleotides of RNA are adenylic acid, uridylic acid, guanylic acid, and cytidylic acid.

nucleotide pair   a hydrogen-bonded pair of purine-pyrimidine nucleotide bases on opposite strands of a double helical DNA molecule. Normally adenine pairs with thymine and guanine pairs with cytosine; also called *complementary base pairs. See* **deoxyribonucleic acid.**

268

nucleotide pair substitution   the replacement of a given nucleotide pair by a different pair, usually through a transition or a transversion, *both of which see.*

nucleus   the spheroidal structure present in most cells which contains the chromosomes. *See* **nuclear envelope**.

nuclide   a species of atom characterized by the constitution of its nucleus. This is specified by the number of protons and neutrons it contains.

nude mouse   a congenitally hairless and athymic strain of laboratory mouse caused by homozygosity for an autosomal recessive gene *(nu/nu)*. Such mice are virtually devoid of T lymphocytes and therefore must be reared in an essentially germ-free environment. *See* **gnotobiosis**.

null allele   an allele that produces no functional product and therefore usually behaves as a genetic recessive. For example, in the human ABO blood group system, the recessive allele (I^O) products no detectable antigen, either in homozygous condition (blood group O) or in heterozygous condition with allele I^A (blood gorup A) or with allele I^B (blood group B).

null cells   lymphocytes which lack the typical cell surface markers of T and B lymphocytes.

null hypothesis method   the standard hypothesis used in testing the statistical significance of the difference between the means of samples drawn from two populations. The null hypothesis states that there is no difference between the populations from which the samples are drawn. One then determines the probability that one will find a difference equal to or greater than the one actually observed. If this probability is 0.05 or less, the null hypothesis is rejected, and the difference is said to be significant.

nulliplex   *See* **autotetraploidy**.

nullosomic   lacking both members of a pair of chromosomes.

number of degrees of freedom   *See* **degrees of freedom**.

numerical taxonomy   the classification of related organisms using a multitude of characteristics, each one of which is given equal weight. The degree of similarity between the organisms is calculated using a digital computer which treats the data collected for all characteristics and determines the similarities, taking all possible pairs of the organisms in question.

nu particles   *See* **nucleosomes**.

nurse cells   cells in the insect ovary that function to nourish the oocyte. In insects with polytrophic meroistic ovaries, the oocyte and its accompanying nurse cells are descendants of a single cell, the germinal cytoblast. In *Drosophila melanogaster,* this cell divides four times to give a clone of 16 interconnected cells. One (either 1^4 or 2^4) becomes the oocyte, and the others, nurse cells. The nuclei of the 15 nurse cells undergo endomitotic cycles of DNA replication and are active in transcribing RNA, which enters the cytoplasm. Nurse cell cytoplasm continually streams into the oocytes, which grows at their expense. *See* **insect ovary types**.

nutritional mutant   a mutation converting a prototroph into an auxotroph.

nutritive chord   *See* **insect ovary types**.

nymph   a wingless, sexually immature stage of a hemimetabolous insect.

nystagmus   a jerky twitching of the eye.

nurse cells

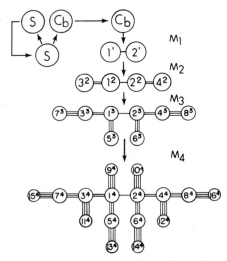

A diagram of the steps in the production of a clone of 16 cystocytes. The stem cell (S) divides into 2 daughters: one behaves like its parents; the other differentiates into a cystoblast (Cb). This cell, by a series of 4 mitoses ($M_1 - M_4$), each followed by incomplete cytokinesis, produces a branching chain of 16 interconnected cells. Cystocytes (represented by open circles) are given superscript designations (1,2,3,4) depending on whether they belong to the 1st, 2nd, 3rd, or 4th generation. The area in each circle is proportional to the volume of the cell. The ring canals connecting adjacent cells are labeled according to the division at which they were produced.

nystatin   a polyene antibiotic (*q.v.*).

O

O 1. Ordovician. 2. oxygen.

O antigens polysaccharide antigens of the cell walls of enterobacteria such as *Escherichia* or *Salmonella;* in contrast to the polysaccharide K antigens of bacterial capsules or the protein H antigens of bacterial flagella.

oat *See Avena sativa.*

obligate parasite an organism that cannot live in the absence of its host.

Oc abbreviation for an ochre suppressor (*q.v.*). For example, if mutation 18 in the *his* gene is a supressor-sensitive Oc type, it would be written his18(Oc). *See* **Am.**

Occam's razor the principle that the simplest sufficient hypothesis is to be preferred, even though others are possible. *See* **parsimony.**

oceanic island an island that has risen from the sea. *See* **continental island.**

ocellus one of the simple eyes located near the compound eyes of an insect; an eyespot in many invertebrates.

ochre codon a triplet of mRNA nucleotides (UAA) usually not recognized by any tRNA molecules; one of three stop codons that normally signal termination of translation. *See* **amber codon, opal codon.**

ochre mutation one of a group of mutations resulting in abnormally short polypeptide chains. Because of a base substitution a codon specifying an amino acid is converted to UAA, which signals chain termination. UAA appears to be the codon primarily used for chain termination in *Escherichia coli. See* **amber mutation, nonsense mutation.**

ochre suppressor any mutant gene coding for a mutant tRNA whose anticodon can respond by the insertion of an amino acid into the stop codon UAA as well as or instead of its standard codon.

octad a fungal ascus containing eight linear ascospores; produced in some ascomycete species when the tetrad of meiospores undergoes a mitotic division following meiosis. *See* **ordered tetrad.**

octopine *See* **opine.**

ocular albinism a hereditary eye disease of humans occurring in two forms, one inherited as an autosomal recessive and the other as an X-linked recessive. In females heterozygous for X-linked ocular albinism, the retinas show a mosaic pattern of pigment distribution due to the random inactivation of X chromosomes at an early developmental stage. In males, the prevalence of X-linked ocular albinism is 1/50,000. *See* **dosage compensation.**

OD optical density. *See* **Beer-Lambert law.**

Oenothera lamarckiana the evening primrose, the species first studied genetically by Hugo De Vries about 1886. During meiosis, plants of this and related species, such as *O. grandiflora,* have their chromosomes arranged in rings rather than pairs. The evolution of this atypical cytogenetic behavior, the result of the accumulation of reciprocal translocations, has been extensively studied.

Ohno's hypothesis the proposal advanced by S. Ohno that the unique regulatory features of the X chromosomes dictate the evolutionary conservation of the primordial X-linkage group among mammals. Thus, if any gene locus is found to be X-linked in humans, it is likely to be X-linked in all other mammalian species.

oil-immersion objective the objective lens system used for highest resolution with the light microscope. The space between the coverslip over the object to be examined and the lens is filled with a drop of oil of the same refractive index as the glass.

Okazaki fragments *See* **replication of DNA.**

Olduvai Gorge a site in Tanzania, Africa, rich in hominid fossils.

Oligocene an epoch in the Tertiary period lasting from 23 to 37 million years before present. *See* **geologic time divisions.**

oligo dA (oligo dT) a homopolymer chain of deoxyriboadenylate (or deoxyribothymidylate) of unspecified length, but generally 100–400 residues.

oligogene a gene producing a pronounced phenotypic effect as opposed to a polygene (*q.v.*) which has an individually small effect.

oligomer a molecule made up of a relatively few monomeric subunits.

oligomycin a polyene antibiotic (*q.v.*).

oligonucleotide a polymer made up of a few (2–10) nucleotides.

oligopyrene sperm *See* **sperm polymorphism.**

oligosaccharide a polymer made up of a few (2–10) monosaccharide units.

oligospermia an abnormally low concentration of sperm in the semen.

ommatidium a facet of the insect compound eye.

ommochrome a product of tryptophan metabolism which given rise to pigments, particularly the eye pigments of certain animals. *See* **xanthommatin.**

oncogene a gene that induces uncontrolled cell proliferation. Cellular and viral oncogenes differ in that only cellular oncogenes contain introns. *See* **Rous sarcoma virus.**

oncogene hypothesis a proposal put forth in 1969 by R. I. Huebner and G. I. Todaro that carcinogens of many sorts act by inducing the expression of retrovirus genes already resident in the target cell. It is now known that while cells from different species harbor genes homologous to retrovirus oncogenes, the cellular genes were the progenitors of the viral oncogenes. The cellular genes are now called *proto-oncogenes* (*q.v.*), and they evidently function in the normal physiology of cells from evolutionarily diverse species.

oncogenic virus   a virus that can transform the cells it infects so that they proliferate in an uncontrolled fashion.

oncolytic   capable of destroying cancer cells.

oncornavirus   an acronym for *onco*genic *RNA virus*. See **retrovirus.**

one gene-one enzyme hypothesis   the hypothesis that a large class of genes exist in which each gene controls the synthesis or activity of but a single enzyme. Since enzymes and other proteins whose syntheses are controlled by more than one gene are now known, this hypothesis has been replaced by the **one gene-one polypeptide hypothesis** (*q.v.*).

one gene-one polypeptide hypothesis   the hypothesis that a large class of genes exist in which each gene controls the synthesis of a single polypeptide. The polypeptide may function independently or as a subunit of a more complex protein.

one step growth experiment   an experiment first performed by E. L. Ellis and M. Delbrück in which phage were absorbed on *E. coli,* incubated, and the concentration of free phage particles determined at periodic intervals. The phage concentration remained constant for a time, then rose sharply, and then plateaued to give a "one step" growth curve. They concluded that after infection there is a latent period, followed by the rupture of the host cell and release of the progeny phage.

ONPG   *o*-nitrophenyl galactoside; an unnatural substrate for the enzyme beta-galactosidase, which cleaves it into galactose and *o*-nitrophenol; the latter is a yellow substance, easily assayed spectrophotometrically. ONPG has been extensively used to determine enzyme activity associated with mutants of the *lac* operon (*q.v.*) in *E. coli.* Unlike IPTG (*q.v.*), ONPG is not an inducer of the operon, so these two substances are often used in combination.

ontogeny   the development of the individual from fertilization to maturity.

oocyte   the cell that upon undergoing meiosis forms the ovum.

oocyte assay of mRNA   a technique developed by J. B. Gurdon which involves the microinjection of nanogram amounts of preparations containing mRNAs into *Xenopus* oocytes. Here the exogenous mRNAs are efficiently translated, showing that the oocyte lacks specificity with respect to the type of mRNA it can process.

oogenesis   the formation of eggs, including the meiotic behavior of oocytes, vitellogenesis, and the formation of egg membranes.

oogonium   a mitotically active cell that serves as the source of oocytes.

oolemma   the plasma membrane of the ovum.

ooplasm   the cytoplasm of an oocyte.

ootid nucleus   one of the four haploid nuclei formed by the meiotic divisions of a primary oocyte. Three of the nuclei are discarded as polar nuclei and the remaining functions as the female pronucleus. *See* **oriented meiotic division.**

opal codon the nonsense mRNA codon UGA. In contrast to amber (Am) and ochre (Oc), no symbol for opal has been agreed upon.

opaque-2 a mutant strain of corn that produces proteins rich in lysine. Mutants of this kind are of potential use in combating kwashiorkor (*q.v.*).

open population one freely exposed to gene flow (*q.v.*).

open reading frame regions in a DNA molecule where successive nucleotide triplets potentially can be read as codons specifying amino acids and where the sequence of these triplets is not interrupted by stop codons. *Contrast with* **closed reading frame.**

operational definition a definition in terms of properties significant to a given experimental situation, without consideration of whether there may be more fundamental characteristics of that which is defined.

operator a chromosomal region capable of interacting with a specific repressor, thereby controlling the functioning of adjacent cistrons. *See* **regulator genes.**

operon a unit consisting of *adjacent cistrons* that function coordinately under the control of an operator gene. *See lac* **operon, regulator genes.**

operon network a collection of operons and their associated regulator genes which interact in the sense that the products of structural genes in one operon serve to activate or suppress another operon by acting as repressors or effectors.

opine a compound, specifically synthesized by crown gall plant cells, that can be used by agrobacteria as specific growth substances. Examples are nopaline [N-α-(1,3-dicarboxyl-propyl)-L-arginine] and octopine [N-α-(D-1-carhboxyethyl)-L-arginine].

opisthe the posterior daughter organism produced in a transverse division of a protozoan.

opportunism a theory that (1) all potential modes of existence will eventually be tried by some group and all potential niches will eventually become occupied, and (2) organisms evolve only as historical conditions permit and not according to what would theoretically be best.

opportunistic species a species specialized to exploit newly opened habitats because of its ability to disperse for long distances and to reproduce rapidly.

opsin a species-specific protein that combines with retinene to form rhodopsin.

opsonin any substance that promotes cellular phagocytosis. When antibodies bind to antigens by their Fab portions (*see* **immunoglobulin**), the shape of the molecule changes to expose the Fc region. Scavenger cells such as macrophages have Fc receptors on their surfaces. Thus, phagocytic cells can bind to and engulf antigen-antibody complexes. Neutrophils and macrophages have receptors for certain activated complement components. Thus antigen-antibody-complement complexes also enhance phagocytosis through immune adherence. IgG antibodies are much more effective opsonins than IgM in the absence of complement, but IgM antibodies are more effective opsonins in the presence of complement.

optical antipodes enantiomers *(q.v.).*

optical density *See* **Beer-Lambert law.**

optical isomers molecular isomers which in solution cause the rotation of the plane of a beam of plane-polarized light passed through the solution. The rotation is due to the asymmetry of the molecule. Molecules with this property are given the prefix *d* or *l* depending on whether the plane is rotated to the right (dextro) or to the left (levo).

orange G an acidic dye often used in cytochemistry.

$$
\begin{array}{c}
\text{OH} \quad \text{H} \\[2pt]
\text{H}\;\;\text{H} \qquad\qquad \text{C}=\text{C} \\[2pt]
\text{C}-\text{C} \qquad\qquad\qquad\qquad \text{CH} \\[2pt]
\text{HC}\qquad\text{C}-\text{H}=\text{N}-\text{C}\qquad \\[2pt]
\text{C}=\text{C}\qquad\qquad \text{C}-\text{C} \\[2pt]
\text{H}\;\;\text{H} \qquad\qquad\qquad\qquad \text{CH} \\[2pt]
{}^{-}\text{O}_3\text{S}-\text{C} \\[2pt]
\text{C}-\text{C} \\[2pt]
\text{H}\qquad \text{SO}_3^{-}
\end{array}
$$

orcein a dye used in cytology. *See* **aceto-orcein.**

$$
\begin{array}{c}
\text{CH}_3 \\
\text{C} \\
\text{HC}\qquad\text{CH} \\
\text{HO}-\text{C}\qquad\text{C}-\text{OH} \\
\text{C} \\
\text{H}
\end{array}
$$

ordered octad *See* **ordered tetrad.**

ordered tetrad a linear sequence of four haploid meiotic cells (or pairs of each of four haploid cells produced by a post meiotic division) within a fungal ascus. This physical arrangement allows identification of chromatids participating in crossover events. Drawing A illustrates that, in a tetrad heterozygous for alleles controlling ascospore pigmentation, single crossovers between these genes and the centromere will generate spores showing 2-2-2-2 and 2-4-2 segregation patterns. Drawing B illustrates that in *Neurospora* such patterns are observed, together with non-crossover asci showing 4-4 distributions.

Ordovician a period in the Palezoic era lasting from 435 to 570 million years before present. *See* **geologic time divisions.**

organ a part of an animal or plant that forms a structural and functional unit (a heart, a flower, etc.).

organ culture the maintenance or growth of organ primordia or the whole or parts of an organ *in vitro* in a way that may allow further differentiation or the preservation of architecture or function or both. *See also* **in *vivo* culturing of imaginal discs.**

organelle a complex cytoplasmic structure of characteristic morphology and function, such as a mitochrondrion or plastid.

organic **1.** pertaining to organisms (dead or alive) or to the chemicals made by them. **2.** chemical compounds based on carbon chains or rings. They may also contain oxygen, hydrogen, nitrogen, and various other elements.

organicism *See* **compositionism.**

organizer a living part of an embryo which exerts a morphogenetic stimulus upon another part, bringing about its determination and morphological differentiation.

organogenesis the formation of organs.

Oriental designating one of the six biogeographic realms (*q.v.*) of the globe, including the

ordered tetrad

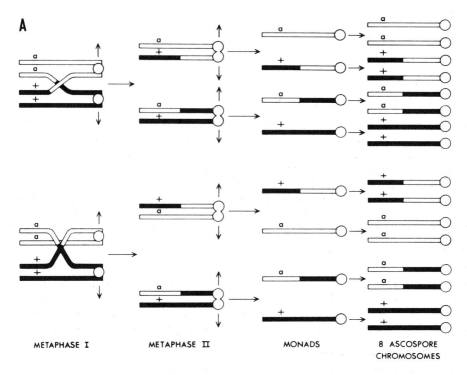

METAPHASE I   METAPHASE II   MONADS   8 ASCOSPORE CHROMOSOMES southern coast of Asia east of the Persian Gulf, the peninsula of India south of the Himalayas, eastern India, south China, Sumatra, Borneo, Java, Celebes, and the Philippines.

oriented meiotic division   an oocyte meiotic division, as in *Drosophila,* where the spindles are oriented in single file with their long axes perpendicular to the egg surface. The nucleus farthest from the surface functions as the oocyte pronucleus. Thus chromosomes differentially distributed to the other nuclei (discentrics, for example) are eliminated in the polar nuclei.

Origin of Species   an abbreviated name for the most famous book by Charles Darwin that documented the phenomenon of evolution and elaborated a theory to explain its mechanism. The full title of the book was *On the Origin of Species by means of Natural Selection, or the Preservation of Favored Races in the Struggle for Life.* The first edition was published in 1859, and no biological treatise written before or since has produced an impact upon society equal to it. The 1,250 copies of the first edition were sold out the first day. *The Origin* went through six editions, the last in 1872.

ori site   a 422 base-pair segment of the *E. coli* chromosome where replication is initiated.

ornithine cycle   a cyclic series of reactions in which potentially toxic, nitrogenous products from protein catabolism are converted to urea which is innocuous. In the cycle diagrammed, ammonia is removed from the system and used in the conversion of ornithine to citrulline. Aspartic acid enters the cycle, and its amino group is incorporated into arginosuccinic acid before it can form ammonia. Arginosuccinic acid is converted to arginine, and the fumaric acid released enters the citric acid cycle (*q.v.*). Urea splits off arginine and regenerates ornithine. In man, mutants are known which block the cycle at any one of its steps, as shown in the diagram on page 278. Blocking produces disorders that include: ornithine transcarbamylase deficiency, from blocking of step 1; citrullinuria (condensing enzyme deficiency), step 2; arginosuccinic aciduria (arginosuccinase deficiency), step 3; lysine intolerance (inhibition of arginase by excess lysine), step 4.

ornithophily   pollination by birds.

orphan drugs   pharmaceuticals developed to treat diseases that afflict relatively few people.

orphan viruses   viruses found in the digestive and respiratory tracts of healthy people; hence they are nonpathogenic (orphan = without an associated disease). *See* **reoviruses**.

orphons   dispersed, single pseudogenes (*q.v.*) derived from tandemly repeated families or gene clusters, such as those for histones or hemoglobins. Orphons may serve as a reservoir of sequences which can evolve new functions, and have probably been important factors in the evolution of higher organisms. Not to be confused with viruses called "orphans." *See* **orphan viruses**.

ortet   the original single ancestor of a clone.

orthochromatic dye   a dye that stains tissues a single color in contrast to a metachromatic dye (*q.v.*).

orthochromatin   conventional, stable chromatin, in contrast to parachromatin (*q.v.*).

orthogenesis   a discredited theory that a steady trend of evolution in a given direction over a prolonged period of time, affecting related groups of organisms, is due to mystic forces rather than to natural selection.

orthopteran   an organism belonging to the Orthoptera, an order of the Hemimetabola containing cockroaches, locusts, grasshoppers and similar insects.

orthoselection   continuous selection on the members of a lineage over a long time, causing

ornithine cycle

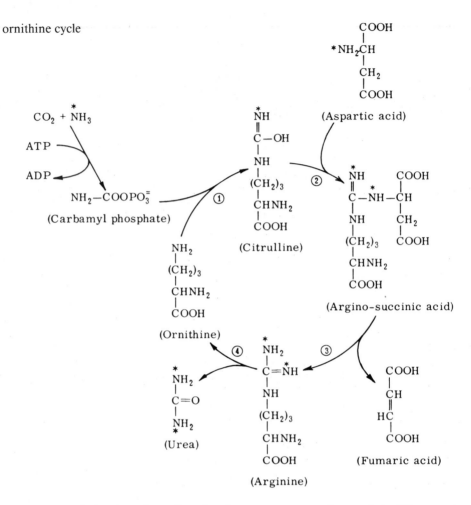

continued evolution in a given direction that may create an impression of "momentum" or "inertia" in evolutionary trends.

orthotopic transplantation   the transplanation of grafts between identical sites in such a way that the graft maintains its normal orientation.

Oryctolagus cuniculus   the rabbit, a mammal belonging to the Lagomorpha commonly reared in the laboratory and the subject of intensive genetic research. An extensive collection of mutations is available influencing a wide variety of morphological and physiological traits.

Oryza sativa   rice *(q.v.)*.

Oryzias latipes   the medaka, a fresh-water fish common to Japan, Korea, and China. It is easily maintained in the laboratory and was the first fish in which Mendel's laws were shown to be valid. Y-linked inheritance was first demonstrated in the medaka and the guppy.

osmium tetroxide, OsO_4, a compound often used as a fixative in electron microscopy.

osmosis   diffusion of a solvent through a semipermeable membrane separating two solu-

278

tions of unequal solute concentrations. The direction of solvent flow tends to equalize the solute concentrations.

otu mutation   a sex-linked female sterile gene in *Drosophila melanogaster* that is remarkable in that different mutant alleles can produce quite different ovarian pathologies. One class produces *ovarian tumors*, hence the *otu* symbol. Another class is characterized by germaria that lack oogonia. A third class generates oocytes that grow slowly and fail to complete development. The accompanying nurse cells contain polytene chromosomes, and the largest of these have 8,000 times the haploid amount of DNA.

Ouchterlony technique   a gel-diffusion, antibody-antigen precipitation test that depends upon horizontal diffusion from two or more opposite sources. An agar slab is prepared and two or more wells are cut in it. One (A) is filled with a aqueous suspension of antibody molecules, while each of the other wells (B and C in the illustration) is filled with a different antigen preparation. The antigen and antibody molecules diffuse toward each other and eventually interact, forming curved precipitation lines. In example I, the antigens in wells B and C are different. In II, well B contains a single antigen and well C two antigens, one identical to that in B. Named after Orjan Ouchterlony, who developed the methodology.

I

II

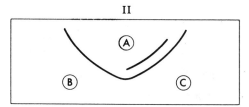

Oudin technique   a gel-diffusion, antibody-antigen precipitation test that depends on simple vertical diffusion in one dimension. A gel column is prepared containing a homogeneous distribution of antibody molecules. Above this is layered an aqueous suspension of antigen molecules. As these diffuse into the gel a moving zone of antigen-antibody precipitate is formed. If several antigens and antibodies are present, separate zones of interacton will be seen. Named after Jacques Oudin, who developed the methodology.

outbreeding   the crossing of genetically unrelated plants or animals; crossbreeding.

outcross   *See* **outbreeding.**

out-group   a species or higher monophyletic taxon that is examined in the course of a phylogenetic study to determine which of two homologous character states may be inferred to be apomorphous. The most critical out-group comparison involves the sister group of the taxon under study.

outlaw gene   a gene favored by selection despite its disharmoneous effects on other genes in the same organism. *See* **meiotic drive.**

ovariectomy   the surgical removal of one or both ovaries.

ovariole   one of several egg tubes constituting the ovary of most insects.

ovary   the female gonad in animals or the ovule-containing region of the pistil of a flower.

overdominance   the phenomenon of heterozygotes having a more extreme phenotype than either homozygote; monohybrid heterosis. Overdominance generally refers to the situation in with AA' individuals are more fit than AA or $A'A'$ individuals.

overlapping code   a hypothetical genetic code, first proposed by George Gamow, in which any given nucleotide is shared by two adjacent codons. The genetic code used in biological systems was later shown to be non-overlapping. *See* **overlapping genes** for the few exceptions.

overlapping genes   genes whose nucleotide sequences overlap to some degree. The overlap may involve control genes (e.g., tryptophan operator and promoter regions in *E. coli*) or structural genes (e.g., in bacteriophage ϕX174, gene E lies entirely within gene D, but they are translated in different reading frames).

overlapping inversion   a compound chromosomal inversion caused by a second inversion that includes part of a previously inverted segment.

overwinding   positive supercoiling of DNA, resulting in further tension in the direction of winding of the two strands of a duplex about each other.

ovicide   any compound that destroys eggs, especially a compound that destroys insect eggs.

oviduct   the tube carrying eggs from the ovary to the uterus.

ovine   referring to members of the sheep family, especially the domestic sheep species *Ovis aries*.

oviparous   laying eggs in which the embryo develops outside the mother's body and eventually hatches. *See* **ovoviviparous, viviparous.**

oviposition   the laying of eggs by a female insect.

ovipositor   an organ at the hind end of the abdomen in female insects, through which eggs are laid.

Ovis aries   the domestive sheep. The blood group genetics of this species has been intensively investigated. *See* **sheep.**

ovisorption   the resorption of oocytes.

ovogenesis   oogenesis (*q.v.*).

ovotestis   the organ of some hermaphroditic animals which functions both as ovary and testis; the gonad of an animal that undergoes consecutive sexuality (*q.v.*).

ovoviviparous   bringing forth young that develop from eggs retained within the maternal body, but separated from it by the egg membranes. Many fish, reptiles, molluscs, and insects are ovoviviparous. *See* **oviparous, viviparous.**

ovulation   the release of a ripe egg from the mammalian ovarian follicle, frequently at the stimulus of a pituitary hormone.

ovule   the structure found in seed plants which develops into a seed after the fertilization of an egg cell within it. *See* **nucellus.**

ovum   an unfertilized egg cell.

oxidation   the gain of oxygen, loss of hydrogen, or loss of electrons by a compound.

oxidation-reduction reactions   chemical reactions in which electrons are transferred from a reductant to an oxidant; as a consequence of the transfer, the reductant is oxidized and the oxidant is reduced.

oxidative phosphorylation   the enzymatic phosphorylation of ADP to ATP, which is coupled to the electron transport chain (*q.v.*). Thus respiratory energy is transformed into phosphate-bond energy.

oxygen   the second most abundant of the biologically important elements. Atomic number 8; atomic weight 15.9994; valence 2^-; most common isotope ^{16}O.

oxyhemoglobin   oxygenated hemoglobin.

oxytocin   a polypeptide secreted by the hypothalamus and stored in the neurohypophysis. Oxytocin causes smooth-muscle contraction and may help terminate pregnancy.

P

p 1. commonly used to represent phosphoric acid molecules, as in ppGpp. 2. referring to the small arm of a chromosome. 3. proton.

P 1. Permian. 2. probability. 3. phosphorus. 4. phosphate (when combined in an abbreviation such ADP, ATP. 5. symbol for the panmictic index (*q.v.*).

P₁ the symbol denoting the immediate parents of the F_1 generation. The symbols P_2 and P_3 are used to designate the grandparental and great-grandparental generations, respectively, if one starts from the F_1 and works backward.

³²P a radioactive isotope of phosphorous, widely used to label nucleic acids; it emits a strong beta particle and has a half life of 14.3 days.

P$_i$ inorganic phosphate.

pachynema *See* **meiosis.**

pachytene stage *See* **meiosis.**

packing ratio the ratio of DNA length to the unit length of the fiber containing it.

paedogenesis precocious sexual maturity in the larval stage of some animals. In the axolotyl, for example, the gill-breathing, water-dwelling larva can mature sexually and reproduce without undergoing metamorphosis to the lung-breathing, land-dwelling adult form. *See Ambystoma mexicanum.*

paedomorphosis an evolutionary phenomenon in which adult descendants resemble the youthful stages of their ancestors; the opposite of "recapitulation," in which the early stages of descendants resemble the adult stages of their ancestors. Paedomorphosis may be produced either by the acceleration of sexual development (progenesis) or by retarded somatic development.

Paeonia californica the California peony, a species with naturally occurring translocation complexes.

PAGE polyacrylamide gel electrophoresis. *See* **electrophoresis, polyacrylamide gel.**

pair bonding an intimate and long lasting association between a male and a female of the same species, generally facilitating the cooperative rearing of their offspring.

282

pairing   synapsis.

pairing segments   the segments of the X and Y chromosomes which synapse and cross over. The remaining segments, which do not synapse, are called the differential segments.

Palearctic   designating one of the six biogeographic realms (*q.v.*) of the globe, including Eurasia except Iran, Afghanistan, the Himalayas, and the Nan-ling Range in China, Africa north of the Sahara, Iceland, Spitzbergen, and the islands north of Siberia.

Paleocene   an epoch in the Tertiary period lasting from 55 to 65 million years before present. *See* **geologic time divisions.**

Paleolithic   that phase of human history prior to the cultivation of plants during which tools were manufactured, food was obtained by hunting, fishing, or collecting wild nuts and fruits. The Paleolithic culture lasted from about 500,000 years ago up to the beginning of the Neolithic stage about 10,000 years ago.

paleontology   the study of extinct forms of life through their fossils, as opposed to neontology.

paleospecies   the successive species in a phyletic lineage which are given ancestor and descendant status according to the geological strata in which they appear. *See* **anagenesis.**

Paleozoic   the era extending from 230 to 570 million years before present; the age of invertebrates. *See* **geologic time divisions.**

Paley's watch   an argument developed by William Paley (1743–1805) for the existence of God based upon the common-sense notion that a watch is too complicated to have originated by accident; it presents its own evidence of having been purposely designed. This argument is commonly applied to living organisms by creationists.

palindrome   a sequence of deoxyribonucleotide base pairs that reads the same (5′ to 3′) on complementary strands; tandem inverted repeats; example:

$$5'\ \text{AATGCGCATT}\ 3'$$
$$3'\ \text{TTACGCGTAA}\ 5'$$

Palindromes serve as recognition sites for restriction endonucleases, RNA polymerase, and other enzymes.

palynology   the study of both living and fossil spores, pollen grains, and other microscopic propagules.

Pan   a genus containing the chimpanzee, *Pan troglodytes.*

Panagrellus redivivus   a free-living nematode whose developmental genetics is under study, primarily for comparison with *Caenorhabditis elegans* (*q.v.*). It differs from *Caenorhabditis* in having XX females and XO males. Its cell lineage (*q.v.*) is known in part.

pancreozymin   a hormone secreted by the duodenum which causes secretion of pancreatic enzymes.

pandemic   designating a disease simultaneously epidemic in human populations in many parts of the world.

panethnic   referring to a hereditary disease which is found in a variety of ethnic groups.

pangenesis   a defunct theory of development, popular in Darwin's time, proposing that small particles (pangenes) from various parts of the body distill into the gametes, resulting in a blending of parental characteristics in their offspring.

panhypopituitarism   *See* **pituitary dwarfism.**

panicle *See* **raceme.**

panmictic index (P) a measure of the relative heterozygosity. $1 - P = F$, Wright's inbreeding coefficient (*q.v.*).

panmictic unit a local population in which mating is completely random.

panmixia panmixis.

panmixis random mating, as contrasted with assortative mating (*q.v.*).

panoistic ovary *See* **insect ovary types.**

pantothenic acid vitamin B_5, a subunit of coenzyme A (*q.v.*).

$$HO-CH_2-\underset{\underset{CH_3}{|}}{\overset{\overset{CH_3}{|}}{C}}-\underset{\underset{H}{|}}{\overset{\overset{OH}{|}}{C}}-\overset{\overset{O}{\|}}{C}-\underset{\underset{H}{|}}{N}-CH_2-CH_2-COOH$$

papain a proteolytic enzyme isolated from the latex of the papaya plant.

paper chromatography *See* **chromatography.**

Papilio glaucus the swallowtail butterfly, a species extensively studied in terms of isolating mechanisms (*q.v.*).

papillary pattern the pattern of dermal ridges on the finger tips and palms.

papilloma a benign cutaneous neoplasm, a wart.

papovavirus a group of animal DNA viruses (including SV4O and polyoma) responsible for papillomas of the rabbit, dog, cattle, horse, and human.

parabiotic twins artificial "Siamese twins" produced by joining two animals surgically. Since their blood circulations will eventually anastomose, one can study the transmission of humoral agents from one "twin" to the other.

paracentric inversion an inversion (*q.v.*) which does not include the centromere.

Paracentrotus lividus a common sea urchin extensively used in studies of molecular developmental genetics. *See* **echinoderm.**

parachromatin a hypothesized type of chromatin that is sensitive to changes in the cellular environment that specifically accompany the developmental process. Controlling elements reside in parachromatin.

paracrystalline aggregate a regular linear arrangement of stacked molecules.

paradigm a scientific hypothesis set up as a model or example; a known occurence which serves as an example for a more general phenomenon.

paraffin section a section of tissue cut by a microtome after embedding in a paraffin wax; the classical method of preparing tissues for microscopical study.

paragenetic referring to a chromosomal change that influences the expression of a gene rather than its structure. *See* **position effect, Lyonization.**

parallel evolution the occurrence of the same or a similar trend independently evolved in two or more lineages; the lineages are usually, although not necessarily, related to one another.

Paramecium a common genus of ciliate protozoa found in stagnant ponds. Common species are *P. aurelia, P. caudatum,* and *P. bursaria.* These are favorite species for the study of nuclear-cytoplasmic interactions.

Paramecium aurelia A cigar-shaped, heavily ciliated protozoan, 100–150 μm long, living in still or running fresh waters. *Paramecium aurelia* consists of a group of 14 syngens. Each syngen is genetically isolated from every other syngen and is also biochemically unique. However, all syngens are so similar morphologically that they have not been given individual species names. Each syngen has two mating types. Mating types were first discovered in syngen 1. *See* **conjugation, killer paramecia.**

parameter the value of some quantitative characteristic in an entire population. For example, the mean height of all males 20 years of age or older in a pigmy tribe. *Compare with* **statistic.**

parametric statistics. *See* **statistics.**

paramutation a phenomenon discovered in corn where one allele influences the expression of another allele at the same locus when the two are combined in a heterozygote. The first allele is referred to as "paramutagenic," the second as "paramutable." The paramutable allele behaves like an unstable hypomorph when the paramutagenic allele is present.

paranemic spiral a spiral made up of two parallel threads coiled in opposite directions. The two threads can be easily separated without uncoiling. *See* **plectonemic spiral.**

paranoia a variety of schizophrenia characterized by suspiciousness, delusions of persecution, and megalomania.

parapatric referring to populations or species that occupy adjacent areas with a narrow zone of overlap within which hybridization commonly occurs.

parapatric speciation a mode of gradual speciation in which new species arise from populations that maintain genetic contact during the entire process by a narrow zone of overlap. Also called *semigeographic speciation. Compare with* **allopatric speciation, alloparapatric speciation.**

paraphyletic group an artifiical group which includes a common ancestor and some, but not all, of its descendants. Placement in the group is based on the common possession of ancestral or primitive characters.

Parascaris equorum a nematode (commonly called the horse thread-worm) studied by early cytologists because it exhibited chromosome diminution (*q.v.*).

parasexuality any mechanism that achieves nonmeiotic recombination. In molds, for example, diploid nuclei arise from rare fusions of two genetically unlike nuclei in heterokaryons. Somatic crossing-over occurs, and eventually haploid nuclei showing new combinations of genes are formed from the diploids. *See* **sexduction, transduction, transformation.**

parasitism a symbiotic association that benefits one member (the parasite) but is harmful to the other (the host).

parathyroid hormone a hormone controlling calcium and phosphorus balance, synthesized by the cells of the parathyroid gland.

paratope   the site within an immunoglobulin Fab that specifically interacts with an antigenic determinant (epitope).

Parazoa   a subdivision of the animal kingdom containing organisms that lack tissue organized into organs, and have an indeterminate shape. *See* **classification**.

parenchyma   tissues composed of thin-walled cells which fit loosely together, leaving intercellular spaces.

parental ditype   *See* **tetrad segregation types.**

parity   the fact of having born children. A woman of parity 0 has born no live children although she may have been pregnant one or more times. A woman of parity 1 has given birth only once. However, the number of children produced at this birth may have exceeded one. *See* **multiparous**

paroral cone   a protuberance in the oral region of a conjugating ciliate which juts into the body of the partner. The haploid nucleus residing in the paroral cone survives and all others degenerate. *See* **conjugation.**

pars amorpha   *See* **nucleolus.**

parsimony principle   the principle of simplicity (Occam's razor) is a rule of logic and science first promulgated by the scholastic William of Occam in 1347. It states that when there are several possible explanations of a phenomenon (i.e., conflicting hypotheses), one selects as most probable that explanation (hypothesis) which is the simplest, i.e., has the fewest assumptions and is most consistent with the data at hand. In phylogenetic analysis, the rule of parsimony is used to select between conflicting hypotheses of geneological relationships as portrayed by cladograms that depict different arrangements of derived character states. The parsimony criterion leads us to select that cladogram that has the fewest conflicting characters.

pars intercerebralis   that medial region in the insect forebrain containing neurosecretory cells.

Parthenium argentatum   the guayule, a desert plant, reproducing sexually or asexually depending upon its genotype and studied by geneticists accordingly.

parthenocarpy   the natural or artificially induced formation of fruit without seeds because of either (1) lack of pollination, (2) lack of fertilization, or (3) death of the embryo at an early stage of development. Edible bananas, for example, set fruit by parthenocarpy. Wild bananas are diploid (N = 11) and reproduced normally, whereas the commercially grown, edible bananas are triploid and both male and female are sterile.

parthenogenesis   the development of an individual from an egg without fertilization. *See* **artificial parthenogenesis, arrhenotoky, gynogenesis, thelytoky.**

partial denaturation   an incomplete unwinding of the DNA double helix; GC-rich regions are more resistant to thermal disruption because three hydrogen bonds form between G and C, whereas only two form between A and T. *See* **deoxyribonucleic acid.**

partial diploid   *See* **merozygote.**

partial dominance   semidominanace (*q.v.*).

particulate inheritance   the Mendelian theory that genetic information is transmitted from one generation to another in the form of discrete units, so that the biological inheritance of offspring is not a solution in which the parental information is blended.

parturition   the act of giving birth to young.

Pascal's pyramid   *See* **binomial distribution.**

PAS procedure   periodic acid Schiff procedure (*q.v.*).

passage   subculture (*q.v.*).

passenger   in recombinant DNA research, a DNA segment of interest that will be spliced into a DNA vehicle for subsequent cloning.

passive cutaneous anaphylaxis   an *in vivo* passive transfer test for detecting the presence of antibody responsible for immediate hypersensitivity reactions.

passive equilibrium   an unstable equilibrium resulting from selective neutrality of alleles at a genetic locus, as occurs in Hardy-Weinberg equilibrium; also called *neutral equilibrium.*

passive immunity   the immunity against a given disease produced by injection into a host of serum containing antibodies formed by a donor organism that possesses active immunity to the disease. *See* **active immunity.**

Patau's syndrome   trisomy 13. *See* **D₁ trisomy syndrome.**

paternal-X inactivation   the method of dosage compensation found in marsupials where the paternal-X chromosome is inactivated in the somatic cells of females. *See* **random-X inactivation.**

path coefficient analysis   a method invented by Sewall Wright for analyzing regular and irregular breeding systems.

pathogenic   producing disease or toxic symptoms.

patroclinous   designating an offspring that resembles the male more closely than the female parent. The sons of an attached-X female *Drosophila* are patroclinous in terms of their sex-linked genes.

P blood group   a system of human blood groups depending on an antigen on the red cells specified by the dominant gene *P*. *See* **blood groups.**

pBR322   a plasmid cloning vector that grows under relaxed control in *E. coli*. It contains ampicillin- and tetracycline-resistance genes and several convenient restriction endonuclease recognition sites.

pea   *See* **Pisum sativum.**

Peckhammian mimicry   *See* **mimicry.**

Pecten irradians   *See* **Pelecypoda.**

pectic acid   a polymer made up of galacturonic subunits.

pectin   a polysaccharide material within the cell wall and middle lamella. Pectin consists of pectic acid many of whose COOH groups have been methylated.

pedigree   a diagram setting forth the ancestral history or genealogical register. Symbols commonly used in such diagrams are illustrated below.

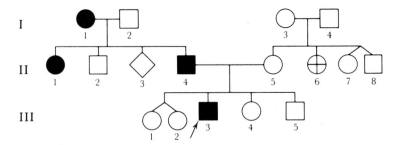

A sample pedigree. Females are symbolized by circles and males by squares. Individuals showing the trait are symbolized by solid figures. Offspring are presented beneath the parental symbols in order of birth from left to right. The arrow points to the propositus. The sex of individual II-3 is unknown. II-6 died at an early age and consequently her phenotype relative to the trait in question was unknown. II-7 and II-8 were dizygotic twins, whereas III-1 and III-2 were identical twins. Other symbols commonly encountered are shown below.

Individual died without leaving offspring	
Heterozygotes for an autosomal recessive gene	
Heterozygous female carrier of a sex-linked recessive gene	
Abortion or stillbirth; sex unknown	
Consanguinous marriage or mating	

pedigree selection   artificial selection of an individual to participate in mating based upon the merits of its parents or more distant ancestors.

pedogenesis   *See* **paedogenesis.**

Peking man   an extinct subspecies of primitive man known from fossil remains found near Peking. Now classified as *Homo erectus pekinensis,* but formerly referred to as *Sinanthropus pekinensis.*

pelargonin   a scarlet-colored anthocyan giving color to the petals of geraniums and asters.

(G = glucose)

Pelargonium zonale   the geranium. Classical studies on the non-Mendelian inheritance of chloroplast mutants were performed on this species.

Pelecypoda   the class containing the bivalve molluscs. The cytogenetics and quantitative genetics of certain species have been studied because of their economic importance. These include the American oyster, *Crassostrea virginica,* the clam, *Mercenaria mercenaria,* the mussel, *Mytilus edulis,* and the scallop, *Pecten irradians.*

P elements   transposable elements in *Drosophila melanogaster* that are responsible for one type of hybrid dysgenesis (*q.v.*). P elements have been cloned in *E. coli* plasmids. When DNA molecules carrying P elements are microinjected into *Drosophila* embryos some P elements integrate into the germ line chromosomes and are transmitted to the progeny of the injected flies. *See* **transposable element.**

P element transformation   the transfer of specific DNA segments into germ line cells of *Drosophila* using the transposable P element to carry exogenous DNA fragments.

Pelger-Huet anomaly   *See* **Pelger nuclear anomaly.**

Pelger nuclear anomaly   a hereditary abnormality in man involving the nuclear morphology of the polymorphonuclear leukocytes. A similar syndrome occurs in the rabbit. Homozygotes show chondrodystrophy in the rabbit, but not in man.

pellagra   a disease caused by deficiency of niacin.

Pelomyxa   a genus of amoebae. *P. carolinensis* is a favorite species for nuclear transplantation studies. *P. palustris* is of interest, since it lacks mitochondria and instead harbors anaerobic bacteria in a permanent symbiotic relationship.

peltate   shield-shaped.

penetrance   the proportion of individuals of a specified genotype that show the expected phenotype under a defined set of environmental conditions. For example, if all individuals carrying a dominant mutant gene show the mutant phenotype, the gene is said to show complete penetrance.

penicillin   any of a family of antibiotics derived from the mold *Penicillium notatum* and related species. Penicillins inhibit the synthesis of the wall of the bacterial cell. Bacteria poisoned with penicillin soon outgrow their cell walls and lyse. Different penicillins differ only in the side chain symbolized by R in the formula. *See* **ampicillin.**

$$R-O-NH-CH-CH \quad C(CH_3)_2$$

(chemical structure of penicillin)

penicillin enrichment technique   *See* **penicillin selection technique.**

penicillin selection technique   a method for isolating an auxotrophic mutant from a wild-type culture of bacteria by adding penicillin to minimal medium. Penicillin interferes with cell wall development, causing growing wild-type cells to rupture. However, nongrowing auxotrophic mutant cells are not killed. After one hour, about 99% of the wild-type cells have lysed, releasing their pool of metabolites into the medium. The culture must be filtered to remove these metabolites, because the auxotrophic mutants would use them for growth and be subject to penicillin-induced lysis; the filtering also removes the penicillin. Alternatively, the enzyme penicillinase can be used to destroy the penicillin. Surviving auxotrophs are then supplied with enriched medium. These cells produce the colonies that are harvested.

Penicillium notatum   an antibiotic-producing ascomycete fungus.

Pennsylvanian   an epoch in the Carboniferous period lasting from 280 to 310 million years before the present. *See* **geologic time divisions.**

pentabarbital   Nembutal (*q.v.*).

pentadactyl   having 5 digits.

pentosephosphate pathway   a pathway of hexose oxidation which is an alternative to the glycolysis-citric acid cycle (*q.v.*). The pathway involves an interconversion of phosphates of 7-, 6-, 5-, 4-, and 3-carbon sugars, and of a sugar lactone and sugar acids in a cycle that effects the complete oxidation of glucose to CO_2 and H_2O with the formation of 36 molecules of ATP. *See* pages 292–293.

penultimate   next to last.

pepo   a specialized berry with a hard rind (pumpkins and cucumbers are examples).

pepsin   a proteolytic enzyme from the gastric mucosa which functions at low *p*H.

peptidase   an enzyme catalyzing the hydrolytic cleavage of peptide bonds (*q.v.*).

peptide   a compound formed of two or more amino acids.

peptide bond   a covalent bond between two amino acids formed when the amino group of one is bonded to the carboxyl group of the other and water is eliminated.

$$-N-H \quad + \quad HO-C- \quad \longrightarrow \quad -N-C- \quad + \quad H_2O$$

with $-N-H$ having an H below, $HO-C-$ having $\|$ and O below, and $-N-C-$ having H and O below.

peptidyl transferase   *See* **translation.**

peptidyl—tRNA binding site   *See* **translation.**

peptone   a partially hydrolyzed protein.

perennation   the survival of plants from growing season to growing season with a period of reduced activity in between.

perennial   a plant that continues to grow from year to year.

perfect flower   *See* **flower.**

perfusion   the introduction of fluids into organs by injection into their arteries.

pericarp   the wall of the ovary after it has matured into a fruit; it may be dry and hard (as in the case of a nut) or fleshy (as in the case of a berry).

pericentric inversion   an inversion that includes the centromere.

periclinal   referring to a layer of cells running parallel to the surface of a plant part. *See* **anticlinal.**

periclinal chimera   a plant made up of two genetically different tissues, one surrounding the other.

perikaryon   that portion of the cell body of a neuron surrounding the nucleus as distinguished from the axon and dendrites.

perinuclear cisterna   the fluid-filled reservoir enclosed by the inner and outer membranes of the nuclear envelope.

periodic   recurring at regular intervals.

periodic acid Schiff procedure   a staining procedure for demonstrating polysaccharides. Abbreviated PAS. *See* **Schiff's reagent.**

periodicity   in molecular genetics, the number of base pairs per turn of the DNA double helix.

periodic table   an arrangement of the chemical elements (*q.v.*) in order of increasing atomic number. Elements with similar properties are placed one under the other, yielding groups and families of elements. Such a table was first drawn up by D. I. Mendeleev.

	IA	IIA	IIIB	IVB	VB	VIB	VIIB		VIII		IB	IIB	IIIA	IVA	VA	VIA	VIIA	INERT GASES
	1 H 1.00797																1 H 1.00797	2 He 4.0026
1																		
2	3 Li 6.939	4 Be 9.0122											5 B 10.811	6 C 12.01115	7 N 14.0067	8 O 15.9994	9 F 18.9984	10 Ne 20.183
3	11 Na 22.9898	12 Mg 24.312											13 Al 26.9815	14 Si 28.086	15 P 30.9738	16 S 32.064	17 Cl 35.453	18 Ar 39.948
4	19 K 39.102	20 Ca 40.08	21 Sc 44.956	22 Ti 47.90	23 V 50.942	24 Cr 51.996	25 Mn 54.9380	26 Fe 55.847	27 Co 58.9332	28 Ni 58.71	29 Cu 63.54	30 Zn 65.37	31 Ga 69.72	32 Ge 72.59	33 As 74.9216	34 Se 78.96	35 Br 79.909	36 Kr 83.80
5	37 Rb 85.47	38 Sr 87.62	39 Y 88.905	40 Zr 91.22	41 Nb 92.906	42 Mo 95.94	43 Tc (99)	44 Ru 101.07	45 Rh 102.905	46 Pd 106.4	47 Ag 107.870	48 Cd 112.40	49 In 114.82	50 Sn 118.69	51 Sb 121.75	52 Te 127.60	53 I 126.9044	54 Xe 131.30
6	55 Cs 132.905	56 Ba 137.34	57 †La 138.91	72 Hf 178.49	73 Ta 180.948	74 W 183.85	75 Re 186.2	76 Os 190.2	77 Ir 192.2	78 Pt 195.09	79 Au 196.967	80 Hg 200.59	81 Tl 204.37	82 Pb 207.19	83 Bi 208.980	84 Po (210)	85 At (210)	86 Rn (222)
7	87 Fr (223)	88 Ra (226)	89 ‡Ac (227)															

† Lanthanum series

58 Ce 140.12	59 Pr 140.907	60 Nd 144.24	61 Pm (147)	62 Sm 150.35	63 Eu 151.96	64 Gd 157.25	65 Tb 158.924	66 Dy 162.50	67 Ho 164.930	68 Er 167.26	69 Tm 168.934	70 Yb 173.04	71 Lu 174.97

‡ Actinium series

90 Th 232.038	91 Pa (231)	92 U 238.03	93 Np (237)	94 Pu (242)	95 Am (243)	96 Cm (247)	97 Bk (247)	98 Cf (249)	99 Es (254)	100 Fm (253)	101 Md (256)	102 No (253)	103 Lw (257)

*( ) Numbers in parentheses are the mass numbers of the most stable or best-known isotope. Atomic weights are based on carbon 12, conforming to the conventions adopted in 1961 by the International Union of Pure and Applied Chemistry.

Reproduced from *A Sourcebook for the Biological Sciences,* second edition, by Evelyn Morholt, Paul F. Brandwein, and Alexander Joseph, © 1958, 1966 by Harcourt, Brace and World, Inc., and reprinted with their permission.

peripheral   the part nearest the surface.

peristalsis   waves of muscular contractions that pass along tubular organs and serve to mix the contents and move them posteriorly.

perithecium   the rounded or flask-shaped, fruiting body of certain ascomycete fungi and lichens. The mature fruiting body of *Neurospora,* for example, contains about 300 ascus sacs.

peritoneal sheath   a network of anastomosing muscle fibers which holds together the ovarioles of an insect ovary.

peritrichous   designating bacteria having flagella all over their surfaces.

permeability   the extent to which molecules of a given kind can pass through a given membrane.

permease   a membrane-bound protein in bacteria that is responsible for transport of a specific substance in or out of the cell; sometimes referred to as *transport proteins.* In *E. coli,* lactose permease actively transports lactose into the cell.

Permian   the most recent of the Paleozoic periods. *See* **geologic time divisions.**

permissible dose   *See* **maximum permissible dose.**

pentosephosphate pathway

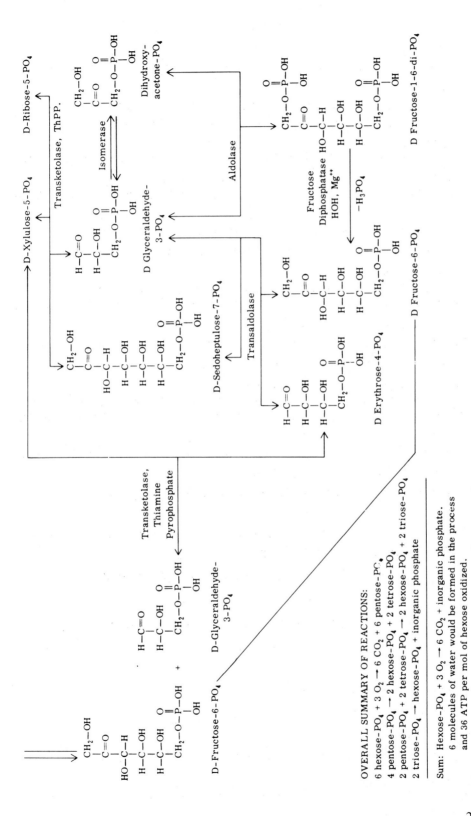

OVERALL SUMMARY OF REACTIONS:

6 hexose-PO_4 + 3 O_2 → 6 CO_2 + 6 pentose-PO_4

4 pentose-PO_4 → 2 hexose-PO_4 + 2 tetrose-PO_4

2 pentose-PO_4 + 2 tetrose-PO_4 → 2 hexose-PO_4 + 2 triose-PO_4

2 triose-PO_4 → hexose-PO_4 + inorganic phosphate

Sum: Hexose-PO_4 + 3 O_2 → 6 CO_2 + inorganic phosphate.
       6 molecules of water would be formed in the process
       and 36 ATP per mol of hexose oxidized.

permissive cells cells in which a particular virus may cause a productive infection (i.e., the production of progeny viruses). Cells in which infection is not productive are called *nonpermissive cells.* Some DNA tumor viruses may cause a productive infection in cells of one species and a tumor in cells of another species.

permissive conditions environmental conditions under which a conditional lethal mutant (e.g., a temperature-sensitive mutant) can survive and produce wild-type phenotype.

permissive temperature *See* **temperature sensitive mutation.**

Peromyscus a genus containing about 40 species of mice native to Central and North America. The genetic data accumulated so far have been primarily related to the deermouse, *P. maniculatus.* More limited data exist for *P. boylei, P. leucopus,* and *P. poliono-tus.* Studies of biochemical variation and cytogenetics are presently the areas of most genetical research within the genus.

peroxisomes membrane-bound intracellular organelles containing at least four enzymes involved in the metabolism of hydrogen peroxide. They are thought to be important in purine degradation, photorespiration, and the glyoxylate cycle. Sometimes called *microbodies.*

petites dwarf colonies of *Saccharomyces cerevisiae.* Such yeasts are slow growing because of mutations affecting mitochondria. *Segregational* petites contain mutated nuclear genes that result in mitochondrial defects. In the case of *cytoplasmic* or *vegetative* petities, it is the mitochondrial genome that contains mutations or deletions of various sizes. In situations where all mitochondrial-encoded gene products are missing, the yeast forms *promitochondria.* These have a normal outer membrane, but the inner membrane contains poorly developed cristae. In spite of the fact that the organelle cannot function in oxidative phosphorylation, it contains nuclear-encoded proteins such as DNA and RNA polymerases, all the enzymes of the citric acid cycle, and many inner membrane proteins.

Petri dish a round, shallow, covered glass container in which microorganisms or dividing eukaryotic cells are cultured on a nutrient gel (usually agar). *See* **plaque.**

pH *See* **hydrogen ion concentration.**

phaeomelanin one of the pigments found in the coats of mammals. It is derived from the metabolism of tyrosine and is normally yellow in color. The amount of this pigment inserted into the hairs is quantitatively and qualitatively controlled by the *agouti* locus. *See* **melanin.**

phage bacteriophage.

phage conversion *See* **prophage-mediated conversion.**

phage cross a procedure requiring the multiple infection of a single bacterium with phages that differ at one or more genetic sites. Upon lysis of the host, recombinant progeny phage are recovered which carry genes derived from both parental phage types. *See* **Visconti-Delbrück hypothesis.**

phage induction the stimulation of prophage to enter the vegetative state, accomplished by exposing lysogenic cells to ultraviolet light. Hydrogen peroxide, x-rays, and nitrogen mustard (*q.v.*) also act as inducing agents.

phagocyte a cell that incorporates particles from its surroundings by phagocytosis.

phagocytosis the engulfment of solid particles by cells; the ingestion of microorganisms by leukocytes.

phagolysosome an organelle formed by the fusion of a phagosome (*q.v.*) and a lysosome (*q.v.*).

phagosome a membrane-bounded cytoplasmic particle produced by the budding off of localized invaginations of the plasmalemma. Recently phagocytosed particles are segregated within the cell in phagosomes.

phanerogam an outmoded term referring to a plant belonging to the Spermatophyta (*q.v.*). *See* **cryptogam.**

phantom a volume of material approximating the density and effective atomic number of tissue. Radiation dose measurements are made within or on a phantom as a means of determining the radiation dose within or on a body under similar exposure conditions.

pharmacogenetics the area of biochemical genetics dealing with genetically controlled variations in responses to drugs.

phase contrast microscope light rays passing through an object of high refractive index will be retarded in comparison with light rays passing through a surrounding medium with a lower refractive index. The retardation or *phase change* for a given light ray is a function of the thickness and the index of refraction of the material through which it passes. Thus in a given unstained specimen, transparent regions of different refractive indices retard the light rays passing through them to differing degrees. Such phase variations in the light focused on the image plane of the light microscope are not visible to the observer. The phase contrast microscope is an optical system that converts such phase variations into visible variations in light intensity or contrast. The phase microscope therefore allows cytologists to observe the behavior of living, dividing cells.

phaseolin a glycoprotein that constitutes up to 50% of the storage protein in the cotyledons of the bean *Phaseolus vulgaris.* It is encoded by a family of about 10 genes.

Phaseolus the genus that includes *P. aureus,* the Mung bean; *P. limensis,* the lima bean; *P. vulgaris,* the red kidney bean.

phene a phenotypic character controlled by genes.

phenetic a taxonomic classification based upon phenotypic characteristics without regard to phylogenetic relationships.

phenocopy the alteration of the phenotype, by nutritional factors or the exposure to environmental stress during development, to a form imitating that characteristically produced by a specific gene. Thus rickets due to a lack of vitamin D would be a phenocopy of vitamin-D-resistant rickets.

phenocritical period the period in the development of an organism during which an effect produced by a gene can most readily be influenced by externally applied factors.

phenogenetics developmental genetics.

phenogram a branching diagram linking taxons by estimates of overall similarity based on evidence from a sample of characters. Characters are not evaluated as to whether they are primitive or derived.

phenogroup *See* **haplotype.**

phenomic lag phenotypic lag.

phenon a set of organisms grouped together by methods of numerical taxonomy (*q.v.*).

phenotype the observable properties of an organism, produced by the genotype in conjunction with the environment.

phenotypic lag the delay of the expression of a newly acquired character. Mutations may appear in a bacterial population several generations after the administration of a mutagen. Phenotypic lag may be due to any of the following reasons: (1) The mutagen may inactivate the gene at once, but this inactivity may not become apparent until the products previously made by the gene are diluted to a sufficient degree. A number of cell divisions occur before the concentration of these products falls below some critical level. (2) the "mutagen" may itself be inactive. It may undergo a series of reactions to yield a compound which is the true mutagen. The latent period would then be the time required for those reactions to take place. (3) The mutagen may cause the gene to become unstable. At a later time it will return to a stable wild type or mutant state. (4) The microbe may be multinucleate, and a mutation may occur in but one nucleus. The latent period then would be the time required for nuclear segregation.

phenotypic mixing the production of a virus with a phenotype that does not match its genotype. During the assembly of a virus, nucleic acid and protein components are drawn randomly from two pools. In a host infected simultaneously by mutant and wild-type viruses, progeny phages are assembled without regard to matching the coat components to the genes in the nucleic acid core. Thus discrepancies sometimes arise, and a virus is produced with coat proteins that are not specified by its genome.

phenotypic sex determination control of gonad development by nongenetic stimuli. For example, the incubation temperature of fertilized eggs determines the type of sexual development in some turtle species.

phenotypic variance the total variance observed in a character. *See* **genetic variance, variance.**

phenylalanine *See* **amino acid.**

phenylketonuria a hereditary disorder of amino acid metabolism in humans, inherited as an autosomal recessive. Homozygotes cannot convert phenylalanine to tyrosine due to a lack of the liver enzyme phenylalanine hydroxylase. Brain dysfunction characteristic of the disease can be avoided by early dietary restriction of phenylalanine. Prevalence is 1/11,000. Abbreviated PKU.

phenylthiocarbamide *See* **taste blindness.**

phenylthiourea phenylthiocarbamide (*q.v.*).

pheoplasts a brown or yellow plastid of brown algae, diatoms, and dinoflagellates.

pheromone a chemical exchanged between members of the same species which affects behavior. Examples of such pheromones are the sex attractants, alarm substances, aggregation-promotion substances, territorial markers, and trail substances of insects.

Philadelphia chromosome an aberrant chromosome observed in metaphase figures from bone marrow cells of patients suffering from chronic myelogenous leukemia. The chromosome aberration was at first thought to involve a terminal deletion of chromosome 22. It is now clear that the Philadelphia chromosome is a translocation in which approximately one half the long arm of chromosome 22 is moved to a terminal position on another chromosome. About 90% of the time the long arm of chromosome 9 is the recipient.

philopatry   a mode of dispersal in which propagules remain at or very near their point of origin.

phi *X*174 (ϕ*X*174)   a single-stranded DNA virus of *Escherichia coli*.

phloem   the vascular tissue that conducts nutrient fluids in vascular plants.

phocomelia   absence of the proximal portion of a limb or limbs, the hands or feet being attached to the trunk by a single bone. A genetic form occurs in humans that is inherited as an autosomal recessive. Phocomelia may also be caused by exposure of the developing embryo to the drug thalidomide (*q.v.*).

phosphate bond energy   the energy liberated as one mole of a phosphorylated compound undergoes hydrolysis to form free phosphoric acid. *See* **ATP.**

phosphodiester   any molecule containing the linkage

$$R-O-\overset{\overset{\displaystyle O}{\|}}{\underset{\underset{\displaystyle O^-}{|}}{P}}-O-R'$$

where R and R′ are carbon-containing groups, O is oxygen, and P is phosphorus. This type of covalent chemical bond involves the 5′ carbon of one pentose sugar (ribose or deoxyribose) and the 3′ carbon of an adjacent pentose sugar in RNA or DNA chains. *See* **deoxyribonucleic acid.**

phosphodiesterase I   a 5′ exonuclease that removes, by hydrolysis, 5′ nucleotides from the 3′ hydroxy-terminus of oligonucleotides.

phospholipid   a lipid containing phosphate esters of glycerol or sphingosine.

phosphorescence   *See* **luminescence.**

phosphorus   an element universally found in small amounts in tissues, a component of nucleic acids. Atomic number 15; atomic weight 30.4735; valence 5^+; most abundant isotope ^{31}P; radioisotope ^{32}P (*q.v.*).

phosphorylation   the combination of phosphoric acid with a compound.

phosphovitin   a phosphoprotein of molecular weight 40,000. Two molecules of phosphovitin and one of lipovitellin comprise the basic subunit of the amphibian yolk platelet. *See* **vitellogenin.**

photoactivated cross-linking   a technique for crosslinking a nucleic acid (e.g., a tRNA chain) to a polypeptide chain with which it is functionally associated (e.g., its cognate synthetase) by irradiating the synthetase-tRNA complex with ultraviolet light. The technique is used to locate the points of intimate contact between the two molecules. *See* **RNase protection.**

photoautotroph   an organism that can produce all of its nutritonal and energy requirements using only inorganic compounds and light.

photoelectric effect   a process in which a photon ejects an electron from an atom so that all the energy of the photon is absorbed in separating the electron and in imparting kinetic energy to it. *Contrast with* **Compton effect.**

photographic rotation technique   a technique used to establish the symmetry of a structure (such as a virus) observed in an electron micrograph. The micrograph is printed *n* times, the enlarging paper being rotated 360°/*n* between successive exposures. Structures with *n*-fold radial symmetry show reinforcement of detail; whereas the micrograph will show no reinforcement when tested for *n* − 1 or *n* + 1 symmetry.

photolysis   decomposition of compounds by radiant energy, especially light.

photomicrography   the technique of making photographs through a light microscope.

photon   a quantum of electromagnetic energy.

photoperiodism   the response of organisms to varying periods of light. In plants, for example, the photoperiod controls flowering. *See* **phytochrome.**

photophosphorylation   the addition of phosphate to AMP and ADP through the energy provided by light during photosynthesis.

photoreactivating enzyme   an exonuclease catalyzing a photochemical reaction which removes UV-induced thymine dimers from DNA.

photoreactivation   reversal of injury to cells caused by ultraviolet light, accomplished by post-irradiation exposure to visible light waves. *See* **thymine dimer.**

photoreceptor   a biological light receptor. *See* **cone, ommatidium, rod.**

photosynthesis   the enzymatic conversion of light energy into chemical energy in green plant cells resulting in the formation of carbohydrates and oxygen from carbon dioxide and water.

phototropism   a tropism in which light is the stimulus.

phragmoplast   a differentiated region of the plant cell which forms during late anaphase or early telophase between the separating groups of chromosomes. The phragmoplast contains numerous microtubules which function to transport material used by the developing cell plate. Once the cell plate forms it divides the phragmoplast into two parts, and subsequently the cell plate is transformed into the middle lamella of the mature cell wall.

Phycomyces   a genus belonging to the same family as the common bread molds *Mucor* and *Rhizopus.* Two species, *P. nitens* and *P. blakesleeanus,* are favorites for genetic studies.

phyletic evolution   the gradual transformation of one species into another without branching; anagenesis; vertical evolution.

phyletic speciation   *See* **phyletic evolution.**

phylogenetic tree   a diagram that portrays the hypothesized genealogical ties and sequence of historical ancestor/descendant relationships linking individual organisms, populations, or taxa. When species are considered they are represented as line segments, and points of branching correspond to subsequent speciation events. When possible the lineage is presented in relation to a geological time scale.

phylogenic   pertaining to phylogeny.

phylogeny   the relationships of groups of organisms as reflected by their evolutionary history.

phylum   *See* **classification.**

Physarum polycephalum   *See* **Myxomycota.**

physical map   a map of the linear order of genes on a chromosome with units indicating their distances determined by methods other than genetic recombination (e.g., nucleotide sequencing, overlapping deletions in polytene chromosomes, electron micrographs of heteroduplex DNAs, etc.).

physiological saline   an isotonic, aqueous solution of salts used for temporarily maintaining living cells.

physiology   the study of the dynamic processes of living organisms.

phytochrome   the molecule responsible for the photoperiodic control of flowering. During the day a form of phytochrome that absorbs light at the far-red end of the spectrum accumulates in plants. This form of the pigment inhibits flowering in short-day plants (*q.v.*) and stimulates flowering in long-day plants *(q.v.)*. During darkness this compound reverts to a red-absorbing form which is stimulatory to the flowering of short-day plants and inhibitory to long-day plants (*q.v.*). Whether a plant belongs to the short- or long-day class is genetically controlled.

phytohemagglutinin   a lectin (*q.v.*) extracted from the red kidney bean, *Phaseolus vulgaris,* that agglutinates human erythrocytes and stimulates lymphocytes to undergo mitosis.

phytohormone   a plant hormone. *See* **auxin, cytokinin, gibberellin**.

phytopathology   the study of plant diseases.

phytophagous   plant-eating.

phytotron   a group of rooms used for growing plants under controlled, reproducible, environmental conditions.

pI   isoelectric point.

pico   a prefix meaning 10^{-12}. Used with units of measurement in the metric system.

picornavirus   a group of extremely small RNA viruses. The name is derived from the prefix *pico* (meaning small) + *RNA* + *virus*. The polio virus belongs to this group.

piebald   designating an animal, especially a horse, having patches of black and white or of different colors. A pinto horse is piebald.

Pieris   a genus of small butterflies extensively studied by ecological geneticists.

pig   *See* **swine**.

pigeon, domestic breeds   the homing pigeon, the carneaux, the dragoon, the white Maltese, the white king, the fantail, the pouter, the tumbler, the roller, the Jacobin, the barb, the carrier pigeon, and the ptarmigan. The species is *Columbia livia domestica.*

pile   *See* **nuclear reactor**.

pilin   the protein from which a pilus is constructed.

pillotinas   large symbiotic spirochaetes living in the hind gut of termites. These spirochaetes are of interest because they contain microtubules, and the presence of microtubules in spirochaetes supports the symbiotic theory of the origin of undulipodia. *See* **motility symbiosis**.

pilus (*plural* **pili**)   a filamentous, hollow appendage extending from the surface of a bacterial cell. Some pili serve as conjugation tubes; e.g., on "male" (F$^+$ and Hfr) cells, the F pilus (or sex pilus) functions as a conjugation tube through which donor DNA is transmitted to a recipient, "female" (F$^-$) cell. *See* **F factor**.

pin   the type of flower characterized by long styles and low anthers found among distylic species such as seen in the genus *Primula*. *See* **thrum**.

pinocytosis   the engulfment of liquid droplets by a cell through the production of pinosomes (*q.v.*).

pinosome   a membrane-bounded cytoplasmic vesicle produced by the budding off of localized invaginations of the plasma membrane. Recently pinocytosed fluids are segregated within the cell in pinosomes.

pioneer the first plant or animal species to become established in a previously uninhabited area.

pistil the female reproductive organ of the flower (*q.v.*), consisting of ovary, style, and stigma.

pistillate designating a flower having one or more pistils and no stamens.

Pisum sativum the garden pea, Mendel's experimental organism.

pitch the number of base pairs in one complete revolution of a DNA double helix.

pith parenchymatous tissue in the center of roots and stems.

Pithecanthropus erectus currently called *Homo erectus erectus* (*q.v.*).

Pitressin a trademark for vasopressin.

pituitary dwarfism in humans, a form of ateliosis inherited as an autosomal recessive. There seems to be three types. In *primordial dwarfism,* growth hormone (*q.v.*) alone is deficient, whereas in *panhypopituitarism* there is a deficiency of all anterior pituitary hormones. In *Laron dwarfism,* excessive amounts of growth hormone are present, suggesting that the defect involves the growth hormone receptors.

pituary gland the master endocrine gland which lies beneath the floor of the brain, within the skull, of vertebrates. *See* **adenohypophysis, neurohypophysis.**

pK dissociation constant.

PKU phenylketonuria (*q.v.*).

placebo an inactive substance given to certain patients randomly chosen (without their knowledge) from a group. A new medicinal compound is administered to the other patients. The effectiveness of the compound is determined by comparing the progress of the treated patients with those receiving the placebo.

placenta an organ consisting of embryonic and maternal tissues in close union through which the embryo of a viviparous animal is nourished.

Plantae the plant kingdom. Its members are all eukaryotes made up of cells containing green plastids. *See* **classification, Eukaryotes.**

plaque a clear, round area on an otherwise opaque layer of bacteria or tissue-cultured cells where the cells have been lysed by a virulent virus. In the example shown on p. 301, the Petri dish contains a gel in which a nutrient broth is suspended. Covering the surface of the medium is a lawn of bacteria. These arose from a layer of 1×10^8 bacteria which was deposited upon the agar surface. Subsequently a small number of viruses was spread over these cells. The holes represent points at which a phage particle was present. The initial phage particle was adsorbed to a bacterium and after a short period the bacterium lysed, releasing new phage. These new phage in turn attacked neighboring bacteria and produced more phage. The process continued until holes visible to the naked eye appeared. Different phages can sometimes be recognized by the morphology of the plaques they produce. The numerous small plaques in the figure are from T_6 bacteriophage, while the four large plaques are from phage T_7.

plaque assay a technique for counting the number of complete, infective phage in a culture by an appropriate dilution to ensure that no more than one phage can infect a given host cell, followed by counting the number of plaques that develop on a bacterial lawn.

plaque forming cells in immunology, antibody secreting cells which can cause a hemolytic plaque with the aid of complement on a lawn of erythrocytes. The term is also applied

plaque

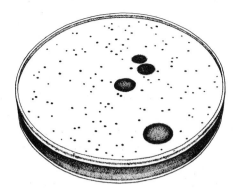

to cells in certain assay systems where the cell killing that creates the plaque is cell mediated, rather than antibody dependent.

plasma   *See* **blood plasma.**

plasmablasts   highly proliferative cells that are developmental intermediates between small B lymphocytes and immunoglobulin-secreting mature plasma cells.

plasma cell   a terminally differentiated immunoglobulin-secreting cell of the B lymphocyte lineage.

plasmacytoma   synonomous with myeloma (*q.v.*).

plasmagene   a self-replicating cytoplasmic gene within an organelle or a symbiont of an eukaryotic cell (e.g., the gene for paramecin, *q.v.*), or in a plasmid of a bacterial cell (e.g., F factor).

plasmalemma   plasma membrane.

plasma lipoproteins   a multicomponent complex of proteins and lipids circulating in the plasma. Plasma lipoproteins are grouped by their densities into four classes: high density lipoproteins (HDL), low density lipoproteins (LDL), intermediate density lipoproteins (IDL), and very low density lipoproteins (VLDL). LDLs are about 25% protein and 75% lipid, and about half of the lipid is cholesterol. LDLs serve as the major cholesterol transport system in human plasma. *See* **familial hypercholesterolemia.**

plasma membrane   the membrane surrounding a cell. *See* **unit membrane.**

plasma protein   any of the dissolved proteins of vertebrate blood plasma which are responsible for holding fluid in blood vessels by osmosis.

plasma transferrins   beta globins that bind and transport iron to the bone marrow and tissue storage areas. Numerous heritable transferrin variants are known.

plasmid   an extrachromosomal genetic element found in a variety of bacterial species that generally confers some evolutionary advantage to the host cell (i.e., resistance to antibiotics, production of colicins, etc.). Plasmids are double-stranded, closed DNA molecules ranging in size from 1-200 kb. Plasmids whose replication is coupled to that of the host so that only a few would be present per bacterium are said to be under *stringent control.* Under *relaxed control,* the number of plasmids per host cell may be from 10 to 100.

plasmid cloning vector   a plasmid used in recombinant DNA experiments as acceptors of foreign DNAs. Plasmid cloning vectors are generally small and replicate in a relaxed fashion. They are marked with antibiotic resistance genes and contain recognition sites for restriction endonucleases in regions of the plasmid that are not essential for its replication. The most widely used plasmid cloning vector is pBR322 (*q.v.*).

plasmid conduction   the process whereby a conjugative plasmid can help a nonmobilizable plasmid to be transferred from a donor to a recipient cell. Nonmobilizable plasmids cannot prepare their DNA for transfer, but can become mobilized by recombination with a conjugative plasmid to form a single transferrable DNA molecule. *See* **plasmid donation, relaxation complex.**

plasmid donation   the process by which a nonconjugative plasmid is transferred from a donor to a recipient cell via the effective contact function provided by a conjugative plasmid. In *E. coli,* for example, the ColE1 plasmid does not have genes for the establishment of effective contact, but an F plasmid in the same cell is conjugative and can provide this function.

plasmid engineering   *See* **recombinant DNA research.**

plasmid fusion   *See* **replicon fusion.**

plasmid incompatibility group   a group of closely related plasmids that are sensitive to the same repressor. Different plasmids belonging to the same plasmid incompatibility group cannot be stably maintained in the progeny of a single cell.

plasmin   an enzyme, cleaved from plasminogen, that hydrolyzes fibrin. *See* **blood clotting.**

plasminogen   a blood proenzyme that is activated by cleavage of a single arg-val peptide bond to form the functional enzyme plasmin (*q.v.*).

plasmodesmata   cytoplasmic threads that form delicate protoplasmic connections between adjacent plant cells.

plasmodium   a multinucleate mass of cytoplasm lacking internal cell boundaries. The term is used to refer to the amoeboid stage in the life cycle of Sporozoa (*q.v.*) and to the vegetative stage of Myxomycota (*q.v.*).

Plasmodium   *See* **malaria.**

plasmogamy   the fusion of protoplasts of two haploid cells without the fusion of their nuclei, as in certain fungi.

plasmolysis   shrinkage of the plant cell protoplasm away from its cellulose wall when placed in hypertonic solutions, due to osmotic withdrawal of water from its large central vacuole.

plasmon   all extrachromosomal hereditary agents considered collectively.

plasmosome   nucleolus (*q.v.*).

plastid   a self-replicating cytoplasmic organelle of plant cells, such as a chloroplast, chromoplast, elaioplast, or leukoplast.

plastoquinone   a group of quinones (*q.v.*) involved in the transport of electrons during photosynthesis in chloroplasts.

plate   1. a flat, round dish (Petri plate) containing agar and nutrients for the culture of bacteria. **2.** to spread or inoculate cells on the surface of semi-solid medium in such a culture dish. **3.** a geological plate. *See* **plate tectonics.**

platelets   anucleate, oval, colorless corpuscles present in blood. Platelets, which are 1/3–1/2 the size of erythrocytes, originate from projections pinched off the surface of megakaryocytes (*q.v.*).

plate tectonics   movement of large segments of the earth's crust (plates) as a consequence of convection currents in the hot interior. Together with the phenomenon of sea floor spreading, these geological events provide mechanisms that help explain biogeographical distributions of plants and animals.

plating efficiency *See* **absolute plating efficiency, relative plating efficiency.**

platyrrhine referring to members of the broad-nosed primate infraorder Platyrrhini that includes the New World monkeys. *Compare with* **catarrhine.**

playback experiment an experiment designed to recover a DNA strand that has been saturated with RNA (*see* **RNA driven hybridization**), and then using it in a further reassociation reaction to show that its $C_0t_{1/2}$ (*q.v.*) corresponds to that expected of nonrepetitive DNA.

plectonemic spiral a spiral in which two parallel threads coil in the same direction about one another and cannot be separated unless uncoiled. The component strands of a DNA duplex are plectonemically coiled. *Compare with* **paranemic spiral.**

pleiotropy the phenomenon of a single gene being responsible for a number of distinct and seemingly unrelated phenotypic effects.

Pleistocene the ice age, lasting from 10,000 B.C. to the beginning of the Pliocene (*q.v.*). One of the two epochs making up the Quaternary period (*q.v.*). See **geologic time divisions.**

Pleistocene refuge one of the favorable areas south of the ice fields where species survived periods of glaciation.

pleomorphic having more than one form or shape.

plesiomorphic character an original, underived, or primitive character. *Contrast with* **apomorphic character.**

Pleurodeles salamanders of two species belonging to this genus, *P. waltlii* and *P. poireti,* have been studied both genetically and cytologically. Working maps of the oocyte lampbrush chromosomes of both species are available.

pleuropneumonia-like organisms a group of bacteria that do not form cell walls. PPLOs are included in the phylum Aphragmabacteria (*see* **classification**). One PPLO, *Mycoplasma pneumoniae,* is the cause in humans of an atypical pneumonia.

Pliocene the most recent epoch in the Tertiary period lasting from 2 to 5 million years before the present. *See* **geologic time divisions.**

-ploid a combining form used in cytology and genetics to designate a particular multiple of the chromosome set of the nucleus of an organism, as 16-ploid, 32-ploid, etc.

ploidy *See* **polyploidy.**

plumage pigmentation genes a group of genes controlling pigmentation of chicken feathers. Pigment will not be produced unless the gene *C* is present. A second gene *I* which inhibits pigment formation is located on a different chromosome. The White Leghorn breed of chickens has the genotype *IICC,* whereas the White Plymouth Rock is *iicc.*

pluralism a belief that evolution is driven by many forces, not just by natural selection.

pluripotent pertaining to an embryonic tissue that has a large number of possible developmental fates because determination has not yet taken place.

plus and minus techniques *See* **DNA sequencing techniques.**

plus (+) and minus (−) viral strands   1. in a single-stranded RNA virus, a plus strand is one having the same polarity as viral mRNA and containing codon sequences that can be translated into viral protein. A minus strand is a noncoding strand that must be copied by an RNA-dependent RNA polymerase to produce a translatable mRNA. 2. in a single-stranded DNA virus, a plus strand is one contained in the virus particle or any strand having the same base sequence. A minus strand has a base sequence complementary to the plus strand; mRNA can be transcribed from the minus strand.

pneumococcal transformation   *See* **transformation.**

pod corn   *Zea mays tunicata,* a primitive variety of corn characterized by kernels, each of which is enclosed in a husk.

podophyllin   a mitotic poison isolated from the plant *Podophyllum peltatum.*

Podospora anserina   a fungus belonging to the same subclass as *Neurospora* and *Sordaria* (*q.v.*). Studies on the genetic control of sexual and vegetative reproduction in *Podospora* are particularly important.

point mutation   as defined in molecular genetics, a mutation caused by the substitution of one nucleotide for another.

Poisson distribution   a function that assigns probabilities to the sequence of outcomes of observing no events of a specified type, one event, two events, and so on without limit. Events following a Poisson distribution are completely randomized. Suppose that within a defined spatial or temporal region one looks at the events occurring in nonoverlapping subregions. A Poisson distribution for the number of events observed in the total region implies that the events occurring in the subregions do not affect each other. A Poisson distribution will not be found if the events are correlated positively (in the case of clumping) or negatively (in the case of mutual repulsion). The Poisson is specified by the average number of events per observation, and its mean and variance are equal. The formula of the function is $P_i = (m^i e^{-m})/i!$ where m is the mean number of events; i ! is the factorial i $\cdot$ (i−1) $\cdot$ (i−2)... 2 $\cdot$ 1; e is the base of natural logarithms, and i is the number for which the probability P_i is given. If m = 1.2, for example, the distribution is as follows:

Number observed	0	1	2	3	4	5	6	7	Above 7
Probability	0.3012	.3614	.2169	.0867	.0260	.0062	.0012	.0002	.0002

Many natural distributions follow a Poisson, including the number of radioactive disintegrations of a radioisotope in a fixed period of time, or the number of larvae of a particular invertebrate species captured by towing a plankton net through a specified volume of sea water.

pokeweed mitogen   a lectin (*q.v.*) extracted from rhizomes of the pokeweed, *Phytolocca americana,* that stimulates the proliferation of lymphocytes, particularly mouse T and B lymphocytes.

pol I, II, III   *See* **DNA polymerase.**

polar   referring to water soluble chemical groups such as a hydrophilic side chain of an amino acid.

polar body   the minute cell produced and discarded during the development of an oocyte. A polar body contains one of the nuclei derived from the first or second division of meiosis, but has practically no cytoplasm.

polar fusion nucleus   in plants, the product of the fusion of the two polar nuclei. This,

304

after fusing with a male nucleus, gives rise to the tripoid endosperm nucleus. *See* **double fertilization.**

polar gene conversion   a phenomenon in which a gradient of conversion frequencies exists from one end of a gene to the other; sites closer to one end of a gene usually have higher conversion frequencies than do those further from that end.

polar granules   RNA-rich granules in the posterior ooplasm of the insect egg. In *Drosophila* only pole cells engulfing such granules can subsequently form germ cells.

polarity gradient   the quantitative effect of a polarity mutation in one gene on the expression of later genes in the operon. The effect is a function of the distance between the nonsense codon and the next chain initiation signal.

polarity mutant   a mutant gene that is able to reduce the rate of synthesis of the proteins that normally would be produced by wild-type alleles of the genes lying beyond it on the chromosome. Such genes exert their effect during the translation of a polycistronic message (*q.v.*). *See* **regulator genes, translation.**

polarization microscope   a compound light microscope used for studying the anisotropic properties of objects and for rendering objects visible because of their optical anisotropy.

polar nuclei   *See* **pollen grain.**

polaron   a chromosomal segment within which polarized genetic recombination takes place by gene conversion.

polar tubules   microtubules of the spindle apparatus that originate at the centriolar or polar regions of the cell. *See* **chromosomal tubules.**

pole cell   one of the cells that are precociously segregated into the posterior pole of the insect embryo before blastoderm formation. Among these cells are the progenitors of the germ cells.

Polish wheat   *Triticum polonicum* (N = 14). *See* **wheat.**

pollen grain   a microspore in flowering plants that germinates to form the male gametophyte (pollen grain plus pollen tube) which contains three haploid nuclei. One of these fertilizes the ovum, a second fuses with the two polar nuclei to form the 3N endosperm, and the third (the vegetative nucleus) degenerates once double fertilization has been accomplished.

pollen mother cell   microsporocyte.

pollen-restoring gene   a gene that permits normal microsporogenesis to occur in the presence of a cytoplasmic male sterility factor.

pollen tube   the tube formed from a germinating pollen grain which carries male gametes to the ovum.

pollination   the transfer of pollen from anther to stigma. *See* **pollen grain.**

polocyte   the small degenerate sister cell of the secondary oocyte. This cell generally divides into two polar bodies, which disintegrate.

polyacrylamide gel   a gel prepared by mixing a monomer (acrylamide) with a cross-linking agent (N,N′-methylenebisacrylamide) in the presence of a polymerizing agent. An insoluble three-dimensional network of monomer chains is formed. In water the network becomes hydrated. Depending upon the relative proportions of the ingredients it is possible to prepare gels with different pore sizes. The gels can then be used to separate biological molecules like proteins of a given range of sizes.

polyacrylamide gel electrophoresis *See* **electrophoresis.**

polyadenylation enzymatic addition of several adenine nucleotides to the 3′ end of mRNA molecules as part of the processing that primary RNA transcripts undergo prior to transport from the nucleus to the cytoplasm. The added segment is referred to as a "poly-A tail."

polyandry the state of having more than one male mate at one time.

poly-A tail *See* **polyadenylation.**

polycentric chromosome polycentromeric chromosome. *See* **centromere.**

polycentromeric chromosome *See* **centromere.**

polycistronic message a giant messenger RNA molecule specifying the amino acid sequence of two or more proteins produced by adjacent cistrons in the same operon. *See* **transcription unit**.

polyclonal an adjective applied to cells or molecules arising from more than one clone; e.g., an antigenic preparation (even a highly purified one) elicits the synthesis of various immunoglobulin molecules. These antibodies would react specifically with different components of the complex antigen molecule. Thus the antibody preparation generated by such an antigen would be polyclonal in the sense that it would contain immunoglobulins synthesized by different clones of B lymphocytes.

polyclone *See* **compartmentalization.**

polycomplex structures, observed in certain insects, within oocyte nuclei, formed by the fusion of components from synaptonemal complexes (*q.v.*) which have detached from the diplotene chromosomes.

polycystic kidney disease cystic fibrosis (*q.v.*).

polycythemia a disease in man characterized by overproduction of red blood cells.

polydactyly the occurrence of more than the usual number of fingers or toes.

polyembryony the formation of multiple embryos from a zygote by its fission at an early developmental stage. Monozygotic twins constitute the simplest example of polyembryony. Monozygotic quadruplets are commonly formed by armadillos. In certain parasitic wasps as many as 2000 embryos can be formed by polyembryony from a single zygote.

polyene antibiotic any antibiotic active against fungi, but not bacteria (nystatin and oligomycin are examples).

polyestrous mammal *See* **estrous cycle.**

polyethylene glycol   a chemical used to promote the fusion of tissue-cultured cells, as in the production of a hybridoma (*q.v.*).

polygamy   polandry and/or polygyny. *Compare with* **monogamy.**

polygene   one of a group of genes that together control a quantative character. *See* **quantitative inheritance.**

polygenic character   a quantitatively variable phenotype dependent on the interaction of numerous genes.

polyglucosan   a polymer such as glycogen made up of a chain of glucose units.

polygyny   the mating of a male with more than one female. *Compare with* **monogamy, polyandry.**

polymer   a macromolecule composed of a covalently bonded collection of repeating sub-units or monomers linked together during a repetitive series of similar chemical reactions. Each strand of DNA is a linear polymer of nucleotide monomers. A linear polypeptide chain is a polymer of amino acid monomers.

polymerase   any enzyme that catalyzes the formation of DNA or RNA molecules from deoxyribonucleotides and ribonucleotides, respectively (e.g., DNA polymerase, RNA polymerase).

polymerization   the formation of a polymer from a population of monomeric molecules.

polymorphism   the existence of two or more genetically different classes in the same inter-breeding population (Rh-positive and Rh-negative humans, for example). The polymorphism may be transient, or the proportions of the different classes may remain the same for many generations. In the latter case the phenomenon is referred to as *balanced polymorphism*. If the classes are located in different regions, *geographic polymorphism* exists.

polymorphonuclear leukocyte   *See* **granulocyte.**

polyneme hypothesis   the concept that a newly formed chromatid contains more than one DNA duplex. *Contrast with* **unineme hypothesis.**

polynucleotide   a linear sequence of joined nucleotides.

polynucleotide kinase   an enzyme which phosphorylates the 5′ hydroxyl termini produced by endonucleases (*q.v.*).

polynucleotide phosphorylase   an enzyme that links ribonucleotides together in a random fashion. It is used to produce artificial mRNA molecules.

polyp   **1.** the sedentary form of a coelenterate. **2.** a small stalked neoplasm projecting from a mucous surface (for example, an intestinal polyp).

polypeptide   a polymer made up of amino acids linked together by peptide bonds (*q.v.*).

polyphasic lethal   a mutation characterized by two or more lethal phases separated by developmental periods in which it produces no deaths.

polyphenism   the occurrence of several phenotypes in a population which are not due to genetic differences between the individuals in question.

polypheny   pleiotropy (*q.v.*).

polyphyletic group   a group of species classified together, some members of which are descended from different ancestral populations. *Contrast with* **monophyletic group.**

polyploid   designating an individual having more than two sets of chromosomes.

polyploidy a chromosome number that is a multiple of the normal diploid number, resulting from the replication within a nucleus of complete chromosome sets without subsequent nuclear division.

polyprotein a cistronic product that is post-translationally cleaved into several independent proteins. For example, an enkephalin precursor protein contains six copies of met-enkephalin and one copy of leu-enkephalin. *See* **enkephalins.**

polyribonucleotide phosphorylase *See* **polynucleotide phosphorylase.**

polyribosome polysome.

polysaccharide a carbohydrate formed by the polymerization of many monosaccharide units. Starch, cellulose, and glycogen (*q.v.*) are examples of polysaccharides.

polysome a multiple structure containing a group of ribosomes held together by a molecule of messenger RNA. A contraction of polyribosome.

polysomy the reduplication of some but not all of the chromosomes of a set beyond the normal diploid number. A metafemale *Drosophila* is polysomic (trisomic for the X).

polyspermy the penetration of more than one sperm into one ovum at the time of fertilization.

Polysphondylium pallidum *See* **Acrasiomycota.**

polytene chromosome a giant cable-like chromosome consisting of many identical chromatids lying in parallel. The chromatin is hypercoiled in localized regions, and since the chromatids are in register, a pattern of bands is produced vertical to the long axis of the chromosome. Polytene chromosomes are found within a limited number of organisms. They are present in the macronucleus anlage of some ciliates, in the synergids and antipodal cells of the ovules of certain angiosperms, and in various tissues of dipterans. The *Drosophila* salivary gland chromosomes (*q.v.*) have been studied most extensively. *See* *Anopheles,* **Balbiani ring,** *Calliphora erythrocephala,* *Chironomus,* *Culex pipiens,* *Glyptotendipes barbipes,* *otu* **mutation,** *Rhynchosciara,* *Sciara,* *Smittia.*

polyteny the interphase replication of each chromosome to produce giant chromosomes that are polytene (many-stranded) cables.

polythetic group a group of organisms that share a large number of features, but no single feature is either essential for group membership or is sufficient to make an orgnanism a member of the group.

polytopic pertaining to the distribution of subspecies in two or more geographically discontinuous areas.

polytrophic meroistic ovary *See* **insect ovary types.**

polytypic species a species subdivided into a number of specialized races.

pome a fleshy, many-seeded fruit such as the apple or pear, in which the enlarged end of the flower-stalk forms much of the flesh.

Pompe disease a hereditary glycogen storage disease in humans arising from a deficiency of the lysosomal enzyme α-1,4-glucosidase, due to a recessive gene on chromosome 17. Prevalence of the disease is 1/100,000.

Pongidae the family of primates containing all anthropoid apes.

popcorn *See* **corn.**

population a local (geographically defined) group of conspecific organisms sharing a common gene pool; also called a *deme.*

population biology the study of the patterns in which organisms are related in space and time. Such disciplines as ecology, taxonomy, ethology, population genetics, and others that deal primarily with the interactions of entire organisms are included under this term.

population cage a special cage in which *Drosophila* populations can be reared for many generations. The cage is designed so that samples of the population can be conveniently withdrawn and food supplies can be replenished.

population genetics the study of the genetic composition of populations. Population geneticists try to estimate gene frequencies and detect the selective influences that determine them in natural populations. They also build mathematic models to elucidate the interaction of factors such as selection, population size, mutation, and migration upon the fixation and loss of linked and unlinked genes.

porcine referring to members of the pig family, especially the domestic pig *Sus scrofa*.

porphyrin any of a class of organic compounds in which four pyrrole nuclei are connected in a ring structure usually associated with metals (like iron or magnesium). Porphyrins form parts of the hemoglobin, cytochrome, and chlorophyll molecules. *See* **heme.**

position effects the change in the expression of a gene accompanying a change in the position of the gene with respect to neighboring genes. The change in position may result from crossing over or from a chromosomal aberration. Position effects are of two types: the *stable* (S) type and the *veriegated* (V) type. S-type position effects are also called *cis-trans* position effects. S-type position effects involve cistrons that possess at least two mutated sites separable by intragenic recombination. In the cis configuration ($m^1 m^2/+ +$) a normal phenotype is observed, whereas in the trans configuration ($m^1 +/+ m^2$) a mutant phenotype is produced. A reasonable explanation for such an observation would be that the mRNA transcribed from a $+ +$ chromatid would function normally, whereas the mRNAs transcribed from $m^1 m^2$, $m^1 +$, or $+ m^2$ chromatids would not. V-type position effects generally involve the suppression of activity of a wild-type gene when it is placed in contact with heterochromatin because of a chromosome aberration. Under some conditions, the gene may escape suppression, and consequently the final phenotype may be variegated, with patches of normal and mutant tissues.

positive assortitive mating *See* **assortative mating.**

positive control control by a regulatory protein that must bind to an operator before translation can take place.

positive eugenics *See* **eugenics.**

positive gene control enhancement of DNA transcription through binding of specific expressor molecules to promoter sites. For example, the binding of CAP-cAMP complexes to promoters of bacterial genes involved in catabolism of sugars other than glucose facilitates binding of RNA polymerase to these operons when glucose is absent. *See* **glucose-sensitive operons.** *Compare with* **negative gene control.**

positive interference the interaction between crossovers such that the occurrence of one exchange between homologous chromosomes reduces the likelihood of another in its vicinity. *Compare with* **negative interference.**

positive supercoiling *See* **supercoiling.**

positron a particle of the atomic nucleus equal in mass to the electron and having an equal but opposite (positive) charge.

post coitum after mating.

posterior the direction away from the head of the animal.

postmating isolating mechanisms   *See* **postzygotic isolating mechanisms.**

postmeiotic fusion   a method for restoring diploidy in eggs produced by parthenogenesis, involving union of two identical haploid nuclei formed by a mitotic division of the egg nucleus.

postmortem   occurring after death.

postoperative   occurring after surgical operation.

postpartum   occurring after childbirth.

postreductional disjunction   referring to the separation of alleles at particular heterozygous loci during the first meiotic division. If the loci are represented by A and A', in the case of postreductional disjunction the two chromatids that enter one sister nucleus have one A and one A' allele, whereas in the case of prereductional disjunction both have A alleles or both A' alleles.

post-transcriptional modification   any alterations made to primary RNA transcripts before leaving the nucleus as mature mRNA. Such modifications include removal of introns, splicing of exons, addition of cap (*q.v.*), or a poly-A tail (*q.v.*); also called *nuclear processing of RNA.*

post-translational modifications   alterations to polypeptide chains after they have been synthesized: e.g., removal of the formyl group from methionine in bacteria, acetylation, hydroxylation, phosphorylation, attachment of sugars or prosthetic groups, oxidation of cysteines to form disulfide bonds, cleavage of specific regions that convert proenzymes to enzymes, etc.

postzygotic isolation mechanism   any factor that tends to reduce or prevent interbreeding between genetically divergent populations or species, but functioning after fertilization has occurred; includes hybrid inviability, hybrid sterility, and hybrid breakdown.

potassium   an element universally found in small amounts in tissues. Atomic number 19; atomic weight 39.102; valence 1^+; most abundant isotope ^{39}K; radioisotope ^{42}K; half life 12.4 hours; radiations emitted—beta particles and gamma rays.

potato   *Solanum tuberosum.* Together with corn, wheat, and rice it is one of the four most valuable of the world's crops.

potentiometer   an instrument for the measurement of electrical potentials or electromotive forces.

Potorous tridactylis   the rat kangaroo, a marsupial favorable for chromosomal studies because of the small number and individuality of its chromosomes.

poult   a young turkey.

poultry breeds   Plymouth Rock, New Hampshire, White Leghorn, Blue Andalusian, Rhode Island Red, Rhode Island White, Australorp, and Orpington. The species is *Gallus gallus domestica.*

PP   inorganic pyrophosphate.

P particle   *See* **kappa.**

pp60c-src   the protein kinase encoded by the *c-src* gene (*q.v.*).

PPLO   *pleuropneumonia-like organism* (*q.v.*).

ppm   parts per million.

ppt   precipitate.

pp60v-src   the protein encoded by the oncogene of the Rous sarcoma virus. It is a phosphoprotein of molecular weight 60 kd, hence the *pp60* in the name; the *v-src* indicates that it is encoded by viral gene *src*. The molecule is a protein kinase that phosphorylates tyrosine subunits in cellular proteins, particularly those that form the adhesion portions of the plasmalemma.

preadaptation   a new character or function arising from a mutation in an organism which becomes a useful adaptation after an environmental change, often occurring many generations later.

pre-adoptive parents   *See* **germinal choice.**

Precambrian   the geological eon between the Phanerozoic and the Hadean eons. *See* **geologic time divisions.**

precursor ribosomal RNA   *See* **pre-ribosomal RNA.**

predacious   behaving as a predator.

predator   an animal that kills and consumes other animals.

preferential association   an immunological theory that specific viral antigens interact more strongly with certain allelic products of the major immunogene complex than with others. This preferential association may make the virus more immunogenic, and hosts with the strongly interacting allelic product would tend to be more immune to viral infection than would those with weakly interacting allelic products.

preformation   *See* **epigenesis.**

prehensile   adapted for grasping.

premating isolating mechanisms   *See* **prezygotic isolation mechanism.**

premature initiation   a second initiation of replication occurring before the first is completed; a phenomenon observed in bacteria grown in a complex nutrient broth or in some phage species that make replicas very rapidly.

prematurely condensed chromosomes   interphase chromosomes that are experimentally forced to undergo rapid condensation to metaphase dimensions. This is done by fusing an interphase cell with a cell in mitosis. The interphase cell is induced to enter mitosis, and its chromosomes contract accordingly.

pre-messenger RNA   the giant RNA molecule transcribed from a structural gene. It contains both introns and exons, but the introns are removed and the exons spliced together before the molecule leaves the nucleus. *See* **nuclear processing of RNA.**

prepattern   a morphogenetic pattern superimposed upon a population of cells arranged in a two-dimensional array. Specific types of differentiation are stimulated in certain cells located in defined areas. In *Drosophila* the gene *eyeless-dominant* produces a mutant sex comb (*q.v.*) prepattern.

preprimosome   *See* **primosome.**

prepupal period   the period between puparium formation and the eversion of the imaginal discs of the insect.

prereductional disjunction   *See* **postreductional disjunction.**

pre-ribosomal RNA   the giant RNA molecule transcribed from a ribosomal RNA gene (*q.v.*). In *Drosophila,* it is 38S, in *Xenopus* it is 40S, and in HeLa cells it is 45S. After transcription, pre-RNA is cleaved one or more times to generate the 5.8S, 18S, and 28S rRNAs that become components of ribosomes.

presence and absence theory   the discarded theory that dominance indicates the presence of the particular gene concerned and recessiveness its absence.

presumptive   indicating the presumed fate of an embryonic tissue in a normal development. For example, if a tissue is "presumptive neural tube," this means that in the course of normal development it will become neural tube tissue.

prezygotic isolation mechanism   any factor that tends to reduce or prevent interbreeding between members of genetically divergent populations or species and functioning before fertilization occurs; includes ecological, temporal, ethological, and other isolating factors.

Pribnow box   a segment upstream from the startpoint of prokaryotic structural genes to which the sigma subunit of the RNA polymerase binds. The segment is 6 bp long, and the nucleotides most commonly found are TATAAT; named after its discoverer David Pribnow. *See* **canonical sequence, Hogness box, promoter.**

primaquine-sensitivity   *See* **glucose-6-phosphate dehydrogenase deficiency, malaria.**

primary constriction   a centromere.

primary culture   a culture started from cells, tissues, or organs taken directly from the organism.

primary immune response   *See* **immune response.**

primary ionization   the ionization produced by the primary particles passing through matter as contrasted to the "total ionization," which includes the "secondary ionizations" of delta rays (*q.v.*).

primary nondisjunction   sex chromosomal nondisjunction in diploid orgnanisms with the XX, XY system of sex determination. In the homogametic sex, gametes are produced with two X chromosomes or none. In the heterogametic sex, primary nondisjunction during the first meiotic division produces gametes with no sex chromosome (O) or with an X and a Y. Primary nondisjunction during the second meiotic division produces XX and O or YY and O gametes.

primary protein structure   *See* **protein structure.**

primary sex ratio   the ratio of male to female zygotes at conception.

primary sexual character   an organ that functions in producing gametes; the ovaries and the testes.

primary speciation   the splitting of one species into two, usually resulting from natural selection favoring different gene complexes in geographically isolated populations.

primary transcript   an RNA molecule as it was initially transcribed from DNA. In eukaryotic cells, a primary transcript usually contains introns (noncoding regions) as well as exons (coding regions). *See* **post-transcriptional modification.**

primase   in *E. coli,* the product of *dnaG* gene, responsible for initiation of precursor fragment synthesis in the lagging strand during discontinuous replication. Primase makes the RNA primer which is subsequently elongated by DNA polymerase III. The primase in *E. coli* consists of a single polypeptide of 60,000 d. Unlike RNA polymerase, primase is not inhibited by rifampicin (*q.v.*) and can polymerize deoxyribonucleotides as well as ribonucleotides *in vitro. See* **dna mutations, replication of DNA.**

primate   a mammal belonging to the order Primates, which includes man, the apes, and monkeys. *See* **classification.**

primed   in immunology, sensitization by contact of competent lymphocytes with antigen to which they are programmed to respond.

primed synthesis technique   a method for nucleotide sequencing involving enzymatically controlled extension of a primer DNA strand. *See* **DNA sequencing techniques.**

primer DNA   single-stranded DNA required for replication by DNA polymerase III in addition to primer RNA (*q.v.*).

primer RNA   a short RNA sequence attached by RNA polymerase to the 5′ ends of many newly synthesized DNA chains, serving as required primers onto which DNA polymerase III adds deoxyribonucleotides during DNA replication. Primers are later enzymatically removed (e.g., by DNA polymerase I) and the gaps closed by ligase. *See* **replication of DNA.**

primitive   at an early stage in the evolutionary history of a given group; morphologically similar to an organism at such an early stage.

primordial dwarfism   *See* **pituitary dwarfism.**

primordium   the early cells that serve as the mitotic progenitors of an organ during development.

primosome   a complex of proteins (including primase) required for the priming action that initiates synthesis of each Okazaki fragment in eukaryotic DNA replication. The complex minus primase is called a *preprimosome. See* **replication of DNA.**

Primula   a genus of cowslips and primrose whose population genetics has been extensively studied.

prion   an infectious protein known to be the cause of scrapie, a transmissible, degenerative disease of the nervous system of sheep and goats. The word prion is a contraction of "protein" and "infection." Prions were formerly called "slow viruses," but now are known to be devoid of nucleic acid and therefore are neither viruses nor viroids.

pro   proline. *See* **amino acid.**

probability of an event   the long-term frequency of an event relative to all alternative events, and usually expressed as decimal fraction. Probabilities range between zero (if the event never occurs) and 1 (if the event always occurs and no alternative event ever occurs). In some cases we know a probability *a priori,* as in the case of a coin-toss. In the long run, the coin will come up tails with a frequency of 0.5. More often a probability must be estimated by averaging the results of many trials.

proband   propositus (*q.v.*).

proband method   a method in human genetics for comparing the proportion in families of children in which a proband shows a specific trait with the proportion expected if the trait were inherited as a single gene. For example, if one considers a group of families, each with both parents heterozygous for a recessive gene and each with two children, the proportion of affected children is 57%, not 25%. This is because the families are chosen in the first place through an affected child, and all sibships in which just by chance no affected individuals occured have been left out. Thus there is an *ascertainment bias* which loads the results in favor of the trait.

probe   in molecular biology, any biochemical (usually labeled with radioactive isotopes or tagged in other ways for ease in identification), used to identify or isolate a gene, a gene product, or a protein. For example, a radioactive mRNA hybridizing with a single-strand of its DNA gene, a cDNA hybridizing with its complementary region in a chromosome, or a monoclonal antibody combining with a specific protein.

procaryote   *See* **prokaryote.**

Procaryotes   *See* **Prokaryotes.**

processed gene an eukaryotic pseudogene (*q.v.*) lacking introns and containing a poly-A segment near the downstream end, suggesting that it arose by some kind of reverse copying from processed nuclear RNA into double-stranded DNA; also called *retrogene.*

processing 1. post-transcriptional modifications of primary transcripts. 2. antigen processing involves partial degradation by macrophages (and, in some cases, coupling with RNA) before the immunogenic units appear on the macrophage membrane in a condition that is stimulatory to cognate lymphocytes.

processive enzyme an enzyme that remains bound to a particular substrate during repititions of the catalytic event.

proctodone a hormone thought to be secreted by cells of the anterior intestine of insects which terminates diapause (*q.v.*).

procumbent designating a plant stem that lies on the ground for all or most of its length (as in the case of vines).;

productive infection viral infection of a cell that produces progeny via the vegetative or lytic cycle.

productivity fertility. In *Drosophila* the term is used specifically to refer to the number of progeny surviving to the adult stage among those produced per mated parental female in a specified time interval.

proenzyme a zymogen (*q.v.*).

proflavin an acridine dye that can function as a mutagen causing reading frame shifts (*q.v.*).

progeny the offspring from a given mating; members of the same biological family with the same mother and father; siblings.

progeny test the evaluation of the genotype of a parent by a study of its progeny under controlled conditions.

progeria a hereditary, premature aging disease of humans. Death usually occurs before age 14. Tissue-cultured cells from patients with progeria have a markedly reduced replicative life span and have a reduced ability to repair damaged DNA. Also called *Hutchinson-Gilford syndrome.*

progesterone a steroid hormone secreted by the corpus luteum and the placenta which maintains pregnancy. The formula is on p. 315.

progestogens hormones involved with ovulation.

prognosis a forecast of the course and termination of a disease.

proinsulin a protein synthesized and processed by the beta cells of the pancreas. The molecule contains both the A and B peptides of insulin (*q.v.*) and an intervening C peptide progesterone

H₃C, O ...

H_3C

The progesterone structural formula with labeled atoms: H_3C, C, O, H_2C, CH_3, CH, CH_2, H_2C, CH_3, C, C, CH_2, H_2C, C, H, CH, H, C, C, CH_2, O, C, H, C, H_2

containing 30 amino acids. Specific proteases cleave the precursor at two points, releasing the connecting peptide and the intact insulin molecule.

prokaryon   synonomous with nucleoid (*q.v.*).

prokaryote   member of the superkingdom Prokaryotes (*q.v.*).

Prokaryotes (*also* **Procaryotes**)   the superkingdom containing all microorganisms that lack a membrane-bound nucleus containing chromosomes. Cell division involves binary fission. Centrioles, mitotic spindles, microtubules, and mitochondria are absent. This superkingdom contains one kingdom, the Monera (*q.v.*). *See* **classification, Eukaryotes.**

prolactin   lactogenic hormone (*q.v.*).

proline   *See* **amino acid.**

promiscuous DNA   DNA segments that have been transferred between organelles, such as mitochondria and chloroplasts, or from a mitochondrial genome to the nuclear genome of the host as a result of transpositional events happening millions of years ago. An example would be a section of mitochondrial DNA present in the nuclear genome of *Strongylocentrotus purpuratus* (*q.v.*).

promitochondria   aberrant mitochondria characteristically found in yeasts grown under anaerobic conditions. Promitochondria have incomplete inner membranes and lack certain cytochromes. *See* **petites.**

promoter   **1.** a region on a DNA molecule to which an RNA polymerase binds and initiates transcription. In an operon, the promoter is usually located at the operator end, adjacent but external to the operator. The nucleotide sequence of the promoter determines both the nature of the enzyme that attaches to it and the rate of RNA synthesis. *See* **down promoter mutations, regulator gene, up promoter mutations. 2.** a chemical which, while not carcinogenic itself, enhances the production of malignant tumors in cells that have been exposed to a carcinogen.

pronase   an enzyme from *Streptomyces* which digests mucoproteins.

pronucleus   the haploid nucleus of an egg, sperm, or pollen grain.

proofreading   any mechanism for correcting errors in replication, transcription, or translation that involves monitoring of individual units after they have been added to the chain; also called *editing.*

pro-oocyte   one of the two cystocytes containing four ring canals which form synaptonemal complexes in *Drosophila melanogaster*. Upon entering the vitellarium the anterior pro-oocyte switches to the nurse cell developmental pathway, leaving the posterior cell as the oocyte. *See* **nurse cells.**

propagule   any kind of reproductive particle; a useful term when one wishes to avoid specific statements about sexual or asexual reproduction, about gametes or spores, etc.

properdin pathway   *See* **complement.**

prophage   in lysogenic bacteria, the structure that carries genetic information necessary for the production of a given type of phage and confers specific hereditary properties on the host.

prophage-mediated conversion   the acquisition of new properties by a bacterium once it becomes lysogenized. A prophage, for example, confers upon its bacterial host an immunity to infection by related phages. Lysogenized bacteria also often show changes in their antigenic properties or in the toxins they produce.

prophase   *See* **mitosis.**

propositus (*female,* **proposita)**   the clinically affected family member through whom attention is first drawn to a pedigree of particular interest to human genetics.

prosimian   a member of the most primitive primate suborder, the Prosimii, containing tree shrews and tarsiers.

Prosobranchiata   one of the three subdivisions of the mollusk class Gastropoda. *See* **classification.**

prospective significance   the normal fate of any portion of an embryo at the beginning of development.

prostaglandin   a group of naturally occurring, chemically related, long-chain fatty acids that exhibit a wide variety of physiological effects (contraction of smooth muscles, lower blood pressure, antagonism of certain hormones, etc.). The first prostaglandin was originally isolated from the prostate gland (hence the name), but they are now known to be produced by many tissues of the body.

prosthetic group   that portion of a complex protein which is not polypeptide in character. Usually the prosthetic group is the active site of such a protein. The heme groups of hemoglobin are examples of prosthetic groups.

protamines   basic proteins that replace histones in sperm cells of many animal species. *See* **nucleoprotein.**

protan   *See* **color blindness.**

protandry   **1.** the maturation of the pollen-bearing organs before the female organs on a monoecious plant. **2.** sequential hermaphroditism in animals, with the male stage preceding the female stage (*compare with* **protogyny**). **3.** the appearance of male animals earlier in the breeding season than females.

protanomaly   *See* **color blindness.**

protanopia   *See* **color blindness.**

protease   an enzyme that digests proteins.

protein   a molecule composed of one or more polypeptide chains, each composed of a linear chain of amino acids covalently linked by peptide bonds. *See* **amino acid, peptide bond, protein structure, translation.**

protein kinase   an enzyme that attaches phosphate groups to serine, threonine, or tyrosine molecules built into proteins.

proteinoid   an amino acid polymer with a molecular weight as high as 10,000 d formed under "pseudoprimeval conditions" by heating to 70°C a dry mixture containing phosphoric acid and 18 amino acids. Such proteinoids are acted upon by proteolytic enzymes and have nutritive value for bacteria, but are nonantigenic.

protein structure   The *primary* structure of a protein refers to the number of polypeptide chains in it, the amino acid sequence of each, and the position of inter- and intrachain disulfide bridges. The *secondary* structure refers to the type of helical configuration possessed by each polypeptide chain resulting from the formation of intramolecular hydrogen bonds along its length. The *tertiary* structure refers to the manner in which each chain folds upon itself. The *quaternary* structure refers to the way two or more of the component chains may interact.

protein synthesis   *See* **translation.**

proteolytic   causing the digestion of proteins into simpler units.

proter   the anterior daughter organism produced by the transverse division of a protozoan.

Proterozoic   pertaining to a subdivision of the Precambrian era in which sponges, protozoa, and certain primitive invertebrates evolved. *See* **geologic time divisions.**

prothallus (prothalium)   the independent gametophyte of a horsetail or fern. *See* **classification** (Phylum Tracheophyta).

prothetely   an experimentally induced abnormality in which an organ appears in advance of the normal time, because of a partially inhibited metamorphosis: for example, the formation of pupal antennae on a caterpillar.

prothoracic gland   a gland located in the prothorax of insects which secretes ecdysone (*q.v.*). *See* **ring gland.**

prothoracicotropic hormone   a peptide hormone produced by neurosecretory cells in the dorsum of the insect brain which stimulates the prothoracic gland (*q.v.*) to synthesize and secrete ecdysones. Symbolized PTTC.

prothrombin   an inactive form of thrombin. *See* **blood clotting.**

protist   an informal term used to refer to any single-celled organism.

protocooperation   population or species interaction favorable to both, but not obligatory for either one.

Protoctista   one of the five kingdoms of living organisms. It contains the eukaryotic microorganisms and their immediate descendants, i.e., the nucleated algae, flagellated water molds, slime molds, and protozoa. *See* **classification, Eukaryotes.**

protogyny   sequential hermaphroditism with the ovary functioning before the testis. *Compare with* **protandry.**

protomers   single polypeptide chains (either identical or nonidentical) of a multimeric protein.

protomitochondria   *See* **petites.**

proton   an elementary particle of the atomic nucleus with a positive electric charge (equal numerically to the negative charge of the electron) and a mass of 1.0073 mass units.

proto-oncogene   a cellular gene that functions in controlling the normal proliferation of cells. It can be converted into an oncogene as the result of (1) a somatic mutational event, or (2) recombination with a viral genome. *See* *myc,* **Rous sarcoma virus.**

protoplasm   the substance within the plasma membrane of a cell; the nucleus and surrounding cytoplasm.

protoplast   the organized living unit of a plant or bacterial cell consisting of the nucleus (or nucleoid), cytoplasm, and surrounding plasma membrane, but with the cell wall left out of consideration. Protoplasts can be generated experimentally; e.g., the walls of *E. coli* cells can be removed by lysozyme treatment. Aphragmabacteria (*see* **classification**) lack cell walls and in this sense are protoplasts.

protoplast fusion   a mechanism for achieving genetic transformation by joining two protoplasts or joining a protoplast with any of the components of another cell.

Protostomia   a subdivision of Bilateria containing species in which the mouth arises from the blastopore. *See* **classification, Deuterostomia.**

prototroph   **1.** an organism that is able to subsist on a carbon source and inorganic compounds. For most bacteria, the carbon source could be a sugar; green plants use carbon dioxide. **2.** a microbial strain that is capable of growing on a defined minimal medium; wild type strains are usually regarded as prototrophs.

Protozoa   the subkingdom containing all unicellular animals.

provirus   **1.** a virus that is integrated into a host cell chromosome and is transmitted from one cell generation to another without causing lysis of the host. **2.** more specifically, a duplex DNA sequence in an eukaryotic chromosome (corresponding to the genome of an RNA retrovirus) that is transmitted from one cell generation to another without causing lysis of the host. Such proviruses are often associated with transformation of cells to the cancerous state.

proximal   toward or nearer to the place of attachment (of an organ or appendage).

Prunus   the genus that includes *P. amygdalus,* the almond; *P. armeniaca,* the apricot; *P. avium,* the cherry; *P. domestica,* the plum; *P. persica,* the peach.

pseudoalleles   genes that behave as alleles in the allelism test but can be separated by crossing over. *See* **cis, trans configurations.**

Pseudocoelomata   a subdivision of the Protostomia containing species having a body cavity that is not lined with peritoneum. The space is formed by dispersion of mesenchyme. *See* **classification.**

pseudocopulation   the mode of pollination in certain orchids in which structures of the flower closely resemble a female insect, and the male insects attempting copulation serve to transfer pollen from one flower to another.

pseudodominance   the phenotypic expression of a recessive allele on one chromosome as a consequence of deletion of the dominant allele from the homologue.

pseudoextinction   disappearance of a taxon by virtue of its being evolved by anagenesis into another taxon.

pseudogamy   the parthenogenetic development of an ovum following stimulation (but not fertilization) by a male gamete or gametophyte.

pseudogene   a gene bearing close resemblance to a known gene at a different locus, but rendered nonfunctional by additions or deletions in its structure that prevent normal transcription and/or translation. Pseudogenes are usually flanked by direct repeats of 10 to 20 nucleotides; such direct repeats are considered to be a hallmark of DNA insertion. *See* **hemoglobin genes.**

pseudohermaphroditism   a condition in which an individual has gonads of one sex and secondary sexual characters of the other sex or of both sexes. Pseudohermaphrodites are designated as male or female with reference to their sex chromosome constitution or the type of gonadal tissue present.

Pseudomonas   a genus of Gram-negative, motile bacteria, each with a single polar flagellum. Two species have been the subjects of genetic study, *P. aeruginosa* and *P. putida*. Lysogeny (*q.v.*) is common in *P. aeruginosa,* rare in *P. putida*.

pseudotumor   an aggregation of blackened cells in *Drosophila* larvae, pupae, and adults of certain genotypes. Such "tumors" result from encapsulation during the larval stage of certain tissues by hemocytes and subsequent melanization of these masses.

pseudouridine   *See* **rare bases.**

pseudovirion   a synthetic virus consisting of the protein coat from one virus and the DNA from a foreign source. *See* **transduction.**

P site   *See* **translation.**

psoralens   photosensitive cross-linking reagents that act specifically on base-paired regions of nucleic acids. *See* **trimethylpsoralen.**

^{32}**P suicide**   inactivation of phages due to the decay of radiophosphorus molecules incorporated into their DNA.

psychosis   a generic term covering any behavioral disorder of a far-reaching and prolonged nature.

PTC   **1.** plasma thromboplastin component. **2.** phenylthiocarbamide. *See* **blood clotting, taste blindness.**

pteridine   a molecule belonging to a family of organic compounds having the basic structure shown. *See* **drosopterin, sepiapterin.**

pteridophytes   the ferns, horsetails, club-mosses, and other vascular spore-bearing plants.

pteroylglutamic acid   folic acid (*q.v.*).

pterygote   an insect belonging to a division that includes all winged species. Some pterygotes (e.g., fleas) are wingless, but they are believed to have been derived from winged ancestors. *See* **apterygotes, classification.**

PTTH   prothoracicotropic hormone (*q.v.*).

Pu   abbreviation for any purine (e.g., adenine or guanine).

puff   *See* **chromosomal puff.**

pullet   a female chicken less than one year old. *See* **cockerel.**

pulse-chase experiment   an experimental technique in which cells are given a very brief exposure (the pulse) to a radioactively labeled precursor of some macromolecule, and then the metabolic fate of the label is followed during subsequent incubation in a medium containing only the nonlabeled precursor (the chase).

pulvillus   the last segment of the foot in an insect. It has a pad with a claw on either side.

punctuated equilibrium model of evolution   *See* **evolution.**

Punnett square   the checkerboard method commonly used to determine the types of zygotes produced by a fusion of gametes from the parents. The results allow the computation of genotypic and phenotypic ratios. Named after R. C. Punnett, its inventor.

pupal period   the developmental period between pupation and eclosion.

puparium formation   the formation of a pupal case by the tanning of the skin molted from the last instar larval insect.

pupation   that stage in the metamorphosis of the insect signaled by the eversion of the imaginal discs.

pupiparous   pertaining to certain dipterous females that lay young ready to pupate. The tsetse fly is an example.

purebred   derived from a line subjected to inbreeding (*q.v.*).

pure culture   a culture that contains only one species of organism.

pure line   a strain of an organism that is homozygous because of continued inbreeding.

purine   a class of organic bases found in nucleic acids.

puromycin   an antibiotic which, because of its structural resemblance to the terminal aminoacylated adenosine group of aminoacyl tRNA, becomes incorporated into the growing polypeptide chain and causes the release of incompleted polypeptide chains (which are terminated with a puromycin residue) from the ribosome.

P value   probability value. A decimal fraction showing the number of times an event will occur in a given number of trials. *See* **probability.**

Py   abbreviation for any pyrimidine (e.g., thymine, cytosine, uracil).

pycnosis the contraction of the nucleus into a compact, strongly staining mass, taking place as the cell dies.

pycocin *See* **bacteriocin.**

plyoric stenosis the constriction of the valve between the stomach and intestine, a congenital disorder of high heritability.

pyrenoid a small, round protein granule surrounded by a starch sheath found embedded in the chloroplasts of certain algae and liverworts.

pyridoxal phosphate the coenzyme of both amino acid decarboxylating enzymes and transaminating enzymes.

pyridoxine vitamin B_6.

pyrimidine a class of organic bases found in nucleic acids.

pyrimidine dimer the compound formed by UV irradiation of DNA whereby two thymine residues, or two cytosine residues, or one thymine and one cytosine residue occupying adjacent positions in the polynucleotide strand become covalently joined. *See* **thymine dimer.**

pyronin Y a basic dye often used in cytochemistry. In 2 M magnesium chloride at *p*H 5.7 pyronin Y stains only undegraded RNA. *See* **methyl green.**

pyrrole molecules ring-shaped compounds containing one nitrogen and four carbon atoms which are components of porphyrin (*q.v.*) molecules.

Q

q *See* **symbols used in human cytogenetics.**

Q Quaternary (*q.v.*).

Q$_{10}$ temperature coefficient; the increase in a reaction or other process (expressed as a multiple of the initial rate) produced by raising the temperature 10° C.

Q$_{02}$ oxygen uptake (in microliters per milligram dry weight per hour). The term is used in the measurements of respiratory rates.

Qa in the mouse, a series of loci located very close to the major histocompatibility complex (H-2) whose products are expressed on the surfaces of some lymphocyte classes and subclasses.

Q-bands *See* **chromosome banding techniques.**

Q beta an RNA phage from which an RNA replicase has been isolated.

Q beta replicase a viral RNA-directed RNA polymerase that is highly selective for its own template.

quadrivalent a meiotic association of four homologous chromosomes.

quadruplex *See* **autotetraploidy.**

quail *See* **Coturnix coturnix japonica.**

quantasome a photosynthetically active particle found in the grana of chloroplasts. Each quantosome is an oblate ellipsoid with axes of about 100 Å and 200 Å. Chlorophyll (*q.v.*) is localized within quantasomes.

quantitative character a character showing quantitative inheritance (beef and milk production in cattle, egg production in hens, DDT resistance in *Drosophila,* stature, weight, and skin pigmentation in man).

quantitative inheritance phenotypes that are quantitative in nature and continuous in distribution are referred to as quantitative characters (*q.v.*). During their genetic transmission, there is an absence of clear-cut segregation into readily recognizable classes showing typical Mendelian ratios. An often used example is ear length in maize. If crosses are made between individuals from lines showing large quantitative differences in ear length, the offspring are intermediate. When F$_1$ individuals are crossed, the F$_2$ population has a mean that is very similar to the F$_1$ mean, but some individuals produce ears as long or as short as the grandparents.

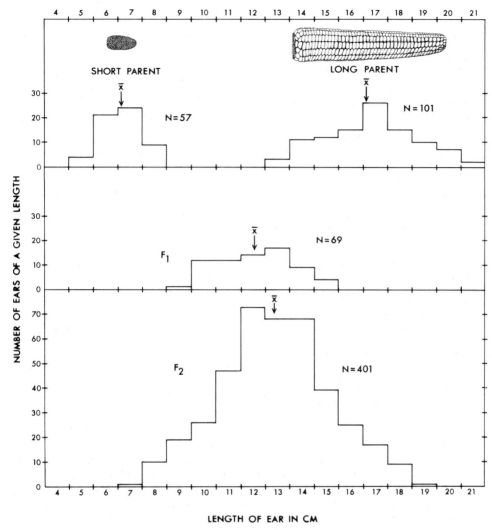

Quantitative inheritance in maize. Distribution of ear lengths in parents, F_1 and F_2 generations of a cross between Tom Thumb popcorn and Black Mexican sweet corn. (After Emerson and East, 1913)

Such results are explained by the "multiple factor hypothesis," which assumes that the quantitative character depends upon the cumulative action of multiple genes (or polygenes), each on a separate chromosome, and each producing a unitary effect. In the corn example, a simple model would employ three genes, each existing in two allelic forms. Each capital letter gene might be responsible for three units of "growth potential," and each small letter gene, for one unit. Thus the capital letter genes are all interchangeable in the sense that each produces the same phenotypic effect, and the same is true for the small letter genes. The long and short eared parental individuals would be AABBCC and aabbcc, respectively, and their offspring would be AaBbCc. These would show little variability, because all plants would be genetically identical. The segregation of the alleles in the F_2 population would produce 27 different genotypic classes, and the cumulative action of the genes would generate 7 phenotypic classes. The most common genotype (making up 1/8th of the total population) would be AaBbCc, genetically identical to the F_1 plants. But there would also be plants of genotype AABBCC and aabbcc (each making up 1/64th of the

323

population) and these would be phenotypically and genetically identical to the grandparents. There would also be individuals with various intermediate ear lengths of genotypes AABBCc, aabbcC, AAbbCC, etc.) and the result would be an F_2 population with a mean equivalent to the F_1, but with a distribution whose width depended on the number of segregating alleles. By comparing the variances of the F_1 and F_2 populations, one can estimate the number of segregating gene pairs responsible for the trait. *See* **Wright's polygenic estimate.**

quantum according to the quantum theory, energy is radiated in discrete quantities of definite magnitude called quanta and absorbed in a like manner.

quantum speciation the rapid evolution of new species, usually within small, peripheral isolates, with founder effects and genetic drift playing important roles. *See* **evolution.**

quarantine a period of forced isolation because of a contagious disease.

quartan malaria *See* **malaria.**

quartet a group of four nuclei or of four cells arising from the two meiotic divisions.

Quaternary the most recent of the Cenozoic periods lasting from the beginning of the Pleistocene to the present. *See* **geologic time divisions.**

quaternary protein structure *See* **protein structure.**

Quercus the genus of oaks including: *Q. alba,* the white oak; *Q. coccinea,* the scarlet oak; *Q. palustris,* the pin oak; *Q. suber,* the cork oak.

quick-stop mutants of *E. coli* that immediately cease replication when the temperature is increased to 42°C.

quinacrine an acridine derivative used in the treatment of certain types of cancer and malaria. It is also used as a fluorochrome in chromosome cytology.

quinone a compound belonging to a class of molecules that function in biological oxidation-reduction systems.

q.v. which see. An abbreviation for Latin, *quod vide.*

R

r **1.** reproductive potential. **2.** ring chromosome; *see* **symbols used in human genetics. 3.** roentgen. **4.** correlation coefficient; *see* **correlation.**

R **1.** a chemical radical. Used to show the position of an unspecified radical in a generalized structural formula of a group of organic compounds. **2.** a drug-resistant plasmid conferring resistance to one or more antibiotics on bacteria in which it resides.

rabbit *see Oryctolagus cuniculus.*

race a phenotypically and/or geographically distinctive subspecific group, composed of individuals inhabiting a defined geographical and/or ecological region, and possessing characteristic phenotypic and gene frequencies that distinguish it from other such groups. The number of racial groups that one wishes to recognize within a species is usually arbitrary but suitable for the purposes under investigation. *See* **subspecies.**

raceme an inflorescence as in the hyacinth, in which the flowers are borne on pedicles arising from the rachis. A branched raceme, such as may be seen in oats, rice, wheat, and rye, is called a panicle.

rachis *See* **spike.**

rad *r*adiation *a*bsorbed *d*ose; an energy equivalent to 100 ergs per gram of irradiated tissue.

Radiata a subdivision of the Eumetazoa containing species characterized by radial symmetry. *See* **classification.**

radiation the emission and propagation of energy through space or a medium in the form of waves. When unqualified, radiation usually refers to electromagnetic radiations (radio waves, infrared, visible light, ultraviolet, x rays, and gamma rays), and, by extension, ionizing particles.

radiation absorbed dose the quantity of radiation that liberates 100 ergs of energy in a gram of matter.

radiation chimera an experimentally produced animal containing hemopoeitic cells of a genotype different from that of the rest of the organism. Recipients receive a single dose of radiation that kills the stem cells of the bone marrow and much of the differentiated hemopoeitic tissue. Very shortly thereafter, they receive an intravenous inoculation of bone marrow or fetal liver cells from non-irradiated donors. The injected stem cells home to the recipient's bone marrow sites and begin repopulating them, and ultimately they replace the recipient's hemopoeitic tissues.

radiation dosage *See* **dose.**

radiation-induced chromosomal aberration a chromosomal aberration (*q.v.*) induced through breakage caused by ionizing radiation. In the table below are shown the origin and mitotic behavior of a variety of radiation-induced aberrations. Original break positions are indicated by short diagonal lines.

CHROMOSOME ABERRATIONS				CHROMATID ABERRATIONS			
TYPE	INTERPHASE	PRE-METAPHASE	ANAPHASE	TYPE	PROPHASE	PRE-METAPHASE	ANAPHASE
CHROMOSOME BREAK	*(diagram)*	*(diagram)*	*(diagram)*	CHROMATID BREAK	*(diagram)*	*(diagram)*	*(diagram)*
INTERSTITIAL DELETION	*(diagram)*	*(diagram)*	*(diagram)*	ISO-CHROMATID BREAK	*(diagram)*	*(diagram)*	*(diagram)*
EXCHANGES — INTRACHANGE (GROSS)	*(diagram)*	*(diagram)*	*(diagram)*	EXCHANGES — INTRACHANGE	*(diagram)*	*(diagram)*	*(diagram)*
INTERCHANGES ASYMMETRICAL	*(diagram)*	*(diagram)*	*(diagram)*	INTERCHANGES ASYMMETRICAL	*(diagram)*	*(diagram)*	*(diagram)*
SYMMETRICAL		*(diagram)*	*(diagram)*	SYMMETRICAL		*(diagram)*	*(diagram)*

After K. Sax, 1940. *Genetics 25:* 41–68. As modified by N.H. Giles, 1954, in *Radiation Biology.* Vol. I: p. 716, A. Hollaender, editor. McGraw-Hill, New York.

radiation sickness a syndrome characterized by nausea, vomiting, diarrhea, psychic depression, and death following exposure to lethal doses of ionizing radiation. The median lethal radiation dose for man is between 400 and 500 r. Such a dose leaves only about 0.5% of the body's reproducing cells still able to undergo continued mitosis. Since each cell continues to function normally in the physiological sense, death is not immediate. Damage shows up first in tissues with a high mitotic rate (the blood cell forming tissues of the bone marrow, for example). Death occurs when the surviving cells are unable to restore by mitosis the needed numbers in time to maintain the physiological functioning of the various vital tissues.

radical scavenger a molecule with a high affinity for free radicals. If a radical scavenger is added to a biological system prior to irradiation it may act as a protective agent.

radical substitution replacement of one amino acid by another with a different polarity or water solubility.

326

radioactive decay the disintegration of the nucleus of an unstable nuclide accompanied by the spontaneous emission of charged particles and/or photons.

radioactive isotope an isotope with an unstable nucleus that stabilizes itself by emitting ionizing radiations.

radioactive series a succession of nuclides, each of which transforms by radioactive disintegration into the next until a stable nuclide results.

radioactivity the spontaneous disintegration of certain nuclides accompanied by the emission of one or more types of radiation, such as alpha particles, beta particles, and gamma photons.

radioautograph autoradiograph (*q.v.*).

radioautographic efficiency autoradiographic efficiency (*q.v.*).

radioautography autoradiography (*q.v.*).

radiobiology a branch of biology that deals with the effects of radiation on biological systems.

radiogenic element an element derived from another element by atomic disintegration.

radiograph a shadow image made on photographic emulsion by the action of ionizing radiation. The image is the result of the differential attenuation of the radiation during its passage through the object being radiographed. A chest x-ray negative is a radiograph.

radioimmunoassay a highly sensitive technique for the quantitative determination of antigenically active substances that are present in very small amounts, such as hormones. The concentration of an unknown, unlabeled antigen is determined by comparing its inhibitory effect on the binding of radioactively labeled antigen to specific antibody with the inhibitory effect of known standards. Symbolized RIA.

radiological survey the evaluation of the radiation hazards incident to the production, use, or existence of radioactive materials or other sources of radiation under a specified set of conditions.

radiomimetic chemical a chemical that mimics ionizing radiations in terms of damage to nucleic acids. Radiomimetic compounds include **sulfur mustards, nitrogen mustards,** and **epoxides,** *all of which see.*

radioresistance the relative resistance of cells, tissues, organs, or organisms to the injurious action of radiation. Ultraviolet-resistant bacteria, for example, can excise ultraviolet-induced thymine dimers from their DNA.

radiotracer *See* **radioactive isotope, labeled compound.**

ramet an individual member of a clone. *Compare with* **ortet.**

Rana frogs of this genus have been used widely in research. The leopard frog, *R. pipiens,* is the most common species bred in the laboratory, and many of its mutations have been recovered and analyzed. Mutant strains are also available for *R. sylvatica, R. esculenta,* and *R. temporaria. R. esculenta* is the only anuran for which working maps of the lampbrush chromosomes are available.

r and K selection theory a theory in population ecology that attempts to establish whether environmental conditions favor the maximization of r (the intrinsic rate of natural increase) or of K (the carrying capacity of the environment). When populations can expand without food reserves limiting their growth, then r selection is in control. When food reserves limit population size, K selection takes over, and increase in one genotype must be at the expense of another. r selection operates in ecological situations were food reserves fluctuate drastically, and species are favored that reproduce rapidly and produce large numbers of offspring. K selection operates in populations that are close to the environmental carrying capacity, and species are favored that reproduce slowly and generate a few offspring that are well adapted to a relatively stable environment.

random assortment   *See* **assortment.**

random genetic drift   genetic drift.

random mating   a population mating system in which every male gamete has an equal opportunity to join in fertilization with every female gamete, including those gametes derived from the same individuals (if the species is monoecious or hermaphroditic); panmixis.

random primers   randomly generated oligodeoxyribonucleotides, some of which anneal to complementary sequences in the template nucleic acid and serve as primers in reactions involving reverse transcriptase.

random sample   a sample of a population selected so that all items in the population are equally likely to be included in the sample.

random-X inactivation   the method of dosage compensation (*q.v.*) found in eutherian mammals. *See* **paternal-X inactivation.**

Ranunculus   a genus of flowering plants including the buttercup. The cytology of meiosis has been extensively studied in species from this genus.

Raphanobrassica   a fertile allotetraploid developed by Karpechenko from a cross between the radish *(Raphanus sativus)* and the cabbage *(Brassica oleracea).*

rapidly reannealing DNA, rapidly reassociating DNA   repetitious DNA (*q.v.*).

rapid-lysing (*r*) mutants   mutants of T-even phage that enhance the rate at which *E. coli* host cells are lysed; on a bacterial lawn, *r*-plaques are larger than wild-type plaques (r^+).

rare bases   the purine and pyrimidine nucleoside phosphates found in transfer RNA molecules. These include ribothymidine, 5,6 dihydrouridine, 5-ribosyluracil (pseudouridine), inosine, 1-methylinosine, 1-methylguanosine, and N^2-dimethylguanosine (see p. 329).

rare earth   any of the series of very similar metals ranging in atomic number from 57 through 71. *See* **periodic table.**

Rassenkreis   *See* **circular overlap, polytypic species.**

rat   *See* **Rattus.**

rat kangaroo   *See* ***Potorous tridactylis.***

Rattus   the genus of rats including *R. norvegicus,* the brown rat and *R. rattus,* the black rat. The white laboratory rat is an albino form of *R. norvegicus.*

Rauscher leukemia virus   a retrovirus isolated from the plasma of leukemic mice.

R bands   *See* **chromosome banding techniques.**

RBC   red blood cell.

RBE   relative biological effectiveness (*q.v.*).

rDNA   **1.** in general, any DNA regions that code for ribosomal RNA components. **2.** specifically, a tandem cluster of eukaryotic rRNA genes with a sufficiently atypical base composition to allow its isolation directly from sheared genomic DNA.

rare bases

inosine

1-methylinosine

N²-dimethylguanosine

1-methylguanosine

ribothymidine

pseudouridine

5,6 dihydrouridine

rDNA amplification   The genes for rRNA are preferentially replicated during oogenesis in amphibia. In *Xenopus laevis,* for example, there are 2,000 rDNA repeats integrated into the chromosomes of the oocyte. However, there are 2,000,000 DNA repeats distributed among about 1,000 extrachromosomal nucleoli that lie near the periphery of the nucleus of each diplotene oocyte. These amplified genes arose from single copies of the chromosomal rDNA, and during pachynema they replicated extrachromosomally by a rolling circle (*q.v.*) mechanism. These extrachromosomal nucleoli function to transcribe the rRNAs stored in the growing oocyte. Amplification of rDNA also occurs commonly in insects with panoistic ovaries and in the macronuclei of protozoa, such as *Tetrahymena. See* **insect ovary types, Miller trees.**

reactor   nuclear reactor.

reading   the unidirectional process by which mRNA sequences are decoded (translated) into amino acid sequences (polypeptide chains).

reading frame   the sequence of codons (adjacent groups of three nucleotides) by which translation may occur. For example, a segment of mRNA 5'AUCCGAGGC3' could be translated in three reading frames, 5'AUC... or 5'UCC... or 5'CCG..., dependent upon the location of the start codon.

reading frame shift   Certain mutagens (acridine dyes, for example) intercalate themselves between the strands of a DNA double helix. During subsequent replication the newly formed complementary strands may have a nucleotide added or subtracted. A cistron containing an additional base or missing a base will transcribe a messenger RNA with a *reading frame shift*. That is, during translation the message will be read properly up to the point or loss or addition. Thereafter, since the message will continue to be read in triplets, all subsequent codons will specify the wrong amino acids (and some may signal chain termination). *See* **translation, nonsense mutation, amino acid, acridine orange, acriflavin.**

reading mistake   the incorrect placement of an amino acid in a polypeptide chain during protein synthesis.

readthrough   **1.** transcription beyond a normal terminator sequence in DNA, due to occasional failure of RNA polymerase to recognize the termination signal or due to the temporary dissociation of a termination factor (such as rho in bacteria) from the terminator sequence. **2.** translation beyond the chain-terminator (stop) codon of a mRNA, as occurs by a nonsense suppressor (*q.v.*) tRNA.

reannealing   in molecular genetics, this term refers to the joining of single-stranded DNA molecules that have complementary base sequences to form duplex molecules. Reannealing and annealing (*q.v.*) differ in that the DNA molecules in the first case are from the same source and in the second case from different sources. *See* **reassociation kinetics, mouse satellite DNA, Alu family, repetitious genes.**

reassociation   in DNA, the pairing of complementary strands that results in duplex formation.

reassociation kinetics   a technique devised by Roy J. Britten and David E. Kohne to measure the rate of reassociation of complementary strands of DNA derived from a single source. The DNA under study is fragmented into pieces several hundred base pairs in length and then disassociated into single strands by heating. Subsequently the temperature is lowered, and the rate of reannealing (*q.v.*) is monitored. Reassociation of DNA is followed in the form of a cot curve, which plots the fraction of molecules that have reannealed against the log of cot. Cot values are defined as $C_0 \times t$, where C_0 is the initial concentration of single-stranded DNA in moles of nucleotides per liter and t is the reannealing time in seconds. Typical cot curves are shown on page 331. DNAs reannealing at low cot values $(10^{-4}-10^{-1})$ are composed of highly repetitive sequences, DNAs reannealing at cot values between 10^0 and 10^2 are moderately repetitive, and DNAs reannealing at higher cot values are nonrepetitive. *See* **repetitious DNA, mouse satellite DNA, Alu family.**

reassortant virus   a virion consisting of DNA from one virus and protein from another viral species; e.g., through genetic engineering, a hybrid virus has been made containing genes from the human influenza virus and capsid proteins that provoke immunity, but also containing avian influenza virus genes that slow the rate of viral replication.

recapitulation   the theory (now profoundly modified) that an individual during its development passes through stages resembling the adult forms of its successive ancestors. The concept is often stated, "ontogeny recapitulates phylogeny." *Contrast with* **paedomorphosis.**

RecA protein   the product of the RecA locus of *E.coli,* which acts as a protease in the SOS response (*q.v.*) and also functions in recombination-repair.

receptor element   *See* **controlling elements.**

reassociation kinetics

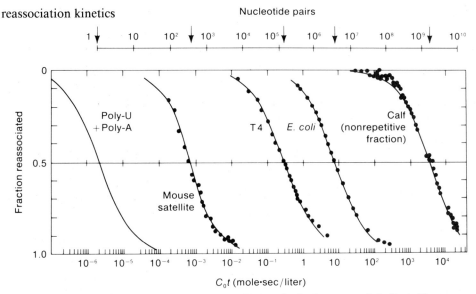

For each of the DNA samples tested, the number of base pairs in the genome is indicated by an arrow on the logarithmic scale at the top of the graph. The poly-U + poly-A sample is a double helix of RNA, with one strand containing only A and the other strand only U. The mouse satellite DNA is a fraction of nuclear DNA in mouse cells that differs in its physical properties from the bulk of the DNA. The calf DNA represents only those sequences that are present in single copies per haploid genome. The denatured DNA samples were fragmented by mechanical shear to chain lengths of about 400 nucleotides and incubated at a temperature near 60°C. The fraction reassociated was measured by the decrease in UV absorption as double strands formed. (After R. J. Britten and D. E. Kohne, "Repeated Sequences in DNA," Science 161:529, 1968. Copyright 1968 by the American Association for the Advancement of Science.)

recessive complementarity   *See* **complementary genes.**

recessive gene   in diploid organisms a gene which is phenotypically manifest in the homozygous state but is masked in the presence of its dominant allele. Usually the dominant gene produces a functional product, while its recessive allele does not. Therefore the normal phenotype is produced if the dominant allele is present (in one or two doses per nucleus), and the mutant phenotype appears only in the absence of the normal allele (i.e., when the recessive gene is homozygous). By extension, the terms dominant and recessive are used in the same sense for heterokaryons and merozygotes.

recessive lethal   an allele which kills the cell or organism that is homozygous or hemizygous for it.

reciprocal crosses   crosses of the forms A ♀ X B ♂ and B ♀ X A ♂, where the individuals symbolized by A and B differ in genotype or phenotype or both. Reciprocal crosses are employed to detect **sex linkage, maternal inheritance,** or **cytoplasmic inheritance,** *all of which see.*

reciprocal genes   complementary genes (*q.v.*).

reciprocal hybrids   hybrid offspring derived from reciprocal crosses of parents from different species.

reciprocal recombination   in the gametes of dihybrids, the production of new linkage arrangements that are different from those of the maternal and paternal homologues. For example, if the nonallelic mutants *a* and *b* were present in the coupling configuration *AB/*

331

ab, crossovers would generate the reciprocal recombinant gametes *Ab* and *aB* in equal amounts.

reciprocal translocation   *See* **translocation.**

rec⁻ mutant   a class of mutations characterized by defective recombination. Such mutants are also radiation-sensitive, which suggests that enzymes functioning during the naturally occurring breakage and rejoining characterizing meiotic crossing over may also repair damage caused by mutagens.

recoil energy   the energy imparted to the positively charged ion formed during the radioactive transmutation of an atom. A high-energy beta particle is emitted concurrently.

recombinant   the new individual or cell arising as the result of recombination.

recombinant DNA technology   techniques for uniting two heterologous DNA molecules, usually as a result of *in vitro* ligation of DNAs from different organisms. Recombinant DNA molecules are commonly produced by experiments in genetic engineering, as exemplified by inserting a human gene coding for insulin into a bacterial plasmid for cloning. Synonymously called *gene splicing, molecular cloning,* and *plasmid engineering.*

recombinant inbred (RI) lines   inbred lines, each derived independently from an F_2 generation produced from crossing two unrelated, inbred, progenitor lines. Each RI line has a characteristic combination of genes with a different pattern of alternative alleles at multiple loci. This technique has been used in mice to fix chance recombinants in a homozygous state in a group of strains derived from two unrelated but highly inbred progenitor strains.

recombinant joint   the edge of a heteroduplex region where two recombining DNA molecules are connected.

recombinant RNA technology   techniques that unite foreign RNA molecules or splice different RNAs from the same species. For example, a heterologous RNA sequence can be constructed by ligation of two or more different RNA molecules with T4 RNA ligase.

recombination   the occurrence of progeny with combinations of genes other than those that occurred in the parents, due to independent assortment or crossing over.

recombination frequency   the number of recombinants divided by the total number of progeny. This frequency is used as a guide in assessing the relative distances between loci on a genetic map.

recombination nodules   electron-dense organelles seen attached to synaptonemal complexes. These presumably play a role in crossing-over.

recombination suppression   *See* **crossover suppressor.**

recon   the smallest unit of DNA capable of recombination.

record of performance   a record of an animal with respect to certain economically important characteristics. Such data are used by livestock breeders in the artificial selection and development of improved breeds.

recurrence risk   the risk that a genetic defect that has appeared once in a family will appear in a child born subsequently.

recurrent parent   backcross parent.

red blood cell   erythrocyte. *See* **hemoglobin, sickle-cell anemia.**

reductase   an enzyme responsible for reduction in an oxidation-reduction reaction.

reduction   the loss of oxygen, gain of hydrogen, gain of electrons. *See* **oxidation.**

reduction divisions *See* **meiosis.**

reductionism a philosophy that each phenomenon in the natural world can be understood from a knowledge of its component parts. *See* **mechanistic philosophy.**

redundant cistrons cistrons frequently repeated on a chromosome. Examples are the cistrons in the nucleolus organizer coding for the ribosomal RNA molecules.

redundant code *see* **degenerate code.**

redundant DNA *See* **repetitious DNA.**

refractile capable of refracting light.

refractive index the ratio of the velocity of light in a vacuum to its velocity in a given substance.

regeneration the replacement by an organism of tissues or organs that have been lost.

regression in psychiatry, the reversal of psychosexual development. The patient may adopt a behavior pattern characteristic of childhood or infancy. Regression is a primary feature of schizophrenia.

regression coefficient the rate change of the dependent variable with respect to the independent variable. The change in mutation frequency per unit change in radiation dose, for example, would be determined by the regression coefficient of the regression line (*q.v.*).

regression line a line that defines how much an increase or decrease in one factor may be expected from a unit increase in another. *See* **line of best fit, scatter diagram.**

regressive evolution changes in the genetic structure of a population that result in loss of complexity, as exemplified by the loss of pigment and eyes in some cave-dwelling animals.

regulation the power of an embryo to continue normal or approximately normal development or regeneration in spite of experimental interference by ablation, implantation, transplantation, etc.

regulative development embryonic development in which the fates of all parts of the embryo are not fixed before fertilization. In such development an ablated part can be repaired, or even separated blastomeres can form identical twins. *See* **mosaic development.**

regulator element *See* **controlling elements.**

regulator gene a gene whose primary function is to control the rate of synthesis of the products of other distant genes. The regulator gene ($_rG$) controls the synthesis of a protein repressor (R) which inhibits the action of an operator gene ($_oG$) and thus turns off the operon it controls.

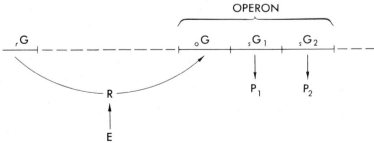

In the diagram, the horizontal line represents a chromosome upon which four genes reside. The left gene can be distant from the other three closely linked genes (in fact, $_rG$ can be on a different chromosome). Genes $_sG_1$ and $_sG_2$ are structural genes or cistrons of the conventional sort which produce specific proteins P$_1$ and P$_2$, respectively, through the formation of specific messenger RNA molecules. The repressor is present in exceedingly small amounts. It possesses two sites, one of which can attach to the operator and one of which can bind an effector (E) molecule. Once bound to E, however, the repressor changes shape and cannot attach to the operator. The effector molecule is generally a substrate of an enzyme produced by $_sG_1$ or $_sG_2$. Thus the system is an inducible one, since synthesis of P$_1$ and P$_2$ proceeds only in the presence of E. *See* **allosteric effect, constitutive mutations, derepression, inducible system, repressible system.**

regulatory sequence   a DNA sequence involved in regulating the expression of the structural gene(s) in the common operon. Examples include attenuators, operators, and promoters.

regulon   a noncontiguous group of genes under control of the same regulator gene.

reiterated genes   genes which are present in multiple copies that are clustered together on specific chromosomes. Ribosomal RNA genes, transfer RNA genes, and histone genes are examples of such tandem multigene families.

rejection   in immunology, destruction of a cell or tissue graft by the immune system of the recipient, directed against antigens on the graft that are foreign to the recipient.

relational coiling   the loose plectonemic coiling of two chromatids about one another. The two chromatids cannot separate until uncoiling is completed.

relative biological effectiveness   the ratio of the doses of different ionizing radiations required to produce the same biological effect.

relative plating efficiency   the percentage of inoculated cells which give rise to colonies, relative to a control where the absolute plating efficiency is arbitrarily sets as 100. *See* **absolute plating efficiency.**

relaxation complex   a group of three proteins tightly bound to some *E. coli* supercoiled plasmids that convert supercoiled DNA to a nicked open circle. When heated or treated with alkali, proteolytic enzymes, or detergents, one of these proteins nicks one strand at a specific site, thereby relaxing the supercoil to a nicked open circular form. During relaxation, the two smaller proteins are released, but the largest protein becomes covalently attached to the 5'-P end of the nick. Nicking plays a role in transfer of the plasmid during conjugation. The site of the nick establishes the *transfer origin.*

relaxation of selection   cessation of selection in an experimental situation.

relaxation proteins   *See* **helix-destabilizing proteins.**

relaxed control   *See* **plasmid.**

relaxed molecule   absence of supercoiling in molecules such as circular DNA.

relaxed mutants   mutants of *E.coli* that do not exhibit the stringent response to starvation for amino acids or other nutrients.

release factors   specific proteins that read termination codons and cause the release of the finished polypeptide.

releaser   in ethology (*q.v.*), the particular physical attributes or behavior of one animal that stimulates another animal to perform a specific response.

relic   surviving beyond others of a kind, as a species that persists in one region after becoming extinct elsewhere or a surviving species of a group of which others are extinct.

relic coil   the relaxed spirals often seen in prophase chromosomes. These coils are believed to be left over from the tightly coiled condition that the chromosome maintained at the previous metaphase.

relief   in photography, etc., to produce the effect of a third dimension in a two-dimensional image, as by the use of shadows.

rem   *r*oentgen *e*quivalent *m*an, the dose of ionizing radiation that has the same biological effect as one *rad* of x rays. It is equal to *rads* $\times$ *rbc* (relative biological effectiveness).

renal   referring to the kidney.

renaturation   the return of a protein or nucleic acid from a denatured state to its native three-dimensional configuration.

Renner complex   a group of chromosomes (and the genes within them) that are distributed as a unit generation after generation. Such complexes are found in species belonging to the genera *Oenothera* and *Rhoeo,* for example, where whole sets of chromosomes are involved in a series of interchanges. At first meiotic metaphase one sees rings of bivalents, rather than independent tetrads.

reovirus   a virus whose name is derived from *r*espiratory and *e*nteric *o*rphan virus. Reoviruses have been found in humans, but their relation to any disease is uncertain. The term "orphan" is used for viruses "without a disease." They are one of the few viruses that contain double-stranded RNA.

rep   *r*oentgen *e*quivalent *p*hysical (*q.v.*).

repair synthesis   enzymatic excision and replacement of regions of damaged DNA as when UV-induced thymine dimers are removed. *See* **cut and patch repair.**

rep DNA   repetitious DNA (*q.v.*).

repeated gene families   synonomous with multiple gene families (*q.v.*).

repeating unit   the length of a nucleotide sequence that is repeated in a tandem cluster.

repeats   small tandem duplications (*q.v.*).

repetition frequency   the number of copies of a given DNA sequence present in the haploid genome.

repetitious DNA   nucleotide sequences occurring repeatedly in chromosomal DNA. Analysis of reassociation kinetics (*q.v.*) reveals that repetitious DNA can belong to the highly repetitive or middle repetitive categories. The highly repetitive fraction contains sequences of several nucleotides repeated millions of times. It is a component of constitutive heterochromatin (*q.v.*). Middle repetitive DNA consists of segments 100–500 bp in length repeated 100 to 10,000 times each. This class of rep DNA contains the genes transcribed into rRNAs and tRNAs.

repetitive DNA   repetitious DNA (*q.v.*).

repetitive genes   *See* **multigene families.**

replacement sites   positions within a gene at which point mutations alter the amino acid specification.

replacement vector   *See* **lambda cloning vehicle.**

replica plating   a technique used to produce identical patterns of bacterial colonies on a series of petri plates. A petri plate containing bacterial colonies is inverted, and its surface is pressed against a cylindrical block covered with velveteen. In this way 10 to 20% of the bacteria are transferred to the fabric. Subsequently bacteria-free plates are inverted and pressed against the velveteen disc to pick up samples of the colonies. About eight replicas may be printed from a single pad in this way. If the medium contained in each of the secondary plates differs in its selective properties, then the hundreds of different bacterial clones transferred may be scored simultaneously for their responses to a given agent.

replicase   *See* **RNA replicase.**

replication   a duplication process requiring copying from a template.

replication-defective virus   a virus defective in one or more genes essential for completing the infective cycle.

replication eye (bubble)   an eye-shaped replicated DNA region within a longer, unreplicated region. *See* **D loop.**

replication fork   *See* **replication of DNA.**

replication of DNA   during DNA replication the two strands of the duplex molecule separate to form a *replication fork.* DNA polymerase (*q.v.*) then adds complementary nucleotides starting at the 3′ end. The strand that is continuously replicated in this way is referred to as the *leading strand.* The other strand is replicated discontinuously in short pieces. These are called *Okazaki fragments* after R. Okazaki, who identified them. The fragments are later connected by DNA ligase to form the *lagging strand. See* **primase.**

replicative forms   double-stranded nucleic acid molecules seen at the time of the replication of single-stranded DNA and RNA viruses.

replicator   *See* **replicon.**

replicon   a genetic element that behaves as an autonomous unit during DNA replication. In bacteria, the chromosome functions as a single replicon, whereas eukaryotic chromosomes contain hundreds of replicons in series. Each replicon contains a segment to which a specific RNA polymerase binds and a replicator locus at which DNA replication commences. The polymerase makes an RNA primer called an initiator.

replicon fusion   the joining of two complete replicating systems, mediated by a transposon. If two plasmids are joined (one of which carries a transposon), the process is called *plasmid fusion.*

replisome   a structure containing DNA polymerase and other enzymes which assembles at the replication fork of a bacterial chromosome to initiate DNA synthesis.

rep protein   a helicase (*q.v.*). identified in the *rep* mutant strain of *E. coli* which hydrolyzes ATP while forcing the strands of the DNA helix apart.

representation   *See* **abundance.**

repressible enzyme   an enzyme whose rate of production is decreased when the intracellular concentration of certain metabolities is increased.

repressible system   a regulatory system in which the product of a regulator gene (the repressor) blocks transcription of the operon only if it first reacts with an effector molecule (called the repressing metabolite). Thus mRNA synthesis occurs only in the absence of the effector. *See* **regulator gene.**

repressing metabolite   *See* **repressible system.**

repression   **1.** the inhibition of transcription or translation when a repressor protein binds to an operator locus on DNA or to a specific site on a mRNA. **2.** the cessation of synthesis of one or more enzymes when the products of the reactions they catalyze reach a critical concentration.

repressor   a protein (synthesized by a regulator gene) that binds to an operator locus and blocks transcription of that operon. *See* **regulator gene.**

reproduction probability   the average number of children of patients with a specific hereditary disease in relation to the average number of children of comparable individuals who do not have the hereditary disease. The reproduction probability is a measure of the selective disadvantage of a hereditary disease.

reproductive death   the suppression of the proliferative ability of a cell which otherwise would divide indefinitely.

reproductive isolation   the absence of interbreeding between members of different species. *See* **isolating mechanisms.**

reproductive potential   the theoretical logarithmic rate of population growth when unimpeded by environmental limitations; also known as the biotic potential; symbolized *r*.

reproductive success   for a given individual, the number of offspring that survive to reproduce.

repulsion   *See* **coupling, repulsion configurations.**

RER   *r*ough surfaced *e*ndoplasmic *r*eticulum. (*q.v.*).

RES   *r*eticulo-*e*ndothelial *s*ystem.

residual genotype   background genotype (*q.v.*).

residual homology   in species hybrids, the homology remaining between those chromosomes that have been derived from a common ancestral chromosome and have since diverged through the accumulation of mutations. *See* **homoeologous chromosomes.**

resistance factor   a class of episomes that confer antibiotic resistance to the recipient bacterium. *See* **R plasmid.**

resistance transfer factor (RTF)   *See* **R plasmid.**

resolvase   an enzyme catalyzing the site-specific recombination between two transposons present as direct repeats in a *cointegrate structure* (*q.v.*).

resolving power   the ability of any magnifying system to reveal fine detail. This ability is often measured as the minimum distance between two lines or points at which they are resolved as two rather than as a single blurred object. The maximum resolving power of the light microscope is about 0.2 μm; that of the electron microscope, about 0.5 nm.

respiration   the aerobic, oxidative breakdown and release of energy from fuel molecules.

respiratory pigment   a substance that combines reversibly with oxygen, thus acting as a carrier of it (hemoglobin, for example).

respiratory quotient (RQ)   the ratio of the volume of carbon dioxide expired to the volume of oxygen inspired. The RQ is used to estimate the type of fuel being utilized by the organism. The theoretical RQ values for carbohydrates, fats, and proteins are 1.0, 0.7, and 0.8, respectively.

responder   in immunology, an animal capable of mounting an immunological response to a particular antigen.

resting cell (nucleus)   any cell (nucleus) not undergoing division. The cell (nucleus) is nevertheless very active metabolically.

restitution   the spontaneous rejoining of experimentally induced, broken chromsomes to produce the original configuration.

restitution nucleus   a nucleus containing double the expected number of chromosomes owing to a failure of the mitotic apparatus to function properly.

restricted transduction   *See* **transduction.**

restriction   ability of a bacteriophage to infect bacteria belonging to certain strains, but not others. *See* **DNA restriction enzyme.**

restriction allele   *See* **DNA restriction enzyme.**

restriction and modification model   a theory propsed by W. Arber to explain host-controlled restriction of bacteriophage growth. According to this model the DNA of the bacterium contains specific nucleotide sequences that are recognized and cleaved by the restriction endonucleases carried by that cell. The bacterium also contains methylases that methylate these sequences. This chemical modification thus protects the DNA of the bacterium from its own endonucleases. However, these serve to degrade foreign DNA introduced by phages.

restriction endonucleases   any one of many enzymes that cleave foreign DNA molecules at specific recognition sites. Restriction endonucleases are coded for by genes called restriction alleles. The enzymes are named by a symbol that indicates the bacterial species from which they were isolated, followed by a Roman numeral which gives the chronological order of discovery when more than one enzyme came from the same source. Some restriction endonucleases, the organism from which they were isolated, and their target nucleotide sequences are shown below. The arrows indicate the cleavage sites.

Bam HI *Bacillus amyloliquefaciens*
$$5' \ldots \text{G-G-A-T-C-C} \ldots 3'$$
$$3' \ldots \text{C-C-T-A-G-G} \ldots 5'$$

Eco RI *Escherichia coli RY13*
$$5' \ldots \text{G-}\overset{*}{\text{A}}\text{-A-T-T-C} \ldots 3'$$
$$3' \ldots \text{C-T-T-A-A-G} \ldots 5'$$

Hin dII *Hemophilus influenzae Rd*   C $\downarrow$ A
$$5' \ldots \text{G-T-(T)-(G)-}\overset{*}{\text{A}}\text{-C-3'}$$
$$3' \ldots \text{C-A-(A)-(C)-T-G-5'}$$
G $\uparrow$ T

$\overset{*}{\text{A}}$ is N^6-methyladenine; (T)(G) signifies that either base can occupy that position. Note that *Bam* HI and *Eco* RI cleave the strands of DNA at specific sites four nucleotides apart. Such staggered cleavage yields DNA fragments with protruding 5' termini. Such ends are said to be "sticky" or "cohesive" because they will hydrogen bond to complementary 3'

338

ends. As a result, the end of any DNA fragment produced by an enzyme, such as *Eco* RI, can anneal with any other fragment produced by that enzyme. This property allows splicing of foreign genes into *E. coli* plasmids. Enzymes like *Hin* dII produce flush or blunt ended fragments.

restriction fragment   a fragment of a longer DNA molecule digested by a restriction endonuclease.

restriction fragment length polymorphisms   variations occurring within a species in the length of DNA fragments generated by a specific endonuclease. Such variations are generated by mutations that create or abolish recognition sites for these enzymes. For example, restriction endonuclease mapping of human structural genes for beta hemoglobin chains has shown that patients with the sickle cell mutation produce abnormal restriction fragments. Since restriction enzyme analyses can be performed on DNA from amniotic fluid cells, RFLPs provide a new approach to prenatal diagnoses.

restriction map   a diagram portraying a linear array of sites on a DNA segment at which one or more restriction endonuclease enzymes cleave the molecule.

restrictive conditions   any environmental condition (e.g., temperature or type of host) under which a conditional mutation either cannot grow or expresses the mutant phenotype. *See* **temperature sensitive mutation.**

restrictive transduction   *See* **transduction.**

reticulate evolution   the netlike lineage relation seen for a series of related allopolyploid species. The cross links represent places where hybridization has occurred and allotetraploid species have arisen. Reticulate evolution is common in plants. *See* **dendritic evolution.**

reticulocyte   an immature erythrocyte at an active stage in the synthesis of hemoglobin.

reticulo-endothelial system   a network of phagocytic cells residing in the bone marrow, spleen, and liver of vertebrates, where they free the blood or lymph of foreign particles.

retinene   a light-absorbing, carotenoid pigment which is a derivative of vitamin A. Retinene combines with opsin to form rhodopsin.

$$\mathrm{CH_3} \diagdown \quad \diagup \mathrm{CH_3}$$

$$
\begin{array}{c}
\mathrm{CH_3}\diagdown \quad \diagup \mathrm{CH_3} \\
\mathrm{C} \\
\diagup \quad \diagdown \\
\mathrm{H_2C} \quad \mathrm{C}\!-\!\mathrm{CH}\!=\!\mathrm{CH}\!-\!\underset{\underset{\mathrm{CH_3}}{|}}{\mathrm{C}}\!=\!\mathrm{CH}\!-\!\mathrm{CH}\!=\!\mathrm{CH}\!-\!\underset{\underset{\mathrm{CH_3}}{|}}{\mathrm{C}}\!=\!\mathrm{CH}\!-\!\underset{\underset{\mathrm{H}}{|}}{\mathrm{C}}\!=\!\mathrm{O} \\
\mathrm{H_2C} \quad \mathrm{C}\diagup\mathrm{CH_3} \\
\diagdown \quad \diagup \\
\mathrm{CH_2}
\end{array}
$$

retinitis pigmentosa   an autosomal recessive genetic disease of the human retina, causing progressive blindness.

retinoblastoma   a malignant neoplasm composed of primitive retinal cells usually occurring in children less than three years old. Retinoblastoma occurs in hereditary and nonhereditary forms. In the case of hereditary retinoblastoma, patients are generally affected bilaterally. The gene involved (*Rb*) is located on the long arm of chromosome 13. In contrast to oncogene activation or alteration, *Rb* functions to suppress tumor formation, since the loss or inactivation of both alleles results in the disease.

retinol   vitamin A (*q.v.*).

retrogene   *See* **processed gene.**

retroregulation   the ability of downstream DNA sequences to regulate translation of an mRNA.

retroviruses   RNA viruses that utilize reverse transcriptase (*q.v.*) during their life cycle. This enzyme allows the viral genome to be transcribed into DNA. The name *retrovirus* alludes to this "backwards" transcription. The transcribed viral DNA is integrated into the genome of the host cell where it replicates in unison with the genes of the host chromosome. The cell suffers no damage from this relationship unless the virus carries an oncogene (*q.v.*). If it does, the cell will be transformed into a cancer cell. Viruses belonging to this group are subdivided into (1) fowl viruses (such as the Rous sarcoma virus), (2) mouse leukemia and sarcoma viruses (Maloney and Rauscher viruses), (3) mouse mammary tumor viruses (Bittner virus), and (4) feline leukemia and sarcoma viruses. Also called leukoviruses or oncornaviruses.

reversals   situations in phylogenetic analysis in which a derived character state changes (reverses) to a preexisting character state. For example, among the vertebrates, limbs are considered to be derived from a limbless state; but snakes are thought to have evolved from tetrapod (four-legged) ancestors and in the process have lost their limbs.

reverse loop pairing   *See* **inversion.**

reverse mutation   a change in a mutant gene which restores its ability to produce a functional protein.

reverse selection   selection in an experimental situation for a trait opposite to the one selected earlier (for example, selection first for increased numbers of thoracic bristles in *Drosophila,* then for decreased numbers).

reverse transcriptase   RNA dependent DNA polymerase (*q.v.*).

reverse transcription   DNA synthesis from an RNA template, mediated by reverse transcriptase. *See* **RNA dependent DNA polymerase.**

reversion   reverse mutation (*q.v.*).

revertant   **1.** an allele that undergoes reverse mutation. **2.** an organism bearing such an allele.

reverted Bar   *See* **Bar.**

reynard   a male fox.

R$_f$ in paper chromatography, a ratio given by the distance traveled by the solute divided by the distance traveled by the solvent. For a given solute molecule the R$_f$ varies with the solvent, and therefore the solvent must be specified for any R$_f$ value.

RF   **1.** replicative form; **2.** recombination frequency.

R factor   resistance factor (*q.v.*).

rhesus factor   *See* **Rh factor.**

rheumatoid factor   a distinctive gamma globulin commonly present in the serum of patients with rheumatoid arthritis.

Rh factor   an antigen occurring on the erythrocytes of certain human beings. The Rh system actually contains several antigens. The most important one was first found in the Rhesus monkey—hence the name. Persons of genotype *r/r* produce no antigen and are classified as Rh-negative. *R/R* and *R/r* individuals produce the antigen. A pregnant mother who is Rh-negative but is carrying an Rh-positive child may produce antibodies against the child *in utero,* causing the child to develop a hemolytic disease called erythroblastosis fetalis. The Rh genetic locus is on the distal end of the long arm of chromosome 1. *See* **blood group.**

Rhizobium a genus of nitrogen fixing bacteria that live as symbionts in the root nodules of legumes. *See* **classification,** Monera.

rhizoid a unicellular or multicellular filament that attaches some nonvascular plants and the gametophytes of some vascular plants to the substrate.

rhizome a creeping underground stem that serves as a means of perennation and vegatative propagation. The iris, for example, produces rhizomes.

Rhizopoda the phylum of protoctists containing singled celled amoebas. *See Amoeba proteus,* **classification.**

Rhodesian man a subspecies of *Homo sapiens* that lived in the middle and upper Pleistocene. Known from fossils found in southern Africa.

rhodoplast the red plastid of red algae.

rhodopsin the visual pigment of vertebrates.

rho factor an oligomeric protein in *E. coli* that attaches to certain sites on its DNA to assist in termination of transcription.

RhoGAM the trade name for anti-Rh gamma globulin used in the prevention of Rh hemolytic disease. *See* **Rh factor.**

rho(ρ)-independent terminators *E. coli* DNA sequences recognized by RNA polymerase as transcription termination signals in the absence of rho (ρ) factors.

Rhynchosciara A genus of primitive flies belonging to the Sciaridae. Two species, *R. angelae* and *R. hollaenderi,* can be reared in the laboratory and have been extensively studied because of the gigantic polytene chromosomes found in various larval tissues (such as the salivary glands, Malpighian tubules, and anterior mid gut).

RIA radioimmunoassay (*q.v.*).

riboflavin vitamin B$_2$, a subunit of flavin adenine dinucleotide and flavin mononucleotide.

riboflavin phosphate flavin mononucleotide (*q.v.*).

ribonuclease the enzyme that hydrolyzes ribonucleic acid.

ribonuclease A an endoribonuclease that specifically attacks pyrimidines at the 3′ phosphate group and cleaves the 5′ phosphate linkage to the adjacent nucleotide. The end products of digestion are pyrimidine 3′ phosphates and oligonucleotides with pyrimidine 3′ phosphate termini.

ribonuclease P   a bacterial enzyme whose catalytic activity depends upon an RNA molecule. Historically, ribonuclease P was the first RNA found to exhibit the characteristics of a true enzyme (i.e., it acts catalytically but is not changed or consumed during the reaction). The enzyme cleaves precursor transfer RNA molecules to produce mature tRNAs. Ribonuclease P is composed of about five times more RNA than protein by weight. The RNA component alone can perform the catalytic function, whereas the protein component alone cannot. This enzyme has been found in *E. coli* and in *Bacillus subtilis.*

ribonuclease T1   an endoribonuclease that specifically attacks the 3′ phosphate groups of guanosine nucleotides and cleaves the 5′ phosphate linkage to the adjacent nucleotide. The end products of digestion are guanosine 3′ phosphates and oligonucleotides with guanosine 3′ phosphate termini.

ribonucleic acid (RNA)   any of a family of polynucleotides characterized by their component sugar (ribose) and one of their pyrimidines (uracil). RNA molecules are single stranded and have lower molecular weights than DNAs. There are three classes of RNAs: (1) messenger RNA (*q.v.*), (2) ribosomal RNA (*see* **ribosome**), and (3) transfer RNA (*q.v.*).

ribonucleoprotein   a complex macromolecule containing both RNA and protein and symbolized by RNP.

ribonucleotide   an organic compound that consists of a purine or pyrimidine base bounded to ribose, which in turn is esterified with a phosphate group. Such monomers are polymerized to form RNA.

ribose   a five-carbon sugar.

ribosomal precursor RNA   *See* **preribosomal RNA.**

ribosomal protein   *See* **ribosome.**

ribosomal RNA (rRNA)   *See* **ribosome.**

ribosomal RNA genes   rRNA genes reside as tandem repeating units in the nucleolus organizer regions of eukaryotic chromosomes. Each unit is separated from the next by a nontranscribed spacer. Each unit contains three cistrons coding for the 28S, 18S, and 5.8S rRNAs. The transcriptional polarity of the unit is 5′-18S-5.8S-28S-3′. Ribosomal RNA genes are often symbolized by rDNA. *See* **preribosomal RNA, rDNA amplification.**

ribosomal stalling   *See* **attenuation.**

ribosome   one of the ribonucleoprotein particles, 10-20 mμ in diameter, which are the sites of translation. Ribosomes consist of two unequal subunits bound together by magnesium ions. Each subunit is made up of roughly equal parts of RNA and protein. Each ribosomal subunit is assembled from one molecule of ribosomal RNA which is noncovalently bonded to 20 to 30 smaller protein molecules to form a compact, tightly coiled particle. In eukaryotes, the rRNAs of cytoplasmic ribosomes are formed by cistrons localized in the nucleolus organizer region of chromosomes. At least four classes of ribosomes exist which can be characterized by the sedimentation constants of their component rRNAs. Animal ribosomes also contain a 5.8S rRNA, which is hydrogen bonded to the 28S rRNA and is derived from the same intermediate precursor as the 28S rRNA.

SEDIMENTATION COEFFICIENT (S)

Ribosomal Source	Ribosome	Large Subunit	Small Subunit	Large rRNA	Small rRNA
Bacteria, chloroplasts	70	50	30	23	16
Plant cytoplasm	80	~60	~40	25	16
Animal cytoplasm	80	~60	~40	29	18
Mitochondria	81	61	47	25	19

5-ribosyluracil pseudouridine. *See* **rare bases.**

ribothymidine *See* **rare bases.**

ribozyme an RNA segment that has the intrinsic ability to break and form covalent bonds. For example, Cech *et al.* have shown in *Tetrahymena thermophila* that the intron in the rRNA precursor molecule can excise itself from the molecule, form a circle, and ligate the exons.

rice *Oryza sativa*. Together with wheat, corn, and potatoes it is one of the world's four most important crops.

Ricinus communis the castor bean. The genetics of sex determination has been extensively studied in this species.

rickets a deficiency disease of growing bone due to insufficient vitamin D in the diet. *See* **vitamin D-resistant rickets.**

Rickettsia a genus of very small bacteria belonging to the phylum Omnibacteria (*see* **classification**). Rickettsias are obligate intracellular parasites responsible for a variety of human diseases (typhus and Rocky Mountain spotted fever are examples). The parasites are generally carried to man via an anthropod vector such as a louse or tick.

rifampicin the most commonly used of the rifamycins (*q.v.*).

rifamycins a group of antibiotic molecules produced by *Streptomyces mediterranei* that interfere with the beta subunit of prokaryotic RNA polymerases and thereby inhibit initiation of transcription.

right splicing junction the boundary between the right (3′) end of an intron and the left (5′) end of an adjacent exon in mRNA; also called the *acceptor splicing site*.

ring canals canals connecting sister cystocytes in the *Drosophila* egg chamber. Such canals are surrounded by a ring of protein material. They function to allow a stream of cytoplasm to flow from nurse cells to oocyte. *See* **nurse cells.**

ring chromosome 1. an aberrant chromosome with no ends. 2. a ring-shaped chromosomal association seen during diakinesis in normal metacentric tetrads with two terminal chiasmata.

Ringer solution a physiological saline containing sodium, potassium, and calcium chlorides used in physiological experiments for temporarily maintaining cells or organs alive *in vitro*. Ringer solution is sometimes simply designated as *"ringer."*

ring gland a gland lying above the hemispheres (h) of the brain of the larval *Drosophila*. Its lateral extremities encircle the aorta (a) like a ring, hence its name. The gland contains three endocrine tissues: the corpus allatum (ca), the prothoracic gland (pg), and the corpus cardiacum (cc). The diagrams below show the ring gland viewed from the side (A) and from above (B). Other symbols are as follows: n_1 = afferent nerve to the corpus cardiacum from the brain; n_2 = efferent nerve from the corpus cardiacum; n_3 = nerve from corpus cardiacum to the corpus allatum; o = oesophagus; vg = ventral ganglion.

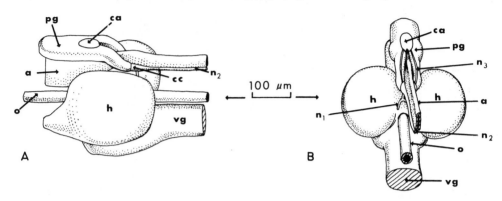

RK a symbol used in *Drosophila* studies to indicate *rank* or valuation of a given mutant. For example, RK1 indicates the best and most used mutants, with sharp classification, excellent viability, and accurate genetic localization. RK5 mutants show poor penetrance, low viability, and their chromosomal loci are not accurately determined.

rII locus the segment of the chromosome of T_4 bacteriophage that was the first to be subjected to fine structure genetic mapping by S. Benzer.

R loop during molecular hybridization, the single-stranded antisense (coding) strand of DNA that is prevented from reannealing because its complementary sense (anticoding) strand is base-paired with an mRNA exon as a heteroduplex.

R-loop mapping a technique for visualizing under the electron microscope the complementary regions shared by a specific eukaryotic RNA and a segment of one strand of a DNA duplex. The RNA-DNA hybrid segment displaces one of the DNA strands, causing it to form a *loop,* hence the name of the technique. Double-stranded regions appear thicker than single-stranded regions in electron micrographs. Introns cannot hybridize with mature mRNA (from which introns have been removed); thus, one intron results in two R-loops; two introns yield three R-loops, etc.

No intervening sequence   **One intervening sequence**

r **mutants** rapid-lysing phage mutants.

344

RNA   ribonucleic acid.

RNA coding triplets   *See* **amino acid, start codon, stop codon.**

RNA-dependent DNA polymerase   an enzyme that synthesizes a single strand of DNA from deoxyribonucleoside triphosphates, using RNA molecules as templates. Such enzymes occur in oncogenic RNA viruses. This class of enzymes, also known as *reverse transcriptases,* can be used experimentally to make complementary DNA (cDNA) from purified RNA. The functioning of these polymerases contradicts the central dogma (*q.v.*) in the sense that the direction of information exchange between DNA and RNA is reversed.

RNA-driven hybridization   an *in vitro* technique that uses an excess of RNA molecules to ensure that all complementary sequences in single-stranded DNA undergo molecular hybridization. *See* **DNA-driven hybridization.**

RNA gene   a DNA segment coding for one of the various types of non-messenger RNA (rRNA, 5S RNA, or tRNA).

RNA phage   an RNA bacteriophage such as MS2 and Qβ.

RNA polymerase   an enzyme that transcribes an RNA molecule from one strand (the template strand) of a DNA molecule. Two kinds of RNA polymerases are known in pro-karyotes: one produces the RNA primer required for DNA replication; the other transcribes all three types of RNA (mRNA, tRNA, rRNA). In eukaryotes, each type of RNA is transcribed by a different RNA polymerase. The holoenzyme from *E. coli* consists of a core (with subunits α, β, and β') and a sigma (σ) subunit that recognizes Pribnow sequences (*q.v.*) in promoter regions of genes. Once the core enzyme has started transcription, the σ factor can move to another core enzyme and help it initiate transcription of the same gene or a different gene at its correct starting position. Substances such as the antibiotics rifa-mycin and actinomycin D interfere with the transcriptional activity of RNA polymerases.

RNA primer   *See* **primer RNA.**

RNA processing   *See* **nuclear processing of RNA, post-transcriptional modification.**

RNA puff   *See* **chromosomal puff.**

RNA replicase   an RNA-dependent RNA polymerase. *See* **MS2** and **Qβ.**

RNase   any enzyme hydrolyzing RNA.

RNase protection   a technique for locating the points of effective contact between a nucleic acid chain and a cognate polypeptide chain; the complex (e.g., tRNA and its cog-nate aminoacyl-tRNA synthetase) is treated with a group of RNases that digest all of the RNA except those regions in contact with the synthetase. *See* **photoactivated crosslinking.**

RNA splicing   the removal of non-coding regions from a large precursor RNA molecule, and the nucleotide sequences transcribed from non-adjacent DNA segments are then joined together to produce a smaller mature RNA. *See* **nuclear processing of RNA.**

RNP   ribonucleoprotein.

rNTP   ribonucleoside 5′-triphosphates.

Robertsonian translocation   synonym for whole-arm fusion (*q.v.*).

rod   one of the elongate, unicellular photoreceptors in the vertebrate eye, involved with vision in dim light. Rods do not discriminate color differences.

roentgen   the quantity of ionizing radiation that liberates 2.083×10^9 ion pairs in a cubic centimeter of air (at 0°C and at a pressure of 760 mm of mercury) or approximately two ion pairs per cubic micron of a substance such as protein (which has a density of 1.35). A gram of tissue exposed to 1 roentgen of gamma rays absorbs about 93 ergs.

roentgen equivalent physical (rep)   the amount of ionizing radiation that will result in the absorption in tissue of 93 ergs per gram.

rogue   a variation from the standard variety, usually inferior.

rolling circle   a model mechanism for the replication of DNA molecules, so named because the growing point can be imagined as rolling around a circular template strand. The circular DNA is shown below in A. In B, a nick opens one strand, and the free 3'-OH end is extended by DNA polymerase. The newly synthesized strand displaces the original parental strand as it grows (C, D). By E, the polymerase has completed one revolution, and by F, two revolutions. The result is a molecule containing three unit genomes, one old and two new. The displaced strand can then serve as a template for a complementary strand. This mechanism is used to generate concatemeric duplex molecules (e.g., phage lambda, amplified rDNA in amphibian oocytes, etc.).

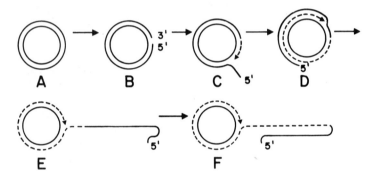

Romalea microptera   the lubber grasshopper. Meiosis has been extensively studied in this species.

Roman numerals:

I	II	III	IV	V	VI	VII	VIII	IX	X
1	2	3	4	5	6	7	8	9	10
XX	XXX	XL	L	LX	LXX	LXXX	XC	IC	C
20	30	40	50	60	70	80	90	99	100
CC	CCC	CD	D	DC	DCC	DCCC	CM	XM	M
200	300	400	500	600	700	800	900	990	1000

Larger numbers, used occasionally for dates, combine thus: 1978—MCMLXXVIII.

root cap   a cap of cells covering the apex of the growing point of a root and protecting it as it is forced through soil.

root hair   a tubular outgrowth of an epidermal cell of a root which functions to absorb water and nutrients from the soil.

root nodules   a small swelling on roots of legumes (*q.v.*) produced as a result of infection by symbiotic nitrogen-fixing bacteria. *See* **Rhizobium.**

Rosa   the genus that includes the rose species, which have been extensively hybridized. Commercially grown species include *R. centifolia, R. damascena, and R. multiflora.*

346

Rose chamber   a closed culture vessel permitting long-term observation of explanted cells under phase microscopy. The fluid culture medium may be periodically renewed without disturbing the growing cells.

rostral   directed toward the front end of the body.

R_0t the product of RNA concentration and the time of incubation in an RNA-driven hybridization; the analog of C_0t values used to describe DNA-driven hybridization reactions.

rotational base substitution   A break is induced by radiation at a corresponding point in both complementary strands of a DNA molecule. The bond broken is between the base and the sugar molecule to which it is attached. Thus two complementary bases (held together by hydrogen bonds) are detached from their backbones. If the pair rotates before it is reinserted in the molecule, the resulting transversional mutation would be termed a rotational base substitution. Many radiation-induced point mutations may result in this way.

rotation technique   *See* **photographic rotation technique.**

rough surfaced endoplasmic reticulum   *See* **endoplasmic reticulum.**

Rous sarcoma virus   an avian retrovirus discovered by P. Rous. The genome of the RSV contains four genes: *gag,* which codes for the protein of the capsid; *pol,* which codes for a reverse transcriptase *(q.v.)*; *env,* which codes for the protein that forms spikes on the envelope of the viral membrane, and *src.* This is an oncogene, so named because it induces *src*omas. The *src* gene codes for a protein kinase, *pp60^{v-src}* *(q.v.)*, which is localized in the plasmalemma. Vertebrate cells contain a gene homologous to the *src* gene. To distinguish the two, the viral gene is abbreviated *v-src* and the cellular gene *c-src.* The two genes differ in that *v-src* has an uninterrupted coding sequence, whereas *c-src* contains seven exons separated by six introns. The *c-src* gene is a proto-oncogene *(q.v.)*. *See* **oncogene hypothesis.**

royal hemophilia   classical hemophilia *(q.v.)* transmitted by a defective X chromosome first carried by Queen Victoria of Great Britain and passed on to plague three generations of European royalty.

royal jelly   a liquid nutrient secreted by nurse bees and fed to larvae to produce queens.

R plasmid   an extrachromosomal DNA molecule that confers on bacteria resistance to one or more antibiotics. It consists of two components: the resistance transfer factor (RTF) required for transfer of the plasmid between bacteria, and the r-determinants (genes conferring antibiotic resistance) that are part of transposons *(q.v.)*.

rpm   revolutions per minute.

RQ   respiratory quotient *(q.v.)*.

rRNA   ribosomal RNA. *See* **ribosome.**

rRNA transcription unit   *See* **Miller tree.**

r-selection   *See* **r and K selection theory.**

RSV   Rous sarcoma virus *(q.v.)*.

RTF   resistance transfer factor. *See* **R plasmid.**

rTU   rRNA transcription unit. *See* **Miller tree.**

ruffled edges   *See* **lamellipodia.**

runner   a procumbent shoot that takes root, forming a new plant that eventually is freed from connection with the parent by decay of the runner. The runner serves as a vegetative propagule.

runting disease   a pathological condition seen in young experimental animals inoculated with allogeneic immunocompetent cells that produce a graft-versus-host reaction (*q.v.*).

rut   the period of sexual activity; estrus.

S

s   1. selection coefficient (*q.v.*) 2. standard deviation (*q.v.*) 3. sedimentation coefficient (*q.v.*) 4. second.

S   1. Svedberg unit (*q.v.*) 2. Silurian. 3. sulfur. 4. DNA synthesis phase of the cell cycle (*q.v.*).

^{35}S   a beta-emitting radioactive isotope of sulfur with a half-life of 87.1 days; commonly used to label proteins via their sulfur-containing amino acids cysteine and methionine.

S_1, S_2, S_3, *etc.*   the representation for continued selfing (self-fertilization) of plants. S_1 designates the generation obtained by selfing the parent plant; S_2, the generation obtained by selfing the S_1 plant, etc.

Saccharomyces cerevisiae   bakers yeast, a favorite eukaryotic microorganism for genetic research. Approximately 150 mutant genes have been distributed among its 17 linkage groups. Mitochondrial mutants also have been extensively studied. *See* **petite mutant.**

sagittal plane   A longitudinal plane that separates a bilaterally symmetrical animal into right and left halves is referred to as the median sagittal plane. Any plane parallel to the median sagittal plane is also sagittal.

salivary gland chromosomes   polytene chromosomes found in the interphase nuclei of the salivary gland cells in larval diptera. These chromosomes undergo complete somatic pairing; consequently the mature salivary gland chromosome consists of two homologous polytene chromosomes fused side by side.

salivary gland squash preparation   a rapid method of preparing insect polytene chromosomes for microscopic investigation without sectioning them. The organ is simply squashed in a drop of stain placed between slide and coverslip. *See* **aceto-orcein.**

S allele   *See* **self-sterility genes.**

Salmo   the genus containing various fish species of economic importance; especially *S. salar,* the Atlantic salmon and *S. gairdneri,* the rainbow trout.

Salmonella   together with *Escherichia* and *Shigella,* one of the genera of enteric bacteria containing favorite species for genetic study. *Salmonella typhimurium,* the mouse typhoid bacillus, has been the most studied, but considerable work has also been done with *S. abony, S. pullorum, S. montevideo, S. minnesota,* and *S. typhosa.* Generalized transduction was first shown in *S. typhimurium* using phage P22.

349

saltation **1.** a theory that new species originate suddenly from one or more mutations with large phenotypic effects ("macromutations"); referred to by R. Goldschmidt as "hopeful monsters." **2.** quantum speciation (*q.v.*). *See* **evolution.**

saltatory replication lateral amplification of a chromosome segment to produce a large number of copies of a specific DNA sequence. *See* **gene amplification, rDNA amplification.**

salvage pathways metabolic pathways for the synthesis of nucleosides and nucleotides utilizing preformed purines and pyrimidines or nucleosides. Examples include the conversion of a base to a nucleoside, interconversion of bases and nucleosides, conversion of nucleosides to nucleotides, and interconversions by base alterations. *Compare with* **de novo pathway.**

samesense mutation a point mutation (usually in the third position of a codon) that does not change the amino acid specificity of the codon so altered; a "silent" mutation. *See* **degenerate code.**

sampling error variability due to the limited size of the samples.

Sandhoff disease a lethal hereditary disease in humans due to a recessive gene on autosome 5. The gene codes for the beta subunit of hexosaminidase A. Neurologic deterioration results from storage of gangliosides due to the enzyme deficiency. Symptoms are similar to those of Tay-Sachs disease (*q.v.*).

Sanger-Coulson method *See* **DNA sequencing techniques.**

7, 11.4, 19 S antibodies *See* **immunoglobin.**

saprophyte a plant living on and deriving its nutrition from dead organic matter.

sarcoma a cancer of connective tissue.

sarcomere the repeating unit, about 2.5 μm long, within striated muscle fibers, containing a set of interacting actin and myosin filaments.

sarcosomes the mitochondria of the flight muscles of insects.

sat DNA satellite DNA (*q.v.*).

satellite a distal chromosomal segment separated from the rest of the chromosome by a thin chromatic filament or stalk called the *secondary constriction.*

satellite DNA any fraction of the DNA of an eukaryotic species that differs sufficiently in its base composition from that of the majority of the DNA fragments to separate as one or more bands distinct from the bands containing the majority of the DNA during isopycnic CsCl gradient centrifugation. Satellite DNAs obtained from chromosomes are either lighter (A + T rich) or heavier (G + C rich) than the majority of the DNA. Satellite DNAs are usually highly repetitious. *See* **Alu family, mouse satellite DNA, reassociation kinetics, repetitious DNA.**

saturation hybridization an *in vitro* reaction in which one polynucleotide component is in great excess, causing all complementary sequences in the other polynucleotide component to enter a duplex form. *See* **DNA-driven hybridization reaction, RNA-driven hybridization.**

scaffold *See* **chromosome scaffold.**

scala naturae an antiquated taxonomic scheme, originating with Aristotle, that arranges organisms along a linear ascending scale of perfection and/or complexity; also known as the "scale of being," the "ladder of perfection," the "chain of complexity," etc.

scanning electron microscope *See* **electron microscope.**

scanning hypothesis a theory to explain initiation of translation in eukaryotes according to which a 40S ribosomal subunit attaches at or near the mRNA cap and then drifts in the 3′ direction until it encounters an AUG start codon; at this point an initiation complex is formed and the reading frame becomes established.

scape a leafless flower-bearing stem arising from ground level (as in the dandelion and daffodil).

scarce mRNA *See* **complex mRNA.**

scatter diagram a diagram in which observations are plotted as points on a grid of x and y coordinates to see if there is any correlation (*q.v.*). For example, one might plot for a number of species the LD 50 (*q.v.*) against the average DNA content per somatic cell nucleus. Finding a correlation suggests that the variables may be interrelated. No correlation suggests that the variables chosen have no bearing on one another.

scattering the change of direction of particles or waves as a result of a collision or interaction. Scattering of electrons by the specimen, for example, is responsible for the electron microscopic image.

SCE sister chromatid exchange.

Schiff reagent a reagent that attaches to and colors aldehyde-containing compounds. Used in the PAS and Feulgen procedures (*q.v.*).

$$
\begin{array}{c}
\text{H--C}=\text{C--H} \\
\text{H}_2\text{N--C} \qquad \text{C} \\
\text{H--C--C--H} \qquad \text{SO}_3\text{H} \\
\text{C} \\
\text{H--C--C--H} \qquad \text{H--C--C--H} \\
\text{HO}_3\text{SHN--C} \qquad \text{C} \qquad \text{C} \qquad \text{C--NHSO}_3\text{H} \\
\text{H--C}=\text{C--H} \qquad \text{H--C}=\text{C--H}
\end{array}
$$

Schistocerca gregaria a locust whose population dynamics has been extensively studied.

schistosomiasis also called *bilharziasis* or *snail fever;* a helminthic infection of humans involving 200 million persons in Africa, the Middle East, the Orient, and South America. The causative agent is the blood fluke *Schistosoma mansoni,* and its snail vector is *Biomphalaria glabrata.* The genetic control of susceptibility of *B. glabrata* to *S. mansoni* is a subject of active investigation.

schizogony a series of rapid mitoses without increase in cell size which gives rise to schizonts.

schizont a sporozoan spore arising from schizogony.

schizophrenia a type of mental disease of high heritability, afflicting between 1 and 5% of the population. A controversy exists as to whether schizophrenia is due to an autosomal dominant gene with about 25% penetrance or is polygenically inherited. *See* **catatonia, hebephrenia, paranoia, simple schizophrenia.**

Schizophyllum commune a basidiomycete that has been the subject of considerable genetic research. Two extensive linkage groups have been elucidated.

Schizosaccharomyces pombe a fission yeast, a species suitable for tetrad analysis (*q.v.*).

Schwann cell the cell that enfolds a myelinated nerve fiber. *See* **myelin sheath.**

Sciara a genus of fungus gnats extensively studied in terms of the cytogenetics of chromosome diminution (*q.v.*). The giant polytene chromosomes of the larval salivary glands of *S. coprophila* have been mapped and certain DNA puffs have been extensively studied.

scintillation counter. *See* **liquid scintillation counter.**

scion a portion of a shoot of one plant that is grafted onto a stock of another.

sclerite a hardened area in the integument of an arthropod, usually separated from adjacent sclerites by membranous areas.

scleroproteins the very stable fibrous proteins, present mainly as a surface coverings of animals. Keratin and collagen are examples of scleroproteins.

scrapie *See* **prion.**

scurvy *See* **ascorbic acid.**

scutellum **1.** the single cotyledon of a grass embryo. **2.** a shield-shaped metathoracic tergite of *Drosophila.*

SD **1.** standard deviation. **2.** segregation distortion (*q.v.*).

SDS sodium dodecyl sulfate (q.v.).

SDS-PAGE technique *See* **electrophoresis.**

SE, S.E. standard error (*q.v.*).

Se *Secretor* gene (*q.v.*).

Searle translocation a reciprocal X-autosome translocation in the mouse which exhibits paternal X inactivation in the somatic cells of female heterozygotes. All other X-autosome translocations studied in the mouse show random inactivation of either normal or rearranged chromosome. *See* **Lyon hypothesis.**

seasonal isolation a type of ecological reproductive isolation in which different species become reproductively active at different times; temporal isolation.

sec second.

secondary constriction a thin chromatic filament connecting a chromosomal satellite to the rest of the chromosome.

secondary gametocytes *See* **meiosis.**

secondary immune response *See* **immune response.**

secondary nondisjunction sex chromosomal nondisjunction in an XXY individual resulting in gametes containing either two X chromosomes, one X, one Y, or an X and a Y.

secondary protein structure *See* **protein structure.**

secondary sex ratio the ratio of males to females at birth; in contrast to the primary sex ratio at conception.

secondary sexual character a characteristic of animals other than the organs producing gametes which differs between the two sexes (mammary glands, antlers, external genitalia, etc.). *See* **primary sexual character.**

secondary speciation the fusion through hybridization of two species that were formerly geographically isolated, followed by establishment of a new adaptive norm through natural selection.

second cousin *See* **cousin.**

second division segregation ascus pattern   in ascomycetes, a 2-2-2-2 or 2-4-2 linear order of spore phenotypes within an ascus. These patterns indicate that a pair of alleles (e.g., those controlling spore pigmentation) separated in the second meiotic division because crossing over occurred between that locus and the centromere. *See* **ordered tetrad.**

second law of thermodynamics   *See* **thermodynamics.**

second site mutation   *See* **suppressor mutation.**

secretin   a hormone which stimulates the secretion of pancreatic juice. The epithelial cells of the duodenum release secretin when activated by the acidic contents of the stomach.

secretion   the passage out of a cell of compounds synthesized within it.

Secretor gene   a dominant autosomal gene in man which permits the secretion of the water-soluble forms of the A and B blood-group antigens into saliva and other body fluids. The *Se* gene is not linked to the *I* locus. *See* **A, B antigens.**

sedimentation coefficient (s)   the rate at which a given solute molecule suspended in a less dense solvent sediments in a field of centrifugal force. The sedimentation coefficient is a rate per unit centrifugal field. The s values for most proteins range between 1×10^{-13} sec and 2×10^{-11} sec. A sedimentation coefficient of 1×10^{-13} sec is defined as one Svedberg unit (S). Thus a value of 2×10^{-11} sec would be denoted by 200S. For a given solvent and temperature s is determined by the weight, shape, and degree of hydration of the molecule.

seed   a mature ovule containing an embryo in an arrested state of development, generally with a food reserve.

segmental alloploid   *See* **allosyndeses.**

segmental interchange   a translocation.

segmented genome   a viral genome fragmented into two or more nucleic acid molecules. For example, alfalfa mosaic virus has four different RNA segments, each packaged in a different virion. Successful infection requires that at least one RNA of each type enters the cell. Such a virus is said to be *heterocapsidic.* If all fragments of a segmented genome are present in the same virion (e.g., influenza virus), the virus is said to be *isocapsidic.*

segregational lag   delayed phenotypic expression of an induced mutation in one nucleoid of a multinucleoid bacterium. The lag period is the time required for the fission of the parent to produce a cell containing only the mutant chromosome.

segregational load   the genetic disability sustained by a population due to genes segregating from advantageous heterozygotes to unfit homozygotes.

segregational petites   *See* **petites.**

segregation distortion   a distortion of expected Mendelian ratios in a cross due to dysfunction or lethality in gametes bearing certain alleles. This form of meiotic drive (*q.v.*) is represented in *Drosophila melanogaster* by the segregation distorter (SD) mutation on chromosome 2. SD/sd$^+$ heterozygous males produce both SD and sd$^+$ spermatozoa, but only those carrying SD are functional. SD cannot achieve fixation, however, because it is lethal when homozygous.

segregation of chromosomes   *See* **disjunction.**

segregation of genes   *See* **Mendel's laws.**

segregation ratio distortion   the distortion of the 1:1 segregation ratio produced by a heterozygote. Such distortions may arise because of abnormalities of meiosis which result in an *Aa* individual producing unequal numbers of *A*- and *a*-bearing gametes, or it may arise from *A*- and *a*- bearing gametes being unequally effective in producing zygotes.

353

selection   the process determining the relative share allotted individuals of different geno-types in the propagation of a population. The selective effect of a gene can be defined by the probability that carriers of the gene will reproduce. *See* **artificial selection, directional selection, disruptive selection, group selection, kin selection, stabilizing selection.**

selection coefficient (*s*)   the proportionate reduction in the average gametic contribution to the next generation made by individuals of one genotype relative to those of another genotype (usually the most fit). For example, if the best adapted genotypes are *AA* and *Aa,* and they are not being selected against, $s = 0$, and their fitness = $(1 - s) = 1$. If individuals of genotype *aa* leave only on the average 80% as many progeny (proportionate to their numbers in the population) as the other genotypes, then the selection coefficient against *aa* individuals is 0.2 or 20%, and their fitness is $(1 - 0.2) = 80\%$.

selection differential   the difference between the average value of a quantitative character in the whole population and the average value of those selected to be parents of the next generation. *See* **record of performance.**

selection pressure   the effectiveness of natural selection in altering the genetic composition of a population over a series of generations.

selective advance   the increment in the average value (measured for a quantitative character being selected in a population) from one generation to the next, usually a fraction of the selection differential (*q.v.*).

selective neutrality   a situation in which the phenotypic manifestations of certain mutant alleles are equivalent to that of the wild-type allele in terms of their fitness values. *See* **neutral gene theory, silent mutations.**

selective plating   a method for selectively isolating recombinants. Two different auxotropic mutants are plated upon a minimal medium. Only the recombinant class receiving the normal allele of each mutant can multiply under these conditions.

selective silencing   any mechanism which consistently eliminates plasmagenes of one parent from the zygote, such as the destruction of chloroplasts or chloroplast DNA in some algae and plants or of sperm mitochondria in some animals.

selective system   any experimental technique that aids in the detection and isolation of a specific (usually rare) genotype. *See* **penicillin selection technique.**

selective variant   in microbial genetics a mutation which confers upon the organism the ability to exist under condtions which kill off all organisms not possessing the mutation. Examples of selective variants are mutations conferring resistance to antibacterial agents or the ability to synthesize some essential metabolite lacking in the medium.

selector genes   *See* **compartmentalization.**

self   to undergo self-pollination (*q.v.*) or self-fertilization (*q.v.*).

self-assembly   the spontaneous aggregation of multimeric biological structures involving formation of weak chemical bonds between surfaces with complementary shapes. For example, most of the components of the phage T_4 capsid (head, tail, base plate, and tail fibers) are self-assembled.

self-color   a hair pigment pattern of uniform color over the body; solid color.

self-compatible   said of a plant that can be self-fertilized.

self-fertilization   the fusion of male and female gametes from the same individual.

self-incompatibility   self-sterility (*q.v.*).

selfish DNA functionless DNA regions that exist merely to perpetuate themselves. According to the selfish DNA theory, an eukaryotic organism is a "throwaway survival machine" used by selfish DNA to replicate itself. Spacer DNA, satellite DNA, and some other kinds of repetitive DNA may be examples.; also called *junk DNA.*

self-pollination the transfer of pollen to the stigmas of the same plant.

self-sterility the inability of some hermaphrodites to form viable offspring by self-fertilization.

self-sterility genes genes that prevent the deleterious effects of inbreeding in monecius plants by controlling the rate of growth of the pollen tube down the style. During this process the sporophytic tissue discriminates against gametophytic tissue containing a common sterility gene.

SEM scanning electron microscope. *See* **electron microscope.**

semelparity reproduction that occurs only once in the life of an individual (e.g., annual plants, Pacific salmon). *Compare with* **iteroparity.**

semen a biochemically complex nutrient fluid containing spermatozoa which is transferred to the female during copulation.

semiconservative replication the method of replication of DNA in which the molecule divides longitudinally, each half being conserved and acting as a template for the formation of a new strand.

semidiscontinuous replication a mode of DNA replication in which one new strand is synthesized continuously, while the other is synthesized discontinuously as Okazaki fragments. *See* **replication of DNA.**

semidominance the production of an intermediate phenotype in individuals heterozygous for the gene concerned.

semidwarf a term used to distinguish mutant strains of wheat that are of agricultural importance from the extremely short dwarfs of purely genetic interest. Semidwarfs grow from half to two-thirds the height of, and have greater yields than those of standard varieties.

semigeographic speciation *See* **parapatric speciation.**

semilethal mutation a mutation causing death of more than 50%, but not of all individuals of mutant genotype.

seminiferous tubule dysgenesis Klinefelter syndrome (*q.v.*).

semipermeable membrane any membrane that permits passage of molecules selectively.

semispecies incipient species.

semisterility a situation in which half or more of all zygotes are inviable (as in crosses between translocation heterozygotes).

Sendai virus a virus, first isolated in Japan, that causes an important and widespread infection of laboratory mice; it belongs to the para-influenza group of mixoviruses. The virus is widely used in cell fusion studies. The viruses so modify the surfaces of infected cells that they tend to fuse. Even UV-killed viruses adsorb on host cells and promote their fusion.

senescent aging.

sense codon any of the 61 triplet codons in mRNA that specify an amino acid.

sense strand in duplex DNA, the strand that serves as a template for the synthesis of RNA; also known as the *anticoding strand.*

sensitive developmental period a period during development when there is an enhanced chance that genetic malfunction will bring development to a standstill. In *Drosophila* these sensitive periods correspond to the onset of embryonic, larval, pupal, or adult development, and it is during such periods that many new systems are differentiated and put to immediate test. Gastrulation (*q.v.*) is a sensitive period for amphibians.

sensitive volume **1.** that portion of an ionization chamber which responds to radiation passing through it. **2.** that biological volume in which an ionization must occur to produce a given effect (such as a mutation).

sensitizing agent an agent which, when added to a biological system, increases the amount of damage done by a subsequent dose of radiation.

sepiapterin a pteridine (*q.v.*) pigment that gives the *Drosophila* testis its yellow color. Sepiapterin is a precursor of the drosopterins (*q.v.*).

ser serine. *See* **amino acid.**

SER *s*mooth surfaced *e*ndoplasmic *r*eticulum.

serial homology the resemblance between different members of a single, linearly arranged series of structures within an organism (the vertebrae are an example).

serial symbiosis theory the theory that eukaryotic cells evolved from bacterial ancestors by a series of symbiotic associations. In its most modern form, it suggests that the mitochondria and microtubule organizing systems of present day eukaryotes evolved from bacteria and spirochaetes that lived as symbionts in a line of single-celled eukaryotes that were the ancestors of both fungi and animals. A subline of these protoctists subsequently entered an endosymbiosis with cyanobacteria. These evolved into chloroplasts, and the plant lineage developed from this group.

seriatim in a series; serially.

sericins a group of proteins found in silk (*q.v.*).

sericulture the culture of *Bombyx mori* (*q.v.*) for the purpose of silk production.

serine *See* **amino acid.**

serine proteases a family of homologous enzymes which require the amino acid serine in their active site and appear to use the same mechanism for catalysis. Members include enzymes involved in digestion (trypsin, chymotrypsin, elastase), blood coagulation (thrombin), clot dissolution (plasmin), complement fixation (C1 protease), pain sensing (kallikrein), and fertilization (acrosomal enzymes).

serology the study of the nature, production, and interactions of antibodies and antigens.

serophyte xerophyte (*q.v.*).

serotonin a cyclic organic compound (5-hydroxytryptamine) that causes certain smooth muscles to contract rapidly, and increases capillary permeability. The symptoms of anaphylaxis are due in large part to serotonin released from platelets that accumulate in the capillary bed of the lung. Serotonin also plays an important role in the metabolism of the central nervous system.

$$NH_2-CH_2-CH_2-C \quad C \quad C-OH$$

serotype an antigenic property of a cell (bacterial cell, red blood cell, etc.) identified by serological methods.

serotype transformation *See* **antigenic conversion.**

serrate toothed, like the edge of a saw.

serum the fluid remaining after the coagulation of blood.

sessile fixed to the substratum; sedentary.

seta *See* **chaeta.**

Sewall Wright effect genetic drift (*q.v.*).

sex **1.** a mode of reproduction involving the formation of haploid gametes by meiosis (gametogenesis) and their union to form zygotes (fertilization). **2.** a classification of organisms or parts of organisms according to the kind of gamete produced; larger, nutrient-rich gametes are female; smaller, nutrient-poor gametes are male. Meiosis in some organisms produces morphologically indistinguishable *isogametes,* in which case the sexes are arbitrarily designated "plus" and "minus." An individual that produces both male and female gametes is *monoecious* (plant) or *hermaphroditic* (animal).

sex cells gametes.

sex chromatin a condensed mass of chromatin representing an inactivated X chromosome. Each X chromosome in excess of one forms a sex chromatin body in the mammalian nucleus. *See* **Barr body, late replicating X chromosome.**

sex chromosomes the homologous chromosomes that are dissimilar in the heterogametic sex.

sex comb a row of bristles arranged like the teeth of a comb on the forelegs of *Drosophila melanogaster* males. *See* **gynandromorph, prepattern.**

sex-conditioned character a phenotype that is conditioned by the sex of the individual. For example, a sex-conditioned, autosomal gene may behave as a dominant in males and as a recessive in females. Furthermore, in the homozygous female the condition may be expressed to a minor degree. Human baldness is an example of a sex-conditioned character.

sex determination the mechanism in a given species by which sex is determined. In many species sex is determined at fertilization by the nature of the sperm that fertilizes the egg. Y-bearing sperm produce male zygotes; X-bearing sperm, female zygotes. In man the Y chromosome is the masculinizing agent, and feminization will occur only when it is absent. *See* **Klinefelter syndrome, Turner syndrome.**

sexduction the process whereby a fragment of genetic material from one bacterium is carried with the sex factor *F* to a second bacterium.

sex factor *See* **fertility factor.**

sex hormone any hormone produced by or influencing the activity of gonads: e.g., gonadotropins, estrogens, androgens. Sex hormones are responsible for development of certain secondary sexual characteristics (e.g., growth of facial hair and muscular development in men).

sex index in *Drosophila,* the number of X chromosomes per set of autosomes (e.g., a male has a sex index of 0.5, a female is 1.0, a metamale 0.33, and a metafemale 1.5).

sex-influenced character sex-conditioned character.

sex-limited character a phenotype expressed in only one sex, although it may be due to a sex-linked or autosomal gene. Examples: the recessive, female sterile genes of *Drosophila;* the genes influencing milk and egg production in farm animals.

sex linkage a special case of linkage occurring when a gene that produces a certain phenotypic trait (often unrelated to primary or secondary sexual characters) is located on the X chromosome. The result of this situation is that in certain crosses the phenotypic trait in question may be observed only in individuals of the heterogametic sex, differences between reciprocal crosses (*q.v.*) may also be observed, and the trait will be observed much less frequently among members of the homogametic sex. Genes residing on the Y chromosome will influence only the heterogametic sex. In *Drosophila* Y-linked genes control spermatogenesis.

sex pilus *See* **F pilus.**

sex ratio the relative proportion of males and females of a specified age distribution in a population.

sex ratio organisms spiroplasmas (*q.v.*) responsible for male specific lethality in certain *Drosophila* species. SROs are transmitted in the ooplasm, and they kill male embryos. The Y chromosome is not associated with SRO-induced male lethality, and therefore only embryos with two X chromosomes can withstand the infection. Sex ratio spiroplasmas are referred to as sex ratio spirochaetes in the earlier literature.

sex reversal the change from functioning as one sex to functioning as the other sex. The change may be a normal occurrence (*See* **consecutive sexuality**) or it may be experimentally or environmentally induced.

sexual differentiation the process by which sexual determination is phenotypically expressed through the proper development of sexual organs and characteristics.

sexual dimorphism *See* **dimorphism.**

sexual isolation ethological isolation (*q.v.*).

sexual reproduction reproduction involving fusion of haploid gamete nuclei, which result from meiosis.

sexual selection a theory first proposed by Charles Darwin that in certain species there occurs a struggle between males for mates and that characteristics enhancing the success of those bearing them would have value and be perpetuated irrespective of their general value in the struggle for existence. In the current literature, sexual selection is generally subdivided into *intrasexual* and *epigamic* selection. Intrasexual selection is what Darwin had in mind and involves the power to conquer other males in battle. In epigamic selection, the female is the active selective agent, making a choice from among a field of genetically variable males of her own species. The results of epigamic selection may be seen in the elaborate sexual displays of certain male birds.

sexuparous   producing offspring by sexual reproduction. The term is used to describe the sexual phase in a species that alternates sexual reproduction with parthenogenesis. This phenomenon occurs in aphids, for example.

shearing   in molecular biology, the process whereby a DNA sample is broken into pieces of fairly uniform size (e.g., by subjecting them to shearing forces in a Waring blender).

sheep   common breeds are Merino, Rambouillet, Dorset, Debouillet, Lincoln, Leicester, Cotswold, Romney, Corridale, Columbia, Romeldale, Panama, Montadale, Polwarth, Targee, Hampshire, Shropshire, Southdown, Suffolk, Cheviot, Oxford, Tunis, Ryeland, and Blackface Highland. The species is *Ovis aries.*

shift   *See* **translocation.**

shifting balance theory   a theory, proposed by Sewell Wright, which maintains that biological evolution proceeds most rapidly when subpopulations of a species remain isolated for a time sufficient to acquire distinctive adaptations, followed by reestablishment of gene flow, the broadening of genetic diversity, and the enhancement of evolutionary flexibility.

Shigella dysenteriae   the dysentery bacillus. Many *E. coli* phages also attack this species.

Shine-Dalgarno (SD) sequence   part or all of the mRNA leader sequence AGGAGG preceding the AUG initiation codon; it is complementary to a sequence at the 3′ end of 16S rRNA and thereby serves as a binding site for ribosomes; named after the discoverers J. Shine and L. Dalgarno.

Shope papilloma virus   an icosahedral particle about 53 nm in diameter containing a double-stranded DNA molecule weighing 5×10^6 d. The virus produces papillomas in rabbits.

short-day plant   a plant with a flowering period that is accelerated by daily exposures to light of less than twelve hours.

short-period interspersion   a genomic pattern in which moderately repetitive DNA sequences (each about 300 bp in length) alternate with nonrepetitive sequences of about 1,000 bp.

shotgun experiment   the random collection of a sufficiently large sample of cloned fragments of the DNA of an organism to create a cloned "gene library" for that species from which cloned molecules of interest can later be selected.

shrub   a perennial, low, woody plant.

sib   a shortened form for sibling.

siblings   brothers and/or sisters; the offspring of the same parents.

sibling species   species that are almost identical morphologically but are reproductively isolated; also called *cryptic species.*

sibmating   a brother-sister mating.

sibship   all the siblings in a family.

sickle-cell anemia   a generally fatal form of hemolytic anemia seen in individuals homozygous for an autosomal, codominant gene H^S. The erythrocytes of such individuals contain an abnormal hemoglobin, Hb^S. These cells undergo a reversible alteration in shape when the oxygen tension of the plasma falls slightly, and they assume elongate, filamentous, and sickle-like forms. Such red cells show a greatly shortened life span, since they tend to clump together and are rapidly destroyed. About 0.2% of the Black babies born in the United States suffer from sickle cell anemia.

sickle-cell trait the benign condition shown by individuals carrying both the normal gene, H^A, and H^S. The erythrocytes of such individuals produce both Hb^A and Hb^S. Such heterozygotes are healthy, and their erythrocytes can be caused to sickle only under conditions where the oxygen concentration is drastically reduced. H^A/H^S individuals suffer far less severely from *Plasmodium falciparum* infections than do H^A/H^A individuals. This malaria plasmodium enters the erythrocyte and lives by engulfing cytoplasm into its food vacuole. Parasites are unable to feed efficiently upon H^A/H^S cells because Hb^S molecules are insoluble compared to Hb^A and cause a great increase in the viscosity of the cytoplasm. About 9% of the Black population of the United States shows the sickle cell trait. *See* **hemoglobin, hemoglobin S, malaria.**

siderophilin plasma transferrin (*q.v.*).

Sigma (Σ) the summation of all quantities following the symbol.

sigma (σ) factor a polypeptide subunit of the RNA polymerase of *Escherichia coli*. This molecule by itself has no catalytic function, but it serves to recognize specific binding sites on DNA molecules for the initiation of RNA transcription. *See* **Pribnow box, RNA polymerase.**

sigma virus a virus that confers CO_2 sensitivity upon *Drosophila melanogaster*.

sigmoid curve an S-shaped curve.

signal hypothesis the notion that the N-terminal amino acid sequence of a secreted polypeptide is critical for attaching the nascent polypeptide to membranes.

signal sequence synonym for leader sequence (*q.v.*).

significance of results. If the probability values are equal to or less than 0.05 but greater than 0.01, the results are said to be *significantly different;* probability values $\leq 0.01 >$ 0.001 are *highly significant;* and those ≤ 0.001 are called *very highly significant* by convention. *See* **P value.**

silent allele an allele that has no detectable product.

silent mutation a gene mutation that has no consequence at the phenotypic level; i.e., the protein product of the mutant gene functions just as well as that of the wild-type gene. Functionally equivalent amino acids may sometimes substitute for one another (e.g., leucine might be replaced by another nonpolar amino acid such as isoleucine). *See* **samesense mutation.**

silk the cocoon filament spun by the fifth instar larva of *Bombyx mori*. Each cocoon filament contains two cylinders of fibroin, each surrounded by three layers of sericin. Fibroin is secreted by the cells of the posterior portion of the silk gland. These cells undergo 18–19 cycles of endomitotic DNA replication before they begin transcribing fibroin mRNAs. The fibroin gene is present in only one copy per haploid genome. It resides on chromosome 23 and is about 18 kb long. The fibroin gene is fundamentally an extensive array of 18 bp repeats coding for Gly-Ala-Gly-Ala-Gly-Ser. The similarity of the fibroin gene to a satellite DNA suggests that the gene grew to its current size and continues to evolve by unequal crossing-over. The sericin proteins receive their name because of the abundance of serine, which makes up over 30% of the total amino acids. There are at least three sericins and one of these is encoded by a gene on chromosome 11. All sericins are secreted by the cells from the middle region of the silk gland. Several mutations are known that influence silk production. Some (*Fib* and *Src-2*) represent mutations in the cistrons coding for fibroin and sericin molecules. Others (*Nd, Nd-s,* and *flc*) seem to have defects in the intracellular transport and secretion of fibroin.

silk worm   *See **Bombyx mori**.*

Silurian   a period in the Paleozoic era lasting from 435 to 440 million years before the present. *See **geologic time divisions**.*

simian virus   any of a group of viruses that attack nonhuman primates.

simian virus 40   a DNA virus that readily infects cultured primate cells. SV40 is lytic in monkey cells, but temperate in mouse cells, causing occasional neoplastic transformations. The virus replicates in the nuclei of host cells and may become stably integrated into the host genome. The virus genome consists of a circular DNA molecule containing 5227 base pairs. The complete nucleotide sequence for this virus has been determined, and it contains both conventional genes, overlapping genes, and split genes. *See **enhancer**.*

simple schizophrenia   schizophrenia (*q.v.*) marked by extreme indifference by the patient to his surroundings, introversion, and general listlessness.

simple-sequence DNA   satellite DNA. (*q.v.*).

simplex   *See **autotetraploidy**.*

Sinanthropus pekinensis   a group of extinct hominids, originally found near Peking, China, but no longer considered a distinct species; now included in the species *Homo erectus*. *See **Peking man**.*

single-copy plasmids   plasmids maintained in bacterial cells in a ratio of one plasmid for each host chromosome.

single-event curve   a dose-response curve in radiobiology which gives a linear relation when the log of survival is plotted against radiation dose. *See **multiple event curve, target theory**.*

single-strand assimilation   the process whereby a single strand of DNA displaces its homologous strand in a duplex, forming a D loop (*q.v.*). The reaction is mediated by RecA protein (*q.v.*) and is involved in recombination and heteroduplex formation.

single-stranded DNA binding protein   in *E. coli*, a tetrameric protein of 74,000 d molecular weight that binds to the single-stranded DNA generated when a helicase (*q.v.*) opens the double helix. This stabilizes the single-stranded molecule and prevents reannealing or the formation of intrastrand hydrogen bonds.

single-strand exchange   pairing of one strand of duplex DNA with a complementary strand in another DNA molecule, displacing its homologue in the other duplex.

sire   the male parent in animal breeding.

sister chromatid exchange   crossing over between the sister chromatids of a meiotic tetrad or between the sister chromatids of a duplicated somatic chromosome.

sister chromatids   identical nucleoprotein molecules joined by a centromere. *See **chromatid**.*

sister group   a species or higher monophyletic taxon that is hypothesized to be the closest genealogical relative of a given taxon. Sister taxa are derived from an ancestral species not shared by any other taxon.

site   the position occupied by a mutation within a cistron.

site-specific recombination a process in which two specific (not necessarily homologous) DNA sequences are joined: e.g., during phage integration/excision or resolution of cointegrate structures during transposition.

sliding filament model a theoretical mechanism that explains muscle contraction by the making and breaking of cross bridges between adjacent thick (myosin) and thin (actin) filaments. A similar model is postulated to explain the shortening of the microtubules of the spindle apparatus, whereby separation of chromosomes or chromatids is effected during cell division.

slime molds *See* **Acrasiomycota, Myxomycota.**

slow component in a reassociation reaction, the last component to reassociate, usually consisting of nonrepetitive (unique) DNA sequences.

slow stop mutants temperature sensitive *dna* mutants of *E. coli* that complete the current round of DNA replication when placed at the restrictive temperature, but cannot initiate another round of replication.

Slp sex-limited protein; a serum protein that is normally found only in male mice and encoded by a gene in the major histocompatibility complex (H-2).

small angle x-ray diffraction the technique used in the analysis of widely spaced repetitions, such as the groups of atoms that form monomeric subunits of a polymer. *See* **x-ray crystallography, large angle x-ray diffraction.**

small cytoplasmic RNAs the cytoplasmic counterparts of small nuclear RNAs (*q.v.*); found in small ribonucleoprotein particles (*q.v.*) in their native state.

small nuclear RNAs a large number of discrete, small RNA species (100–300 bp) that vary in abundance from about 10^5 to 10^6 molecules per cell. These RNAs exist within small ribonucleoprotein particles found only in the nucleus and are called snRNPs (snurps). Some snRNAs may be involved in splicing (*q.v.*). *See* **U1RNA.**

Smittia a genus of chironomid possessing giant polytene chromosomes (*q.v.*) and hence subjected to cytological study.

smooth endoplasmic reticulum *See* **endoplasmic reticulum.**

smut **1.** a fungus disease of cereals characterized by black masses of spores. **2.** any parasitic fungus of the order Ustilagnales which causes smut disease.

snapdragon *See* **Antirrhinum majus.**

sneak synthesis *See* background constitutive synthesis.

snRNA a small nuclear RNA (*q.v.*).

snRNPs small nuclear ribonucleoproteins. *See* **small nuclear RNAs.**

S1 nuclease an endonuclease from *Aspergillus oryzae* that selectively degrades single-stranded DNA to yield 5′ phosphoryl mono- or oligonucleotides.

snurps *See* **small nuclear RNAs.**

social evolution a continued increase in the complexity of human society resulting from the selection, transmission, and utilization of the useful information gained in each generation.

sodium an element universally found in small amounts in tissues. Atomic number 11; atomic weight 22.9898; valence 1^+; most abundant isotope ^{23}Na; radioisotopes: ^{24}Na, half-life 15 hours, radiations emitted—beta particles and gamma rays; ^{22}Na, half-life 2.6 years, radiations emitted—positrons and gamma rays.

sodium dodecylsulfate $CH_3(CH_2)_{11}OSO_3Na$ (sodium lauryl sulfate), an anionic detergent used in the SDS-PAGE method of protein fractionation. *See* **electrophoresis.**

sodium pentobarbital   *See* **Nembutal.**

solenoid structure   a supercoiled DNA structure produced during nuclear chromosomal condensation in eukaryotes.

Solo man   an extinct subspecies of primitive man known from fossil remains found near the Solo River in Java. Currently classified as *Homo erectus soloensis.*

solution hybridization   liquid hybridization (*q.v.*).

soma   the somatic cells of a multicellular organism in contrast to the germ cells.

somatic cell   any cell of the eukaryotic body other than those destined to become sex cells. In diploid organisms, most somatic cells contain the 2N number of chromosomes; in tetraploid organisms, somatic cells contain the 4N number, etc.

somatic cell genetic engineering   correction of genetic defects in somatic cells by genetic engineering: e.g., insertion of genes for insulin production into defective pancreatic cells. Such correction would not be hereditary.

somatic cell genetics   the genetic study of asexually reproducing body cells, utilizing cell fusion techniques, somatic assortment, and somatic crossing over.

somatic cell hybrid   a hybrid cell resulting from cell fusion (*q.v.*).

somatic crossing over   crossing over during mitosis of somatic cells which leads to the segregation of heterozygous alleles.

somatic doubling   the doubling of the diploid chromosome set. Such doubling may be induced experimentally by applying the alkaloid colchicine in a lanolin paste to somatic tissues that are undergoing mitosis.

somatic mutation   a mutation occurring in any cell that is not destined to become a germ cell. If the mutant cell continues to divide, the individual will come to contain a patch of tissue of genotype different from the cells of the rest of the body.

somatic pairing   the conjoining of the homologous chromosomes in somatic cells, a phenomenon seen in dipterans. The fact that the polytene chromosomes of *Drosophila* undergo somatic pairing makes possible the identification of chromosomal rearrangements, the mapping of deficiencies, and, as a result, the cytological localization of genes.

somatostatin   a peptide hormone, present in hypothalamic extracts, that inhibits the release of growth hormone (somatotropin) by the anterior lobe of the pituitary gland, and of insulin and glucagon by the pancreas. The gene for this 14-residue peptide has been chemically synthesized, spliced into a plasmid, and cloned in *E. coli.* Biologically active somatostatin has been recovered from these clones.

somatotropin   growth hormone.

somites   a series of segmentally arranged masses of muscle on either side of the spinal cord of the vertebrate embryo.

Sordaria fimicola   an ascomycete fungus often used in studies of gene conversion (*q.v.*).

sorus   a cluster of spores or sporangia, such as occur on the underside of fern fronds.

SOS boxes   the operator sequences in *E. coli* DNA that are recognized by a repressor called the LexA protein. This protein represses several loci involved in DNA repair functions. *See* **SOS response.**

SOS response   an error-prone mechanism of repairing damaged DNA in *E. coli* by the coordinated induction of several enzymes. Damaged DNA somehow activates an enzyme called RecA protease, and this protease cleaves a protein called LexA repressor. Many genes involved in repair functions become activated when this repressor is cleaved. *See* **SOS boxes.**

Southern blotting   a technique, developed by E.M. Southern, for transferring electrophoretically resolved DNA segments from an agarose gel to a nitrocellulose filter paper sheet via capillary action. Subsequently the DNA segment of interest is probed with a radioactive, complementary nucleic acid, and its position is determined by autoradiography. A similar technique, referred to as *northern blotting,* is used to identify RNAs. For example, an electropherogram containing a multitude of different mRNAs could be probed with a radioactive cloned gene. In cases where proteins have been separated electrophoretically, a specific protein on an electropherogram can be identified by the *western blotting* procedure. In this case the probe is a radioactively labeled antibody raised against the protein in question.

sow   the adult female of swine.

spacer DNA   untranscribed segments of eukaryotic and some viral genomes flanking functional genetic regions (cistrons). Spacer segments usually contain repetitive DNA. The function of spacer DNA is not presently known, but it may be important for synapsis. *See* **transcribed spacer.**

spawn   to deposit eggs.

spay   to remove the ovaries.

special creation   a nonscientific philosophy asserting that each species has originated through a separate act of divine creation by processes that are not now in operation in the natural world.

specialized   **1.** an organism having a narrow range of tolerance for one or more ecological conditions. **2.** a species having a relatively low potential for further evolutionary change; the opposite of generalized.

specialized transduction   *See* **transduction.**

speciation   **1.** the splitting of an ancestral species into daughter species that co-exist in time; horizontal evolution or speciation; cladogenesis. **2.** the gradual transformation of one species into another without an increase in species number at any time within the lineage; vertical evolution or speciation; phyletic evolution or speciation.

species   **1.** biological (genetic) species: reproductively isolated systems of breeding populations. **2.** paleospecies (successional species): distinctly different appearing assemblages of organisms as a consequence of species transformation (*q.v.*). **3.** taxonomic (morphological; phenetic) species: phenotypically distinctive groups of coexisting organisms. **4.** microspecies (agamospecies): asexually reproducing organisms (mainly bacteria) sharing a common morphology and physiology (biochemistry). **5.** biosystematic species (ecospecies; coenospecies): populations that are isolated by ecological factors rather than ethological isolation (*q.v.*).

species group   superspecies (*q.v.*).

species selection   a form of group selection (*q.v.*) in which certain species (produced by cladogenesis) continue the cladogenic process and others become extinct.

species transformation   the transformation of a species (A) into another (species B) during the passage of time. Species transformation does not increase the number of species, since species A and B do not co-exist in time. *See* **speciation.**

specific activity   the ratio of radioactive to non-radioactive atoms or molecules of the same kind. Sometimes given as the number of atoms of radioisotope per million atoms of stable element. Also expressed in curies per mole.

specific immune suppression   an immune response in which the initial exposure to a particular antigen results in the loss of the ability of the organism to respond to subsequent exposures of that antigen, but not to different antigens.

specific ionization   the number of ion pairs per unit length of path of the ionizing radiation in a given medium (per micron of tissue, for example).

specificity   selective reactivity between substances: e.g., between an enzyme and its substrate, between a hormone and its cell-surface receptor, or between an antigen and its corresponding antibody.

specificity factors   proteins that temporarily associate with the core component of RNA polymerase and determine to which promoters the enzyme will bind (e.g., the sigma factor, *q.v.*). *See* **anti-specificity factor.**

specimen screen   the support for sections to be viewed under the electron microscope consisting of a disc made of copper or gold mesh.

spectrophotometer   an optical system used in biology to compare the intensity of a beam of light of specified wave length before and after it passes through a light-absorbing medium. *See* **microspectrophotometer**

spelt   *Triticum spelta* (N = 21), the oldest of the cultivated hexaploid wheats, grown since the latter days of Roman Empire. *See* **wheat.**

S period   *See* **cell cycle.**

sperm   a spermatozoan or spermatozoa.

spermateleosis   spermiogenesis (*q.v.*).

spermatheca   the organ in a female or a hermaphrodite which receives and stores the spermatozoa donated by the mate.

spermatid   one of four haploid cells formed during meiosis in the male. Spermatids without further division transform into spermatozoa, a process known as spermiogenesis (*q.v.*).

spermatocyte   a diploid cell that undergoes meiosis and forms four spermatids. A primary spermatocyte undergoes the first of the two meiotic divisions and gives rise to two secondary spermatocytes. Each of these divides to produce two haploid spermatids.

spermatogenesis   an inclusive term covering both male meiosis and spermiogenesis.

spermatogonia   mitotically active cells in the gonads of male animals which are the progenitors of primary spermatocytes.

Spermatophyta   in older taxonomies the division of the plant kingdom containing the contemporary dominant flora. Spermatophytes are characterized by the production of pollen tubes and seeds. All angiosperms and gymnosperms are included in the Spermatophyta. *See* **classification.**

spermatozoan   a haploid male gamete produced by meiosis.

sperm bank   a depository where samples of human semen are stored in liquid nitrogen at

−196°C; when needed, perhaps years later, a sample can be thawed and used in artificial insemination.

spermiogenesis   the formation of sperm from the spermatids produced during the meiotic divisions of spermatocytes.

sperm polymorphism   the production of normal and aberrant sperm during spermatogenesis. The normal sperm are called *eupyrene,* those containing subnormal numbers of chromosomes are *oliogopyrene,* and those lacking a nucleus altogether are *apyrene.* Apyrene and oligopyrene sperm are formed by certain snails (*Viviparus malleatus* is an example) and moths *(Bombyx mori),* but the function of these abnormal gametes is unknown.

Sphaerocarpus   a genus of liverworts. Classic tetrad analyses have been made on species from this genus.

S phase   *See* **cell cycle.**

spherocytic anemia   a hemolytic anemia in humans characterized by erythrocytes of spherical shape.

spheroplast   a protoplast (*q.v.*) to which some cell wall remnants are attached. For example, a rod-shaped bacterium treated with lysozyme becomes spherical because the enzyme removes peptidoglycan components that give rigidity to the cell wall.

sphingomyelin   a molecule belonging to a family of compounds that occur in the myelin sheath of nerves. All sphingomyelins contain sphingosine, phosphorylcholine, and a fatty acid group.

$$CH_2-(CH_2)_{11}CH_3$$

$$
\begin{array}{l}
H \\
C=C \longleftarrow \text{Sphingosine} \\
| \quad \backslash \\
H-C-OH \quad H \\
| \\
H-C-NH-CO-R \longleftarrow \text{Fatty Acid Group} \\
| \\
\quad\quad O \\
\quad\quad \| \\
CH_2-O-P-O-CH_2-CH_2-\overset{+}{N}\!\!<^{CH_3}_{CH_3}\!\!-CH_3 \\
\quad\quad O_- \\
\end{array}
$$

Phosphorylcholine

sphingosine   an amino dialcohol component of the sphingolipids, which are abundant in the brain.

$$
CH_3-(CH_2)_{12}-CH=CH-\underset{OH}{\overset{H}{\underset{|}{\overset{|}{C}}}}\!-\!\underset{NH_2}{\overset{H}{\underset{|}{\overset{|}{C}}}}\!-\!\underset{OH}{\overset{}{\underset{|}{CH_2}}}
$$

spike   an inflorescence, such as the catkin of the pussy willow, in which the flowers arise directly from a central axis, the rachis. *See* **raceme.**

spikelet   in grasses, a secondary spike bearing few flowers.

spindle   a collection of microtubules responsible for the movement of eukaryotic chromosomes during nuclear division. *See* **mitosis.**

spindle attachment region   (*also* **spindle fiber attachment, spindle fiber locus**)   centromere (*q.v.*).

spindle fiber   one of the microtubular filaments of a spindle.

366

spindle poison *See* **colchicine.**

spiracle the external opening of an insect trachea.

spiral cleavage a type of embryonic development seen in invertebrates such as annelids and molluscs. The first and second divisions of the zygote are in vertical planes but at right angles to one another, producing a quartet of blastomeres. The next divisions are horizontal, cutting off successive quartets. However, each quartet is slightly displaced from the one above, giving a spiral appearance to the embryo. The direction of the spiral is genetically determined.

spirochete (*also* **spirochaete**) bacteria that are nonflagellated, spiral, and move by flexions of the body. *See* **classification** (Spirochaetae), ***Treponema.***

spiroplasmas helical, motile bacteria that resemble spirochaetes. Unlike spirochaetes, spiroplasmas lack a cell wall, and they are therefore included in the Aphragmabacteria (*see* **classification**). Spiroplasmas are responsible for certain plant diseases and cause male-specific lethality among the progeny of female *Drosophila* carrying them. *See* **sex ratio organisms.**

splenomegaly enlargement of the spleen.

splice junctions segments containing a few nucleotides that reside at the ends of introns and function in excision and splicing reactions during the processing of transcripts from split genes. The sequence at the 5′ end of any intron transcript is called the *donor junction* and the sequence at the 3′ end the *acceptor junction.* U1 RNA (*q.v.*) contains a segment complementary to the donor and acceptor junctions of intron transcripts.

splicing   **1.** RNA splicing: the removal of introns and the joining of exons from eukaryotic primary RNA transcripts to create mature RNA molecules of the cytoplasm. **2.** DNA splicing: *See* **recombinant DNA research.**

splicing homeostasis a phenomenon in which a maturase (*q.v.*) helps to catalyze the excision of an intron from its own primary RNA transcript. In so doing, the maturase destroys its own mRNA and thereby limits its own level of activity.

split genes genes containing coding regions (exons) that are interrupted by non-coding regions (introns). This type of genetic organization is typical of most eukaryotic genes and some animal viral genomes, but introns are not found in prokaryotic organisms.

sp. n. new species.

spontaneous generation *See* **biopoesis.**

spontaneous mutation a naturally occurring mutation.

spontaneous reaction exergonic reaction (*q.v.*).

sporangium a structure housing asexual spores.

spore   **1.** Sexual spores of plants and fungi are haploid cells produced by meiosis. **2.** Asexual spores of fungi are somatic cells that become detached from the parent and can either germinate into new haploid individuals or can act as gametes. **3.** Certain bacteria respond to adverse growth conditions by entering a spore stage until more favorable growth conditions return. Such spores are metabolically inert and exhibit a marked resistance to the lethal effects of heat, drying, freezing, deleterious chemicals, and radiation.

spore mother cell a diploid cell which by meiosis gives rise to four haploid spores.

sporogenesis the production of spores.

sporophyte the spore-producing, 2N individual. In the higher plants the sporophyte is the conspicuous plant. In lower plants like mosses the gametophyte is the dominant and conspicuous generation.

Sporozoa   a class of parasitic protoctists in the phylum Apicomplexa that reproduce sexually with an alternation of generations (*q.v.*). Both haploids and diploids undergo schizogony (*q.v.*) to produce small infective spores. All species of *Plasmodium* belong to the Sporozoa. *See* **classification, malaria.**

sporulation   **1.** the generation of a bacterial spore. **2.** production of meiospores by fungi and many other eukaryotic organisms.

spreading position effect   the situation in which a number of genes in the vicinity of a translocation or inversion seem to be simultaneously inactivated.

38, 40, 45S pre-ribosomal RNAs   *See* **pre-ribosomal RNA.**

src   the oncogene of the Rous sarcoma virus (*q.v.*). *See* ***c-src.***

30, 40, 47, 50, 60, 61S ribosomal subunits   *See* **ribosome.**

70, 80, 81S ribosomes   *See* **ribosome.**

4S RNA   transfer RNA (*q.v.*).

5S RNA   an RNA molecule composed of 120 nucleotides that is a component of the large ribosomal subunit. It is transcribed independently of the ribosomal RNA gene (*q.v.*). In man the 5S RNA locus is near the telomere of the short arm of chromosome 1. In *Drosophila melanogaster* it is on 2R at 56 E-F. *See* **Drosophila salivary gland chromosomes.**

5.8S RNA   an RNA molecule composed of 120 nucleotides that is a component of the large ribosomal RNA genes (*q.v.*) of the nucleolus organizer and functioning as a component of the large ribosomal subunit. *See* **ribosomal RNA gene.**

5S RNA genes   genes that are transcribed into 5S RNAs. Such genes occur in tandemly linked clusters in all eukaryotes. In *Xenopus laevis* 5S RNA genes account for 0.5% of the entire genome. There are three separate 5S RNA multigene families. Two of these, the *major oocyte* and *trace oocyte* families, are expressed only in oocytes, while a third, *somatic* 5S DNA, is expressed in all types of somatic cells. The major oocyte, trace oocyte, and somatic 5S DNAs are present in 20,000, 1,300, and 400 copies, respectively per haploid genome.

16, 18, 19, 23, 25, 28S rRNAs   *See* **pre-ribosomal RNA, ribosomal RNA genes, ribosome.**

SSC   sister-strand crossover. *See* **sister-chromatid exchange.**

ssDNA   single-stranded DNA.

stabilizing selection   normalizing selection (*q.v.*).

stable equilibrium   an equilibrium state of alleles at a genetic locus to which the population returns following temporary disturbances of the equilibrium frequencies. For example, a locus with overdominance should form a stable equilibrium as long as selection favoring heterozygotes remains constant.

stable isotope   a nonradioactive isotope of an element.

stacking   **1.** the planar alignment of adjacent flattish nitrogen bases in a DNA double helix. **2.** stacking of dye molecules on RNA to yield metachromasy (*q.v.*).

stage micrometer   a glass slide on which are etched parallel lines usually exactly 10 microns apart. A stage micrometer is used for standardizing an ocular micrometer or for determining the magnification of photomicrographs and illustrations made with a drawing tube (*q.v.*).

staggered cuts the result of breaking two strands of duplex DNA at different positions near one another, as occurs by action of many restriction endonucleases (*q.v.*).

stallion an adult male horse, donkey, or zebra.

stamen the pollen-bearing organ of the angiosperm flower. It consists of a filament bearing a terminal anther. *See* **flower.**

standard deviation (s) a measure of the variability in a population of items. The standard deviation of a sample is given by the equation $s = \sqrt{\dfrac{\Sigma(x - \bar{x})^2}{N - 1}}$, where N is the number of items in the sample and $\Sigma(x - \bar{x})^2$ is the sum of the squared deviations from the mean.

standard error (S.E.) a measure of variation of a population of means. S.E. = $s/\sqrt{N - 1}$, where N = the number of items in the population and s = standard deviation.

standard type the most common form of an organism.

Stanford-Binet test used to gauge intelligence, it consists of a series of questions and problems grouped for applicability to ages up to sixteen years. Some questions require verbal recognition and others recognition of form and manual skills. The subject's performance is expressed in terms of his mental age. *See* **intelligence quotient.**

Staphylococcus aureus the pathogenic bacterium causing "staph" infections.

starch the storage polysaccharide of most plants. It is a polymer made up of α-D-glucose molecules.

start codon a group of three adjacent ribonucleotides (AUG) in an mRNA coding for the methionine in eukaryotes (formylated methionine in bacteria) that initiates polypeptide formation; also called an *initiation codon. See* **genetic code.**

startpoint in molecular genetics, the base pair on DNA that corresponds to the first nucleotide incorporated into the primary RNA transcript by RNA polymerase.

startsite synonym for startpoint (*q.v.*).

stasigenesis referring to a period during the paleontological history of a lineage during which little or no significant evolutionary change occurred.

stasipatric speciation speciation resulting from the dispersion of a favorable chromosomal rearrangement that yields homozygotes which are adaptively superior in a particular part of the geographical range of the ancestral species.

stasis in evolutionary studies, the persistence of a species over a span of geological time without significant change.

stationary phase   a period of little or no growth that follows the exponential growth phase (*q.v.*) in a culture of microorganisms or in a tissue culture.

statistic   the value of some quantitative characteristic in a sample from a population. *Compare with* **parameter.**

statistics   the scientific discipline concerned with the collection, analysis, and presentation of data. The analysis of such data depends upon the application of probability theory. Statistical inference involves the selection of one conclusion from a number of alternatives according to the result of a calculation based on observations. *Parametric* methods in statistical analysis assume that the data follow a defined probability distribution (for example, a **normal, binomial,** or **Poisson distribution,** *all of which see*), and the results of the calculations are valid only if the data are so distributed. The Student's *t* test (*q.v.*) is an example of a parametric procedure. *Nonparametric* methods in statistical inference are free from assumptions as to the shape of the underlying probability distribution. The Mann-Whitney rank sum test and the sign test are examples of nonparametric procedures. *See* **analysis of variance, chi-square method, Gaussian curve, null hypothesis, Student's *t* test.**

status quo hormone   synonym for allatum hormone (*q.v.*).

steady-state system   a system the components of which seem unchanging because material is entering and leaving the system at identical rates.

stem cell   **1.** one of the mitotically active somatic cells that serve to replenish those that die during the life of the metazoan organism. **2.** one of the mitotically active germinal cells that produce a continuing supply of gametes.

stem structure   in molecular biology, the base-paired (unlooped) segment of a single-stranded RNA or DNA hairpin (*q.v.*).

stereochemical structure   the three-dimensional arrangement of the atoms in molecules.

stereoisomers   molecules that have the same structural formula, but which differ in the spatial arrangement of dissimilar groups bonded to a common atom.

steric   relating to stereochemical structure (*q.v.*).

sterile   **1.** unable to reproduce. **2.** free from living microorganisms; axenic.

sterilization   **1.** elimination of the ability to reproduce. **2.** the process of killing or removing all living microorganisms from a sample.

sternite   a ventral sclerite (*q.v.*).

steroid   a lipid belonging to a family of saturated hydrocarbons containing seventeen carbon atoms arranged in a system of four fused rings. The hormones of the gonads and adrenal cortex, the bile acids, vitamin D, digitalis, and certain carcinogens are steroids.

steroid receptor   a cytoplasmic receptor protein that can bind to a specific steroid hormone. The receptor-hormone complex then moves into the nucleus and binds to a specific DNA site to regulate gene activity.

sterol   a compound with the general chemical ring structure of a steroid, but with a long side-chain and an alcohol group. Cholesterol (*q.v.*) is an example of a sterol.

sticky ends   complementary single-stranded projections from opposite ends of a DNA duplex or from different duplex molecules that are terminally redundant. Sticky ends allow the splicing of hybrid molecules in recombinant DNA experiments. Many restriction endonucleases (*q.v.*) create sticky ends by making staggered cuts (*q.v.*) in a palindromic restriction site. Also called *cohesive ends.*

stigma   the receptive surface usually at the apex of the style of a flower on which compatible pollen grains germinate.

stillbirth   the birth of a dead fetus.

stochastic process   a process that can be visualized as consisting of a series of steps at each of which the movement made is random in direction.

stock   1. that part of a plant, usually consisting of the root system together with part of the stem, onto which is grafted a scion. 2. an artificial mating group, as, for example, a laboratory stock of mutant *Drosophila.*

stolon   a horizontally growing plant stem that produces adventitious roots when in contact with the soil.

stop codon   a ribonucleotide triplet signaling the termination of the translation of a protein chain (UGA, UAG, UAA).

strain   an intraspecific group of organisms possessing only one or a few distinctive traits, usually genetically homozygous (pure-breeding) for those traits, and maintained as an artificial breeding group by humans for domestication (e.g., agriculture) or for genetic experimentation. There is no clear distinction between the terms *strain* and *variety,* but the latter is generally applied when the differences between such intraspecific groups is substantial.

strand displacement   a replication mechanism, used by certain viruses, in which one DNA strand is displaced as a new strand is being synthesized.

stratigraphic time divisions   geologic time divisions. (*q.v.*).

streak plating   a technique of spreading microorganisms over the surface of a solidified medium for the purpose of isolating pure cultures.

Streptocarpus   the genus containing the Cape primroses. The inheritance of flower pigmentation has been thoroughly studied in various species in this genius. *See* **anthocyanins.**

Streptococcus pyogenes   the pathogenic bacterium causing "strep" infections.

streptolydigins   a group of antibiotics that, when bound to the beta subunit of bacterial RNA polymerase, prevent transcriptional elongation.

Streptomyces   a genus of saprophytic bacteria inhabiting the soil. Many species produce useful antibiotics, such as streptomycin and the tetracyclines. One hundred or so genetic markers have been identified in the most studied species, *S. coelicolor.*

streptomycin   an antibiotic produced by *Streptomyces griseus,* which binds to the 30S subunit of the bacterial ribosome and leads to faulty translation of the advancing messenger tape. *See* **ribosome, translation.**

streptomycin suppression   seen in bacterial mutants with an altered ribosomal protein (S12). This enables them to initiate polypeptide synthesis in the presence of streptomycin, and it also reduces the extent of misreading induced by that antibiotic. Such cells are converted from streptomycin-sensitive to streptomycin-resistant.

streptomycin

$$HN=C \begin{matrix} NH_2 \end{matrix}$$

(streptomycin chemical structure with labeled atoms: NH2, HN=C, NH, H, NH, C, NH2, H3C, CH, OH, C, C, OH, OH, C, HO, H, OH, H, C, O, HO C, O, CH2OH, C, H, H3C—NH, H, C, C, H, OH, H)

streptonigrin an antibiotic produced by *Streptomyces flocculus* which causes extensive chromosomal breakage.

(streptonigrin chemical structure with labeled atoms: CH3O, O, C, H, C, C, C, CH, C, C, C, N, H2N, C, N, C, C—COOH, O, H2N—C, C—CH3, C, C, HO—C, CH, CH3O—C, CH, C, OCH3)

stress fibers bundles of parallel aligned microfilaments underlying the plasma membrane of cultured eukaryotic cells. Stress fibers contain actin, alpha-actinin, myosin, and tropomyosin, and they appear to be involved with cell motility.

stringent control *See* **lambda cloning vehicle.**

stringent response the cessation of tRNA and ribosome synthesis by bacteria under poor growth conditions.

stroma the proteinaceous background matrix of a chloroplast or mitochondrion.

stromatolites living or fossil microbial mats dominated by cyanobacteria and fine sediment (usually calcium carbonate) trapped by these photosynthetic microbial communities. The oldest stromatolites are more than three billion years old. These are among the oldest known fossils.

Strongylocentrotus purpuratus a common sea urchin used in studies of molecular developmental genetics. *See* **echinoderm.**

strontium[90] a radioisotope of strontium with a half-life of 28 years generated during the explosion of nuclear weapons. ^{90}Sr is one of the major sources of radiation due to fallout.

structural change chromosomal aberration (*q.v.*).

structural gene a DNA sequence coding for RNA or a protein; regulatory genes are structural genes whose products control the expression of other genes.

structural protein any protein that substantially contributes to shape and structure of cells and tissues: e.g., the actin and myosin components of muscle filaments, the proteins of the cytoskeleton, collagen, etc.

struggle for existence the phrase used by Darwin to describe the competition between animals for environmental resources such as food or a place to live, hide, or breed. Darwin wrote in *On the Origin of Species,* "I use the term *struggle for existence* in a large and metaphorical sense . . . including (which is more important) not only the life of the individual, but success in leaving progeny."

stud an animal kept for breeding, as a horse or a mare.

Student *t* test a statistical method used to determine the significance of the difference between two sample means. The method was developed by the British stastician W.S. Gosset, who used the pseudonym "Student" in his publications. *See* page 480 for *t* distribution.

style a slender column of tissue arising from the top of the ovary and through which the pollen tube grows.

subculture a culture made from a sample of a stock culture of an organism transferred into a fresh medium.

subdioecy a sexual state of certain plants in which some unisexual individuals show imperfect sexual differentiation.

sublethal gene *See* **subvital mutation.**

submetacentric chromosome a chromosome that appears J-shaped at anaphase because the centromere is nearer one end than the other.

subspecies **1.** a taxonomically recognized subdivision of a species. **2.** geographically and/or ecologically defined subdivisions of a species with distinctive characteristics. *See* **race.**

substitutional load the cost to a population in genetic deaths of replacing an allele by another in the course of evolutionary change.

substitution vector *See* **lambda cloning vector.**

substrate **1.** the specific compound acted upon by an enzyme. **2.** substratum.

substrate race a local race of organisms selected by nature to agree in coloration with that of the substratum.

substratum the ground or other surface upon which organisms walk, crawl, or are attached.

subtertian malaria *See* **malaria.**

subvital mutation a gene that significantly lowers viability, but causes the death before maturity of less than 50% of those individuals carrying it. *Contrast with* **semilethal mutation.**

succulent a plant having watery or juicy tissues.

sucrose the sugar of commerce, a disaccharide composed of glucose and fructose.

sucrose gradient centrifugation *See* **centrifugation separation.**

Sudan black B a commonly used lysochrome.

sue **mutations** *See* **suppressor-enhancing mutations.**

sugar *See* **carbohydrate.**

sulfanilamide an antimetabolite widely used to suppress bacterial growth.

sulfatide lipidosis an autosomal recessive disease in man due to a defect in the production of the lysosomal enzyme arylsulfatase A. The clinical symptoms are paralysis, blindness, and dementia, leading to death during childhood.

sulfur an element universally found in small amounts in tissues. Atomic number 16; atomic weight 32.064; valence 2^-, 4^+, 6^+; most abundant isotope ^{32}S; radioisotope ^{35}S (*q.v.*).

sulfur-containing amino acids: cysteine, cystine, methionine. *See* **amino acids.**

sulfur mustard the first chemical mutagen to be discovered. *See* **nitrogen mustard.**

supercoiling the coiling of a covalently closed circular duplex DNA molecule upon itself so that it crosses its own axis. A supercoil is also referred to as a *superhelix.* The B form of DNA is a right-handed double helix. Winding of the DNA duplex in the same direction as that of the turns of the double helix is called *positive supercoiling.* Twisting of a duplex DNA molecule in a direction opposite to the turns of the strands of the double helix is called *negative supercoiling.*

374

superdominant overdominant (*q.v.*).

superfemale metafemale (*q.v.*).

superficial near the surface.

supergene a chromosomal segment protected from crossing over and so transmitted from generation to generation as if it were a single recon.

superhelix *See* **supercoiling.**

superinfection the introduction of such a large number of viruses into a bacterial culture that each bacterium is attacked by several phages.

superior in human anatomy the direction toward the head; above.

supermale metamale (*q.v.*).

supernatant the fluid lying above a precipitate in a centrifuge, following the centrifugation of a suspension.

supernumerary chromosome a chromosome present, often in varying numbers, in addition to the characteristic invariable complement of chromosomes. *See* **B chromosomes.**

superovulation the simultaneous release of more than the normal number of eggs from an ovary. This can be induced artificially by hormone treatment in cattle and other livestock for embryo transfer (*q.v.*) to surrogate mothers.

superrepression an uninducible state for a gene usually attributed to (1) a defective operator locus to which a functional repressor protein cannot bind, or (2) a mutant regulatory gene whose repressor protein product is insensitive to the inducer substance; a phenomenon that causes a gene to be permanently "turned off."

superspecies a complex of related allopatric species (often called a species group). Such species are grouped together because of their morphological similarities. That the superspecies constitutes a natural grouping is demonstrated by finding in the genus *Drosophila* that whenever species hybrids are produced under laboratory conditions the parents are from the same species group.

supersuppressor a mutation that can suppress the expression of specific alleles of mutations at a number of different chromosomal sites; usually a nonsense suppressor.

supervital mutation a mutation that increases the viability of individuals bearing it above the wild-type level.

suppression **1.** the restoration of a lost or aberrant genetic function (*see* **suppressor mutation**). **2.** in immunology, a specific or non-specific state of induced non-responsiveness in the immune system. *See* **immunological suppression, suppressor T cell.**

suppressor-enhancing mutations genetic alterations that boost the activity of both temperature-sensitive as well as temperature-insensitive suppressors.

suppressor mutation **1.** an *intergenic* suppressor mutation may code for an abnormal tRNA whose anticodon reads the mutated codon either in the sense as the normal codon or as an acceptable substitute for the original meaning. **2.** an *intragenic* suppressor restores the original reading frame after a frameshift. Also called a *second site mutation.*

suppressor T cell a subpopulation of T lymphocytes (designated T_s) whose function is to suppress the response of other lymphocytes to a particular antigen.

surrogate mother a woman (or other female mammal) that receives an embryo transplant from another donor.

survival of the fittest   the corollary of Darwin's theory of natural selection; namely, that as a result of the elimination by natural selection of those individuals least adapted to the environment, those that ultimately remain are the fittest.

survival value   the degree of effectiveness of a given phenotype in promoting the ability of that organism to contribute offspring to the future populations.

Sus   the genus that includes *S. scrofa,* the European wild hog, and *S. scrofa domesticus,* the domestic swine. *See* **swine.**

SV 40   simian virus 40 (*q.v.*).

Svedberg   *See* **sedimentation coefficient.**

sweepstakes route   a potential migration pathway along which species disperse with difficulty. Chance events play a large role in colonization of new areas along this route. For example, birds blown far out to sea by a storm may accidentally land on an oceanic island and colonize it, but it is unlikely that this will happen a second time for that species.

sweet corn   *See* **corn.**

swine   any of a number of domesticated breeds of the species *Sus scrofa.* Popular breeds include Berkshire, Chester White, Duroc, Hampshire, Hereford, Ohio Improved Chester, Poland China, Spotted Poland China, Tamworth, Yorkshire.

switch gene   a gene that causes the epigenotype to switch to a different developmental pathway.

switching sites   break points at which gene segments combine in gene rearrangements.

swivelase   *See* **gyrase, topoisomerase.**

symbiont   an organism living in a mutually beneficial relationship with another organism from a different species: e.g., the coexistence of algae and fungi in lichens.

symbiosis   any interactive association between two or more species living together. *See* **commensalism, mutualism, parasitism.**

symbiotic theory of the origin of undulipodia   the theory proposed by L. Margulis that the ancestral eukaryote acquired undulipodia (*q.v.*) as the result of a motility symbiosis with spirochaetes.

symbols used in human cytogenetics   A-G, the chromosome groups; 1–22, the autosome numbers; X, Y, the sex chromosomes; p, the short arm of a chromosome; q, the long arm of a chromosome; ace, acentric; cen, centromere; dic, dicentric; inv, inversion; r, ring chromosome; t, translocation; a plus ($+$) or minus ($-$) when placed before the autosome number or group letter designation indicates that the particular chromosome is extra or missing; when placed after a chromsome arm a plus or minus designation indicates that the arm is longer or shorter than usual; a diagonal (/) separates cell lines when describing mosaicism. Examples: 45,XX, $-$C, 45 chromosomes, XX sex chromosomes, a missing chromosome from the C group; 46, XY, t (Bp$-$;Dq$+$), a reciprocal translocation in a male between the short arm of a B and the long arm of a D group chromosome; inv (Dp$+$, q$-$), a pericentric inversion involving a D chromosome; 2p$+$, an increase in the length of the short arm of a chromosome 2; 46,XX, r, a female with one ring X chromsome; 45,X/ 46, XY, a mosaic of two cell types, one with 45 chromosomes and a single X, one with 46 chromosomes and XY sex chromosomes. *See* **human mitotic chromosomes.**

sympatric speciation   an uncommon process by which populations inhabiting (at least in part) the same geographic range become reproductively isolated.

sympatric species   species whose areas of distribution coincide or overlap.

376

symplesiomorphic character   an ancestral or plesiomorphic character shared by two or more different organisms.

synapomorphic character   a derived or apomorphic character shared by two or more different organisms.

synapsis.   *See* **meiosis.**

synaptonemal complex   a tripartite ribbon consisting of parallel, dense, lateral elements surrounding a medial complex. The lateral elements lie in the central axes of the paired homologous chromosomes of a pachytene bivalent. The medial complex contains a system of interdigitating protein filaments that are oriented perpendicularly to the lateral elements and serve to maintain their parallel configuration during meiotic synapsis. *See* **Gowen's crossover suppressor, meiosis.**

A drawing of a segment of a bivalent as seen under the electron microscope. (c) Chromatin; (cs) central space; (le) lateral element; (sc) synaptonemal complex; (tr) transverse rods of the medial complex.

syncaryon   synkaryon.

syncytium   a mass of protoplasm containing many nuclei not separated by cell membranes.

syndactyl   having webbed digits either as a normal aspect of the species or, in man, pathologically.

syndesis   meiotic chromosomal synapsis.

syndrome   a group of symptoms that occur together, characterizing a disease.

synergid   one of two haploid cells that lie beside the ovum in the embryo sac (*q.v.*). Synergids function to nourish the ovum.

synergism   the phenomenon in which the action of two agents used in combination is more effective than the sum of their individual actions.

synezis   the clumping of chromosomes into a dense knot which adheres to one side of the nucleus. Synezis is a common occurrence during leptonema in microsporocytes.

syngamy   the union of the nuclei of two gametes following fertilization to produce a zygote nucleus.

syngen   *See Paramecium aurelia.*

syngeneic   pertaining to genetically identical organisms such as identical twins or the members of a highly inbred strain. Because syngeneic animals have the same antigens on their tissues, they can exchange skin or organ grafts successfully. *Compare with* **congenic strain, allogeneic.**

synkaryon **1.** the zygote nucleus resulting from the fusion of two gametic nuclei. **2.** the product of nuclear fusion in somatic cell genetic experiments.

synomone *See* **allomone.**

synonym in taxonomy, a different name for the same species or variety.

syntenic genes genes thought to reside on the same chromosome because of their behavior in somatic cell hybridization experiments. A gene that is concurrently lost or retained along with a marker gene during the mitoses of hybrid cells is assumed to reside on the same chromosome as the marker gene.

synthetase an enzyme catalyzing the synthesis of a molecule from two components, with the coupled breakdown of ATP or some other nucleoside triphosphate.

synthetic lethal a lethal chromosome derived from normally viable chromsomes by crossing over.

synthetic linkers short, chemically synthesized DNA duplexes containing sites for one or more restriction endonucleases. Synthetic linkers are used most commonly in the cloning of blunt-ended DNA molecules.

synthetic polyribonucleotides RNA molecules made without a nucleic acid template, by either enzyme action or chemical synthesis in the laboratory. *See* **polynucleotide phosphorylase.**

Systema Naturae the taxonomic treatise first published in 1735 by C. V. Linné (Linnaeus). Sixteen editions were completed during his lifetime. He originated the Linnaean system of binary nomenclature used to this day. The insistence of Linné on the constancy and objective classification of species posed the problem of the method of origin of species.

systematics the study of classification; taxonomy.

systemic generally distributed throughout or affecting the whole organism.

T

t   *See* **symbols used in human cytogenetics.**

t   the Student *t* statistic which is used for testing the difference between the means of two samples. *See* **Student *t* test,** and page 480.

T   thymine or thymidine.

tachyauxesis   *See* **heterauxesis.**

tachytelic evolution   *See* **evolutionary rates.**

tactoid   a paracrystalline aggregate.

tandem duplication   an aberration in which two identical chromosomal segments lie one behind the other. The order of the genes in each segment is the same.

tandem repeat   tandem duplication.

"T" antigen   an antigen found in the nuclei of cells infected or transformed by certain oncogenic viruses such as polyoma. The antigen is thought to be a protein coded for by a virus cistron.

target number   *See* **extrapolation number.**

target organ   the receptor organ upon which a hormone has its effect.

target theory   a theory developed to explain some biological effects of radiation on the basis of ionization occurring in a very small sensitive region within the cell. One or more "hits," that is, ionizing events, within the sensitive volume are postulated to be necessary to bring about the effect. *See* **extrapolation number.**

target tissue   **1.** the tissue against which antibodies are formed. **2.** the tissue responding specifically to a given hormone.

tassel   the staminate inflorescence of corn.

taste blindness   the inability in man to taste the chemical phenylthiocarbamide (PTC). Taste-blind individuals are homozygous for an autosomal recessive gene. PTC tastes bitter to individuals carrying the dominant allele.

TATA box   Hogness box (*q.v.*); pronounced "tah-tah."

tautomeric shift   a reversible change in the location of a hydrogen atom in a molecule which alters it from one to another isomer. Thymine and guanine are normally in *keto* forms, but when in the rare *enol* forms (see diagram below) they can join by three hydrogen bonds with keto forms of guanine or thymine, respectively. Likewise, adenine and cytosine are normally in amino forms, but when in the rare *imino* forms they can join by two hydrogen bonds with amino forms of cytosine or adenine, respectively.

Rare enol form
of thymine (T*)

Thymine

Rare imino form
of adenine (A*)

Adenine

Cytosine

Guanine

Rare imino form
of cytosine (C*)

Rare enol form
of guanine (G*)

From F. J. Ayala and J. A. Kiger, Jr., *Modern Genetics*, The Benjamin/Cummings Publishing Company, Inc., 1980.

tautomerism   the phenomenon in which two isomeric forms of molecules exist in equilibrium.

taxon   the general term for a taxonomic group whatever its rank.

taxonomic category   the rank of a taxon in the hierarchy of classification. *See* **classification.**

taxonomic congruence   the degree to which different classifications of the same organisms postulate the same groupings. When the classifications compared are based on different sources of information (independent sets of data), congruence provides a measure of the degree to which the classifications remain stable as various lines of evidence are considered.

taxonomic extinction   nonsurvival of a taxon, either by extinction or by pseudoextinction (*q.v.*).

taxonomist   a specialist in taxonomy (*q.v.*).

taxonomy   the study of the classification of living things. Classically taxonomy is concerned with description, naming, and classification on the basis of morphology. More recently taxonomists have been concerned with the analyses of patterns of variation in order to discover how they evolved, with the identification of evolutionary units, and with the determination by experiment of the genetical interrelationships between such units, as well as the role of the environment in their formation.

Tay-Sachs disease   a lethal hereditary disease found at relatively high frequency in Ashkenazi Jews of Eastern European ancestry. About 2–4% are heterozygous for the Tay-Sachs gene. Developmental retardation begins in infancy, followed by paralysis, mental deterioration, and blindness. The defective recessive gene resides on autosome 15. The normal allele codes for the alpha subunit of hexosaminidase A. In homozygotes, the enzyme defect results in the accumulation of gangliosides (*q.v.*) in the brain. *See* **hexosaminidase, Sandhoff disease.**

TCA   trichloracetic acid (*q.v.*).

TψC loop   the hairpin loop nearest the 3′ end of tRNA molecules, containing the modified base pseudouridine (ψ). This loop is thought to interact with ribosomal RNA. *See* **transfer RNA.**

T cell   T lymphocyte. *See* **lymphocyte.**

T complex   a region on chromosome 17 of the mouse containing genes affecting tail length and cell surface antigens. Mutations in some of these genes may arrest development at several embryonic stages.

TD50   the average number of cells that must be injected per animal to ensure that 50% of the treated animals will exhibit the effect being tested (usually the development of a tumor).

T-DNA   a group of seven genes (collectively referred to as transferred DNA) of the Ti plasmid (*q.v.*) that integrates into the nuclear DNA of the host plant during tumor induction. T-DNA is always present in crown gall cells of plants. *See Agrobacterium tumefaciens.*

teleology   the explanation of a phenomenon such as evolution by the purposes or goals it serves. Teleological explanations usually invoke supernatural powers and are therefore nonscientific.

teleonomy   the doctrine that the existence in an organism of a structure or function implies that it has conferred an advantage on its possessor during evolution.

telocentric chromosome   a chromosome with a terminal centromere.

telolecithal egg   one in which the yolk spheres are accumulated in one hemisphere. *See* **vegetal hemisphere.**

telomere   one of the two terminal chromomeres of a chromosome.

telophase   *See* **mitosis.**

telotrophic meroistic ovary   *See* **insect ovary types.**

tem   triethylene melamine, an aziridine mutagen (*q.v.*). See formula on p. 382.

TEM   transmission electron microscope. *See* **electron microscope.**

tem

$$H_2C - N \underset{C}{\overset{CH_2}{\diagup}} N \quad N \overset{CH_2}{\diagup} CH_2$$

temperate phage   a non-virulent bacterial virus which infects, but rarely causes lysis. It can become a prophage and thereby lysogenize the host cell.

temperature   the unit of temperature used in science is the degree Celsius or Centigrade (°C) which equals $\frac{1}{100}$th of the difference in temperature between melting ice and boiling water at an atmospheric pressure of 760 mm of mercury. In the Fahrenheit system the interval between the freezing and boiling points of water is divided into 180 degrees, 32°F being the freezing point and 212°F the boiling point. Fahrenheit and Celsius are converted by F = (C× $\frac{9}{5}$) + 32. In the Kelvin system absolute zero (approximately -273°C, -460°F) is taken as zero, and its degrees match the Celsius scale. Thus 373°K would equal the boiling point of water. See nomogram on p. 383.

temperature sensitive mutation   a mutation which is manifest in only a limited temperature range. The product of such a gene generally functions normally, but is unstable above a certain temperature. Thus the mutant when reared at the lower (permissive) temperature is normal, but when placed at the higher (restrictive) temperature shows the mutant phenotype.

template   the macromolecular mold for the synthesis of a negative antitemplate macromolecule. The antitemplate then serves as a mold for the template. Thus the duplication of the template requires two steps. A single strand of DNA serves as a template for the complementary strand of DNA or mRNA.

template RNA   messenger RNA.

template switching   in *E. coli*, a bizarre *in vitro* reaction often accompanying strand displacement in which DNA polymerase I shifts from the original template strand to the displaced strand.

temporal isolation   *See* **seasonal isolation.**

Tenebrio molitor   a cereal beetle commonly called the mealworm.

Teosinte   the closest wild relative of corn. *See* ***Euchlaena mexicana.***

tepa   an aziridine mutagen (*q.v.*).

$$O = P - N \underset{N}{\diagdown} \begin{array}{c} CH_2 \\ CH_2 \end{array}$$

teratocarcinoma   embryonal tumors originating in the yolk sac or gonads of amniotes and

temperature

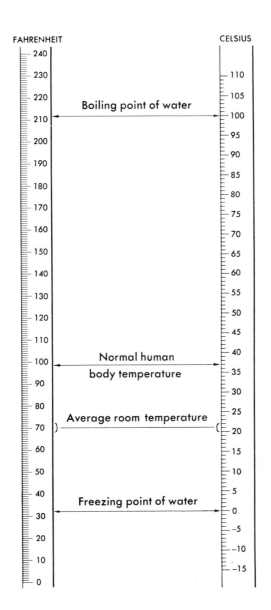

FAHRENHEIT

240
230
220
210
200
190
180
170
160
150
140
130
120
110
100
90
80
70
60
50
40
30
20
10
0

CELSIUS

110
105
100
95
90
85
80
75
70
65
60
55
50
45
40
35
30
25
20
15
10
5
0
–5
–10
–15

Boiling point of water

Normal human
body temperature

Average room temperature

Freezing point of water

capable of differentiating into a variety of cell types. These tumors are used to study the regulatory mechanisms involved in embryological development.

teratogen   any agent that raises the incidence of congenital malformations.

teratoma   a tumor composed of an unorganized aggregation of different tissue types.

tergite   a dorsal sclerite (*q.v.*).

terminal chiasmata   the end-to-end association of homologous chromsome arms resulting from terminalization.

terminal deletion   *See* **deletion.**

terminal fusion   a mechanism for restoring diploidy to an individual originating by parthenogenesis, involving the union of two noninseminated egg nuclei from the same secondary oocyte.

terminalization   *See* **meiosis.**

terminal redundancy   referring to the repetition of the same sequence of nucleotides at both ends of a DNA molecule.

terminal taxa   the groups that occur at the ends of branches in a cladogram.

terminal transferase   a deoxyribonucleotidyltransferase that is used by molecular biologists to add a homopolymer tail, e.g., polydeoxyadenylate, to each end of a vehicle DNA. The enzyme is then used to add poly T tails to a passenger DNA. The passenger and vehicle are then annealed via their complementary termini, ligated, and cloned.

termination codon   a codon that signals the termination of a growing polypeptide chain. *See* **amber mutation, ochre mutation, stop codon.**

termination factors   *See* **release factors.**

terminator sequence   a DNA sequence that provides the signal to stop transcription by RNA polymerase; not to be confused with a termination (stop) codon that serves as a signal to cease translation.

territoriality   the defense by an animal or group of an area against members of the same species.

territory   an area of the habitat occupied by an individual or group. If members belonging to the same species enter the territory, they are attacked as trespassers.

tert.   tertiary; pertaining to the third item or rank in a classification.

Tertiary   a period of the Cenozoic era lasting from 2 to 65 million years before the present. *See* **geologic time divisions.**

tertiary base pairs   the 9 base pairs of a tRNA molecule responsible for its three-dimensional folding. Most of these base pairs are evolutionarily conserved in all tRNA molecules.

tertiary protein structure   *See* **protein structure.**

tesserae   functionally different patches of endoplasmic reticulum, each bearing a characteristic set of enzymes.

test cross   a mating between an individual of unknown genotype, but showing the dominant phenotype for one or more genes, with a tester individual known to carry only the recessive alleles of the genes in question. The test cross reveals the genotype of the tested parent. For example, an individual showing the *A* and *B* phenotypes is crossed to an *aabb* tester. The F_1 contains only individuals of *AB* phenotype. This reveals the genotype of the tested parent to be *AABB.*

tester strain   a multiply recessive strain that provides the genotypically known mate used in a testcross.

testicular feminization   a condition in which XY individuals develop as females. Spermatogenesis does not occur in the testes, which are generally located within the abdomen. The syndrome is due to a defect in androgen receptors. The condition shows X-linked inheritance in both humans and the mouse.

testis (*plural* **testes**)   the gamete-producing organ of a male animal.

testosterone a masculinizing, steroid hormone secreted by interstitial cells of the testis.

test-tube baby the production of a child by *in vitro* fertilization, followed by embryo transplantation to complete gestation in a normal uterus. This may be provided by the biological or surrogate mother.

tetra-allelic referring to a polyploid in which four different alleles exist at a given locus. In a tetraploid $A_1A_2A_3A_4$ would be an example.

tetracyclines a family of antibiotics obtained from various species of *Streptomyces*. Tetracyclines bind to the 30 S subunit of prokaryotic ribosomes and prevent the normal binding of aminoacyl–tRNA at the A site. *See* **translation.**

tetrad 1. four homologous chromatids (two in each chromosome of a bivalent) synapsed during first meiotic prophase and metaphase. *See* **meiosis. 2.** four haploid products of a single meiotic cycle.

tetrad analysis the analysis of crossing over by the study of all the tetrads arising from the meiotic divisions of a single primary gametocyte. To perform such an analysis one must use an organism in which the meiotic products are held together, as for example, in the case of meiospores confined in an ascus sac. Genera suitable for such analyses include *Ascobolus, Aspergillus, Bombardia, Neurospora, Podospora, Saccharomyces, Schizosaccharomyces, Sordaria,* and *Sphaerocarpus.*

tetrad segregation types For a bivalent containing the genes *A* and *B* on one homologue and *a* and *b* on the other, three patterns of chromatid segregation are possible: *AB, AB, ab, ab* (referred to as the parental ditype); *AB, Ab, aB, ab,* where two chromatids are recombinant (the tetratype); and *Ab, Ab, aB, aB,* where all chromatids are recombinant (the nonparental ditype).

Tetrahymena pyriformis a common species of ciliate protozoa. Next to *Paramecium aurelia,* the protozoan for which the most genetic information is available.

tetramer a structure resulting from an association of four subunits. If the subunits are all identical, they form a homotetramer; if the subunits are not all identical, they form a heterotetramer.

tetramine an aziridine mutagen (*q.v.*).

$$H_2C\text{---}CH_2$$

tetramine structure diagram

tetranucleotide theory a defunct theory of DNA structure, formulated about 1910 by P. A. Levene, proposing that the molecule consists of repeating units, each of which contained four nucleotides (one each of adenine, guanine, thymine, cytosine).

tetraparental mouse a mouse developed by artificial fusion of embryonic cells from two genetically different blastulas.

tetraploid having four haploid sets of chromosomes in the nucleus.

tetrapterous having four wings.

tetrasomic having one chromosome in the complement represented four times in each nucleus.

tetratype *See* **tetrad segregation types.**

thalassemias a group of human anemias due to imbalance in the ratio of alpha and/or beta hemoglobin subunits. Since there are four alpha genes per genome, deletions (commonly produced by unequal crossing over) can result in an individual having any number of alpha genes from zero to four. The complete absence of alpha genes produces a condition lethal at or before birth, called *hydrops fetalis.* With only one alpha gene, excess beta chains form a tetramer (β_4), resulting in hemoglobin H disease. Individuals with two or three alpha genes are almost indistinguishable from normal. Incomplete beta chains can be produced by nonsense codons. Deletions in beta genes are commonly produced by unequal crossing over, as are the hybrid chains containing δ and β segments (Hb Lepore) or A, and β segments (Hb Kenya). Beta thalassemia (also called Cooley's anemia) is a hemoglobinopathy in which few functional beta globin chains are made. A point mutation, within an intron that alters the cutting and splicing signal, causes an extra piece of intron RNA to be present in processed mRNA; the extra piece shifts the reading frame and causes translation to stop prematurely, yielding a truncated and nonfunctional beta globin molecule. *See* **hemoglobin genes.**

thalidomide a drug used in Europe during the early 1960s as a tranquilizer. It was safe for adults, but it induced phocomelia (*q.v.*) in babies when their mothers took it during the 3rd to 5th weeks of pregnancy.

thallus a plant body showing no differentiation of root, stem, and leaf as in higher plants.

thelytokous parthenogenesis *See* **thelytoky.**

thelytoky a rare type of parthenogenesis in which unfertilized eggs yield females, and males may be absent entirely. The ova that develop in this way are usually diploid, in which case the offspring are genetically identical with the parent.

Theobroma cacao the cacao tree, source of chocolate.

theobromine a mutagenically active purine analogue. It is the main alkaloid stimulant in chocolate.

thermal denaturization profile *See* **melting profile.**

thermal neutron a fast neutron from uranium fission that has been slowed down by elastic collision with a moderator such as graphite to energies equivalent to those of gas molecules at room temperature (approximately 0.025 eV). The biological effect of thermal neutrons is attributable to the summation of capture and decay radiations. In biological material the reactions ^1H (n, γ) ^2H and ^{14}N (n, p) ^{14}C are the most important sources of tissue ionization. The relative importance of these reactions depends on the size of the organism. Protons from nitrogen capture are the major cause of the biological effects of thermal neutrons in an organism the size of *Drosophila.*

thermodynamics, laws of The first law: energy can be transformed from one form to another but cannot be created or destroyed. The second law: whenever energy is transformed from one kind into another, only a fraction of it is available for doing useful work; the rest remains as the heat energy of the molecules left at the completion of the reaction. The third law: at absolute zero all molecules are perfectly aligned and motionless; their entropy (*q.v.*) is zero.

thermophilic heat loving. Said of bacteria which grow at temperatures between 45°C and 65°C (found in fermenting manure and hot springs).

theta (θ) structure a replication eye (*q.v.*) in a circular replicon; also called a *Cairns molecule* (after John Cairns, its discoverer). *See* **D loop.**

thiamine vitamin B_1, the anti-beriberi factor.

thiamine pyrophosphate a coenzyme of the carboxylases and aldehyde transferases.

thin layer chromatography *See* **chromatography.**

thioglycollic acid treatment a procedure used to rupture disulphide bridges linking adjacent protein chains.

Thio-tepa trade name for triethylenethiophosphoramide (*q.v.*). The formula is on p. 388.

third cousin *See* **cousin.**

Thio-tepa

$$\begin{array}{c} \overset{S}{\underset{||}{H_2C}} \\ \overset{|}{\underset{H_2C}{H_2C}} > N - P - N < \overset{CH_2}{\underset{CH_2}{|}} \\ \overset{|}{\underset{N}{N}} \\ \overset{/}{CH_2} \overset{\backslash}{-} CH_2 \end{array}$$

third law of thermodynamics *See* **thermodynamics.**

thirty-seven percent survival dose the radiation dose at which the number of hits equals the number of targets. The dose at which there is an average of one hit per target. *See* **target theory.**

thr threonine. *See* **amino acid.**

three-point cross a series of crosses designed to determine the order of three, nonallelic, linked genes upon a single chromosome on the basis of their crossover behavior.

three-strand double exchange *See* **inversion.**

threonine *See* **amino acid.**

threshold dose the dose of radiation below which the radiation produces no detectable effect.

threshold effect hypothesis the notion that certain traits with a polygenic basis develop only if the additive effects of contributory alleles exceed a critical value. This hypothesis is often used to explain many all-or-none phenomena with a polygenic mode of inheritance (e.g., susceptibility vs. resistance to certain diseases).

thrombin *See* **blood clotting.**

thrombocyte blood platelet. *See* **platelets.**

thrum the type of flower characterized by short styles and high anthers found among distylic species such as seen in the genus *Primula*. *See* **pin.**

Thy-1 antigen an antigen on the plasma membrane of thymocytes (*q.v.*) which can be used to distinguish them from other lymphocyte groups.

thylakoid *See* **chloroplast.**

thymidine the deoxyriboside of thymine. *See* **nucleoside.**

thymidine kinase an enzyme catalyzing the phosphorylation of thymidine to thymidine monophosphate.

thymidylate kinase an enzyme catalyzing the phosphorylation of thymidine monophosphate and thymidine disphosphate to thymidine diphosphate and thymidine triphosphate, respectively.

thymidylic acid *See* **nucleotide.**

thymine a pyrimidine base found in DNA (formula on p. 389).

thymine dimer two thymine molecules joined as shown in the accompanying illustration (p. 389). The reaction of greatest importance when ultraviolet radiation interacts with DNA is the formation within single polynucleotide chains of bonds between adjacent thymine molecules. Such thymine dimers block future DNA replication.

thymocyte a thymus-derived lymphocyte or T lymphocyte.

thymus an organ lying in the chest of mammals, formed embryologically from gill thymine

thymine dimer

pouches. This organ functions to populate the body with lymphoid cells. It reaches a maximum size about the time of sexual maturity and then atrophies.

thyroglobulin   *See* **thyroid hormones.**

thyroid hormones   thyroxine and triiodothyronine are synthesized by the thyroid gland in response to thyrotropin, a hormone from the anterior pituitary gland. The synthesis of thyroid hormones begins with the selective accumulation of inorganic iodide by the epithelial cells of the gland. The trapped iodide is then oxidized to iodine, and tyrosine residues of the glycoprotein thyroglobulin are then iodinated, converting them to monoiodotyrosine residues (a). The iodotyrosyl residues are then coupled with the elimination of the alanine side chain to form triiodothyronine (b) and thyroxine (c). *See* **cretinism.**

thyroid-stimulating hormone (TSH)   a glycoprotein hormone stimulating secretion by the thyroid. TSH is produced by the adenohypophysis.

thyrotropin   *See* **thyroid hormones.**

thyroxine   *See* **thyroid hormones.**

tiller   a grass side shoot produced at the base of a stem.

timber line   the line in high latitudes and in high elevations in all latitudes beyond which trees do not grow.

time-lapse microcinematography   a technique of photographing living cells under phase contrast with a motion picture camera taking exposures at selected intervals (1 frame per minute, for example) and then projecting the film at a more rapid speed (24 frames per second, for example). Under such circumstances time will be speeded up almost 1500 times, and one can gain a much clearer understanding of the dynamic processes the cells undergo.

Ti plasmid   a *tumor-inducing* (hence the acronym) plasmid found in the bacterium *Agrobacterium tumefaciens* (*q.v.*) that is responsible for crown gall disease of dicotyledonous plants; used as a vector in genetic engineering of plants. The wild-type plasmid produces tumor cells, but it can be modified so that it can carry foreign genes into cells without making the recipient cells tumorous. During tumor induction, a specific segment of the Ti plasmid, called the T-DNA (transferred DNA), integrates into the host plant nuclear DNA.

Tiselius apparatus   electrophoresis apparatus.

tissue   a population consisting of cells of the same kind performing the same function.

tissue culture   the maintenance or growth of tissue cells *in vitro* in a way that may allow further differentiation and preservation of cell architecture or function or both. *Primary cells* are those taken directly from an organism. Treating a tissue with the proteolytic enzyme trypsin dissociates it into individual primary cells that grow well when seeded onto culture plates at high densities. Cell cultures arising from multiplication of primary cells in tissue culture are called *secondary cell cultures.* Most secondary cells divide a finite number of times (*see* **Hayflick number**) and then die. A few secondary cells may pass through this "crisis period" and be able to multiply indefinitely to form a continuous *cell line.* Cell lines have extra chromosomes and are usually abnormal in other respects as well. The immortality of these cells is a feature shared in common with cancer cells.

tissue typing   identification of the major histocompatibility antigens of transplant donors and potential recipients, usually by serological tests. Donor and recipient pairs should be of identical ABO blood group, and in addition should be matched as closely as possible for H antigens in order to minimize the likelihood of allograft rejection.

titer   the amount of a standard reagent necessary to produce a certain result in a titration (*q.v.*).

titration   a method of determing the amount of some substance present in a solution by measuring the amount of a reagent which must be added to cause a defined chemical change.

T lymphocyte   *See* **lymphocyte.**

T_m   the temperature at which a population of double-stranded nucleic acid molecules becomes half-dissociated into single strands; referred to as the *melting temperature* for that system.

TMV   tobacco mosaic virus.

Tn   symbol for transposon.

tobacco   *See* **Nicotiana.**

tobacco mosaic disease   a virus disease of the tobacco plant which produces lesions on the leaves. The causal virus contains a central core of RNA and a coat made up of 2130 protein subunits, each of molecular weight 17,500.

tolerance   *See* **immunological tolerance.**

toluidine blue   a metachromatic basic dye used in cytochemistry.

tomato   *See* ***Lycopersicon esculentum.***

T24 oncogene   a gene isolated from a line of human bladder carcinoma cells. This oncogene appears to be homologous to the oncogene carried by the Harvey murine sarcoma virus (*q.v.*). The change that leads to the activation of the T24 ocogene is due to a single base substitution.

tonofilaments   synonymous with keratin filaments. *See* **intermediate filaments, keratin.**

topoisomerase   an enzyme that can interconvert topological isomers of DNA. These enzymes alter DNA topology by changing the linking number of circular duplex DNAs or by interconverting knotted and catenated forms. Topoisomerase has replaced a number of earlier terms (DNA relaxing enzyme, swivelase, untwisting enzyme, nick-closing enzymes). Topoisomerases are divided into two classes: Type 1 enzymes make a transient break in one strand of the duplex, whereas type 2 enzymes introduce transient double-strand breaks. During the relaxation of DNA by type 1 topoisomerases, one strand of the helix is passed through its complementary strand.

topological isomers   *See* **linking number.**

topology   the study of the properties of geometrical figures that are subjected to deformations such as bending or twisting.

tormogen cell   *See* **trichogen cell.**

tortoise-shell cat   a female cat of genotype *Bb* showing black and orange patches of fur in its coat. The mottling results from random inactivation of one or the other X chromosome during development. *See* **calico cat.**

totipotency   the capacity of a cell to differentiate into all of the cells of the adult organism. A zygote is normally totipotent, but most cells of the embryo become progressively restricted in this capacity as development progresses.

toxin   a poisonous substance elaborated by a microorganism, as well as some fungal, plant, and animal species. An example is alpha amanatin (*q.v.*).

toxoid   a poisonous protein that has been detoxified without harm to its antigenic properties.

T phages   a group of virulent viruses attacking *E. coli.* A long DNA molecule is tightly packed into each polyhedral capsid. Each phage attaches to its host by a hollow tail through which its DNA molecule is inserted into the bacterium. *See* **bacteriophages.**

TPN   triphosphopyridine nucleotide. *See* **nicotinamide-adenine dinucleotide phosphate.**

T_R Triassic (*q.v.*).

trace element an element required only in minute amounts by an organism for its normal health.

tracer *See* **radioactive isotope, labeled compound.**

trachea **1.** the tubular invaginations of the epidermis which serve to convey oxygen to the cells of insects. **2.** the windpipe of land-living vertebrates.

tracheoles one of the terminal branches of the trachea of insects.

traction fiber one of the fibers connecting the centromeres of the various chromosomes to either centriole. *See* **mitotic apparatus.**

trailer a nontranslated sequence at the 3′ end of mRNA following the termination signal, exclusive of the poly-A tail.

trait *See* **character.**

trans *See* **cis, trans configuration.**

transcribed spacer that part of a primary rRNA transcript that is discarded during the formation of functional RNAs of the ribosome (*q.v.*).

transcriptase RNA polymerase (*q.v.*).

transcription the formation of an RNA molecule upon a DNA template by complementary base pairing; mediated by RNA polymerase (*q.v.*).

transcription unit the segment of DNA between the sites of initiation and termination of transcription by RNA polymerase; more than one gene may reside in a transcription unit. A polycistronic message (*q.v.*) may be translated as such in prokaryotes. However, in eukaryotes only one protein molecule can be translated from a given mRNA molecule, because the primary transcripts containing information for more than one polypeptide chain are processed into monocystronic mRNA molecules before leaving the nucleus.

transdetermination change in developmental fate of a cell or group of cells. *See* **in vivo culturing of imaginal discs.**

transduced element the chromosomal fragment transferred during transduction.

transduction the transfer of bacterial genetic material from one bacterium to another using a phage as a vector. In the case of *restrictive* or *specialized transduction* only a few bacterial genes are transferred. This is because the phage has a specific site of integration on the host chromosome, and only bacterial genes close to this site are transferred. In the case of *generalized transduction* the phage can integrate at almost any position on the host chromosome, and therefore almost any host gene can be transferred with the virus to a second bacterium. Transducing phage are usually defective in one or more normal phage functions, and may not be able to replicate in a new host cell unless aided by a normal "helper" phage. *See* **abortive transduction.**

transfection **1.** the process of infection of bacterial protoplasts by viral nucleic acid alone (without the viral protein capsid); it differs from bacterial transformation in that the latter involves intact bacterial cells possessing cell walls, and heritable modifications of cellular properties are required. **2.** the insertion of foreign DNA into cultured eukaryotic cells by exposing them to naked DNA.

transferases enzymes that catalyze the transfer of functional groups between donor and acceptor molecules. The most common molecules transferred are amino, acyl, phosphate, and glycosyl groups.

392

transfer factor a dialyzable extract (lymphokine) from sensitized T lymphocytes that can transfer some types of cell-mediated immunity from one individual to another.

transferred immunity *See* **adoptive transfer.**

transferrins *See* **plasma transferrins.**

transfer RNA (tRNA) an RNA molecule that transfers an amino acid to a growing polypeptide chain during translation (*q.v.*). Transfer RNA molecules are among the smallest biologically active nucleic acids known. For example, an alanine transfer RNA isolated from yeast contains 77 nucleotides and is folded back upon itself and kept in a "clover leaf" configuration by the characteristic pairings of the bases G to C and A to U.

A-C-C-A$_{OH}^{77}$
$_p$G≡C (1)
G≡C
G   U
C≡G
G≡C
U   U
G≡C-A-G-G-C-C
C-U-C-C-G-G
TΨC loop
U-U-A
G

U^d loop
G-A-U-G
C-G-C-G-G
G-G-U^d-A   G-C-G-C   G^d
A
T-Ψ-C

C≡G
U=A
C≡G
C≡G
C≡G
U   Ψ   anticodon
U   I^m loop
I-G-C
$_m$RNA-C-C-G-

All transfer RNAs attach to their amino acids by the 3′ end which contains a terminal adenylic followed by two cytidylic acids. The 5′ end always carries a terminal guanylic acid. Transfer RNAs contain several purines and pyrimidines not generally encountered in other RNA molecules. These *rare bases* (*q.v.*) are formed following transcription, since nuclei are known to contain enzymes that are capable of modifying certain bases on preformed RNA. The site recognized by tRNA synthetase is believed to be located in the neck region adjacent to the dihydrouridine loop (see arrows). The anticodon occupies positions 36–38. In the drawing A,U,C, and G have their usual meanings. The rare bases are symbolized as follows: Ψ, pseudouridylic acid; T, ribothymidylic acid; U^d, dihydrouridylic acid; G^m, methylguanylic acid; G^d, dimethyl guanylic acid; I, inosinic acid; and I^m, methyl inosinic acid. *See* **isoacceptor transfer RNA.**

trans-filter induction *in vitro* inductions using organizer tissues separated from reactive cells by a Millipore filter (*q.v.*). The trans-filter induction system allows one to interrupt induction at any time by removing the inducing cells from the surface of the filter.

transformants cells that show an inherited modification after exposure to a transforming principle or after incorporating plasmid DNA.

transformation in eukaryotic cells, the conversion to a state of unregulated growth in tissue culture, resembling or identical to an *in vitro* tumor.

transformation series the various expressions of a character ordered in a hypothesized sequence from the most primitive, plesiomorphic state to the most derived, apomorphic state. This sequence may be linear (e.g., $A^0 \rightarrow A^1 \rightarrow A^2$) or it may be branched.

transformer (tra) a third chromosomal gene in *Drosophila* which converts homozygous females into sterile males.

transgenic animals animals into which cloned genetic material has been experimentally transferred. In the case of laboratory mice, one-celled embryos have been injected with plasmid solutions, and some of the transferred sequences were retained throughout embryonic development. Some sequences became integrated into the host genome and were transmitted through the germ line to succeeding generations. A subset of these foreign genes expressed themselves in the offspring.

transgressive variation progeny phenotypes outside the range of that which occurs in the parents; usually attributed to polygene segregation.

transient diploid a relatively short stage in the life cycles of predominantly haploid fungi or algae during which meiosis occurs.

transient polymorphism polymorphism existing in a population during the period when an allele is being replaced by a superior one.

transition *See* **base-pair substitution.**

translation the formation of a protein directed by a specific messenger RNA (mRNA) molecule. Translation occurs in a ribosome (*q.v.*). A ribosome begins protein synthesis once the 5′ end of a mRNA tape is inserted into it. As the mRNA molecule moves through the ribosome, much like a tape through the head of a tape recorder, a lengthening polypeptide chain is produced. Once the leading (5′) end of the messenger tape emerges from the first ribosome, it can attach to a second ribosome, and so a second identical polypeptide can start to form. When the 3′ end of the mRNA molecule has moved through the first ribosome the newly formed protein is released, and the vacant ribosome is available for a new set of taped instructions. The assembly of amino acids into a peptide starts at the amino end (N terminus) and finishes at the carboxyl end (C terminus). There are two binding sites for transfer RNA (tRNA) in the ribosome. The P site (peptidyl-tRNA binding site) holds the tRNA molecule that is attached to the growing end of the nascent polypeptide. The A site (aminoacyl-tRNA binding site) holds the incoming tRNA molecule charged with the next amino acid. The tRNAs are held so that their anticodons form base pairs with adjacent complementary codons of the mRNA moving through the ribosome. In the diagram, a mRNA molecule is shown progressing through a ribosome. At time T_0 codon 5 of the messenger tape occupies the P site and codon 6 the A site. About half a second later the mRNA has advanced to the left by one codon. The bond between tRNA I and the 5th amino acid of the nascent polypeptide has been split, and it has been linked to amino acid 6. The atoms involved in the rearrangements are in boldface type. This reaction is catalyzed by *peptide transferase,* an enzyme that is bound tightly to the ribosome. Transfer RNA molecule I is discharged from the ribosome, tRNA II has entered the P site and is now attached to a polypeptide chain six amino acids long. A new tRNA (III) carrying an appropriate anticodon enters the A site. Note that in the diagram letters represent atoms; whereas circled letters represent molecules (i.e., nucleotides or amino acid residues). *See* **elongation factors, initiation factors, leader sequence peptide, N-formyl methionine.**

translational amplification a mechanism for producing large amounts of a polypeptide based upon prolonged mRNA lifetime. Since there is only a single copy of the ovalbumin gene per genome, large numbers of ovalbumin molecules produced by the cells of the chicken oviduct are generated by translational amplification.

translational control the regulation of gene expression through determining the rate at which a specific RNA message is translated.

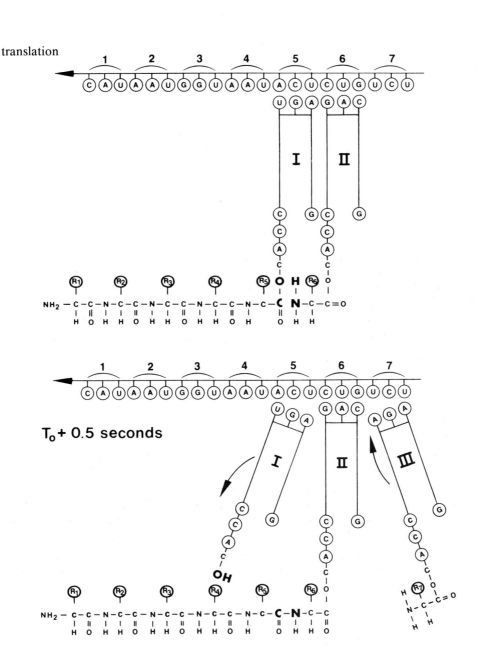

translation

1 2 3 4 5 6 7

T_0 + 0.5 seconds

translocase   a protein that forms a complex with GTP and the ribosome. Translocation of the charged tRNA from the A site to the P site is coupled with the hydrolysis of GTP to GDP and with the release of the translocase. *See* **elongation factors, translation.**

translocation   1. a chromosome aberration which results in a change in position of a chromosomal segment within the genome, but does not change the total number of genes present. An *intrachromosomal translocation* is a three-break aberration that results in the transposition of a chromosomal segment to another region of the same chromosome. Such an aberration is often called a *shift. Interchromosomal translocations* involve interchanges between nonhomologous chromosomes. A *reciprocal translocation* is a two-break aberration that reuslts in an exact interchange of chromosomal segments between two nonhomologous chromosomes and produces two monocentric translocated chromosomes. A *non-*

reciprocal or aneucentric translocation is a two-break aberration that results in a dicentric and an acentric translocated chromosome. A three-break interchromosomal translocation can generate a deficient chromosome and a recipient chromosome containing an intercalated segment of the other nonhomologous chromosome. This would be called an *insertional translocation*.

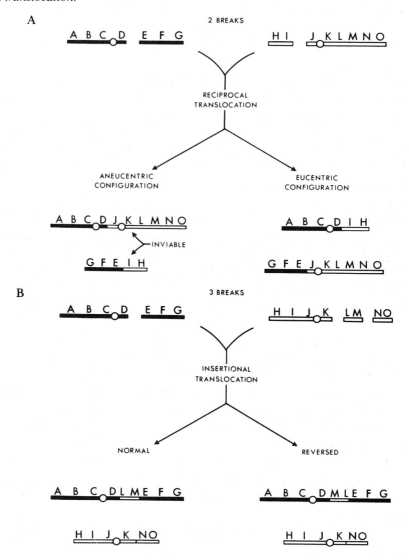

2. the movement of mRNA through a ribosome during translation (*q.v.*). Each translocation exposes a mRNA codon in the A site for base pairing with a tRNA anticodon.

translocation Down syndrome   familial Down syndrome, caused by having three copies of chromosome 21—two as separate chromosomes and one translocated to another chromosome, usually number 14. In families where one parent is a translocation heterozygote, the probability of having a child with Down syndrome is 0.33.

translocation heterozygote   an individual or cell in which two pairs of homologous chromosomes have reciprocally exchanged nonhomologous segments between one member of each pair. Each chromosome pair contains both homologous and nonhomologous segments, i.e., one normal (untranslocated chromosome) and one translocated chromosome. A translocation heterozygote forms a quadrivalent chromosomal association during pachynema (*see* **meiosis**), and the subsequent segregation of the four chromosomes is determined by their centromere orientations.

adjacent disjunction

alternate disjunction

In the case of *adjacent disjunction,* each daughter cell receives a normal and a translocated chromosome. The gametes produced from such cells are inviable because they contain certain genes in duplicate and are deficient for others. In the case of *alternate disjunction,* both translocated chromosomes go to one cell and both normal ones to the other. The gametes that result are viable because they contain all of the genetic material.

translocation mapping   gene mapping by the use of translocation chromosomes as markers. The semisterility that is usually associated with structural translocation heterozygotes is a phenotypic marker that can be used to locate the position of the breakpoint of the translocation relative to other more conventional gene markers.

transmission electron microscope   *See* **electron microscope.**

transmission genetics   that part of genetics concerning the mechanisms involved in the transfer of genes from parents to offspring.

transmutation   the transformation of one element into another accompanying radioactive decay (*q.v.*).

transplantation   transfer of a part of an organism to another organism or to another position in the same organism. In zoology the term is used interchangeably with graft (*q.v.*). In botany, graft is used in the above sense, and transplantation is used in the sense of planting again in a different place.

transplantation antigen   a protein coded by genes in the major (e.g., human HLA, mouse H2) or minor histocompatibility loci and present on most vertebrate cells. Transplantation antigens are targets for T lymphocytes if the graft bears antigens different from that of the host.

transposable elements   a class of DNA sequences that can move from one chromosomal site to another. This movement requires a transposase and a resolvase, that recognize short

397

nucleotide sequences that are repeated in inverted order at both ends of the element. Transposable elements were first identified in maize by Barbara McClintock (*see* **Activator-Dissociation system**) and called *controlling elements*. It is now known that McClintock's *Ac* locus is a 4.6 kb transposable element and that *Ds* is a defective DNA segment, derived from an *Ac* element by a short deletion in its transposase gene. Transposable elements were next detected in bacteria as *insertion sequences* (*q.v.*), and more recently in yeast (*see* **ty elements**) and *Drosophila* (see **copia elements, P elements**).

transposase   an enzyme that catalyzes the insertion of a transposon.

transposition   the process of insertion of a replica of a transposable element at a second site; another replica remains at the original insertion site.

transposons   one kind of transposable element in both prokaryotes and eukaryotes that is immediately flanked by inverted repeat sequences, that in turn are immediately flanked by direct repeat sequences. Transposons usually possess genes in addition to those needed for their insertion (e.g., genes for resistance to antibiotics, sugar fermentation, etc.).

transvection effect   the ability of a gene to influence the activity of an allele on the opposite homologue only when the two holomologous chromosomes are synapsed. Chromosomal rearrangements that prevent synapsis between homologous chromosome regions tend to prevent the expression of the wild-type allele in heterozygotes. In *Drosophila melanogaster,* when trans-heterozygotes for certain mutants of the *bithorax* complex are made structurally heterozygous for chromosomal rearrangements, a cytologically observable disruption in pairing of homologous chromosomes is produced in the *bithorax* region. Under such circumstances the mutant phenotype is made more extreme. Not to be confused with transfection (*q.v.*). *See* **cis, trans configurations.**

transverse axis or section   the plane or section that cuts across the body at right angles to the frontal and sagittal planes.

transversion   *See* **base-pair substitution.**

trend   an apparently directional evolutionary change in a character within a lineage; a chronocline. For example, many mammalian lineages show a tendency to increase in size over part of their evolutionary history. *See* **orthogenesis.**

triallelic   referring to a polyploid in which three different alleles exist at a given locus. In a tetraploid $A_1A_2A_3A_3$ and $A_1A_2A_2A_3$ would be examples.

Triassic   the most ancient of the Mesozoic periods lasting from 195 to 230 million years before present. *See* **geologic time divisions.**

Tribolium   a genus of flour beetles containing *T. castaneum* and *T. confusum,* genetically well known species. At present about 125 mutations are known for *castaneum* and 75 for *confusum.*

tricarboxylic acid cycle   *See* **citric acid cycle.**

trichloracetic acid   a compound commonly used to precipitate proteins during biochemical extractions.

$$Cl-\underset{\underset{Cl}{|}}{\overset{\overset{Cl}{|}}{C}}-COOH$$

trichocyst   the minute protrusible, spindle-shaped organelles in the ectoplasm of many ciliates.

trichogen cell   a large cell which secretes the long, tapering hair of the insect bristle. A smaller tormogen cell forms the circular chitinous socket around the base of the bristle.

trichogyne   a receptive hypha extending from the fruiting body of a fungus such as *Neurospora.*

triethylenethiophosphoramide   a mutagenic, alkylating agent. *See* **Thio-tepa.**

trihybrid cross   a mating that involves differences at three loci; the mated parents are triple heterozygotes.

triiodothyronine   *See* **thyroid hormones.**

trimester   one-third of the human nine-month gestation period; cited as first, second, and third trimesters.

trimethylpsoralen   a low molecular weight, planar molecule that undergoes photochemical reactions with pyrimidines. Trimethylpsoralen molecules intercalate into double-stranded DNA. Upon exposure to ultraviolet light, the molecules attach covalently to pyrimidines forming both monoadducts and interstrand cross links. Trimethylpsoralen has no effects on proteins.

trioecy   a sexual trimorphism in plants that can exist either as male, female, or bisexual individuals.

triparental recombinant   a progeny phage containing marker genes derived from each of three different phages which simultaneously infected the host cell. The observation of triparental recombinants demonstrates that repeated recombinational events occur in the infected cell between replicating nucleic acid molecules derived from all parent viruses. *See* **Visconti-Delbrück hypothesis.**

tripartite ribbons   synaptonemal complexes (*q.v.*).

triplet   a unit of three successive bases in DNA or RNA which codes for a specific amino acid. *See* **amino acid, genetic code, translation.**

triplet code   a code in which a given amino acid is specified by a set of three nucleotides. *See* **amino acid, genetic code, translation.**

triplex   *See* **autotetraploid.**

triploid   an organism having three haploid sets of chromosomes in each nucleus.

Tripsacum   a wild grass that can be hybridized with corn.

tris buffer   a commonly used buffer employing tris (hydroxymethyl) aminomethane as the buffering agent.

triskelion   *See* **endocytotic vesicle.**

trisomic   an organism that is diploid but contains one extra chromosome, homologous with one of the existing pairs, so that one kind of chromosome is present in triplicate. *See* **Down syndrome, metafemale.**

trisomy 21 syndrome   Down syndrome (*q.v.*).

trisomy 18 syndrome   *See* **Edward syndrome.**

tritanomaly   *See* **color blindness.**

tritanopia   *See* **color blindness.**

tritiated thymidine, uridine   *See* **tritium.**

Triticum   the genus containing the various species of wheat (*q.v.*).

tritium the radioactive isotope of hydrogen, with a half-life of 12.46 years. Tritium-labeled thymidine and uridine are often used to tag newly synthesized DNA and RNA, respectively. Tritium is the radioisotope of choice in radioautography because it emits an extremely weak beta particle when it undergoes radioactive decay. In a medium of unit density, the average tritium beta particle will penetrate only 1 μm. Therefore, in autoradiographs of tritium-labeled cells, silver grains of the photographic emulsion will be localized within 1 μm of the decaying atoms. Symbolized ^{3}H.

Triturus a genus of salamanders that are studied for their lampbrush chromosomes (*q.v.*). Species for which working chromosome maps are available include *T. alpestris, T. cristatus, T. helveticus, T. italicus, T. marmoratus,* and *T. vulgaris.* Another favorite, *T. viridescens,* has been renamed *Notophthalmus viridescens* (*q.v.*).

trivalent an association of three homologous chromosomes in meiosis.

tRNA transfer RNA (*q.v.*).

tRNA genes genes that are transcribed into tRNAs. Most tRNA genes are present in multiple copies. For example, the average repetition frequency for each tRNA gene species in yeast, *Drosophila,* humans, and *Xenopus,* has been estimated to be 5, 10, 15, and 200, respectively, per haploid genome.

tRNA isoacceptors those tRNAs that accept the same amino acid, yet differ in primary sequence, either in the anticodon, or in other regions of the molecule, or both. Such tRNA isoacceptors that differ in nucleotide sequences are generally encoded by different genes.

T4 RNA ligase an enzyme from bacteriophage T4 that catalyzes an ATP-dependent covalent joining of 5′ phosphate and 3′ hydroxyl termini of oligoribonucleotides.

tRNA suppressor *See* **nonsense suppressor.**

tRNA synthetase recognition site the site on the tRNA molecule which is bound to the aminoacetyl synthetase. In the case of yeast phenylalanine tRNA this region is located adjacent to the dihydrouridine loop and consists of the nucleotides

$$3' - \text{CUCGA} - 5'$$
$$| \ | \ | \ |$$
$$5' - \text{GAGC} - 3'. \ See \ \textbf{amino acid activation, transfer RNA.}$$

trophoblast the non-embryonic part of the blastocyst, which attaches to the uterine wall and later develops into the fetal portion of the placenta.

trophocyte nurse cell (*q.v.*).

tropomyosin a protein, associated with actin in muscle cells, that participates in regulation of contraction. *See* **isoforms.**

troponin a spherical, 70 kd protein that is bound to actin filaments.

trp tryptophan. *See* **amino acid.**

truncation selection a method employed by breeders in which members of a population are chosen for mating, saving only those whose phenotypic merit (on a quantitative scale) is either above or below a certain value (the truncation point).

Trypanosoma a genus of flagellated protozoa containing the species responsible for African sleeping sickness and Chagas disease.

trypsin a proteolytic enzyme that cleaves peptide chains on the carboxyl side of lysine and arginine residues only; initially secreted from the pancreas in the inactive form of trypsinogen, and activated to trypsin by the enzyme enterokinase in intestinal juice.

tryptophan   *See* **amino acid.**

tryptophan synthetase   the enzyme that catalyzes the union of indole and serine to form tryptophan.

TSH   thyroid-stimulating hormone (*q.v.*).

ts mutation   temperature sensitive mutation (*q.v.*).

T suppressor cell   *See* **suppressor T cell.**

t **test**   Student *t* test (*q.v.*).

tube nucleus   the vegetative nucleus that resides in a growing pollen tube.

tuber   the swollen end of an underground stem that contains stored food material and serves as an organ of vegetative propagation (the potato, the ranunculus, and the tuberous begonia produce tubers).

tubulin   the principal protein component of microtubules. Tubulin is a dimer composed of alpha and beta subunits. The genes coding for each of the subunits show heterogeneity. In *Drosophila,* for example, there are four different alpha genes, each occupying a different locus on chromosome 3. It follows that there are different families of alpha subunits and consequently different classes of microtubules.

tumor   a clump of cells (usually disfunctional) due to abnormal proliferation. Benign tumors are not life-threatening (e.g., warts); malignant tumors are potentially lethal cancers.

tumor specific transplantation antigen   an antigen found on a tumor cell, but not on the normal cells or tissues of the individual in which it arose and which can lead to immune rejection of the tumor *in situ* or if transplanted.

tumor virus   *See* **oncogenic virus.**

Turbatrix aceti   a free-living nematode whose developmental genetics is under study, primarily for comparison with *Caenorhabditis elegans* (*q.v.*). Its cell lineage is known in part.

turbid plaque   *See* **plaque.**

Turner syndrome   a group of abnormalities in humans due to monosomy for the X chromosome. Such individuals are female in phenotype, but are sterile. The ovaries are rudimentary or missing. In the mouse, the equivalent condition (XO) does not have detrimental effects except for a possible shortening of the female breeding life.

turnover   the dynamic replacement of atoms in a tissue or organism without any net change in the total number of atoms.

turnover number   the number of molecules of a substrate transformed per minute by a single enzyme molecule under optimal conditions.

Tween 80   the trademark for polyoxyethylene sorbitan mono-oleate (also called Polysorbate 80); used as an emulsifier and dispersing agent for many purposes.

twins   pairs of individuals produced at one birth. *See* **dizygotic twins, monozygotic twins.**

twin spots   paired patches of tissue genetically different from each other and from the background tissue, produced by mitotic crossing over in an individual of heterozygous genotype during development of the organism. In the *Drosophila* situation diagrammed below, twin spots occur on a female of genotype *y* +/+ *sn*. Most of the thorax and abdomen show wild-type pigment and long, black bristles, but there are adjacent patches of yellow (drawn to appear pale) and singed tissue. These twin spots are clones of cells derived from the reciprocal products of an exchange between the singed gene and the centromere.

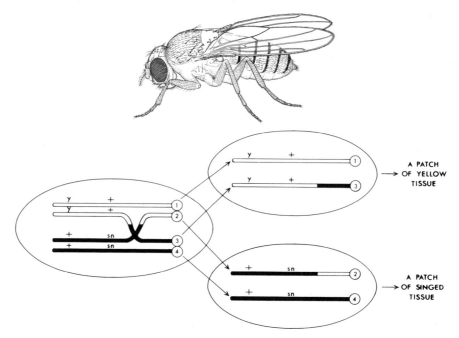

Twin spots or sectors can also result from transposition of a controlling element to an unreplicated recipient site. If the recipient site has been replicated, twin spots result only if the target lies on the chromatid opposite that of the donor site. For example, in maize, transposition of Ds (*see* **Activator-Dissociation system**) usually is accompanied by its disappearance from the donor site (contrary to the mode of transposition in bacteria). Transposition of Ds almost always occurs after the donor element has been replicated. Therefore, twin sectors normally result from transposition because one chromatid has either one or no element and the sister chormatid has two elements.

twisting number   the number of base pairs in duplex DNA divided by the number of base pairs per turn of the helix.

two fold rotational symmetry   the situation in which the two DNA strands of a duplex have the same base sequence when read with the same polarity. For example, the Eco RI *(q.v.)* endonuclease recognizes a DNA segment six base pairs long that has two fold rotational symmetry. *See* **palindrome.**

two genes–one polypeptide chain   participation of more than one gene in the coding of a given polypeptide chain, as occurs in the production of immunoglobulins. Separate genetic elements in the same chromosome coding for the variable, joining, diversifying, and constant regions of the heavy chain are brought together to form a functional gene. *See* **immunoglobulin.**

two-point cross   a genetic recombination experiment involving two linked genes.

two-strand double exchange   *See* **inversion.**

ty elements   transposable elements of the yeast *Saccharomyces cerevisiae*. The symbol *ty* comes from *t*ransposon-*y*east. There are 30–35 ty elements per haploid yeast genome, and each consists of a central region of about 5.6 kb pairs of DNA flanked by direct repeats about 330 base pairs long. *See* **transposable elements.**

type specimen   the individual chosen by the taxonomist to serve as the basis for naming and describing a new species.

typological thinking   the consideration of the members of a population as replicas of a hypothetical type, while disregarding all variations.

typology   the Platonic/Aristotelian views applied to biology, basing classification upon "types" thought to represent the Platonic *eidos* (entities existing in the world of ideas and only imperfectly mirrored in biological organisms).

tyr   tyrosine. *See* **amino acid.**

tyrosinase   an enzyme that converts tyrosine to dopa and oxidizes this to dopa quinone. Dopa quinone molecules undergo spontaneous polymerization to form melanin. *See* **albinism.**

tyrosine   *See* **amino acid.**

tyrosinemia   a hereditary disease in man arising from a deficiency of the enzyme p-hydroxyphenylpyruvic oxidase.

U

U   **1.** uracil or uridine. **2.** uranium.

ubiquinone   coenzyme Q (*q.v.*).

ubiquitin   an acidic protein of 76 residues present in all cells from bacteria to mammals. It probably lies at the surface of eukaryotic nucleosomes, but does not have any discernible effect on nucleosome structure. Ubiquitin appears to be released from chromatin at mitosis. Very little is known of its function aside from its possible role in a cytoplasmic system for protein degradation.

ubiquitous RNA   a class of small nuclear RNAs (snRNA) found among the heterogeneous nuclear RNA component, but not derived from splicing out introns from primary transcripts. Ribosomal 5S and 5.8S rRNAs and transfer RNAs are examples of families of ubiquitous RNA of known function. Most other snRNAs are currently of unknown function, except the family designated 7S, which seems to play a role in secretion of proteins.

UDPG   uridine diphosphate glucose (*q.v.*).

ultracentrifuge   a powerful centrifuge that can attain speeds as high as 60,000 rpm and generate sedimenting forces 500,000 times that of gravity. The instrument is used to sediment macromolecules. *See* **centrifugation separation techniques, centrifuge, sedimentation constant.**

ultramicrotome   an instrument that cuts ultrathin sections (50-100 nm thick) of plastic embedded tissues using knives made from polished diamonds or from triangles of broken plate glass. *See* **knife breaker.**

ultrasonic   pertaining to ultrasound.

ultrasound   sound waves of frequency higher than the audible limit of about 20,000 vibrations per second.

ultra-violet absorption curve   the curve showing the relation between the relative amount of ultra-violet radiation absorbed by a solution of molecules and the wave length of the incident light.

ultra-violet microscope   an optical system utilizing ultra-violet radiation. Since glass filters out UV, quartz transmitting or glass reflecting lenses must be used in the UV microscope. Such a microscope has double the resolving power of the light microscope. Furthermore, if monochromatic UV of a wave length absorbed by nucleic acids (260mμ) is used, nucleic acid-rich structures may be photographed in unstained cells. In combination with a spectrophotometer the UV microcope provides a method for the quantitive estimation of nucleic acids in cells.

ultra-violet radiation   that part of the invisible electromagnetic spectrum just beyond the violet with wave lengths between 1000 and 4000 Å.

umbel   an inflorescence like that of the parsley in which the stems of the individual flowers all arise from the apex of the main axis.

uncoiling of chromomeres   the phenomenon seen during the production of loops by lampbrush chromosomes. This phenomenon suggests that the chromomere is a region where the DNA molecule is tightly coiled and that it can be spun out in preparation for transcription.

uncoupling agent   a substance (such as DNP) that can uncouple phosphorylation of ADP from electron transport.

underdominance   the unusual situation where a heterozygote shows an attribute, such as viability or fertility, that is lower than either homozygote.

underwinding   coiling of a DNA molecule in a left-handed direction, i.e., opposite to that of the double helix; negative supercoiling.

undulipodium   any cellular projection surrounding a cylindrical shaft containing a bundle of eleven microtubules, nine of which form a circle around the central pair, while the cortical microtubules are doublets. Cilia and flagellae (*q.v.*) are examples. *See* **axoneme.**

unequal crossing over   the phenomenon first described in *Drosophila* at the *Bar* locus. Improper pairing of the duplicated segments is thought to occur followed by crossing over. The result is one crossover chromatid with one copy of the segment and another with three copies.

ungulate   a hoofed mammal. Perissodactyla plus Artiodactyla. *See* **classification.**

uniformitarianism   a geological theory that "the present is the key to the past." In other words, the phenomena of volcanism, crustal movements, erosion, glaciation, etc., that can be seen today have been operating through billions of years of earth history, and they are the primary forces that have made the earth what it is today. *Contrast with* **catastrophism.**

unineme hypothesis   the concept that a newly formed chromatid contains only one DNA duplex extending from one end to the other. *Contrast with* **polyneme hypothesis.**

uniovular twins   monozygotic twins (*q.v.*).

uniparental inheritance   a phenomenon in which all offspring of a given mating seem to have received certain phenotypes only from one of the parents (usually the female) regardless of the genotype or phenotype of the other parent; such inheritance is usually the result of macromolecules or organelles stored in the cytoplasm.

unique DNA   a class of DNA determined by C_0t analysis to represent sequences that are present only once in the genome. Most structural genes and their introns are unique DNAs. *See* **reassociation kinetics.**

unisexual flower   a flower having only stamens or only carpels. A plant can bear either one or both kinds of unisexual flowers. *See* **flower.**

unit membrane   the trilamellar membrane seen when the sectioned plasma membrane is viewed under the electron microscope. The membrane, which has a total thickness of about 75 Å, appears as two strata each about 20 Å thick separated by a light interzone about 35 Å wide. *See* **fluid mosaic concept.**

univalent a single chromosome seen during meiosis when bivalents are also present. A univalent has no synaptic mate. An example of a univalent would be the sex chromosome of an XO male.

universal code theory the assumption that the genetic code is used exclusively by all forms of life. This is true with a few exceptions. In yeast mitochondria, CUA codes for threonine instead of leucine; and in mammalian mitochondria, AUA codes for methionine instead of isoleucine, and codons AGA and AGG signal termination instead of coding for arginine. UGA codes for tryptophan instead of signaling termination in mitochondria from both sources.

universal donor an individual with type O blood, who is able to donate blood to O, A, B, or AB recipients. *See* **blood group.**

universal recipient an individual with type AB blood, who can receive blood from AB, A, B, or O donors. *See* **blood group.**

univoltine *See* **voltinism.**

unordered tetrad *See* **nonlinear tetrad.**

unscheduled DNA synthesis DNA synthesis that occurs at some stage in the cell cycle other than the S period. *See* **interphase cycle.**

unstable equilibrium the situation where the equilibrium value for an allele in a population fluctuates because of temporary environmental changes. For example, these may lead to a sudden selection for or against the allele, or a drastic reduction in the size of the population may result in genetic drift away from the former equilibrium value.

unstable mutation a mutation with a high frequency of reversion. The original mutation may be caused by the insertion of a controlling element (*q.v.*), and its exit produces a reversion.

untwisting enzyme *See* **topoisomerase.**

unwinding proteins proteins that bind to, destabilize, and unwind the DNA helix ahead of the replicating fork. *See* **gene 32 protein.**

u orientation *See* **n orientation.**

up promoter mutations mutations in promoter sites that increase the rate of initiation of transcription; promoters with this property are called "high level or strong promoters."

upstream monomeric sequences of a linear polymer proceeding in the opposite direction from gene expression: (1) from the 5′ toward the 3′ end on the sense strand of duplex DNA; (2) from the 3′ toward the 5′ end of mRNA; (3) from the free carboxy toward the free amino end of a polypeptide chain. *Contrast* **downstream.**

uracil a pyrimidine base found only in RNA.

uracil fragments during DNA replication in *E. coli,* polymerases I and III occasionally make mistakes and incorporate dUTP instead of TTP. Several enzymes remove these uracils from both leading and lagging strands, creating "uracil fragments."

urea *See* **ornithine cycle.**

urethane a carcinogen that induces tumorous nodules in the lungs of mammals.

$$NH_2-C-O-CH_2-CH_3$$
$$\overset{\|}{O}$$

uric acid the end-product of nucleic acid catabolism in mammals; the main nitrogenous constituent of the urine of reptiles and birds.

uridine *See* **nucleoside.**

uridine diphosphate galactose *See* **uridine diphosphate glucose.**

uridine diphosphate glucose (UDPG) a compound acting as a coenzyme and substrate to various enzymes. UDPG may be converted to uridine diphosphate galactose by the enzyme epimerase. These coenzymes play essential roles in carbohydrate metabolism.

uridylic acid *See* **nucleotide.**

U1 RNA an RNA molecule 165 nucleotides long that is a component of the most abundant snRNP in mammalian cells. There are about a million of these particles per nucleus. The U1 RNA contains a segment adjacent to its 5′ cap which exhibits complementarity to the sequences at the donor and acceptor splice junctions of split genes. It is believed that the U1 RNA binds to the transcript at both ends of the segment corresponding to an intron causing it to loop into a structure that allows concurrent intron excision and exon splicing. *See* **lupus erythematosis, small nuclear RNAs.**

Ustilago a genus in the *Basidiomycota* (*see* **classification**). These are yeast-like smut fungi, and two species, *U. maydis* and *U. violacea,* have been subjects of genetic research, especially in terms of recombination-defective and radiation-sensitive mutations.

UV ultraviolet radiation (*q.v.*).

UV-induced dimers *See* **thymine dimers.**

UV-reactivation a phenomenon in which survival of an ultraviolet-irradiated lambda phage is greater on an irradiated host than on an unirradiated host. The repair mechanism involved in UV-reactivation utilizes an error-prone replication system of the host such as that of the SOS response (*q.v.*).

V

vaccine a suspension of dead or weakened bacteria or viruses injected into the body to immunize against the same pathogen.

vaccinia virus the DNA virus of cowpox which serves for vaccination against smallpox.

vacuum evaporator the vacuum chamber containing a set of electrodes through which a current can be passed to heat a metal foil placed between the electrodes. The heated metal evaporates, and atoms from it coat the specimen lying below the electrodes. The coating is cast at an angle, and as a consequence the specimen appears in relief when viewed under the electron microscope. A vacuum evaporator equipped with graphite electrodes is also employed to prepare the carbon films that sometimes are used to support ultrathin sections on specimen screens.

val valine. *See* **amino acid.**

valence a number representing the combining or displacing power of an atom; the number of electrons lost, gained, or shared by an atom in a compound; the number of hydrogen atoms with which an atom will combine, or which it will displace.

valine *See* **amino acid.**

valinomycin *See* **ionophore.**

van der Waals forces the relatively weak, short-range forces of attraction existing between atoms or molecules, caused by the interaction of varying dipoles.

var variety.

variable any organismal attribute that may have different values in various situations (e.g., between members of different species, between members of the same species, within an individual at different times, etc.).

variable domain that domain of an immunoglobulin light or heavy chain that has variable amino acid sequences within an individual.

variable region the N-terminal portion of an immunoglobulin chain that binds to antigen.

variance When all values in a population are expressed as plus and minus deviations from the population mean, the variance is the mean of the squared deviations.

variant an individual that is different from an arbitrary standard type (usually wild type) for that species. Variants are not necessarily mutants; for example, many birth defects are simply developmental accidents or environmentally induced. *See* **phenocopy.**

variate   a specific quantitative value of a variable (*q.v.*).

variation   divergence among individuals of a group, specifically a difference of an individual from others of the same species which cannot be ascribed to a difference in age, sex, or position in the life cycle. The variations of evolutionary significance are gene-controlled phenotypic differences of adaptive significance.

variegated position effect   *See* **position effect.**

variegation   **1.** irregularity in the pigmentation of plant tissues due to a variety of causes (virus infection, segregation of normal and mutant plastids, bridge-breakage-fusion-bridge cycles, transposable elements, etc.). **2.** irregularity in the pigmentation of animal tissues or their products (hair, feathers, etc.) due to a variety of causes (X chromosome inactivation, defective embryonic migration of melanocytes, localized physiological conditions, mitotic recombination, etc.).

variety   *See* **strain.**

vasopressin   a peptide hormone secreted by the hypothalamus and stored in the neurohypophysis, which constricts arterioles and promotes resorption of water by the kidney tubules.

vector   **1.** a self-replicating DNA molecule that transfers a DNA segment between host cells; also called a "vehicle." *See* **plasmid cloning vector. 2.** an organism (such as the malaria mosquito) that transfers a parasite from one host to another.

vegetal hemisphere   the surface of the amphibian egg farthest from the nucleus, the yolk-rich hemisphere of the egg.

vegetative   designating a stage or form of growth, especially in a plant, distinguished from that connected with reproduction.

vegetative cell   an actively growing cell, as opposed to one forming spores.

vegetative nucleus   **1.** the macronucleus of a ciliate. **2.** the tube-nucleus of a pollen grain.

vegetative petites   *See* **petites.**

vegetative reproduction   in plants, the formation of a new individual from a group of cells, without the production of an embryo or seed. More generally, asexual reproduction. *See* **agamospermy, apomixis.**

vegetative state   the noninfective state during which a phage genome multiplies actively and controls the synthesis by the host of the materials necessary for the production of infective particles.

vehicle   a plasmid or bacteriophage possessing a functional replicator site, and containing a genetic marker to facilitate its selective recognition, used to transport foreign genes into recipient cells during recombinant DNA experiments; also called a *vector.*

ventral   the side of an animal which is normally directed downwards with reference to gravity.

ventriculus   the midgut; the digestive portion of the insect intestine.

vermilion   a sex-linked, recessive eye color mutation in *Drosophilia melanogaster.* This was the first *Drosophila* mutation to be understood biochemically. Flies lacking the plus allele of *vermilion* have an enzymatic defect that makes them unable to convert tryptophan to formylkynurenine (*q.v.*). If the latter compound is supplied in the larval diet, the adults

410

that develop show the normal eye color. Formylkynurenine is converted through a series of steps to hydroxykynurenine (*q.v.*), and eventually to xanthommatin (*q.v.*).

vermilion plus substance formylkynurenine; so called because the synthesis of this compound in *Drosophila melanogaster* is controlled by the plus or wild-type allele of the gene *vermillion* (*q.v.*).

vernalization the treatment of germinating seeds with low temperatures to effect their flowering. Winter varieties of certain cereals, if vernalized, can be sown in the spring and harvested in the summer.

veronal buffer a barbital buffer commonly used in the preparation of fixatives for electron microscopy.

Veronica a large genus of hardy herbs belonging to the family Scrophulariaceae. Classic studies on the genetic control of self-sterility were performed on this species.

verruca an epidermal tumor of viral origin; a wart.

Versuche über Pflanzen-Hybriden (*Experiments on Plant-Hybridisation*). The title of Mendel's epoch-making paper published in 1866.

vertical classification a system of classification that recognizes taxa corresponding to clades and groups transitional forms with their descendents rather than with their ancestors; the opposite of *horizontal classification* (*q.v.*).

vertical evolution the process whereby an ancestral species changes through time (without splitting) to become distinctively different, and therefore recognized as a new species; phyletic evolution. *See* **speciation.**

vertical transmission passage of genetic information from one cell or individual organism to its progeny by conventional heredity mechanisms (mitosis, meiosis), in contrast to transfer of genetic information by infection-like processes between contemporary individuals (horizontal transmission).

vestigial pertaining to a degenerate or imperfectly developed organ that is no longer functionally useful.

V gene one of many (perhaps hundreds) of genes coding for the variable (N-terminus) region of an immunoglobulin chain.

viability a measure of the number of individuals surviving in one phenotypic class relative to another class, taken as standard, under specified environmental conditions.

vicariance distribution a discontinuous biogeographical distribution of organisms that previously inhabited a continuous range. The current gaps in the distribution were caused by some extrinsic factor (geologic or climatic).

Viceroy butterfly *Basilarchia archippus* (*q.v.*).

Vicia faba the broad bean. A plant often used in cytogenetics because its cells contain a small number (N = 6) of large chromosomes. The semiconservative nature of DNA replication was first demonstrated by analyzing autoradiographs of ^{3}H thymidine-labeled chromosomes from cells of *Vicia* root tips.

villus a finger-like projection extending from an epithelium. Such a villus is composed of many cells. *Contrast with* **microvillus.**

vimentin a 55,000 d cytoskeletal protein commonly found in fibroblasts. In glial cells it is copolymerized with an acidic protein of 50,000 d molecular weight, while in muscle cells it is combined with desmin (*q.v.*).

411

vinblastine a vinca alkaloid (*q.v.*).

vinca alkaloids mitotic poisons isolated from *Vinca rosea.*

vincaleukoblastine a vinca alkaloid (*q.v.*).

Vinca rosea the Madagascar periwinkle, source of the vinca alkaloids.

vincristine a vinca alkaloid (*q.v.*).

vinculin a fibrous protein responsible for anchorage of actin filaments to the inner side of the cell membrane. Vinculin is located in patches called *adhesion plaques* on the cell membrane that are thought to be responsible for intercellular adhesion. Cells infected with Rous sarcoma virus (*q.v.*) produce a kinase that phosphorylates the tyrosine residues of vinculin. It is hypothesized that the phosphorylation of vinculin both destabilizes actin linkages (allowing transformed cells to become rounded) and weakens intercellular adhesion (allowing metastasis).

viral-specific enzyme any enzymes produced in the host cell after viral infection and coded by viral genes.

virion a completed phage particle consisting of a nucleic acid core and a protein coat.

viroid a disease-causing agent of plants consisting of single-stranded RNA molecules typically 270–380 nucleotides long and therefore thousands of times smaller than the most diminutive virus. Viroids are not encapsulated in a protein coat. Viroids do not pass through a DNA stage in their life cycle; they are replicated directly as RNA and are not integrated into the host genome. Typical viroids are the planta macho and apical stunt viroids of tomatoes and the spindle tuber viroid of potatoes.

virulence the realitve ability of an organism to produce disease.

virulent phage a phage which causes lysis of the host bacterium. *Contrast with* **temperate phage.**

virus an ultramicroscopic, obligate, intracellular parasite incapable of autonomous replication. Viruses can only reproduce by entering a host cell and using its translational system. Viruses probably orginated as nucleic acids that escaped from cells. Viruses are generally classified according to the type of nucleic acid they contain and the morphology of the nucleocapsid (*q.v.*). Examples are given below.

NUCLEIC ACID TYPE	CAPSID MORPHOLOGY	EXAMPLES
RNA viruses	helical symmetry	tobacco mosaic virus, influenza viruses, potato virus Y
	cubical symmetry	polio virus, reoviruses, retroviruses
DNA viruses	helical symmetry	smallpox
	cubical symmetry	polyoma virus, ϕX174, Shope papilloma virus
	with head and tail	T phages, λ, P22

virus receptors sites on the cell membrane to which viruses attach. Such sites contain neuraminic acid (*q.v.*).

Visconti-Delbrück hypothesis according to this proposal, bacteriophages multiply upon entering a host, and the replicating units so formed mate repeatedly. Mating occurs in pairs and is at random with respect to the pairing partner. During any given mating cycle a segment of genetic material from one parent can exchange with that from a second parent phage yielding recombinant units.

visibles referring to phenotypically observable mutants, as opposed to lethals, which are scored by the absence of an expected class of individuals in a cross designed to detect induced mutants.

vitalism a philosophy holding that the phenomena exhibited in living organisms are the result of special forces distinct from chemical and physical ones. *See* **mechanistic philosophy.**

vital stain a dye used to stain living cells (Janus green, methylene blue, trypan blue, etc.).

vitamin an organic compound (often functioning as a coenzyme) that is required in relatively minute amounts in the diet for the normal growth of a given organism.

vitamin A the antiblindness vitamin. *See* **retinene.**

vitamin B complex a family of water-soluble vitamins, including thiamin (B_1), riboflavin (B_2), nicotinic acid, pantothenic acid, pyridoxin (B_6), and cobalamin (B_{12}).

vitamin C ascorbic acid, an important regulator of the oxidation-reduction state of protoplasm.

vitamin D calciferol; a fat soluble vitamin required in man for the prevention of rickets. The vitamin affects the absorption and deposition of calcium and phosphate.

413

vitamin D-resistant rickets   a hereditary disease in man inherited as a sex-linked dominant due to a defect in the renal transport mechanism for phosphorus. The condition is indistinguishable from ordinary rickets except that the symptoms occur in the presence of adequate dietary vitamin D.

vitamin E   alpha tocopherol, a vitamin functioning as an antioxidant.

vitamin H   biotin (*q.v.*).

vitellarium   the portion of the insect ovariole posterior to the germarium. Egg chambers complete development within the vitellarium.

vitelline membrane   a membrane that surrounds the ovum. In *Drosophila* the term is used specifically for the membrane that immediately surrounds the oolemma and is formed by the fusion of deposits in the intercellular space between the oocyte and the columnar follicle cells that invest it.

vitellogenesis   the formation of yolk.

vitellogenic hormone   *See* **allatum hormone.**

vitellogenin   a protein synthesized by vitellogenic females and incorporated into the yolk spheres of the developing oocyte. In *Xenopus laevis,* vitellogenins are synthesized by the liver. In *Drosophila melanogaster,* vitellogenins are synthesized by abdominal and thoracic fat bodies and by the columnar follicle cells surrounding the oocyte. *See* **lipovitellin, phosphovitin.**

viviparous   **1.** producing living young rather than eggs. Embryogenesis occurs within the mother's body, as with most mammals. **2.** bearing seeds that germinate within the fruit, as in the mangrove.

Viviparus malleatus   a prosobranchiate snail showing a bizarre type of spermatogenesis with the production of oligopyrene sperm (*q.v.*).

vixen   a female fox.

vol   volume.

voltinism   a polymorphism in terms of whether or not the embryos produced by an insect enter diapause (*q.v.*). For example, in *Bombyx mori* univoltine strains produce only diapause embryos. Bivoltine strains porduce a nonhibernating brood, then diapause embryos.

von Gierke disease   a hereditary glycogen storage disease in humans arising from a deficiency of the enzyme glucose-6-phosphatase. Inherited as an autosomal recessive. Prevalence 1/100,000.

vulgare wheats   *See* **wheat.**

Vulpes vulpes   the red fox, a species bred on a large scale on ranches for its pelt. Numerous mutations influencing fur color are known.

W

Wahlund effect the tendency for a large population that is subdivided into many small panmictic units with different gene frequencies to exhibit a deficiency of heterozygotes and an excess of homozygotes in comparison with expectations from applying the Hardy-Weinberg formula utilizing gene frequencies averaged from all subgroups. The phenomenon, discovered by S. Wahlund, is the result of inbreeding within each subgroup.

Wallace effect the hypothesis put forth by Alfred Russel Wallace that natural selection favors the evolution of mechanisms that ensure the reproductive isolation of sexual populations that have reached the level of elementary biological species. Reproductive isolation prevents the production of sterile hybrids which compete for food reserves.

waltzer one of the many neurological mutants in the laboratory mouse. Homozygotes are deaf and characteristically show a circling and head shaking behavior.

Warburg apparatus an apparatus consisting of a manometer (*q.v.*) to which a Warburg vessel can be attached and immersed in a water bath at constant temperature.

Warburg vessel the chamber in which respiring tissue slices or organisms are placed for determinations in the Warburg apparatus (*q.v.*) of the quantity of gas liberated or consumed.

Waring blender an electric kitchen appliance used to homogenize mixtures, but used in the laboratory to generate the shearing forces required to detach conjugating bacteria, to strip bacteriophages or their "ghosts" from host cell surfaces, to homogenize tissue samples, etc.

warning coloration conspicuous colors or markings on an animal that is poisonous, distasteful, or similarly defended against predators. Such coloration is presumed to facilitate learning, on the part of predators, to avoid the possessor of such markings.

Watson-Crick model *See* **deoxyribonucleic acid.**

wax any esters of fatty acids and long chain monohydroxyalcohols.

weak interactions forces between atoms, such as ionic bonds, hydrogen bonds, and van der Waals forces, which are weak relative to covalent bonds (*q.v.*).

weighted mean the mean obtained when different classes of observations or quantities are given different weights (are multiplied by different factors) in the calculation.

Weismannism the generally accepted concept proposed August Weismann that acquired characters are not inherited and that only changes in the germ plasm are transmitted from generation to generation.

western blotting   *See* **Southern blotting.**

Whatman   the trademark for a family of widely used filter papers.

wheat   the world's most important grain crop, produced by species of the genus *Triticum*. Wheat species are generally grouped together on the basis of their chromosome numbers. The einkorn wheats are diploids (N = 7), the emmer wheats are tetraploids (N = 14), and the vulgare wheats are hexaploids (N = 21). The hexaploids, of which *T. aestivum* is the most common example, contain three genomes, A, B, and D, each of which is comprised of a set of seven pairs of nonhomologous chromosomes. The A, B, and D genomes were derived from *Triticum monococcum, T. searsii,* and *T. tauschii,* respectively. See **einkorn, emmer, durum, spelt, wild einkorn,** and **wild emmer wheat,** *Aegilops squarrosa.*

whelp   an unweaned puppy.

white leghorn   *See* **plumage pigmentation genes.**

white plymouth rock   *See* **plumage pigmentation genes.**

whole-arm fusion, whole arm transfer   *See* **centric fusion.**

whole chromatid conversion   a type of gene conversion in which a 6:2 ratio is observed. *See* **chromatid conversion.**

wild einkorn wheat   *Triticum aegilopoides* (N = 7). *See* **einkorn wheat, wheat.**

wild emmer wheat   *Triticum dicoccoides* (N = 14). *See* **wheat.**

wild type   the most frequently observed phenotype, or the one arbitrarily designated as "normal."

wild type gene   the allele commonly found in nature or arbitrarily designated as "normal."

Williston's rule   during evolution of the serially repeated parts of an organism, the number of such parts tends to diminish as the remaining parts diversify in response to division of labor.

Wilms tumor   a maligant renal tumor of young children associated with a deletion of band p13 on chromosome 11.

Wilson disease   a disease of human beings inherited as an autosomal recessive in which the biliary excretion of copper and its incorporation into ceruloplasmin is impaired. Copper accumulation leads to liver and brain damage. Prevalence 1/75,000.

winter variety   a variety of cereal that must be sown in the autumn of the year preceding that in which it is desired that the plants should flower. If sown in the spring, they will not flower in the same growing season.

wobble hypothesis   a hypothesis developed by F.H.C. Crick to explain how one tRNA may reconize two codons. The anticodon in each tRNA is a base triplet. The first two bases of the mRNA codon pair according to the base pairing rules. The third base in the anticodon, however, has a certain amount of play or wobble which permits it to pair with any one of a variety of bases occupying the third position of different codons. Thus U in the third position would recognize A or G, for example, and transfer RNA with a CUU anticodon would bind to either of two codons (GAA or GAG).

Wolman disease   a hereditary disease in man arising from a deficiency of the lysosomal enzyme acid lipase; due to a recessive gene on chromosome 10.

woodpecker finch *Camarhynchus pallidus,* the famous finch of the Galapagos Islands which uses a tool (a cactus spine) to dislodge insects from bark crevices.

working hypothesis a hypothesis that serves as the basis for future experimentation.

W reactivation *See* **UV-reactivation.**

Wright's inbreeding coefficient (F) the probability that two allelic genes united in a zygote are both descended from a gene found in an ancestor common to both parents. Also the proportion of loci at which an individual is homozygous.

Wright's polygene estimate in the case of quantitative inheritance (*q.v.*), the number of segregating pairs of polygenes (n) can be estimated from the variances computed for the F_1 and F_2 populations. If these are symbolized $s_{F_1}^2$ and $s_{F_2}^2$ respectively, and the means for the high and low P_1 strains are X_h and X_l, respectively, then

$$n = \frac{(X_h - X_l)}{8(s_{F_2}^2 - s_{F_1}^2)}$$

See **quantitative inheritance.**

writhing number the number of times the axis of a DNA molecule crosses itself by supercoiling.

w/v a concentration expressed as a weight per unit volume.

W, Z chromosomes the sex chromosomes of an animal in which the female is the heterogametic sex (*Bombyx mori,* for example). In such cases the W chromosome is female determining and the male is ZZ.

X

x *See* **basic number.**

X "crossed with" (as in A ♀ × B ♂).

X₂ the offspring of an F_1 test cross.

xantha any of many chloroplast mutations in various cereal species. The *xantha* 3 mutant of barley, for example, is characterized by chloroplasts that accumulate excessive numbers of pigment granules and never develop orderly arrays of grana.

xanthommatin the brown ommochrome pigment of the compound eye of *Drosophila*. It is formed by the condensation of two molecules of hydroxykynurenine (*q.v.*).

X chromosome the sex chromosome found in double dose in the homogametic sex and in single dose in the heterogametic sex.

X chromosome inactivation in mammalian development, the repression of one of the two X chromosomes in the somatic cells of females as a method of dosage compensation. At an early embryonic stage in the normal female, one of the two X chromosomes undergoes

418

inactivation, apparently at random. From this point on, all descendent cells will be clonal in that they will have the same X chromosome inactivated as the cell from which they arose. Thus the mammalian female is a mosaic composed of two types of cells—one which expresses only the paternal X chromosome, and another which expresses only the maternal X chromosome. In some cells and tissues, the inactivated X chromosome can be seen as a dense body in the nucleus (referred to as a Barr body or sex chromatin). In abnormal cases where more than two X chromosomes are present, only one X remains active and the others are inactivated. In marsupials, the paternal X is selectively inactivated during female development. Recent studies have shown that certain genes located distally on the short arm of the human X chromosome escape inactivation. *See* **H-Y antigen, Lyon hypothesis.**

xenia referring to the situation in which the genotype of the pollen influences the developing embryo or the maternal tissue of the fruit so as to produce a phenotypically demonstrable effect upon the seed.

xenogeneic transplantation xenoplastic transplantation (*q.v.*).

xenoplastic transplantation the transplantation between individuals of different genera or widely distant species.

Xenopus a genus of African aquatic anurans containing two species, *X. laevis* and *X. borealis,* that are favorites for research in molecular genetics. Studies on the nucleolar mutants of *X. laevis* led to the conclusion that the nucleolus contains about 450 genes that function to transcribe rRNA. The 5S component of rRNA is coded for by genes that cluster near the tips of the long arms of all the chormosomes. *X. laevis* and *X. borealis* have about 24,000 and 9,000 5S rRNA genes, respectively. *See* **rDNA amplificaton.**

xeroderma pigmentosum a group of hereditary diseases inherited as autosomal recessives in which the skin is extremely sensitive to sunlight or ultraviolet light, and death is usually due to skin cancer. Normal skin cells can repair UV damage to DNA by cut-and-patch repair (*q.v.*). Skin cells from patients with xeroderma pigmentosum cannot perform such repair and apparently lack a UV-specific endonuclease. It appears that patients with the symptoms of xeroderma pigmentosum may be homozygous for mutations occurring at any one of at least seven different chromosomal sites.

xerophyte a drought-resistant plant (a cactus, for example).

Xg blood group a blood group defined by an antigen controlled by a gene located distally on the short arm of the human X chromosome.

X-inactivation *See* **X chromosome inactivation.**

Xiphophorus maculatus the platyfish, and *X. helleri,* the swordtail. Pigment cell genetics and the genetics of sex determination have been intensively studied in laboratory strains of these fresh water species.

X linkage the presence of a gene located on the X chromosome; usually termed "sex linkage" (*q.v.*).

XO the symbolic designation of the situation in some heterogametic organisms in which the X chromosome is present and the Y chromosome is absent.

XO monosomy *See* **Turner syndrome.**

x radiation radiations produced when high-speed electrons strike a metallic target. X-rays have wavelengths in the range between ultraviolet and gamma radiation and are ionizing radiations.

x-ray crystallography the use of the diffraction patterns produced by x-ray scattering from crystals to determine the three dimensional structure of the atoms or molecules in the crystal. *See* **large angle x-ray diffraction, small angle x-ray diffraction.**

XXY trisomy *See* **Klinefelter syndrome.**

xylem the vascular tissue which conducts water and mineral salts and provides mechanical support in vascular plants.

XYY trisomy a human karyotype observed in about 1 in 1000 male births. Most adult XYY males are over six feet tall. A few are sterile, and some are mentally retarded or have behavioral disorders. XYY individuals make up a greater than average proportion of the patients in mental-penal institutions. This may be accounted for in part by their diminished intelligence, which may make it easier for them to be apprehended.

Y

Y chromosome   the sex chromosome found only in the heterogametic sex. In *Drosophila,* the Y chromosome bears genes controlling spermiogenesis.

yearling   an animal one year old or in its second year of life. In race horses the term refers to an animal living during the period between January 1st of the year after the year in which is was foaled and the next January 1st.

Y fork   the point at which a DNA molecule is being replicated; the two template strands of the parental molecule separate, forming the arms of a Y-shaped structure. The unreplicated double-stranded DNA distal to the arms forms the base of the Y. *See* **replication of DNA.**

yields symbol   →

Y linkage   genes located on the Y sex chromosome, exhibiting holandric (*q.v.*) inheritance. *See Oryzias latipes.*

yolk   the complex collection of macromolecules and smaller nutrient molecules with which the oocyte is preloaded prior to fertilization. *See* **vitellogenin.**

Y-suppressed lethal   a sex-linked, recessive lethal that causes death of XO *Drosophilia melanogaster* but allows survival of normal males.

Z

Z atomic number, the number of protons plus electrons in the nucleus of the netural atom.

Z chromosome the sex chromosome found in both heterogametic females and homogametic males. *See* **W, Z chromosomes.**

Z DNA *See* **deoxyribonucleic acid.**

Zea mays maize; Indian corn, the plant species for which the most genetic information exists. *See* **corn.**

zeatin the first cytokinin obtained from a plant. Isolated from maize kernels in 1964.

Zebu the Brahman (*q.v.*) breed of cattle.

zero, absolute *See* **absolute zero.**

zero order kinetics the progression of an enzymatic reaction in which the formation of product proceeds at a linear rate with the time. This rate is not increased if additional substrate is added. *See* **first order kinetics.**

zero time binding DNA strands of DNA containing intramolecular repeats that form duplexes at the start of a reassociation reaction.

Zimmermann cell fusion a technique developed by Ulrich Zimmermann in which cells are exposed to a low-level, high-frequency electric field that orients them into chains. A direct current pulse is then used to open micropores in adjoining cell membranes. These micropores allow mixing of the cytoplasms, and the cells may eventually fuse. The Zimmermann technique may also alter the permeability of the plasmalemma so that DNA fragments the size of genes can enter the cell.

zinc   a biological trace element. Atomic number 30; atomic weight 65.37; valence 2^+; most abundant isotope ^{64}Zn, radioisotope ^{65}Zn, half life 250 days, radiation emitted—positrons.

Zinjanthropus boisei   a tool-making, manlike species of ape that existed in Africa some 1.7 million years ago, now classified as *Australopithecus boisei. See* **Olduvai Gorge.**

Zn   zinc

zonal electrophoresis   a technique that makes possible the separation of charged macromolecules and the characterization of each molecule in terms of its electrophoretic mobility. *See* **electrophoresis.**

zoogeographic realms   the divisions of the land masses of the world according to their distinctive faunas. *See* **biogeographic realms.**

zoogeography   the study of the geographical distribution of animals.

ZPG   zero population growth; a population status in which birth and death rates are equivalent.

ZR 515   a synthetic juvenile hormone analogue which mimics the effects of JH and is more resistant to breakdown by the esterases normally found in insect hemolymph. *See* **allatum hormone.**

Z, W chromosomes   *See* **W, Z chromosomes.**

zygonema   *See* **meiosis.**

zygote   the diploid cell resulting from the union of the haploid male and female gametes. The term is also used to refer to the organism that develops from a single celled zygote.

zygotene stage   *See* **meiosis.**

zygotic induction   the induction of vegetative replication in a prophage that is transferred during conjugation to a nonlysogenic F⁻ bacterium.

zygotic lethal   in *Drosophila,* a lethal gene whose effect is apparent in the embryo, larva, or adult, but which does not render inviable any gamete that carries it.

zygotic meiosis   referring to meiosis which immediately follows the formation of a zygote, as in haplonts.

zymogen   the enzymatically inactive precursor of a proteolytic enzyme. Zymogens usually become activated by posttranslational modifications. For example, the zymogen pepsinogen is converted to the digestive enzyme pepsin by cleavage in a particualr peptide sequence.

zymogen granules   enzyme-containing particles elaborated by the cells of the pancreas.

Appendix A
Classification

Classification: the subdivision of organisms into an evolutionary hierarchy of groups. The formal hierarchy proceeding from the largest to the smallest group is Kingdom, Phylum, Class, Order, Family, Genus, and Species. To allow further subdivisions the names Grade or Division are sometimes placed between Kingdom and Phylum, the name Branch is placed between Phylum and Class, the name Cohort between Class and Order, and the name Tribe between Family and Genus. In addition the prefixes Super- and Sub- may be added to any group name.

Classification of living organisms

SUPERKINGDOM Prokaryotes *(q.v.)*
  KINGDOM 1 Monera *(q.v.)*
           PHYLUM Aphragmabacteria *(Mycoplasma)*
           PHYLUM Fermenting bacteria *(Diplococcus, Streptococcus)*
           PHYLUM Spirochaetae *(Spirochaeta)*
           PHYLUM Thiopneutes (Sulfur-reducing bacteria, *Desulfovibrio*)
           PHYLUM Methaneocreatrices (Methane-forming bacteria, *Methanobacillus*)
           PHYLUM Anaerobic photosynthetic bacteria, *(Rhodospirillum)*
           PHYLUM Cyanobacteria (blue-green algae, *Anabaena*)
           PHYLUM Chloroxybacteria *(Prochloron)*
           PHYLUM Nitrogen-fixing aerobic bacteria *(Rhizobium)*
           PHYLUM Pseudomonads *(Pseudomonas)*
           PHYLUM Aeroendospora (aerobic endospore-forming bacteria, *Bacillus*)
           PHYLUM Micrococci *(Staphylococcus)*
           PHYLUM Chemoautotrophic bacteria *(Nitrobacter)*
           PHYLUM Omnibacteria *(Escherichia, Haemophilus, Salmonella, Serratia, Shigella)*
           PHYLUM Actinobacteria *(Actinomyces, Streptomyces)*
           PHYLUM Myxobacteria *(Myxococcus)*
SUPERKINGDOM Eukaryotes *(q.v.)*
  KINGDOM 2 Protoctista *(q.v.)*
           PHYLUM Caryoblastea *(Pelomyxa)*
           PHYLUM Dinomastigota *(Gonovaulax)*
           PHYLUM Rhizopoda *(Amoeba)*
           PHYLUM Chrysophyta *(Ochromonas)*
           PHYLUM Haptophyta *(Prymnesium)*
           PHYLUM Euglenophyta *(Euglena)*

PHYLUM Cryptophyta *(Cyathomonas)*
PHYLUM Zoomastigina (animal flagellates, *Trypanosoma*)
PHYLUM Xanthophyta *(Vaucheria)*
PHYLUM Eustimatophyta *(Vischeria)*
PHYLUM Bacillariophyta (diatoms, *Diatoma*)
PHYLUM Phaeophyta (brown algae, *Fucus, Nereocystis*)
PHYLUM Rhodophyta (red algae, *Polysiphonia*)
PHYLUM Gamophyta (conjugating green algae and desmids, *Spirogyra* and *Micrasterias*)
PHYLUM Chlorophyta (green algae forming flagellated gametes, *Acetabularia, Chlamydomonas, Volvox*)
PHYLUM Actinopoda (radiolarians and heliozoans, *Acanthocystis, Sticholonche*)
PHYLUM Foraminera *(Fusulina, Globigerina)*
PHYLUM Ciliophora (ciliates, *Paramecium, Tetrahymena*)
PHYLUM Apicomplexa (sporozoans, *Plasmodium*)
PHYLUM Cnidosporidia (microsporidians, *Nosema*)
PHYLUM Labyrinthulamycota (slime nets, *Labryrinthula*)
PHYLUM Acrasiomycota (cellular slime molds, *Dictyostelium, Polysphondylium*)
PHYLUM Myxomycota (plasmodial slime molds, *Echinostelium*)
PHYLUM Plasmodiophoromycota *(Plasmidiophora)*
PHYLUM Hypochytridiomycota *(Hypochytrium)*
PHYLUM Chytridiomycota *(Blastocladiella)*
PHYLUM Oomycota *(Saprolegnia)*

KINGDOM 3 Fungi *(q.v.)*
PHYLUM Zygomycota (conjugating fungi, *Mucor, Phycomyces, Pilobolus, Rhizopus*)
PHYLUM Ascomycota (sac fungi, *Neurospora, Saccharomyces*)
PHYLUM Basidiomycota (club fungi, *Agaricus, Schizophyllum, Ustilago*)
PHYLUM Deuteromycota *(Aspergillus, Penicillium)*
PHYLUM Mycophycophyta (lichens, *Cladonia*)

KINGDOM 4 Animalia *(q.v.)*
SUBKINGDOM Parazoa *(q.v.)*
PHYLUM Placozoa *(Trichoplax)*
PHYLUM Porifera (sponges, *Euplectella*)
SUBKINGDOM Mesozoa *(q.v.)*
PHYLUM Mesozoa *(Dicyema)*
SUBKINGDOM Eumetazoa *(q.v.)*
GRADE Radiata *(q.v.)*
PHYLUM Cnidaria (Coelenterates)
CLASS Hydrozoa (hydroids, *Hydra*)
CLASS Scyphozoa (true jelly fish, *Physalia*)
CLASS Anthozoa (corals and sea anemones, *Metridium*)
PHYLUM Ctenophora (comb jellies, *Mnemiopsis*)
GRADE Bilateria *(q.v.)*
SUBGRADE Protostomia *(q.v.)*
SUPERPHYLUM Acoelomata *(q.v.)*
PHYLUM Platyhelminthes (flatworms)
CLASS Turbellaria (planarians)
CLASS Trematoda (flukes, *Schistosoma*)
CLASS Cestoda (tapeworms)
PHYLUM Nemertina (Rhynchocoela) (ribbon worms)
SUPERPHYLUM Pseudocoelomata *(q.v.)*
PHYLUM Acanthocephala (spiny-headed worms)
PHYLUM Entoprocta (entoprocts)
PHYLUM Aschelminthes *(q.v.)*
SUBPHYLUM Rotifera (rotifers)
SUBPHYLUM Gastrotricha (gastrotrichs)
PHYLUM Loricifera (loriciferans)
SUBPHYLUM Kinorhyncha (kinorhynchs)

426

SUBPHYLUM   Priapulida (priapulids)
SUBPHYLUM   Nematoda (round worms, *Ascaris, Caenorhabditis*)
SUBPHYLUM   Nematomorpha (Gordiacea) (horsehair worms)
SUPERPHYLUM   Coelomata (*q.v.*)
  DIVISION   Tentaculata
    PHYLUM   Phoronida (phoronids)
    PHYLUM   Ectoprocta (bryozoa)
    PHYLUM   Brachiopoda (brachiopods)
  DIVISION   Inarticulata (*q.v.*)
    PHYLUM   Sipunculoidea (sipunculids)
    PHYLUM   Mollusca (molluscs)
        CLASS   Amphineura (chitins)
        CLASS   Scaphopoda (tooth shells)
        CLASS   Gastropoda (snails, *Cepaea, Lymnea*)
        CLASS   Pelecypoda (bivalves, *Crassostrea, Mytilus*)
        CLASS   Cephalopoda (squids, octopuses)
  DIVISION   Articulata (*q.v.*)
    PHYLUM   Echiuroidea (echiuroids)
    PHYLUM   Annelida (segmented worms)
        CLASS   Polychaeta (marine worms)
        CLASS   Oligochaeta (earthworms)
        CLASS   Hirudinea (leeches)
    PHYLUM   Tardigrada (water bears)
    PHYLUM   Onychophora (peripatus)
    PHYLUM   Pentastomida (tongue worms) (parasites)
    PHYLUM   Arthropoda
      BRANCH   Chelicerata
        CLASS   Merostomata (king crabs)
        CLASS   Pycnogonida (sea spiders)
        CLASS   Arachnida (scorpions, opilionids, mites, spiders)
      BRANCH   Mandibulata
        CLASS   Crustacea
              ORDER   Branchiopoda (shrimps)
              ORDER   Ostracoda (ostracods)
              ORDER   Copepoda (copepods)
              ORDER   Cirripedia (barnacles)
              ORDER   Malacostraca (lobsters and crabs)
        CLASS   Myriapoda
              ORDER   Diplopoda (millipedes)
              ORDER   Chilopoda (centipedes)
        CLASS   Hexapoda
         SUBCLASS   Entognatha
              ORDER   Collembola (springtails)
              ORDER   Protura (proturans)
              ORDER   Diplura (campodeans)
         SUBCLASS   Insecta
          COHORT   Apterogota (primitively wingless insects)
              ORDER   Archaeognatha (machilids)
              ORDER   Zygentoma (silverfish)
          COHORT   Pterygota (winged insects)
            SUBCOHORT   Paleoptera (extended wing)
              ORDER   Ephemeroptera (Mayflies)
              ORDER   Odonata (dragonflies)
            SUBCOHORT   Neoptera (hinged wing)
             SUPERORDER   Hemimetabola (no pupal stage)
              ORDER   Embioptera (embiids)
              ORDER   Phasmida (stick insects)
              ORDER   Orthoptera (grasshoppers)

ORDER Grylloblatteria (grylloblatids)
ORDER Dictyoptera (roaches, *Blattella*)
ORDER Dermaptera (earwigs)
ORDER Psocoptera (booklice)
ORDER Phthiraptera (sucking lice)
ORDER Hemiptera (true bugs)
ORDER Plecoptera (stone flies)
ORDER Thysanoptera (thrips)
SUPERORDER Holometabola
ORDER Coleoptera (beetles, *Tribolium*)
ORDER Raphidioptera (snakeflies)
ORDER Megaloptera (alder flies)
ORDER Neuroptera (lacewings)
ORDER Hymenoptera *(Apis, Mormoniella, Habrobracon)*
ORDER Trichoptera (caddis flies)
ORDER Lepidoptera (moths, *Bombyx, Ephestia*)
ORDER Mecoptera (scorpion flies)
ORDER Siphonaptera (fleas)
ORDER Diptera *(Aedes, Anopheles, Chironomus, Culex, Drosophila, Glyptotendipes, Lucilia, Musca, Rhynchosciara, Sciara)*
SUBGRADE Deuterostomia (*q.v.*)
PHYLUM Echinodermata
ORDER Crinoidea (sea lilies)
ORDER Asteroidea (starfish)
ORDER Ophiuroidea (brittle stars)
ORDER Echinoidea (sea urchins)
ORDER Holothuroidea (sea cucumbers)
PHYLUM Chaetognatha (arrow worms)
PHYLUM Pogonophora (beard worms)
PHYLUM Chordata (notochord-bearing animals)
SUBPHYLUM Acraniata (*q.v.*)
COHORT Hemichordata (acorn worms and pterobranchs)
COHORT Urochordata (tunicates)
COHORT Cephalochordata (lancelets)
SUBPHYLUM Craniata (*q.v.*)
COHORT Agnatha (jawless vertebrates)
CLASS Cyclostomata (lampreys)
COHORT Gnathostoma (jawed vertebrates)
CLASS Chondrichthyes (elasmobranchs)
CLASS Osteichthyes (bony fish)
SUBCLASS Palaeopterygii (ancient fishes)
ORDER Acipenseriformes (sturgeons)
ORDER Semionotiformes (gars)
SUBCLASS Neopterygyii (modern fishes)
ORDER Salmoniformes (trouts)
ORDER Anguilliformes (eels)
ORDER Cypriniformes *(Oryzias)*
ORDER Cyprinidontiformes *(Fundulus, Lebistes)*
ORDER Perciformes *(Tilapia)*
ORDER Siluriformes (catfishes)
ORDER Elopiformes (tarpons)
ORDER Clupeiformes (herrings)
ORDER Gasterosteiformes (sea horses)
ORDER Pleuronectiformes (flounders)
SUBCLASS Crossopterygii (coelacanths, *Latimeria*)
CLASS Amphibia
ORDER Apoda (caecilians)

ORDER Urodela (salamanders, *Ambystoma, Triturus, Pleurodeles*)

ORDER Anura (frogs and toads, *Rana, Xenopus*)

CLASS Reptilia (turtles, alligators, lizards, snakes)

CLASS Aves (birds)

SUBCLASS Palaeognathae (flightless birds, ostriches, emus)

SUBCLASS Neognathae (modern birds)

ORDER Anseriformes (ducks)

ORDER Galliformes (quail, turkeys, *Gallus*)

ORDER Columbiformes (pigeons)

ORDER Psittaciformes (parrots)

ORDER Passeriformes (song birds)

CLASS Mammalia

SUBCLASS Protheria (egg-laying mammals)

SUBCLASS Metatheria (marsupials)

SUBCLASS Eutheria (placental mammals)

ORDER Insectivora (moles, shrews)

ORDER Chiroptera (bats)

ORDER Edentata (sloths)

ORDER Rodentia (rodents, *Cavia, Chinchilla, Mesocricetus, Mus, Peromyscus, Rattus*)

ORDER Lagomorpha (rabbits, *Oryctolagus*)

ORDER Cetacea (whales)

ORDER Proboscidea (elephants)

ORDER Pinnipedia (seals)

ORDER Perissodactyla (odd-toed ungulates, *Equus*)

ORDER Artiodactyla (even-toed ungulates, *Bos, Ovis*)

ORDER Primates (lemurs, tarsiers, monkeys, apes, *Homo sapiens*)

KINGDOM 5 Plantae

PHYLUM Bryophyta (*q.v.*)

SUBPHYLUM Hepaticae (liverworts)

SUBPHYLUM Anthocerotae (hornworts)

SUBPHYLUM Musci (mosses)

PHYLUM Tracheophyta (vascular plants)

SUBPHYLUM Psilophyta *(Psilotum)*

SUBPHYLUM Lycopodophyta (clubmosses and quillworts)

SUBPHYLUM Sphenopsida (horsetails)

SUBPHYLUM Pteropsida (ferns and seed plants)

SUPERCLASS Filicinae (ferns)

SUPERCLASS Gymnospermae (cone-bearing, seed plants)

CLASS Pteridospermophyta (seed ferns)

CLASS Cycadophyta (cyads)

CLASS Ginkgophyta (ginkgos)

CLASS Coniferophyta (conifers)

CLASS Gnetophyta (gnetophytes)

SUPERCLASS Angiospermae (flowering plants)

CLASS Dicotyledoneae

ORDER Magnoliales (magnolias)

ORDER Rosales (rose, apple, plum, strawberry, etc.)

ORDER Legumiales *(Pisum, Phaseolus)*

ORDER Salicales (willows)

ORDER Fagales (beech, oak, etc.)

ORDER Geraniales (geranium, flax, citrus fruits)

ORDER Cactales (cactuses)

ORDER Scrophulariales *(Antirrhinum, Collinsia)*

ORDER Ranales (buttercups, water lilies)

ORDER Myrtales (myrtle, eucalyptus, *Oenothera*)

ORDER   Cruciales (cabbage, turnip, etc.)
ORDER   Cucurbitales *(Cucurbita, Cucumis)*
ORDER   Caryophyllales (carnations)
ORDER   Gentianales (gentians, olives, lilacs)
ORDER   Primulales (*Primula*)
ORDER   Malvales (*Gossypium,* linden, elm, hemp)
ORDER   Rubiales (coffee, quinine)
ORDER   Saxifragales (current, gooseberry)
ORDER   Umbellales (carrot, parsnip)
ORDER   Sapindales (horse chestnut, maple)
ORDER   Solanales *(Datura, Lycopersicon, Nicotiana)*
ORDER   Celastrales (grapes, ivy)
ORDER   Personales (snapdragon, etc.)
ORDER   Boraginales (heliotrope, etc.)
ORDER   Lamiales (lavender, mint, etc.)
ORDER   Asterales (composites, *Haplopappus*)

CLASS   Monocotyledoneae
ORDER   Palmales (palms)
ORDER   Graminales (grasses, *Hordeum, Oryzea, Triticum, Zea)*
ORDER   Liliales (lilies, tulips, amaryllis, iris)
ORDER   Commelinales *(Tradescantia)*
ORDER   Arales (callas, taros)
ORDER   Zingiberales (bananas)
ORDER   Orchidales (orchids)
ORDER   Bromeliales (pineapple)

Adapted from Lynn Margulis, 1974, *The classification and evolution of Prokaryotes and Eukaroytes. Handbook of Genetics 1*:1–41, Plenum Press, and Lynn Margulis and Karlene V. Schwartz, 1982, *Five Kingdoms: An Illustrated Guide to the Phyla of Life on Earth,* W. H. Freeman.

Appendix B
Domesticated Species

Domesticated species, organisms that have been trained to live with or be of service to humans, include agricultural plants, livestock, household pets, laboratory animals, etc. Below is a listing of the common and scientific names of a variety of economically important domesticated organisms. Species that are given their own entries in the dictionary (i.e., barley, chicken, corn, etc.) are omitted from this list.

alfalfa   *Medicago sativa*
almond   *Prunus amygdalus*
alpaca   *Lama pacos*
amaryllis   *Amaryllis belladonna*
anise   *Pimpinella anisum*
apple   *Malus pumila*
apricot   *Prunus armeniaca*
artichoke   *Cynara scolymus* (globe)
ash   *Fraxinus americana* (white)
asparagus   *Asparagus officinalis*
aster   *Aster novaeangliae* (New England)
avocado   *Persea americana*
balsam fir   *Abies balsamea*
bamboo   *Bambusa vulgaris*
banana   *Musa paradisaica*
bass   *Micropterus salmoides* (large mouth)
bean   *Vicia faba* (broad): *Ricinus communis* (castor); *Phaseolus limensis* (lima); *P. aureus* (Mung); *P. vulgaris* (string)
beech   *Fagus grandifolia* (American); *F. sylvatica* (European)
beet   *Beta vulgaris*
begonia   *Begonia rex*
birch, paper   *Betula papyrifera*
blackberry   Cultivated blackberries are derived chiefly from three species of *Rubus: R. argutus, R. alleghaniensis,* and *R. frondosus.*
blueberry   *Vaccinium corymbosum*
bluegrass   *Poa pratensis*

Brazil nut   *Bertholletia excelsa*
broccoli   *Brassica oleracea italica*
broomcorn   *Sorghum vulgare technicum*
Brussels sprouts   *Brassica oleracea gemmifera*
buckwheat   *Fagopyrum sagittatum*
cabbage   *Brassica oleracea capitata*
camellia   *Camellia japonica*
canary   *Serinus canaria*
cantaloupe   *Cucumis melo cantalupensis*
cardamom   *Elettaria cardamomum*
carnation   *Dianthus caryophyllus*
carp   *Cyprinus carpio*
carrot   *Daucus carota sativa*
cashew nut   *Anacardium occidentale*
cat   *Felis catus*
cattle   *Bos taurus*
cauliflower   *Brassica oleracea botrytis*
cedar, eastern red   *Juniperus virginiana*
celery   *Apium graveolens dulce*
cherry   *Prunus cerasus* (sour), *P. avium* (sweet)
chestnut (European)   *Castenea sativa*
chinchilla   *Chinchilla brevicaudata*
Chinese cabbage   *Brassica rapa*
chive   *Allium schoenoprasum*
chrysanthemum   *Chrysanthemum morifolium*
cinnamon   *Cinnamomum zeylanicum*
clove   *Syzygium aromaticum*

clover   *Trifolium pratense* (red), *T. repens* (white)

cocoa   *Theobroma cacao*

coconut   *Cocos nucifera*

coffee   *Coffea arabica*

coriander   *Coriandrum sativum*

crabapple   *Pyrus ioensis*

cranberry   *Vaccinium macrocarpon*

crocus   *Crocus susianus* (cloth of gold)

cucumber   *Cucumis sativus*

daffodil   *Narcissus pseudo-narcissus*

date palm   *Phoenix dactilifera*

dill   *Anethum graveolens*

Douglas fir   *Pseudotsuga taxifolia*

duck   *Anas platyrhynchos* (mallard)

ebony   *Diospyros ebenum*

eggplant   *Solanum melongena esculentum*

elephant   *Elephas maximus* (Indian)

endive   *Cichorium endivia*

ermine   *Mustela erminea*

fig   *Ficus carica*

flax   *Linum usitatissimum*

fox   *Vulpes vulpes* (silver)

foxglove   *Digitalis purpurea*

geranium   *Pelargonium graveolens*

gerbil   *Merinoes unguiculatis* (Mongolian)

ginger   *Zingiber officinale*

gladiolus   *Gladiolus communis* (one of hundreds of species)

goat   *Capra hircus*

goldfish   *Carassius auratus*

goose   *Cygnopsis cygnoid* (Chinese)

grape   *Vitis vinifera* (common wine)

grapefruit   *Citrus paradisi*

guava   *Psidium guajava*

hazelnut   *Corylus americana*

hemlock   *Tsuga heterophylla* (western)

hemp   *Cannabis sativa* (marijuana), *Agave sis-alana* (sisal)

hickory   *Carya ovata* (shagbark)

holly   *Ibex opaca* (American), *I. aquifolium* (European)

hollyhock   *Althea rosea*

honey locust   *Gleditsia triacanthos*

hop   *Humulus lupulus*

huckleberry   *Gaylussacia baccata* (black)

hyacinth   *Hyacinthus orientalis*

iris   *Iris versicolor grandiflorum* (blue flag)

ivy   *Hedera helix* (English)

jasmine   *Jasminum officinale*

juniper   *Juniperus communis*

kapok   *Ceiba pentandra*

laburnum   *Laburnum anagyroides*

lavender   *Lavandula officinalis*

lemon   *Citrus limon*

lettuce   *Lactuca sativa*

licorice   *Glycyrrhiza glabra*

lily   *Lilium longiflorum eximum* (Easter)

lime   *Citrus aurantifolia*

lingonberry   *Vaccinium vitis-idaea* (also called mountain cranberry or lowbush cranberry)

llama   *Lama glama*

lotus   *Nelumbo lutea* (yellow)

maguey   *Agava cantala* (Manila), *A. atrovirens* (pulque)

mahogany   *Swietenia mahagoni* (West Indian)

mango   *Mangifera indica*

marigold   *Tagetes erecta* (garden)

marten   *Martes americana*

millet   *Setaria italica* (foxtail), *Pennisetum glaucum* (pearl), *Panicum miliaceum* (broomcorn)

monkey   *Macaca mulatta* (Rhesus)

morning-glory   *Ipomoea purpurea*

mushroom   *Agaricus campestris* (meadow)

musk ox   *Ovibos moschatus*

mustard   *Brassica hirta* (white)

narcissus   *Narcissus poeticus*

nasturtium   *Tropaeolum majus*

oak   *Quercus suber* (cork), *Q. alba* (white)

okra   *Hibiscus esculentus*

olive   *Olea europaea*

orange   *Citrus aurantium* (bitter); *C. sinensis* (sweet); the naval orange is a cultivar of this species.

oyster   *Crassostraea virginica*

papaya   *Carica papaya*

parakeet   *Melopsittacus undulatus*

parsnip   *Pastinaca sativa*

pea   *Pisum sativum*

peach   *Prunus persica*

peanut   *Arachis hypogaea*

pear   *Pyrus communis*

pecan   *Carya illinoensis*

peony   *Paeonia officinalis*

pepper   *Capsicum frutescens* (red), *Piper nigrum* (black)

peppermint   *Mentha piperita*

perch   *Perca flavescens* (yellow)

persimmon   *Diospyros kaki* (Japanese)

philodendron   *Philodendron cordatum*

phlox   *Phlox drummondii*

pickerel   *Esox niger* (chain)

pig   *Sus scrofa*

pigeon   *Columba livia*

pike   *Esox lucius* (northern)

pine   *Pinus lambertiana* (sugar)

pineapple   *Ananas comosus*

pistachio nut   *Pistacia vera*

plum   *Prunus domestica*

poinsettia   *Euphorbia pulcherrima*

poppy   *Papaver somniferum* (opium)

potato   *Solanum tuberosum*

pumpkin   *Cucurbita pepo*

quince   *Cydonia oblonga*
rabbit   *Oryctolagus cuniculus* (European)
radish   *Raphanus sativus*
raspberry   *Rubus occidentalis* (black)
redbud   *Cercis canadensis*
reindeer   *Rangifer tarandus*
rhubarb   *Rheum officinale*
rice   *Oryza sativa*
rose   Many species of *Rosa* are grown
   commercially. The most common are *R.
   centifolia, R. damascena,* and *R. multiflora.*
rubber   *Hevea brasiliensis*
rutabaga   *Brassica napobrassica*
rye   *Secale cereale*
salmon   *Salmo salar* (Atlantic)
sandalwood   *Santalum album*
sequoia   *Sequoia gigantia* (big tree), *S.
   sempervirens* (coastal redwood)
sesame   *Sesamum indicum*
sheep   *Ovis aries* (domestic)
snapdragon   *Antirrhinum majus*
sorghum   *Sorghum vulgare*
soybean   *Glycine max*
spearmint   *Mentha spicata*
spinach   *Spinacia oleracea*
spruce   *Picea pungens* (Colorado blue)
strawberry   *Fragaria virginiana* (American),
   F. chiloensis (Chilean), *F. vesca* (European)
strawflower   *Helichrysum bracteatum*

sugar cane   *Saccharum officinarum*
sunflower   *Helianthus annuus*
sweet potato   *Ipomoea batatas*
Swiss chard   *Beta vulgaris cicla*
tangerine   *Citrus reticulata*
tea   *Camellia sinensis (Thea sinensis)*
teak   *Tectona grandis*
thyme   *Thymus vulgaris*
timothy   *Phleum pratense*
trout   *Salvelinus frontinalis* (brook), *S.
   namaycush* (lake), *Salmo gairdneri*
   (rainbow)
tulip   *Tulipa gesneriana*
tulip tree   *Liriodendron tulipifera*
turkey   *Meleagris gallopavo*
turnip   *Brassica rapa*
vanilla   *Vanilla plantifolia*
vicuña   *Vicugna vicugna*
violet   *Saintpaulia ionantha* (African)
walnut   *Juglans regia* (English)
water buffalo   *Bubalus bubalis*
watermelon   *Citrullus vulgaris*
wild rice   *Zizania aquatica*
willow   *Salix babylonica* (weeping)
yak   *Bos grunniens*
yam   *Dioscorea alata*
yucca   *Yucca brevifolia* (Joshua tree)
zinnia   *Zinnia elegans*

Appendix C
Chronology

Genetics, cytology and evolutionary biology have received stimulation from both related and quite independent sciences. In many cases the development of a particular physical instrument or technique has led to a golden age of discovery. Often research in various areas has advanced in nonsynchronous spurts, and consequently it is difficult to develop courses in genetics, cytology, and evolutionary biology from a strictly historical standpoint. The student, however, should have some idea of the chronological order in which certain events having a bearing on these sciences took place. The following chronology will fill this need, even though many experts will complain about the inclusion of some events and the omission of others. Furthermore, a decade from now some of the recent discoveries may be relegated to less prominent positions. The student should keep the following thought in mind when perusing this catalogue. In science a great unifying concept generally does not spring full-blown from the mind of a single individual. Rather when the time is ripe perhaps a dozen authorities may grope about for an explanation, and all may be on the verge of the answer. However, often one scientist may first express the unifying concept in a clear fashion, and as a matter of convenience he or she is the one listed as the progenitor of the idea. In the bibliography following this appendix one can find listed scientific histories and collections of various epoch-making scientific papers. Reading such masterpieces will serve to put flesh upon the skeleton provided by this chronology.

1590	Z. and H. Janssen combine two double convex lenses in a tube and produce the first compound microscope.
1651	W. Harvey puts forward the concept that all living things (including man) originate from eggs.
1657	R. de Graaf discovers follicles in the human ovary, but interprets them incorrectly as eggs.
1665	R. Hooke publishes *Micrographia,* in which he gives the first description of cells.
1668	F. Redi disproves the theory of spontaneous generation of maggots.
1677	A. van Leeuwenhoek observes sperm of man and other mammals.
1694	J. R. Camerarius conducts early pollination experiments and reports the existence of sex in flowering plants.
1735	C. V. Linné (better known today in the Latinized form Linnaeus) publishes the first edition of the *Systema Naturae.* Sixteen editions of this taxonomic work are completed during his lifetime. It is the 10th edition of this book that serves as the starting point for the modern scientific naming of animals, as his *Species Plantarum* is for plants. Linné originated the "Linnean" system of binary nomenclature used to this day. His insistence on the constancy and objective classification of species posed the problem of the method of the origin of species.

1761–67 J. G. Kölreuter carries out crosses between various species of *Nicotiana* and finds that the hybrids are quantitatively intermediate between their parents in appearance. The hybrids from reciprocal crosses are indistinguishable. He concludes that each parent contributes equally to the characteristics of the offspring.

1769 L. Spallanzani demonstrates that the "spontaneous generation" of microorganisms in a nutrient medium can be prevented, provided the vessel is sealed and subjected to the temperature of boiling water for 30 minutes or more. In 1780 he performs artificial insemination experiments with amphibians and demonstrates that physical contact of the egg with spermatic fluid is necessary for fertilization and development.

1798 T. R. Malthus publishes annonymously *An Essay on the Principle of Population*. This essay suggests to Darwin in 1838 the concept of the struggle for existence and the survival of the fittest.

Edward Jenner publishes *An Inquiry into the Causes and Effects of the Variolae Vaccinae, a Disease Discovered in Some of the Western Countries of England, Particularly Gloucestershire and Known by the Name of Cow Pox*. In it he gave the first account of vaccination with cowpox virus to prevent smallpox. He thus establishes the principle of active immunization and initiates the science of immunology.

1809 J. B. de Monet Lamarck puts forward the view that species can change gradually into new species through a constant strengthening and perfecting of adaptive characteristics, and that these acquired characteristics are transmitted to the offspring.

1820 C. F. Nasse describes the sex-linked mode of inheritance of hemophilia in man.

1822–24 T. A. Knight, J. Goss, and A. Seton all independently perform crosses with the pea and observe dominance in the F_1 and segregation of various hereditary characters in the F_2. However, they do not study later generations or determine the numerical ratios in which the characters are transmitted.

1825 F. V. Raspail founds the science of histochemistry by using the iodine reaction for starch.

1827 K. E. von Baer gives the first accurate description of the human egg.

1830 G. B. Amici shows that the pollen tube grows down the style and into the ovule of the flower.

1831 R. Brown notes nuclei within cells.

On December 27th, H.M.S. *Beagle* sets sail from Plymouth for a voyage around the world. It carries as naturalist the 22-year-old Charles Darwin. The *Beagle* reaches the Galapagos Islands, September 15, 1833, and Darwin spends five weeks surveying the plant and animal life. The tremendous abundance of related species suggests to Darwin that the islands were colonized by a few species from the mainland and that these species later evolved various offspring species that became specialized to live in each of the many new environments provided.

1838 The word *protein* first appears in the chemical literature in a paper by G. J. Mulder. The term, however, was invented by J. J. Berzelius.

1838–39 M. J. Schleiden and T. Schwann develop the cell theory, a great generality of biology equaled only by the theory of evolution. Schleiden notes nucleoli within nuclei.

1841 A. Kölliker shows spermatozoa to be sex cells that arise by transformation of cells in the testis.

1845 J. Dzierzon reports that drones hatch from unfertilized eggs, worker and queen bees from fertilized eggs.

1855 R. Virchow states the principle that new cells come into being only by division of previously existing cells.

1856 Gregor Mendel, a monk at the Augustinian monastery of St. Thomas in Brünn, Austria (now Brno, Czechoslovakia), begins breeding experiments with the garden pea, *Pisum sativum*.

1859 Darwin publishes *On the Origin of Species*.

1860 T. A. E. Klebs introduces paraffin embedding.

1861 L. Pasteur disproves the theory of spontaneous generation of micro-organisms.

1865 Gregor Mendel presents the results and interpretations of his genetic studies on the garden pea at the Brünn Society for the Study of Natural Science at their monthly meetings held February 8 and March 8.

1866 G. Mendel's *Versuche über Pflanzenhybriden (Experiments on Plant-Hybridization)* is published and ignored.

1869	F. Miescher isolates nucleoprotein.
1869	F. Galton publishes *Hereditary Genius.* He emphasizes the genetic basis of intelligence and is the first to compare mental traits in twins.
1870	W. His invents the microtome.
1873	A. Schneider gives the first account of mitosis.
1875	O. Hertwig concludes from a study of the reproduction of the sea urchin that fertilization in both animals and plants consists of the physical union of the two nuclei contributed by the male and female parents.
	E. Strasburger describes cell division in plants.
	F. Galton draws attention to the use of twin studies for elucidating the relative importance of hereditary and environmental influence upon behavior.
1876	O. Bütschli describes nuclear dimorphism in ciliates.
1877	H. Fol reports watching the spermatozoan of a starfish penetrate the egg. He was able to see the transfer of the intact nucleus of the sperm into the egg, where it became the male pronucleus.
	E. Abbe begins to publish important contributions to the theory of microscopic optics.
1878	W. Kuhne coins the word *enzyme.*
1879	W. Flemming studies mitosis in the epithelium of the tail fin of salamanders. He shows that nuclear division involves a longitudinal splitting of the chromosome and a migration of the sister chromatids to the future daughter nuclei. He also coins the term *chromatin.*
1881	E. G. Balbiani discovers "cross-striped threads" within the salivary gland cells of *Chironomus* larvae. However, he does not realize that these are polytene chromosomes.
	R. Koch devises the methods used to this day for isolating pure cultures of bacteria.
1882	W. Flemming discovers lampbrush chromosomes and coins the term *mitosis.*
1883	E. van Beneden studies meiosis in a species of the round worm, *Ascaris,* which (fortunately) has a diploid chromosome number of only four. He shows that the gametes contain half as many chromosomes as the somatic tissues and that the characteristic somatic number is re-established at fertilization. He also describes fertilization in mammals.
	W. Roux suggests that the filaments within the nucleus which stain with basic dyes are the bearers of the hereditary factors.
1887	A. Weismann postulates that a periodic reduction in chromosome number must occur in all sexual organisms.
1888	T. Boveri describes the centriole.
	W. Waldeyer coins the word *chromosome* for the filaments referred to by Roux (1883).
1889	F. Galton introduces the quantitative measurement of populations into genetics.
1890	R. Altmann reports the presence of "bioblasts" within cells and concludes that they are "elementary organisms" that live as intracellular symbionts and carry out processes vital to their hosts. Later (1898) C. Benda names these organelles *mitochondria.*
	E. von Behring shows that blood serum from previously immunized animals contains factors that are specifically lethal to the organisms used for the immunization. These factors are now called antibodies.
1896	E. B. Wilson publishes *The Cell in Development and Heredity.* This influential treatise distils the information gained concerning cytology in the half century since Schleiden and Schwann put forth the cell theory.
1898	T. Boveri describes chromatin diminution in *Ascaris megalocephala.*
1899	M. W. Beijerinck demonstrates that tobacco mosaic disease is due to a self-reproducing agent that will pass through bacterial filters and can neither be seen with light microscope nor grown upon bacteriological media. He proposes that this *virus* is a self-reproducing subcellular form of life which represents a hitherto unknown class of organisms.
1900	H. de Vries, C. Correns, and E. Tschermak independently rediscover Mendel's paper. Using several plant species de Vries and Correns had performed breeding experiments that paralleled Mendel's earlier studies and had independently arrived at similar interpretations of their results. Therefore upon reading Mendel's publication they immediately recognized its significance. W. Bateson also stresses the importance of Mendel's contribution in an address to the Royal Society of London.
	K. Pearson develops the chi-square test.
	K. Lansteiner discovers the blood agglutination phenomenon in man.

436

P. Ehrlich proposes that antigens and antibodies bind together because they have structural complementarity.

1901   H. de Vries adopts the term *mutation* to describe sudden, spontaneous, drastic alterations in the hereditary material of *Oenothera*.

T. H. Montgomery studies spermatogenesis in various species of Hemiptera. He concludes that maternal chromosomes only pair with paternal chromosomes during meiosis.

K. Landsteiner demonstrates that humans can be divided into three blood groups: A, B, and C. The designation of the third group was later changed to O.

E. von Behring wins the Nobel Prize for his studies on antiserum therapy.

1902   C. E. McClung notes that in various insect species equal numbers of two types of spermatozoa are formed; one type contains an "accesssory chromosome" and the other does not. He suggests that the extra chromosome is a sex determinant, and he next argues that sex must be determined at the time of fertilization, not just in insects, but perhaps in other species (including man).

T. Boveri studies the development of haploid, diploid, and aneuploid sea urchin embryos. He finds that in order to develop normally, the organism must have a full set of chromosomes, and he concludes that the individual chromosomes must carry *different* essential hereditary determinants.

W. S. Sutton advances the chromosome theory of heredity which proposes that the independent assortment of gene pairs stems from the behavior of the synapsed chromosomes during meiosis. Since the direction of segregation of the homologs in a given bivalent is independent of those belonging to any other bivalent, the genes they contain will also be distributed independently.

F. Hofmeister and E. Fischer propose that all proteins are formed by the condensation of amino acids bound through regularly recurrent peptide linkages.

1902–09   W. Bateson introduces the terms *genetics, alleomorph, homozygote, heterozygote,* F_1, F_2, and *epistatic* genes.

1903   W. Waldeyer defines *centromeres* as chromosome regions with which the spindle fibers become associated during mitosis.

1904   A. F. Blakeslee discovers heteromixis in fungi.

1905   L. Cuénot performs crosses between mice carrying a gene which gives them yellow fur. Since they always produce yellow furred and agouti offspring in a 2:1 ratio, he concludes they are heterozygous. W. E. Castle and C. C. Little show in 1910 that yellow homozygotes die *in utero*. This dominant allele in the agouti series (A^y) is thus the first gene shown to behave as a homozygous lethal.

1906   W. Bateson and R. C. Punnett report the first case of linkage (in the sweet pea).

1907   R. G. Harrison cultures fragments of the central nervous systems of frogs in hemolymph and observes the outgrowth of nerve fibers. In so doing he invents tissue culture.

1908   G. H. Hardy and W. Weinberg, working independently, formulate the so-called Hardy-Weinberg law of population genetics.

1909   G. H. Shull advocates the use of self-fertilized lines in production of commercial seed corn. The hybrid corn program which resulted created an abundance of foodstuffs worth billions of dollars.

A. E. Garrod publishes *Inborn Errors of Metabolism,* the earliest study of the biochemical genetics of man (or any other species).

F. A. Janssens suggests that exchanges between nonsister chromatids produce chiasmata.

C. C. Little initiates a breeding program that produces the first inbred strain of mice (the strains now called DBA).

W. Johannsen's studies of the inheritance of seed size in self-fertilized lines of beans lead him to realize the necessity of distinguishing between the appearance of an organism and its genetic constitution. He invents the terms *phenotype* and *genotype* to serve this purpose, and he also coins the word *gene*.

C. Correns and E. Bauer study the inheritance of chloroplast defects in variegated plants, such as *Mirabilis jalapa* and *Pelargonium zonale*. They find that the inability to form healthy chloroplasts is in some cases inherited in a non-Mendelian fashion.

H. Nilsson Ehle puts forward the multiple factor hypothesis to explain the quantitative inheritance of seed-coat color in wheat.

1910	T. H. Morgan discovers white eye and consequently sex linkage in *Drosophila. Drosophila* genetics begins.
	W. Weinberg develops the methods used for correcting expectations for Mendelian segregation from human pedigree data under different kinds of ascertainment applied to data from small families.
	P. Rous shows that injection of a cell-free filtrate from chicken sarcomas induces new sarcomas in recipient chickens.
1911	T. H. Morgan proposes that the genes for white eyes, yellow body, and miniature wings in *Drosophila* are linked together on the X chromosome.
	W. R. B. Robertson points out that a metacentric chromosome in one orthopteran species may correspond to two acrocentrics in another and concludes that during evolution metacentrics may arise by the fusion of acrocentrics. Whole arm fusions are called *Robertsonian translocations* in his honor.
1912	A. Wegener proposes the continental drift concept.
	F. Rambousek suggests that the "cross-striped threads" within the salivary gland cells of fly maggots are chromosomes.
	T. H. Morgan demonstrates that crossing-over does not take place in the male of *Drosophila melanogaster*. He also discovers the first sex-linked lethal.
1913	Y. Tanaka reports that crossing over does not take place in the female of *Bombyx mori*. In this species the female is the heterogametic sex.
	W. H. Bragg and W. L. Bragg demonstrate that the analysis of x-ray diffraction patterns can be used to determine the three-dimensional atomic structure of crystals.
	A. H. Sturtevant provides the experimental basis for the linkage concept in *Drosophila* and produces the first genetic map.
1914	C. B. Bridges discovers meiotic nondisjunction in *Drosophila*.
	C. C. Little postulates that the acceptance or rejection of transplanted tumors has a genetic basis.
1915	F. W. Twort isolates the first filterable bacterial virus.
	R. B. Goldschmidt coins the term *intersex* to describe the aberrant sexual types arising from crosses between certain different races of the gypsy moth, *Lymantria dispar*.
	J. B. S. Haldane, A. D. Sprunt, and N. M. Haldane describe the first example of linkage in vertebrates (mice).
1916	H. J. Muller discovers interference in *Drosophila*.
1917	F. d'Herelle coins the term *bacteriophage* and develops methods for assaying virus titre.
	O. Winge calls attention to the important role of polyploidy in the evolution of angiosperms.
	C. B. Bridges discovers the first chromosome deficiency in *Drosophila*.
1918	H. Spemann demonstrates that a living part of an embryo can exert a morphogenic stimulus upon another part, bringing about its morphological differentiation (embryonic induction). He thus discovers and names the *organizer*.
	H. J. Muller discovers the balanced lethal phenomenon in *Drosophila*.
1919	T. H. Morgan calls attention to the equality in *Drosophila melanogaster* between the number of linkage groups and the haploid number of chromosomes.
	C. B. Bridges discovers chromosomal duplications in *Drosophila*.
1920	A. F. Blakeslee, J. Belling, and M. E. Farnham describe trisomics in the Jimson weed, *Datura stramonium*.
	H. Kniep describes and correctly interprets the incompatibility system in *Schizophyllum commune* and related basidiomycetes.
1921	F. G. Banting and C. H. Best isolate insulin and study its physiological properties.
	R. B. Goldschmidt publishes the first genetic analysis and discussion of the evolutionary implications of industrial melanism.
	C. B. Bridges reports the first monosomic (haplo-4) in *Drosophila*.
1922	L. V. Morgan discovers attached-X chromosomes in *Drosophila*.
	A. F. Blakeslee, J. Belling, M. E. Farnham, and A. D. Bergner discover a haploid *Datura*.
1923	C. B. Bridges discovers chromosomal translocations in *Drosophila*.
	R. Feulgen and H. Rossenbeck describe the cytochemical test which currently is most used for DNA localization.

O. Winge discovers the first case of monosomy in a plant (wheat).

T. Svedberg builds the first ultracentrifuge.

A. E. Boycott and C. Diver describe "delayed" Mendelian inheritance controlling the direction of the coiling of the shell in the snail *Limnea peregra.*

A. H. Sturtevant suggests that the direction of coiling of the *Limnea* shell is determined by the character of the ooplasm, which is in turn controlled by the mother's genotype.

The XX-XY type of sex determination is demonstrated for certain dioecious plants: for *Elodea* by J. K. Santos, for *Rumex* by H. Kihara and T. Ono, and for *Humulus* by O. Winge.

1925   C. B. Bridges completes his cytogenetic analysis of the aneuploid offspring of triploid *Drosophila,* and he defines the relations between the sex chromosomes and autosomes that control sexual phenotype.

E. M. East and A. J. Mangelsdorf propose the first satisfactory interpretation of the phenomenon of self-sterility in flowering plants.

A. H. Sturtevant analyzes the *Bar* phenomena in *Drosophila* and discovers position effect.

F. Bernstein suggests that the A B O blood groups are determined by a series of allelic genes.

T. H. Goodspeed and R. E. Clausen produce an amphiploid in *Nicotiana.*

1926   E. G. Anderson establishes that the centromere of the X chromosome of *Drosophila* is at the end opposite the locus of *yellow.*

S. S. Chetverikov initiates the genetic analysis of wild populations of *Drosophila.*

J. B. Sumner isolates the first enzyme in crystalline form and shows it (urease) to be a protein.

A. H. Sturtevant finds the first inversion in *Drosophila.*

R. E. Clausen and T. H. Goodspeed describe the first analysis of monosomics in a plant *(Nicotiana).*

1927   K. M. Bauer reports that the rejection of skin grafts does ̣o̩ occur when skin is transplanted from one monozygous twin to another.

J. Belling proposes that interchanges between nonhomologous chromosomes result in ring formation at meiosis.

J. B. S. Haldane suggests that the genes known to control certain coat colors in various rodents and carnivores may be evolutionarily homologous.

J. Belling introduces the aceto-carmine technique for staining chromosome squashes.

B. O. Dodge initiates genetic studies on *Neurospora.*

H. J. Muller reports the artificial induction of mutations in *Drosophila* by x rays.

1928   L. J. Stadler reports the artificial induction of mutations in maize, and demonstrates that the dose-frequency curve is linear.

F. Griffith discovers type-transformation of pneumococci. This lays the foundation for the work of Avery, MacLeod, and McCarthy (1944).

L. F. Randolph distinguishes supernumerary chromosomes from the normal chromosomes of the plant cell. He calls the normal ones "A chromosomes" and the supernumerary ones "B chromosomes."

E. Heitz introduces the term *heterochromatin.*

1929   A. Fleming discovers penicillin.

C. D. Darlington is the first to suggest that chiasmata function to hold homologs together at meiotic metaphase I and so ensure that they pass to opposite poles at anaphase I.

R. C. Tryon demonstrates successful selection for rate of maze learning in the rat.

1930   R. E. Cleland and A. F. Blakeslee demonstrate that the peculiar patterns of the transmission of groups of genes in various *Oenothera* races result from a system of balanced lethal and reciprocal translocation complexes.

R. A. Fisher publishes *The Genetical Theory of Natural Selection.*

K. Landsteiner receives a Nobel Prize for his studies in immunology.

1931   C. Stern, and independently H. B. Creighton and B. McClintock, provide the cytological proof of crossing-over.

S. Wright publishes *Evolution in Mendelian Populations.* This work together with Fisher's book and a series of papers by J. B. S. Haldane published between 1930 and 1932 under the general title *A Mathematical Theory of Natural and Artificial Selection* constitute the mathematical foundation of population genetics.

C. D. Darlington suggests that chiasmata can move to the ends of bivalents without breakage of the chromosomes. This process of *terminalization,* as he called it, is now known to occur in some species but not in others.

B. McClintock shows in maize that if a segment of a chromosome has become relatively inverted, individuals heterozygous for such an inversion often show reversed pairing at pachynema.

1932   M. Knoll and E. Ruska describe the prototype of the modern electron microscope.

1933   T. S. Painter initiates cytogenetic studies on the salivary gland chromosomes of *Drosophila.*

H. Hashimoto works out the chromosomal control of sex determination for *Bombyx mori.*

A. W. K. Tiselius reports the invention of an apparatus which permits the separation of charged molecules by electrophoresis.

B. McClintock demonstrates in maize that a single exchange within the inversion loop of a paracentric inversion heterozygote generates an acentric and a dicentric chromatid.

T. H. Morgan receives a Nobel Prize for his development of the theory of the gene.

1934   M. Schlesinger reports that certain bacteriophages are composed of DNA and protein.

P. L'Héritier and G. Teissier experimentally demonstrate the disappearance of a deleterious gene from populations of *Drosophila melanogaster* maintained in population cages for many generations.

A. Fölling discovers phenylketonuria, the first hereditary metabolic disorder shown to be responsible for mental retardation.

H. Bauer postulates that the giant chromosomes of the salivary gland cells of fly larvae are polytene.

B. McClintock shows that the nucleolus organizer of *Zea mays* can be split by a translocation and that each piece is capable of organizing a separate nucleolus. She thus sets the stage for the later demonstration (1965) that the genes for rRNA are present in multiple copies.

1935   J. B. S. Haldane is the first to calculate the spontaneous mutation frequency of a human gene.

F. Zernicke describes the principle of the phase microscope.

G. W. Beadle and B. Ephrussi and A. Kuhn and A. Butenandt work out the biochemical genetics of eye pigment synthesis in *Drosophila* and *Ephestia,* respectively.

W. M. Stanley succeeds in isolating and crystalizing the tobacco mosaic virus.

C. B. Bridges publishes the salivary gland chromosome maps for *Drosophila melanogaster.*

H. Spemann receives a Nobel Prize for his studies on embryonic induction.

1936   J. Schultz notes the relation of the mosaic expression of a gene in *Drosophila* to its position relative to heterochromatin.

T. Caspersson uses cytospectrophotometric methods to investigate the quantitative chemical composition of cells.

J. J. Bittner shows that mammary carcinomas in mice can be caused by a virus-like factor transmitted through the mother's milk.

A. H. Sturtevant and T. Dobzhansky publish the first account of the use of inversions in constructing a chromosomal phylogenetic tree.

C. Stern discovers somatic crossing over in *Drosophila.*

R. Scott-Moncrieff reviews the inheritance of plant pigments. The major part of this work was done by a group of English geneticists at the John Innes Horitcultural Institute, and these early workers established that gene substitutions resulted in chemical changes in certain flavanoid and carotenoid pigments.

1937   T. Dobzhansky publishes *Genetics and the Origin of Species.*

A. F. Blakeslee and A. G. Avery report that colchicine induces polyploidy.

T. M. Sonneborn discovers mating types in *Paramecium.*

F. C. Bawden and N. W. Pirie show that the tobacco mosaic virus, although being made mostly of protein, also contains a small amount (about 5%) of RNA.

P. A. Gorer discovers the first histocompatibility antigens in the laboratory mouse.

H. Karström points out that the synthesis of certain bacterial enzymes is stimulated when the substrates these enzymes attack are added to the medium. He coins the term "adaptive enzymes" for these and differentiates them from the "constitutive enzymes" that are always formed irrespective of the composition of the medium.

1938   B. McClintock describes the bridge-breakage-fusion-bridge cycle.

T. M. Sonneborn discovers the killer factor of *Paramecium.*

M. M. Rhoades describes the mutator gene *Dt* in maize.

H. Slizynska makes a cytological analysis of several overlapping *Notch* deficiencies in the salivary gland X chromosomes of *Drosophila melanogaster* and determines the band locations of the *w* and *N* genes.

1939   E. L. Ellis and M. Delbrück perform studies on coliphage growth which mark the beginning of modern phage work. They devise the "one-step growth" experiment which demonstrated that after the phage adsorbs onto the bacterium, it replicates within the bacterium during the "latent period," and finally the progeny are released in a "burst."

P. Levine and R. E. Stetson discover maternal immunization by a fetus carrying a new blood group antigen inherited from the father. This antigen is subsequently identified with the human Rh blood group system as the cause of erythroblastosis fetalis.

A. W. K. Tiselius and E. A. Kabat demonstrate that antibodies belong to the gamma class of serum globulins.

1940   W. Earle establishes the strain L permanent cell line from a C3H mouse.

1941   G. W. Beadle and E. L. Tatum publish their classic study on the biochemical genetics of *Neurospora* and promulgate the one gene-one enzyme theory.

J. Brachet and T. Caspersson independently reach the conclusion that RNA is localized in the nucleoli and cytoplasm and that a cell's content of RNA is directly linked with its protein synthesizing capacity.

A. J. P. Martin and R. L. M. Synge develop the technique of partition chromatography and use it to determine the amino acids in protein hydrolyzates.

A. H. Coons, H. J. Creech and R. N. Jones develop immunofluorescence techniques to demonstrate the presence of antibody-reactive sites on specific cells.

1942   R. Schoenheimer publishes *The Dynamic State of Body Constituents* and describes the use of isotopically tagged compounds in metabolic studies. He introduces the concepts of "metabolic pools" and the "turnover" of the organic compounds in cells.

L. J. Stadler and F. M. Uber demonstrate that the effectiveness of ultraviolet light in inducing mutations corresponds in corn pollen to the absorption spectrum of nucleic acid.

G. D. Snell sets out to develop highly inbred strains of mice to study the genes responsible for graft rejection.

1943   The production of radioisotopes is begun at the X-10 reactor at Oak Ridge, Tennessee.

A. Claude isolates and names the microsome fraction and shows that it contains the majority of the RNA of cells.

S. E. Luria and M. Delbrück initiate the field of bacterial genetics when they demonstrate unambiguously that bacteria undergo spontaneous mutation.

1944   O. T. Avery, C. M. MacLeod, and M. McCarty describe the pneumococcus transforming principle. The fact that it is rich in DNA suggests that DNA and not protein is the hereditary chemical.

T. Dobzhansky describes the phylogeny of the gene arrangements in the third chromosome of *Drosophila pseudoobscura* and *D. persimilis.*

E. L. Tatum, D. Bonner, and G. W. Beadle use mutant strains of *Neurospora crassa* to work out the intermediate steps in the synthesis of tryptophan.

1945   R. R. Humphrey demonstrates that the female is the heterogametic sex in urodeles.

E. B. Lewis discovers the cis-trans position effect in *Drosophila.*

R. D. Owen reports that in cattle dizygotic twins are born with, and often retain throughout life, a stable mixture of each other's red blood cells. This chimerism, which results from vascular anastomoses within the chorions of the fetuses, provides the first example of immune tolerance.

1946   A. Claude introduces tissue fractionation techniques based upon differential centrifugation and works out methods for characterizing the fractions biochemically.

Genetic recombination in bacteriophage is demonstrated by M. Delbrück and W. T. Bailey and by A. D. Hershey.

J. Lederberg and E. L. Tatum demonstrate genetic recombination in bacteria.

C. Auerbach and J. M. Robson report that mustard gas induces mutations in *Drosophila.*

J. A. Rapoport demonstrates the mutagenic effectiveness of formaldehyde in *Drosophila.*

J. Oudin develops the gel-diffusion, antigen-antibody precipitation test that bears his name.

Nobel prizes are awarded to H. J. Muller, for his contributions to radiation genetics; to J.

B. Sumner, for crystalizing enzymes; and to W. M. Stanley for his studies on the purification and chemical characterization of viruses.

1948   A. Boivin, R. Vendrely, and C. Vendrely show that in the different cells of an organism the quantity of DNA for each haploid set of chromosomes is constant.

H. K. Mitchell and J. Lein show that tryptophan synthetase is missing in certain mutant strains of *Neurospora*. This finding constitutes the first direct evidence for the one gene-one enzyme theory.

P. A. Gorer, S. Lyman, and G. D. Snell discover the major histocompatibility locus in the mouse. It resides on chromosome 17 and is named H2.

O. Ouchterlony develops the double diffusion antigen-antibody precipitation test that bears his name.

G. D. Snell introduces the term *histocompatibility gene* and formulates the laws of transplantation acceptance and rejection.

1949   B. Ephrussi, H. Hottinger, and A. M. Chimenes discover the *petite* cytoplasmic mutation in yeast.

M. M. Green and K. C. Green demonstrate that the *lozenge* locus in *Drosophila* can be divided into three subloci.

A. Kelner discovers photoreactivation of potential UV damage by visible light in *Saccharomyces*.

J. V. Neel provides genetic evidence that the sickle-cell disease is inherited as a simple Mendelian autosomal recessive.

L. Pauling, H. A. Itano, S. J. Singer, and I. C. Wells show that the H^S gene produces an abnormal hemoglobin, and they introduce the concept of genetically controlled molecular diseases.

M. L. Barr and E. G. Bertram demonstrate that the sex chromatin is morphologically different in the neurons of male and female cats.

1950   B. McClintock discovers the *Ac, Ds* system of transposable elements in maize.

E. Chargaff lays the foundations for nucleic acid structural studies by his analytical work. He demonstrates for DNA that the numbers of adenine and thymine groups are always equal and so are the numbers of guanine and cytosine groups. These findings later suggest to Watson and Crick that DNA consists of two polynucleotide strands joined by hydrogen bonding between A and T and between G and C.

A. Lwoff and A. Gutman study a lysogenic strain of *Bacillus megatherium* and demonstrate that each bacterium harbors a noninfectious form of a virus which gives the host the capacity to generate new phage without the intervention of exogenous bacteriophages. They propose the term "prophage" for this noninfectious phase. Together with L. Siminovitch and N. Kjeldgaard, Lwoff shows that prophage can be induced by ultraviolet light to produce infective virus.

E. Lederberg discovers lambda, the first viral episome of *E. coli*.

H. Latta and J. F. Hartmann introduce glass knives for ultramicrotomy.

1951   G. Gey establishes the human HeLa permanent cell line.

J. Mohr is the first to demonstrate autosomal linkage in humans (between the genes specifying the Lewis and Lutheran blood groups).

C. Stormont, R. D. Owen, and M. R. Irwin describe serological cross-reactions in the multiple allelic *B* and *C* blood group systems of cattle.

Y. Chiba demonstrates the presence of DNA in chloroplasts using the Feulgen-staining cytochemical technique.

N. H. Horowitz and U. Leupold generate large populations of temperature-sensitive mutations to determine what percentage of the genes in *E. coli* and *N. crassa* perform functions that are indispensable. The values obtained were 23% and 46%, respectively.

1952   G. E. Palade publishes the first high resolution electron micrographs of mitochondria.

R. Dulbecco adapts the techniques of bacterial virology to study animal viruses. He counts plaques made by western equine encephalomyelitis virus on monolayers of cells obtained from chick embryos.

D. Mazia and K. Dan isolate the sea urchin mitotic apparatus and start work on its biochemical characterization.

N. D. Zinder and J. Lederberg describe transduction in *Salmonella*.

442

R. Briggs and T. J. King transplant living nuclei from blastula cells into enucleated frogs' eggs. They later demonstrate that transplanted nuclei undergo chromosome changes.

W. Beermann observes stage and tissue specificities in the puffing patterns of polytene chromosomes and suggests that these are the phenotypic reflections of differential gene activities.

F. Sanger and his colleagues work out the complete amino acid sequence for the protein hormone, insulin, and show that it contains two polypeptide chains held together by disulfide bridges.

A. D. Hershey and M. Chase demonstrate that the DNA of phage enters the host, whereas most of the protein remains behind.

D. M. Brown and A. Todd demonstrate that DNA and RNA are 3′-5′ linked polynucleotides.

G. Pontecorvo and J. A. Roper describe mitotic recombination in *Aspergillus nidulans*. Here, unlike the situation in *Drosophila,* recombinant nuclei give rise to a clone that can be propagated for further analysis.

1953   J. D. Watson and F. H. C. Crick propose a model for DNA comprised of two helically intertwined chains tied together by hydrogen bonds between the purines and pyrimidines.

C. C. Lindegren discovers gene conversion in *Saccharomyces.*

A. Howard and S. R. Pelc demonstrate by autoradiography that during the cell division cycle of eukaryotes there exists a period following mitosis during which DNA synthesis does not take place (G_1), a subsequent period of DNA synthesis during which the DNA content of the interphase nucleus is doubled (S), a second growth period (G_2), and then mitosis.

W. Hayes discovers polarized behavior in bacterial recombinations. He isolates the *Hfr H* strain of *E. coli* and shows that certain genes are readily transferred from *Hfr* to F⁻ bacteria, whereas others are not.

R. E. Billingham, L. Brent, and P. B. Medawar show that immunological tolerance can be produced experimentally.

Porter-Blum and Sjöstrand ultramicrotomes become commercially available.

J. B. Finean, F. S. Sjöstrand, and E. Steinmann publish the first electron micrographs of sectioned chloroplasts.

K. R. Porter discovers and names the *endoplasmic reticulum* and identifies it as the source of cytoplasmic basophilia.

G. D. Snell finds that the major histocompatibility complex of the mouse (H-2) is composed of multiple loci.

1954   A. J. Dalton and M. D. Felix give the first detailed description of the ultrastructure of the Golgi complex.

J. Dausset observes that some patients who had received multiple blood transfusions produced antibodies against antigens found on the white blood cells of other individuals but not against those of their own cells. These antibodies defined the first HLA antigens and led to the definition of the human histocompatibility system.

N. K. Jerne proposes that the role of the antigen is to cause the selective stimulation of the cells synthesizing the corresponding antibody before the antigen appeared on the scene.

E. S. Barghoorn and S. A. Tyler report finding fossils of filamentous and coccoid microorganisms in sedimentary rocks over two billion years old. This discovery demonstrates that life existed in the Proterozoic eon.

A. C. Allison provides evidence that individuals heterozygous for the sickle-cell gene are at an advantage in a malarial environment.

1955   M. B. Hoagland obtains cell-free preparations which synthesize protein.

S. Benzer works out the fine structure of the *r* II region of phage T₄ of *E. coli,* and coins the terms *cistron, recon,* and *muton.*

N. K. Jerne proposes the selective theory of antibody formation which postulates that all antibodies are genetically determined. His basic concepts are incorporated into all later clonal selection theories.

H. Fraenkel-Conrat and R. C. Williams reconstitute "hybrid" tobacco mosaic virus from nucleic acid and protein components arising from different sources.

O. Smithies uses starch-gel electrophoresis to identify plasma protein polymorphisms.

R. H. Pritchard studies the linear arrangement of a series of allelic adenine-requiring mutants of *Aspergillus*. He confirms Pontecorvo's conclusion that crossing-over can occur between different alleles of the same gene, provided they are characterized by mutations at different subsites.

C. de Duve and four colleagues describe intracellular vesicles that contain hydrolytic enzymes and name them *lysosomes*.

P. Grabar and C. A. Williams devise the technique of immunoelectrophoresis to analyze complex mixtures of antigenic molecules.

1956   H. B. Kettlewell studies industrial melanism in the Peppered Moth. He demonstrates that moths that are conspicuous in their habitats are indeed eaten by birds more often than inconspicuous forms.

F. Jacob and E. L. Wollman are able experimentally to interrupt the mating process in *E. coli* and show that a piece of DNA is inserted from the donor bacterium into the recipient.

Groups led by S. Ochoa and A. Kornberg succeed in the *in vitro* enzymatic synthesis of polymers of ribonucleotides and deoxyribonucleotides, respectively.

J. H. Tjio and A. Levan demonstrate that the diploid chromosome number for man is 46.

C. O. Miller and his co-workers isolate and determine the chemical structure of kinetin, a substance promoting cell division in plants.

T. T. Puck, S. J. Cieciura, and P. I. Marcus succeed in growing clones of human cells *in vitro*.

G. E. Palade and P. Siekevitz isolate ribosomes.

M. J. Moses and D. Fawcett independently observe synaptonemal complexes in spermatocytes.

A. Gierer and G. Schramm and H. Fraenkel-Conrat demonstrate independently that a chemically pure nucleic acid, namely tobacco mosaic virus RNA, is infectious and genetically competent.

1957   S. A. Berson and R. S. Yalow report the first use of the radioimmunoassay procedure for the detection of insulin antibodies developed by patients in response to the administration of exogenous insulin.

J. H. Taylor, P. S. Woods, and W. L. Hughes are the first to use tritiated thymidine in high resolution autoradiography in experiments that demonstrate the semiconservative distribution of label during chromosome replication in *Vicia faba*.

V. M. Ingram reports that normal and sickle-cell hemoglobin differ by a single amino acid substitution.

A. Todd receives the Nobel Prize for his studies on the structure of nucleosides and nucleotides.

1958   F. Jacob and E. L. Wollman demonstrate that the single linkage group of *E. coli* is circular and suggest that the different linkage groups found in different *Hfr* strains result from the insertion at different points of a factor in the circular linkage group which determines the rupture of the circle.

F. H. C. Crick suggests that during protein formation the amino acid is carried to the template by an adaptor molecule containing nucleotides and that the adaptor is the part which actually fits on the RNA template. Crick thus predicts the discovery of transfer RNA.

P. C. Zamecnik and his colleagues characterize amino acid-transfer RNA complexes.

H. G. Callan and H. G. MacGregor demonstrate that the linear integrity of chromatids of amphibian lampbrush chromosomes is maintained by DNA, not protein.

M. Okamoto and, independently, R. Riley and V. Chapman discover genes that control the pairing of homoeologous chromosomes in wheat.

J. R. Raper and co-workers work out the two-locus, multiple allelic structure of the *A* and *B* incompatibility factors of *Schizophyllum commune*.

F. C. Steward, M. O. Mapes, and K. Mears succeed in rearing sexually mature plants from single diploid cells derived from the secondary phloem of roots of the wild carrot, *Daucus carota*. They conclude that each cell of the multicellular organism has all the ingredients necessary for the formation of the complete organism.

M. Meselson and F. W. Stahl use the density gradient equilibrium centrifugation technique to demonstrate the semiconservative distribution of density label during DNA replication in *E. coli*.

Nobel Prizes are awarded to G. W. Beadle, E. L. Tatum, and J. Lederberg for their contributions to genetics and to F. Sanger for his contributions to protein chemistry.

1959   J. Lejeune, M. Gautier, and R. Turpin show that Down syndrome is a chromosomal aberration involving trisomy of a small telocentric chromosome.

C. E. Ford, K. W. Jones, P. E. Polani, J. C. de Almeida, and J. H. Briggs discover that females suffering from Turner syndrome are XO.

P. A. Jacobs and J. A. Strong demonstrate that males suffering from Klinefelter syndrome are XXY.

S. J. Singer conjugates ferritin with immunoglobulin to produce a labeled antibody which is readily recognized under the electron microscope.

R. L. Sinsheimer demonstrates that bacteriophage phiX174 of *E. coli* contains a single-stranded DNA molecule.

F. M. Burnet improves Jerne's selective theory of antibody formation by suggesting that the antigen stimulates the proliferation of only those cells that are genetically programmed to synthesize the complementary antibodies.

G. M. Edelman resolves immunoglobulin G into heavy and light chains.

A. Lima-de-Faria demonstrates by autoradiography that heterochromatin replicates later than euchromatin.

M. Chèvremont, S. Chèvremont-Comhaire, and E. Baeckeland demonstrate DNA in mitochondria using a combination of autoradiographic and Feulgen-staining techniques.

K. McQuillen, R. B. Roberts, and R. J. Britten demonstrate in *E. coli* that ribosomes are the sites where protein synthesis takes place.

E. Freese proposes that mutation can occur as the result of single, base-pair changes in DNA. He coins the terms *transitions* (where a purine is replaced by the other purine or a pyrimidine by the other pyrimidine) and *transversions* (where a purine is replaced by a pyrimidine and vice versa).

C. Pelling finds selective labeling of puffed regions of polytene chromosomes after they are incubated in a nutrient solution containing ^{3}H uridine.

R. H. Whittaker suggests the grouping of organisms into five kingdoms: the bacteria, the eukaryotic microorganisms, animals, plants, and fungi.

S. Ochoa and A. Kornberg receive Nobel Prizes for their studies on the *in vitro* synthesis of nucleic acids.

1960   P. Nowell discovers phytohemagglutinin and demonstrates it use in stimulating mitoses in human leukocyte cultures.

H. E. Huxley and G. L. Zubay develop the negative staining procedure for electron microscopy of subcellular particles.

P. Siekevitz and G. E. Palade describe the synthesis of secretory proteins on membrane-bound ribosomes.

P. Doty, J. Marmur, J. Eigner, and C. Schildkraut demonstrate that complementary strands of DNA molecules can be separated and recombined.

G. Barski, S. Sorieul, and F. Cornefert report the first successful *in vitro* hybridization of mammalian cells.

U. Clever and P. Karlson experimentally induce specific puffing patterns in polytene chromosomes by injecting *Chironomus* larvae with ecdysone.

P. B. Medawar and F. M. Burnet receive a Nobel Prize for their studies on immunological tolerance.

1961   F. Jacob and J. Monod suggest that ribosomes do not contain the template responsible for the orderly assembly of amino acids. They propose that instead each DNA cistron causes synthesis of an RNA molecule of limited life span which harbors the amino acid sequence information in its nucleotide sequence. This molecule subsequently enters into temporary association with a ribosome and so confers upon it the ability to synthesize a given protein. This messenger RNA is subsequently demonstrated by S. Brenner, F. Jacob, and M. Meselson and by F. Gros, W. Gilbert, H. Hiatt, C. G. Kurland, and J. D. Watson.

M. F. Lyon and L. B. Russell independently provide evidence suggesting that in mammals one X chromosome is inactivated in some embryonic cells and their descendants, that the other is inactivated in the rest, and that mammalian females are consequently X-chromosome mosaics.

V. M. Ingram presents a theory explaining the evolution of the four known kinds of hemoglobin chains from a single primitive myoglobin-like heme protein by gene duplication and translocation.

B. D. Hall and S. Spiegelman demonstrate that hybrid molecules can be formed containing one single-stranded DNA and one RNA molecule which are complementary in base sequence. Their technique opens the way to the isolation and characterization of messenger RNAs.

S. B. Weiss and T. Nakamoto isolate RNA polymerase.

G. von Ehrenstein and F. Lipmann combine messenger RNA and ribosomes from rabbit reticulocytes with amino acid-transfer RNA complexes derived from *E. coli*. Since this cell-free system synthesized a protein similar to rabbit hemoglobin, they concluded that the genetic code is universal.

F. H. C. Crick, L. Barnett, S. Brenner, and R. J. Watts-Tobin show that the genetic language is made up of three-letter words.

F. Jacob and J. Monod put forward the operon concept.

W. Beermann demonstrates that a puffing locus on a *Chironomus* polytene chromosome is inherited in a Mendelian fashion.

M. W. Nirenberg and J. H. Matthaei develop a cell-free system from *E. coli* that incorporates amino acids into protein when supplied with template RNA preparations. They show that the synthetic polynucleotide, polyuridylic acid, directs the synthesis of a protein resembling polyphenylalanine.

M. Meselson and J. J. Weigle demonstrate in phage lambda that recombination involves breakage and reunion (but not replication) of the chromosome.

H. Dintzis shows that the direction of synthesis of the hemoglobin molecule is from amino to carboxyl terminus.

H. Moor, K. Muhlenthaler, H. Waldner, and A. Frey-Wyssling develop the first freeze-fracture procedure that permits ultrastructual observations impossible with conventional sectioning methods.

U. Z. Littauer shows that ribosomes contain only two high molecular weight species of RNA, with sedimentation values of 16S and 23S in bacteria and 18S and about 28S in animals.

C. Tokunaga demonstrates that the *engrailed* gene of *Drosophila melanogaster* causes a shift from one developmental prepattern to a different but related prepattern.

J. P. Waller and J. I. Harris find that bacterial ribosomes contain a large number of different proteins.

1962   H. Ris and W. Plaut show by electron microscopy that chloroplasts contain DNA.

E. Zuckerkandl and L. Pauling calculate the approximate times of derivation of different hemoglobin chains from their common ancestors during eukaryotic evolution.

F. M. Ritossa discovers that the salivary gland chromosomes of *Drosophila buskii* respond to heat shocks by puffing.

The distinction between T and B lymphocytes is shown in publications by J. F. A. P. Miller, R. A. Good *et al.*, and N. L. Warner *et al.*

R. R. Porter uses enzymes to cleave immunoglobulin molecules. He demonstrates that each molecule has two antigen binding portions (F_{ab}) and a crystallizable segment (F_c) that does not bind antigen. He shows that the heavy and light chains are present in 1:1 ratio and suggests the four chain model.

D. A. Rodgers and G. E. McClearn discover differences between mouse strains in alcohol preference.

U. Henning and C. Yanofsky show that amino acid replacements can arise from crossing over within triplets.

J. B. Gurdon reports that a normal fertile frog can develop from an enucleated egg injected with a nucleus from an intestinal cell. This experiment demonstrates that somatic and germinal nuclei are qualitatively equivalent.

Polyribosomes are discovered independently in three laboratories (by A. Gierer, by J. R. Warner, A. Rich, and C. E. Hall, and by T. Staehelin and H. Noll).

A. M. Campbell proposes that episomes become integrated into host chromosomes by a crossover event resembling the exchanges that were previously reported between synapsed ring and rod-shaped chromosomes in eukaryotes.

Nobel prizes are awarded to J. D. Watson, F. H. C. Crick, and M. H. F. Wilkins for their studies on the structure of DNA and to M. F. Perutz and J. C. Kendrew for their studies on the structure of hemoglobin and myoglobin.

1963   B. B. Levine, A. Ojida, and B. Benacerraf publish the first paper on the immune response genes of guinea pigs.

R. Rosset and R. Monier discover 5S RNA.

T. Okamoto and M. Takanami show that mRNA binds to the small ribosomal subunit.

H. Noll, T. Staehelin, and F. O. Wettstein demonstrate the tape mechanism of protein synthesis.

J. G. Gall produces evidence that the lampbrush chromatid contains a single DNA double helix.

B. J. McCarthy and E. T. Bolton use their DNA-agar technique to measure genetic relatedness between diverse species of organisms.

E. Hadorn demonstrates allotypic differentiation in cultured imaginal discs of *Drosophila*.

J. Monod and S. Brenner publish the replicon model.

R. Sager and M. R. Ishida isolate chloroplast DNA from *Chlamydomonas*.

R. H. Epstein, R. S. Edgar, and their collaborators introduce the use of conditional lethal mutations in T_4 phage for studying the action of indispensable genes.

J. Cairns demonstrates by autoradiography that the genophore of *Escherichia coli* is circular and that during its semiconservative replication a single Y-shaped, replicating fork proceeds from a starting point and generates two circular offspring genophores.

E. Margoliash determines the amino acid sequences for cytochrome c derived from a wide variety of species and generates the first phylogenetic tree for specific gene product.

L. B. Russell shows in the mouse that, when an X chromosome containing a translocated autosomal segment undergoes inactivation in somatic cells, the autosomal genes closest to the breakpoint are also inactivated. Thus the X inactivation spreads into the attached autosomal segment.

E. Mayr publishes *Animal Species and Evolution*. This volume provides a synthesis of modern ideas concerning the mechanism of speciation, and it has a profound influence on scientists working in this area.

1964   R. B. Setlow and W. L. Carrier and, independently, R. P. Boyce and P. Howard-Flanders describe the mechanism of excision repair in bacteria.

A. S. Sarabhai, A. O. W. Stretton, S. Brenner, and A. Bolle establish the colinearity of gene and protein product in the case of the protein coating the head of virus T_4 of *Escherichia coli*. C. Yanofsky, B. C. Carlton, J. R. Guest, D. R. Helinski, and U. Henning establish the colinearity of gene and protein product in the case of tryptophan synthetase for *Escherichia coli*.

M. S. Fox and M. K. Allen show that transformation in *Diplococcus pneumoniae* involves incorporation of segments of single-stranded donor DNA into the DNA of the recipient.

E. T. Mertz, L. S. Bates, and O. E. Nelson show that the *opaque-2* mutation modifies the amino acid composition of the mature endosperm, resulting in a striking improvement in the nutritional quality of maize seed.

J. G. Gorman, V. J. Freda, and W. Pollack demonstrate that the sensitization of Rh-negative mothers can be prevented by administration of Rh antibody immediately after delivery of their first Rh-positive baby.

D. J. L. Luck and E. Reich isolate mitochondrial DNA from *Neurospora*. They demonstrate subsequently (1966) that this DNA replicates by the classical semiconservative mechanism.

G. Marbaix and A. Burny isolate a 9S RNA from mouse reticulocytes and suggest that it may be mRNA.

R. Holliday puts forth a model that defines a sequence of breakage and reunion events which must occur during crossing over between the DNA molecules of homologous chromosomes.

J. W. Littlefield develops a method for selecting somatic cell hybrids utilizing HGPRT⁻ and TK⁻ fibroblasts cultured on HAT medium.

D. D. Brown and J. B. Gurdon show that no synthesis of the 18S and 28S rRNAs occurs in *Xenopus* tadpoles homozygous for a deficiency covering the nucleolus organizer.

W. D. Hamilton puts forth the genetical theory of social behavior.

W. Gilbert finds that nascent proteins bind to the large ribosomal subunit, as do the tRNAs.

the genetic code is translated by the sequential reading of triplets of bases starting at a defined point in phage T_4 lysozyme.

H. Wallace and M. L. Birnstiel demonstrate that an anucleolate deletion in *Xenopus laevis* removes more than 99% of the rDNA.

R. C. Lewontin and J. L. Hubby use electrophoretic methods to survey gene-controlled protein variants in natural populations of *Drosophila pseudoobscura*. They demonstrate that between 8 and 15% of all loci in the average individual genome are in the heterozygous condition. Using similar techniques H. Harris demonstrates the existence of extensive enzyme polymorphisms in human populations.

P. Rous receives the Nobel Prize for his studies on oncogenic viruses.

1967   S. Spiegelman, D. R. Mills, and R. L. Peterson report the results of serial transfer experiments in which they select those $Q\beta$ bacteriophage molecules that replicate most rapidly *in vitro*. As these experiments in extracellular evolution progressed, the molecule became smaller as its replication rate increased. By the 74th transfer the replicating RNA molecule was only 20% of its original length and was the smallest known self-duplicating molecule.

H. G. Khorana and his coworkers use polynucleotides with known repeating di- and trinucleotide sequences to solve the genetic code.

K. Taylor, Z. Hradecna, and W. Szybalski show that mRNA can originate from *both* DNA strands.

B. Mintz uses allophenic mice to demonstrate that melanocytes which provide color to the fur of the mouse are derived from 34 cells which have been determined at an early stage in embryogenesis.

J. B. Gurdon transplants somatic nuclei into frog eggs at different developmental states. The synthesis of RNA and DNA of transplanted nuclei changes to the kind of synthesis characteristic of the host cell nucleus.

L. Goldstein and D. M. Prescott perform nuclear transplantations in *Amoeba*. These show there are specific proteins that move from the cytoplasm to the nucleus, and these presumably control the nucleic acid metabolism of the nuclei they enter.

C. B. Jacobson and R. H. Barter report the use of amniocentesis for intrauterine diagnosis and management of genetic defects.

M. Goulian, A. Kornberg, and R. L. Sinsheimer report the successful *in vitro* synthesis of biologically active DNA. The template they presented to the purified *Escherichia coli* DNA polymerase was single-stranded DNA from phi X 174.

M. L. Birnstiel reports the isolation of pure rDNA from *Xenopus laevis*.

T. O. Diener and W. B. Raymer show that the potato spindle tuber disease is caused by a viroid.

M. C. Weiss and H. Green develop the HAT selection procedure and localize the gene for thymidine kinase. This was the first use of somatic cell genetics to localize human genes.

G. Wald receives the Nobel Prize in physiology for his studies on the biochemistry of vision. This work included contributions to the biochemical understanding of inherited color blindness.

1968   R. T. Okazaki and four colleagues report that newly synthesized DNA contains many fragments. These represent short lengths of DNA that are replicated in a discontinuous manner and then spliced together.

J. Morgan, D. P. McKenzie, and X. Le Pinchon develop the concept of plate tectonics to explain continental drift.

Differential synthesis of genes for ribosomal DNA during amphibian oogenesis is reported by J. G. Gall and by D. D. Brown and I. B. Dawid.

M. Kimura proposes the neutral gene theory of molecular evolution.

H. O. Smith, K. W. Wilcox, and T. J. Kelley isolate and characterize the first specific restriction endonuclease (Hind II).

D. Y. Thomas and D. Wilkie demonstrate recombination of yeast mitochondrial genes.

R. P. Donahue, W. B. Bias, J. H. Renwick, and V. A. McKusick assign the Duffy blood group locus to chromosome 1 in man. This is the first gene localized in a specific autosome.

J. A. Huberman and A. D. Riggs demonstrate that mammalian chromosomes contain serially arranged, independently replicating units each about 30 μm long.

E. H. Davidson, M. Crippa, and A. E. Mirsky show that more than 60% of the RNA labeled during oogenesis in *Xenopus laevis* is synthesized during the lampbrush stage and stored during the remaining months of oocyte maturation. This RNA is presumably a long-lived mRNA stored for use in early embryogenesis.

O. Hess and G. Meyer report extensive studies on structural modifications of the Y chromosome in various *Drosophila* species which demonstrate that the Y chromosome contains genes that control stage-specific steps in the development of sperm.

S. A. Henderson and R. G. Edwards demonstrate that the number of chiasmata per oocyte declines with increasing maternal age in the mouse and that the number of univalents increases with age. If the same occurs in human females one would expect (as has been demonstrated) an increase in aneuploid offspring with advancing maternal age.

J. E. Cleaver shows that the repair replication of DNA is defective in patients with xeroderma pigmentosum.

R. W. Holley, H. G. Khorana, and M. W. Nirenberg receive Nobel Prizes for discoveries concerning the interpretation of the genetic code and its function in protein synthesis.

1969   J. Abelson, L. Barnett, S. Brenner, M. Gefter, A. Landy, R. Russell, and J. D. Smith provide proof of the proposed mechanism of nonsense suppression by determining the actual nucleotide sequence of mutant tyrosine transfer ribonucleic acids.

The *in situ* hybridization techniques for the cytological localization of specific nucleotide sequences are developed by J. G. Gall and M. L. Pardue and by H. John, M. L. Birnstiel, and K. W. Jones.

B. C. Westmoreland, W. Szybalski, and H. Ris develop an electron microscopic technique for physically mapping genes in lambda phage. They photograph heteroduplex DNA molecules obtained by annealing of the *l* strand of one parent and the *r* strand of a second parent that has deletions, insertions, substitutions, or inversions.

R. Burgess, A. A. Travers, J. J. Dunn, and E. K. Bautz isolate and identify the sigma factor from RNA polymerase.

O. L. Miller and B. R. Beatty publish electron micrographs showing amphibian genes in the process of transcribing RNA molecules.

J. R. Beckwith (together with five associates) reports the isolation of pure *lac* operon DNA from *Escherichia coli.*

G. M. Edelman (together with five associates) publishes the first complete amino acid sequence for human gamma G_1 immunoglobulin.

Y. Hotta and S. Benzer and W. L. Pak and J. Grossfield independently induce and physiologically characterize neurological mutants in *Drosophila.*

C. Boon and F. Ruddle correlate the loss of particular chromosomes from a somatic hybrid cell line containing both human and mouse chromosomes with the loss of specific phenotypic characters. This approach permits assignment of specific loci to certain human chromosomes.

R. E. Lockard and J. B. Lingrel purify the 9S RNA fraction obtained from polysomes of mouse reticulocytes and show that it directs the synthesis of mouse hemoglobin β chains. They thus confirm the suggestion of Marbaix and Burny (1964).

A. Ammermann reports that in the hypotrichous ciliate *Stylonychia mytilus* macronuclear anlage undergo endomitotic DNA replication to form polytene chromosomes. Subsequently the major portions of these are destroyed, and over 90% of the macronuclear DNA is degraded and excreted into the medium.

R. I. Huebner and G. I. Todaro put forward the oncogene theory.

H. A. Lubs describes a fragile site on the human X chromosome and shows that it is present in mentally retarded males. Subsequent studies show that this locus (Xq27) is associated with a common form of X-linked mental retardation.

M. Delbrück, S. E. Luria, and A. D. Hershey receive a Nobel prize for their contributions to viral genetics.

1970   B. M. Alberts and L. Frey isolate the protein product of gene 32 of phage T_4 and demonstrate that this protein binds cooperatively to single-stranded DNA. They suggest that 32 protein functions to initiate unwinding of the DNA molecule so that replication can begin.

H. G. Khorana (together with twelve associates) reports the total synthesis of the gene for an alanine tRNA from yeast.

J. Yourno, T. Kohno, and J. R. Roth succeed in fusing two bacterial enzymes into one large protein molecule that combines the functions of both. The enzyme fusion was accomplished by fusing the *his D* and *his C* genes in the histidine operon of *Salmonella,* using a pair of frame shift mutations.

D. Baltimore and H. M. Temin report the existence of an RNA-dependent DNA polymerase in two oncogenic RNA viruses (Rauscher mouse leukemia and Rous fowl sarcoma).

M. L. Pardue and J. G. Gall demonstrate that pericentric heterochromatin is rich in repetitious DNA.

D. E. Wimber and D. M. Steffensen localize the 5S RNA cistrons on the right arm of chromosome 2 of *Drosophila melanogaster.*

T. Caspersson, L. Zech, and C. Johansson use quinacrine dyes in chromosomal cytology and demonstrate specific fluorescent banding patterns in human choromosomes.

R. Sager and Z. Ramanis publish the first genetic map of non-Mendelian genes. This group of eight genes reside on a chloroplast chromosome of *Chlamydomonas.*

R. T. Johnson and P. N. Rao induce premature chromosome condensation by fusing mitotically active cells with interphase cells *in vitro.*

1971   M. L. O'Riordan, J. A. Robinson, K. E. Buckton, and H. J. Evans report that all 22 pairs of human autosomes can be identified visually after staining with quinacrine hydrochloride. They demonstrate that the Philadelphia chromosome is a deleted chromosome 22.

Y. Hotta and H. Stern characterize the DNA that is synthesized during meiotic prophase in the lily. Synthesis during zygonema represents the delayed replication of a small fraction of the DNA that failed to replicate during the previous S phase. The DNA synthesized during pachynema has the characteristics of repair replication.

S. H. Howell and Stern demonstrate that an endonuclease present in microspores reaches its highest concentration early in pachynema, the stage when crossing-over is thought to occur.

B. Dudock, C. Di Peri, K. Scileppi, and R. Reszelbach present evidence for the phenylalanyl tRNA synthetase recognition site being adjacent to the dihydrouridine loop.

C. R. Merril, M. R. Geier, and J. C. Petricciani infect fibroblasts cultured from a patient suffering from galactosemia with transducing lambda phage carrying the galactose operon. The cells then make the missing transferase and survive longer in culture than uninfected galactosemic cells.

R. J. Konopka and S. Benzer report recovery of induced clock mutants of *Drosophila.*

D. T. Suzuki, T. Grigliatti, and R. Williamson isolate a temperature-sensitive paralytic mutant of *Drosophila.*

J. E. Manning and O. C. Richards detect circular chloroplast DNA molecules in lysates of *Euglena* chloroplasts.

C. Kung induces and isolates behavioral mutants of *Paramecium aurelia* and shows that many mutants have electrophysiological defects in their plasma membranes.

K. Dana and D. Nathans use restriction endonucleases to cleave the circular DNA of simian virus 40 into a series of fragments and then deduce their physical order.

1972   P. Lobban and D. Kaiser develop a general method for joining any two DNA molecules, employing terminal transferase to add complementary homopolymer tails to passenger and vehicular DNA molecules.

A. F. Zakharov and N. A. Egolina develop the BUDR labelling technique to produce harlequin chromosomes.

G. H. Pigott and N. G. Carr show that ribosomal RNAs from cyanobacteria hybridize with DNA from the chloroplasts of *Euglena gracilis.* This genetic homology provides strong support to the theory that chloroplasts are the descendants of endosymbiotic cyanobacteria.

S. J. Singer and G. L. Nicholson put forth the fluid mosaic model of the structure of cell membranes.

B. Benacerraf and H. O. McDevitt show for the mouse that *Ir* genes are linked with the H2 complex.

Y. Suzuki and D. D. Brown isolate and identify the mRNA for silk fibroin from *Bombyx mori,* and Suzuki, L. P. Gage, and Brown characterize the fibroin gene.

N. Eldredge and S. J. Gould propose the punctuated equilibrium model to explain the tempo of major evolutionary changes.

R. Silber, V. G. Malathi, and J. Hurwitz discover RNA ligase.

D. A. Jackson, R. H. Symons, and P. Berg report splicing the DNA of SV40 virus into the DNA of the lambda virus of *E. coli.* They are thus the first to join the DNAs of two different organisms *in vitro.*

J. Mendlewicz, J. L. Fleiss, and R. R. Fieve demonstrate that manic-depressive psychosis is transmitted by a dominant gene located on the short arm of the X chromosome.

M. L. Pardue, E. Weinberg, L. H. Kedes, and M. L. Birnstiel locate the histone genes on chromosome 2 of *Drosophila melanogaster.*

P. S. Carlson, H. H. Smith, and R. D. Dearing succeed in producing interspecific plant hybrids by parasexual means.

J. Hedgpeth, H. M. Goodman, and H. W. Boyer identify the nucleotide sequence in the DNA of coliphage lambda that is recognized by a specific endonuclease.

G. M. Edelman and R. R. Porter receive a Nobel Prize for their studies on the chemical structure of antibodies.

1973   D. R. Mills, F. R. Kramer, and S. Spiegelman publish the sequence for the 218 nucleotides in a replicating RNA molecule. The molecule is MDV-1, a variant derived from the RNA of Qβ phage exposed to experimental selection and thus forced to undergo "extracellular evolution" to a shorter length. (See first entry for 1967.)

S. H. Kim, G. J. Quigley, F. L. Suddath, A. McPherson, D. Sneden, J. J. Kim, J. Weinzierl, and A. Rich propose a three-dimensional structure for yeast phenylalanine transfer RNA.

R. Kavenoff and B. H. Zimm use a newly developed viscoelastic method for measuring the molecular weights of DNA molecules isolated from cells from different *Drosophila* species. They conclude that a chromosome contains one long molecule of DNA and that it is not interrupted in the centromere region.

P. Debergh and C. Nitsch succeed in culturing haploid tomato plants directly from microspores.

W. G. Hunt and R. K. Selander analyze a zone of hybridization between two subspecies of the house mouse, using gel electrophoresis to trace the boundary.

P. J. Ford and E. M. Southern show for *Xenopus laevis* that different 5S RNA genes are transcribed in the oocyte than in somatic cells.

W. Fiers, W. M. Jou, G. Haegerman, and M. Ysebaert are the first to sequence a gene coding for a protein (the coat protein of the male-specific phage MS2).

B. E. Roberts and B. M. Patterson report the preparation of a wheat germ cell-free system for the *in vitro* translation of experimentally supplied mRNAs.

A. Garcia-Bellido, P. Ripoll, and G. Morata report the developmental compartmentalization of the wing disc of *Drosophila.*

S. N. Cohen, A. C. Y. Chang, H. W. Boyer, and R. B. Helling construct the first biologically functional, hybrid bacterial plasmids by *in vitro* joining of restriction fragments from different plasmids.

1974   J. Shine and L. Dalgarno show that the 3' terminus of *E. coli* 16S rRNA contains a stretch of nucleotides that is complementary to ribosome binding sites of various coliphage mRNAs. They suggest that this region of the 16S rRNA may play a base-pairing role in the termination and initiation of protein synthesis on mRNA.

I. Zaenen, N. van Larebeke, H. Teuchy, M. van Montagu, and J. Schell discover the tumor-inducing plasmid of the crown gall bacterium.

K. M. Murray and N. E. Murray manipulate the recognition sites for restriction endonucleases in lambda phage so that its chromosome can be used as a receptor site for restriction fragments from foreign DNAs. Lambda thus becomes a cloning vehicle.

A. Tissieres, H. K. Mitchell, and U. M. Tracy find that heat shocks result in the synthesis of six new proteins in *Drosophila.* These are also synthesized by tissues that do not have polytene chromosomes.

B. Dujon, P. P. Slonimski, and L. Weill propose a model for recombination and segregation of mitochondrial genomes in *Saccharomyces cerevisae.* According to it, mtDNA molecules are present in the zygote cell in multiple copies. These pair at random, and during any mating cycle a segment from one parent can exhange with that from a second parent mt DNA yielding recombinant units.

R. D. Kornberg proposes that chromatin is built up of repeated structural units of 200 base pairs of DNA and two each of the histones H2A, H2B, H3, and H4. These structures, which are later called nucleosomes, are isolated by M. Noll. A. L. Olins and D. E. Olins publish the first electron micrographs of chromatin spreads from erythrocyte nuclei which show nucleosomes.

B. Ames develops a rapid screening test for detecting mutagenic and possibly carcinogenic compounds.

S. Brenner describes methods for inducing, isolating, and mapping mutations in the nematode *Caenorhabditis elegans.*

A. Claude, C. de Duve, and G. Palade receive Nobel Prizes for their contributions to cell biology.

1975 G. Köhler and C. Milstein perform experiments with mouse cells which show that somatic cell hybridization can be used to generate a continuous "hybridoma" cell line producing a monoclonal antibody.

Molecular biologists from around the world meet at Asilomar, California to write an historic set of rules to guide research in recombinant DNA experiments.

The NIH Recombinant DNA Committee issues guidelines aimed at eliminating or minimizing the potential risks of recombinant DNA research.

M. Grunstein and D. S. Hogness develop the colony hybridization method for the isolation of cloned DNAs containing specific DNA segments or genes.

D. Pribnow determines the nucleotide sequences of two independent bacteriophage T7 promoters, and compares these and other known promoter sequences to form a model for promoter structure and function.

E. M. Southern describes a method for transferring DNA fragments from agarose gels to nitrocellulose filters. The filters are subsequently hybridized to radioactive RNA and the hybrids detected by autoradiography.

W. D. Benton and R. W. Davis describe a rapid and direct method for screening plaques of recombinant λgt phages involving transfer of phage DNA to a nitrocellulose filter and detection of specific DNA sequences by hybridization to complementary labeled nucleic acids.

F. Sanger and A. R. Coulson develop the first rapid method for determining the nucleotide sequences in DNA by primed synthesis with DNA polymerase. A second widely used method is described in 1977 by A. M. Maxam and W. Gilbert.

G. Morata and P. A. Lawrence show in *Drosophila* that the *engrailed* mutation allows cells of the posterior wing compartment to mix with those of the anterior compartment. Therefore the normal allele of this gene functions to define the boundary conditions between the sister compartments of the developing wing.

B. Mintz and K. Illmensee inject XY diploid cells from a malignant mouse teratocarcinoma into mouse blastocyts which then are transferred to foster mothers. Cells derived from the carcinoma appear in both somatic and germ cells of some F_1 males. When these are mated, some F_2 mice contain marker genes from the carcinoma. The experiments demonstrate that the nuclei of teratocarcinoma cells remain developmentally totipotent, even after hundreds of transplant generations during which they functioned in malignant cancers.

S. L. McKenzie, S. Henikoff, and M. Meselson isolate mRNAs for heat shock proteins and show that they hybridize to specific puff sites on the *Drosophila* polytene chromosomes.

L. H. Wang, P. H. Duesberg, K. Beemon, and P. K. Vogt locate within the RNA genome of the Roux sarcoma virus the segment responsible for its oncogenic activity.

R. Dulbecco, H. Temin, and D. Baltimore receive Nobel Prizes for their studies on oncogenic viruses.

1976 H. R. B. Pelham and R. J. Jackson describe a simple and efficient mRNA-dependent *in vitro* translation system using rabbit reticulocyte lysates.

R. V. Dippell shows in *Paramecium* that kinetosomes contain RNA (not DNA) and that RNA (not DNA) synthesis accompanies kinetosome reproduction.

N. Hozumi and S. Tonegawa demonstrate that the DNA segments coding for the variable and constant regions of an immunoglobulin chain are distant from one another in the chromosomes isolated from mouse embryos, but the segments are adjacent in chromosomes isolated from mouse plasmacytomas. They conclude that somatic recombination during the differentiation of B lymphocytes moves the constant and variable gene segments closer together.

W. Y. Chooi shows that ferritin-labeled antibodies raised against proteins (isolated from rat ribosomes) bind to the terminal knobs of fibers extending from Miller trees (isolated from the ovarian nurse cells of *Drosophila*). This observation proves that Miller trees are rRNA transcription units and shows that at least some ribosomal proteins attach to a precursor rRNA molecule before its transcription is completed.

B. G. Burrell, G. M. Air, and C. A. Hutchinson report that phage phi X174 contains overlapping genes.

Formal guidelines regulating research involving recombinant DNA are issued by the National Institutes of Health in the United States.

The first genetic engineering company is formed and named Genentech.

A. Efstratiadis, F. C. Kafatos, A. Maxam, and T. Maniatis are the first to enzymatically generate eukaryotic gene segments *in vitro*. They synthesize double stranded DNA molecules that contain the sequences transcribed into the mRNAs for the alpha and beta chains of rabbit hemoglobin.

J. T. Finch and A. Klug propose that the 300Å threads seen in electron micrographs of fragmented chromatin are formed by the folding of DNA-nucleosome filaments into solenoids.

1977   A. Knoll and E. S. Barghoorn find microfossils undergoing cell division in rocks 3400 million years old. This discovery pushes back the age of life on earth to the lower Archean eon.

C. Jacq, J. R. Miller, and G. G. Brownlee describe the presence of "pseudogenes" within the 5S DNA cluster of *Xenopus laevis* oocytes.

J. C. Alwine, D. J. Kemp, and G. R. Stark prepare diazobenzyloxymethyl (DBM) paper and describe methods for transferring electrophoretically separated bands of RNA from an agarose gel to the DBM paper. Specific RNA bands are then detected by hybridization with radioactive DNA probes, followed by autoradiography. Since this method is the reverse of that described by Southern (1975) in that RNA rather than DNA is transferred to a solid support, it has come to be known as "northern blotting."

F. Sanger and eight colleagues report the complete nucleotide sequence for the DNA genome of bacteriophage phi X174.

M. Leffak, R. Grainger, and H. Weintraub show that "old" histone octamers remain intact during DNA replication and that "new" octamers consist entirely of proteins synthesized during the time of replication.

W. Gilbert induces bacteria to synthesize useful nonbacterial proteins (insulin and interferon).

R. W. Old, H. G. Callan, and K. W. Gross use labeled cloned histone genes from sea urchins to localize, by *in situ* hybridization, histone mRNAs being transcribed on the lampbrush chromosomes of salamander oocytes.

D. S. Hogness, D. M. Glover, and R. L. White report the occurrence of intervening segments in some of the 28S rRNA genes of *Drosophila melanogaster*. Introns are then described for genes that encode proteins, namely, the rabbit beta-globin gene (A. Jeffreys and R. A. Flavell) and the chicken ovalbumin gene (R. Breathnach, J. L. Mandel, and P. Chambon).

J. F. Pardon and five colleagues use neutron contrast matching techniques to demonstrate that in nucleosomes the DNA segment that attaches to the histone octamer is on the outside of the particle.

J. Sulston and H. R. Horvitz work out the postembryonic cell lineages for *Caenorhabditis elegans*.

R. S. Yalow receives a Nobel Prize for developing the radioimmunoassay procedure.

1978   R. M. Schwartz and M. O. Dayhoff compare sequence data for a variety of proteins and nucleic acids from an evolutionary diverse assemblage of prokaryotes, eukaryotes, mitochondria, and chloroplasts. Their computer-generated evolutionary trees identifies the times during evolution when protoeukaryotic organisms entered into symbiosis with mitochondria and chloroplasts (about 2 and 1 billion years ago, respectively).

W. Gilbert coins the terms *introns* and *exons*.

454

The world's first "test-tube baby" is born in Britain as a result of *in vitro* fertilization and subsequent implantation of the embryo to the mother's uterus.

T. Maniatis, R. C. Hardison, E. Lacy, J. Lauer, C. O'Connell, D. Quon, G. K. Sim, and A. Efstratiadis develop a procedure for gene isolation, which involves construction of cloned libraries of eukaryotic DNA and screening of these libraries for individual sequences by hybridization to specific nucleic acid probes.

D. J. Finnegan, G. M. Rubin, M. W. Young, and D. S. Hogness make detailed analyses of dispersed repetitive DNAs in *Drosophila*. Their discoveries represent a new beginning in the understanding of mutability, transposition, transformation, hybrid dysgenesis, and retroviruses in eukaryotes.

V. B. Reddy and eight colleagues publish the complete nucleotide sequence for simian virus 40 and correlate the sequence with the known genes and mRNAS of the virus.

Y. W. Kan and A. M. Dozy demonstrate the value of using restriction fragment length polymorphisms as linked markers for the prenatal diagnosis of sickle cell anemia.

W. Arber, H. O. Smith, and D. Nathans receive Nobel Prizes for the development of techniques utilizing restriction nucleases to study the organization of genetic systems.

1979   J. G. Sutcliffe determines the complete 4,362 nucleotide pair sequence of the plasmid cloning vector pBR322.

J. C. Avise, R. A. Lansman, and R. O. Shade successfully use restriction endonucleases to measure mitochondrial DNA sequence relatedness in natural populations.

The National Institutes of Health relax guidelines on recombinant DNA to allow viral DNA to be studied.

B. G. Barrell, A. T. Bankier, and J. Drouin report that the genetic code of human mitochondria has some unique, non-universal features.

E. F. Fritsch, R. M. Lawn, and T. Maniatis determine the chromosomal arrangement and structure of human globin genes utilizing recombinant DNA technology.

A. Wang, A. Rich, and their colleagues discover the Z form of DNA.

1980   L. Olsson and H. S. Kaplan produce the first human hybridomas that manufacture a pure antibody in laboratory culture.

The United States Supreme Court rules that genetically modified microorganisms can be patented. General Electric company, on behalf of A. Chakrabarty, obtains a patent for a genetically engineered microorganism capable of consuming oil slicks.

J. W. Gordon, G. A. Scangos, D. J. Plotkin, J. A. Barbosa, and F. H. Ruddle produce transgenic mice.

A. R. Templeton provides a new theoretical framework for speciation by the founder principle.

M. R. Capecchi describes a technique for efficient transformation of cultured mamalian cells by direct microinjection of DNA into cells with glass micropipettes.

Nobel Prizes in Physiology and Medicine go to G. D. Snell, J. Dausset, and B. Benacerraf for their contributions to immunogenetics.

P. Berg, W. Gilbert, and F. Sanger receive Nobel Prizes in Chemistry for their contributions to the experimental manipulation of DNA.

1981   R. C. Parker, H. E. Varmus, and J. M. Bishop demonstrate that the tumorigenic properties of the Rous sarcoma virus are due to a protein encoded by the *v-src* gene. Cells from various vertebrates contain a homologous gene, *c-src*. The two genes differ in that *v-src* has an uninterrupted coding sequence, whereas *c-src* contains seven exons separated by six introns.

L. Margulis publishes *Symbiosis in Cell Evolution*. Here she summarizes the evidence for the theory that organelles such as mitochondria, chloroplasts, and kinetosomes evolved from prokaryotes that lived as endosymbionts in the ancestors of modern day eukaryotes.

R. Lande proposes a new model of speciation based on sexual selection on polygenic traits. This model results in a revival of interest in sexual selection.

T. E. Wagner first demonstrates the successful transfer of functional genetic material between mammalian species; a rabbit blood protein is formed in mice.

J. D. Kemp and T. H. Hall transfer the gene of a major seed storage protein (phaseolin) from beans to the sunflower via a plasmid of the crown gall bacterium *Agrobacterium tumefaciens,* creating a "sunbean."

M. D. Edge assembles 514 paired nucleotides to make a synthetic gene for interferon, the longest gene to be pieced together in the laboratory thus far.

The complete nucleotide sequence and genetic organization of the human mitochondrial genome are worked out by a research group composed of S. Anderson, B. G. Barrell, F. Sanger, and eleven other scientists.

M. E. Harper and G. F. Saunders demonstrate that single copy genes can be mapped on human mitotic chromosomes utilizing an improved *in situ* hybridization technique.

J. Banerji, S. Rusconi, and S. Schaffner show that the transcription of the beta-globin gene is enhanced hundreds of times when this gene is linked with certain SV40 nucleotide sequences that they name "enhancer sequences."

M. Chalfie and J. Sulston identify among the touch-insensitive mutants of *Caenorhabditis elegans* five genes that affect a specific set of six sensory neurons.

T. R. Cech, A. J. Zaug, and P. J. Grabowski report that certain introns can function as ribozymes.

K. E. Steinbeck, L. McIntosh, L. Bogorad, and C. J. Arntzen demonstrate that the resistance of a weed, *Amaranthus hybridus*, to triazine herbacides is controlled by a chloroplast gene that encodes a polypeptide to which the herbacide binds. Resistant strains of the weed produce a modified gene product that fails to bind triazine.

1982   Eli Lilly International Corporation is the first to market a drug made by recombinant DNA techniques. The drug is human insulin, sold under the trade name "Humulin."

G. Scott and colleagues demonstrate that interferon produced by recombinant DNA techniques blocks virus action associated with the common cold.

E. P. Reddy, R. K. Reynolds, E. Santos, and M. Barbacid report that the genetic change which leads to the activation of an oncogene carried by a line of human bladder carcinoma cells is due to a single base substitution in this gene. The result is the incorporation of valine instead of lysine in the 12th amino acid of the protein encoded by the oncogene.

P. M. Bingham, M. G. Kidwell, and G. M. Rubin show that P strains of *Drosophila* contain 30–50 copies per genome of a transposable P element. This is the cause of hybrid dysgenesis. Then A. C. Spradling and Rubin demonstrate that cloned P elements, when microinjected into *Drosophila* embryos, become integrated into germ line chromosomes and that P elements can be used as vectors to carry DNA fragments of interest into the *Drosophila* germ line.

J. G. Sutcliffe, R. J. Milner, F. E. Bloom, and R. A. Lerner report the isolation of identifier sequences unique to genes expressed in neural tissues.

A. Klug receives a Nobel Prize for the development of crystallographic electron microscopy and for structual elucidation of biologically important nucleic-acid complexes.

1983   E. A. Miele, D. R. Mills, and F. R. Kramer succeed in constructing the first recombinant RNA molecule, involving the insertion of a synthetic foreign decaadenylic acid into a variant of the RNA genome of a small bacterial virus called Qβ via the action of the Qβ replicase.

A bacterial enzyme called ribonuclease P, whose enzymatic activity depends on its RNA complement alone, is discovered by S. Altman, N. Pace and their colleagues. This is the first case in which an RNA molecule is shown to have such properties.

H. J. Jacobs (and six colleagues) report the presence of promiscuous DNA in the sea urchin.

R. Doolittle and six colleagues show that the gene responsible for the oncogenic activity of the simian sarcoma virus is derived from a gene that encodes a growth factor normally synthesized by blood platelets.

I. S. Greenwald, P. W. Sternberg, and H. R. Horvitz demonstrate that the *lin*-12 mutant of *Caenorhabditis* functions as a developmental control gene.

B. McClintock receives the Nobel Prize for her discovery of transposable genetic elements.

1984   N. K. Jerne, G. Köhler, and C. Milstein receive the Nobel Prize in medicine for their contributions to immunology.

Index of Scientists Listed in the Chronology

Bolton, E. T., 1963
Bonner, D., 1944
Boon, C., 1969
Boveri, T., 1888, 1898, 1902
Boyce, R. P., 1964
Boycott, A. E., 1923
Boyer, H. W., 1972, 1973
Brachet, J., 1941
Bragg, W. H., 1913
Bragg, W. L., 1913
Breathnach, R., 1977
Brenner, S., 1961, 1963, 1964, 1965, 1969, 1974
Brent, L., 1953
Bridges, C. B., 1914, 1917, 1921, 1923, 1925, 1935
Briggs, J. H., 1959
Briggs, R., 1952
Britten, R. J., 1959, 1966
Brown, D. D., 1964, 1968, 1972
Brown, D. M., 1952
Brown, R., 1831
Brownlee, G. G., 1965, 1977
Buckton, K. E., 1971
Burgess, R., 1969
Burnet, F. M., 1959, 1960
Burny, A., 1964 (1969)
Burrell, B. G., 1976
Butenandt, A., 1935
Bütschli, O., 1876

Cairns, J., 1963
Callan, H. G., 1958, 1977
Camerarius, J. R., 1694
Campbell, A. M., 1962
Capecchi, M. R., 1980
Cappecchi, M., 1966
Carlson, P. S., 1972
Carlton, B. C., 1964
Carr, N. G., 1972
Carrier, W. L., 1964
Caspersson, T., 1936, 1941, 1970
*Castle, W. E., 1905 (1910)
Cech, T. R., 1981
Chakrabarty, A., 1980
Chalfie, M., 1981
Chambon, P., 1977
Chang, A. C. Y., 1973
Chapman, V., 1958
Chargaff, E., 1950
Chase, M., 1952
Chetverikov, S. S., 1926
Chèvremont, M., 1959
Chèvremont-Comhaire, S., 1959
Chiba, Y., 1951
Chimenes, A. M., 1949

Chooi, W. Y., 1976
Cieciura, S. J., 1956
Clark, A. J., 1965
Claude, A., 1943, 1946, 1974
Clausen, R. E., 1925, 1926
Cleaver, J. E., 1968
Cleland, R. E., 1930
Clever, U., 1960
Cohen, S. N., 1973
Coons, A. H., 1941
Cornefert, F., 1960
Correns, C., 1900, 1909
Coulson, A. R., 1975
Craig, L., 1965
Creech, H. J., 1941
Creighton, H. B., 1931
Crick, F. H. C., 1953, 1958, 1961, 1962, 1966
Crippa, M., 1968
Cuénot, L., 1905

Dahm, K. H., 1966
Dalgarno, L., 1974
Dalton, A. J., 1954
Dan, K., 1952
Dana, K., 1971
Darlington, C. D., 1929, 1931
Darwin, C., 1831, 1859
Dausset, J., 1954, 1980
Davidson, E. H., 1968
Davis, R. W., 1975
Dawid, I. B., 1968
Dayhoff, M. O., 1978
de Almeida, J. C., 1959
Dearing, R. D., 1972
Debergh, P., 1973
de Duve, C., 1955, 1974
de Graff, R., 1657
Delbrück, M., 1939, 1943, 1946, 1969
de Vries, H., 1900, 1901
d'Herelle, F., 1917
Diener, T. O., 1967
Dintzis, H., 1961
Di Peri, C., 1971
Dippell, R. V., 1976
Diver, C., 1923
Dobzhansky, T., 1936, 1937, 1944
Dodge, B. O., 1927
Donahue, R. P., 1968
Doolittle, R., 1983
Doty, P., 1960
Dozy, A. M., 1978
Dreyer, W. J., 1965
Drouin, J., 1979
Dudock, B., 1971
Duesberg, P. H., 1975
Dujon, B., 1974

Dulbecco, R., 1952, 1975
Dunn, J. J., 1969
Dzierzon, J., 1845

Earle, W., 1940
East, E. M., 1925
Edelman, G. M., 1959, 1969, 1972
Edgar, R. S., 1963, 1966
Edge, M. D., 1981
Edwards, R. G., 1968
Efstratiadis, A., 1976, 1978
Egolina, N. A., 1972
Ehrlich, P., 1900
Eigner, J., 1960
Eldredge, N., 1972
Ellis, E. L., 1939
Emrich, J., 1966
Ephrussi, B., 1935, 1949
Epstein, R. H., 1963
Evans, H. J., 1971

Farnham, M. E., 1920, 1922
Fawcett, D., 1956
Felix, M. D., 1954
Feulgen, R., 1923
Fiers, W., 1973
Fieve, R. R., 1972
Finch, J. T., 1976
Finean, J. B., 1953
Finnegan, D. J., 1978
Fischer, E., 1902
Fisher, R. A., 1930
Flavell, R. A., 1977
Fleiss, J. L., 1972
Fleming, A., 1929
Flemming, W., 1879, 1882
Fol, H., 1877
Fölling, A., 1934
Ford, C. E., 1959
Ford, P. J., 1973
Fox, M. S., 1964
Fraenkel-Conrat, H., 1955, 1956
Freda, V. J., 1964
Freese, E., 1959
Frey, L., 1970
Frey-Wyssling, A., 1961
Fritsch, E. F., 1979

Gage, L. P., 1972
Gall, J. G., 1963, 1968, 1969, 1970
Galton, F., 1869, 1875, 1889
Garcia-Bellido, A., 1973
Garrod, A. E., 1909
Gautier, M., 1959

Gefter, M., 1969
Geier, M. R., 1971
Gey, G., 1951
Gierer, A., 1956, 1962
Gilbert, W., 1961, 1964, 1966, 1977, 1978, 1980
Glover, D. M., 1977
Goldschmidt, R. B., 1915, 1921
Goldstein, L., 1967
Good, R. A., 1962
Goodman, H. M., 1972
Goodspeed, T. H., 1925, 1926
Gordon, J. W., 1980
Gorer, P. A., 1937, 1948
Gorman, J. G., 1964
Goss, J., 1822–24
Gould, S. J., 1972
Goulian, M., 1967
Grabar, P., 1955
Grabowski, P. J., 1981
Grainger, R., 1977
Green, H., 1967
Green, K. C., 1949
Green, M. M., 1949
Greenwald, I. S., 1983
Griffith, F., 1928
Grigliatti, T., 1971
Gros, F., 1961
Gross, K. W., 1977
Grossfield, J., 1969
Grunstein, M., 1975
Guest, J. R., 1964
Gurdon, J. B., 1962, 1964, 1967
Gutman, A., 1950

Hadorn, E., 1963
Haegerman, G., 1973
Haldane, J. B. S., 1915, 1927, 1931, 1935
Haldane, N. M., 1915
Hall, B. D., 1961
Hall, C. E., 1962
Hall, T. H., 1981
Hamilton, W. D., 1964
Hardison, R. C., 1978
Hardy, G. H., 1908
Harper, M. E., 1981
Harris, H., 1965, 1966
Harris, J. I., 1961
Harrison, R. G., 1907
Hartmann, J. F., 1950
Haruna, I., 1965
Harvey, W., 1651
Hashimoto, H., 1933
Hayes, W., 1953
Hayflick, L., 1965
Hedgpeth, J., 1972

Lacy, E., 1978
Lamarck, J. B. de Monet, 1809
Lande, R., 1981
Landsteiner, K., 1900, 1901, 1930
Landy, A., 1969
Lansman, R. A., 1979
Latta, H., 1950
Lauer, J., 1978
Lawn, R. M., 1979
Lawrence, P. A., 1975
Lederberg, E., 1950
Lederberg, J., 1946, 1952, 1958
Leffak, M., 1977
Lein, J., 1948
Lejeune, J., 1959
LePinchon, X., 1968
Lerner, R. A., 1982
Leupold, U., 1951
Levan, A., 1956
Levine, B. B., 1963
Levine, P., 1939
Lewis, E. B., 1945
Lewontin, R. C., 1966
L'Héritier, P., 1934
Lima-de-Faria, A., 1959
Lindegren, C. C., 1953
Lingrel, J. B., 1969
Linné, C. V. (Linnaeus), 1735
Lipmann, F., 1961
Littauer, U. Z., 1961
*Little, C. C., 1905 (1910), 1909, 1914
Littlefield, J. W., 1964
Lobban, P., 1972
Lockard, R. E., 1969
Lubs, H. A., 1969
Luck, D. J. L., 1964
Luria, S. E., 1943, 1969
Lwoff, A., 1950, 1965
Lyman, S., 1948
Lyon, M. F., 1961

MacGregor, H. G., 1958
MacLeod, C. M., 1944
Malathi, V. G., 1972
Malthus, T. R., 1798
Mandel, J. L., 1977
Mangelsdorf, A. J., 1925
Maniatis, T., 1976, 1978, 1979
Manning, J. E., 1971
Mapes, M. O., 1958
Marbaix, G., 1964 (1969)
Marcus, P. I., 1956
Margoliash, E., 1963
Margulies, A. D., 1965
Margulis, L., 1981
Marmur, J., 1960
Martin, A. J. P., 1941

Matthaei, J. H., 1961
*Maxam, A. M., 1975, 1976 (1977)
Mayr, E., 1963
Mazia, D., 1952
McCarthy, B. J., 1963
McCarty, M., 1944
McClearn, G. E., 1962
McClintock, B., 1931, 1933, 1934, 1938, 1950,
  1983
McClung, C. E., 1902
McDevitt, H. O., 1972
McIntosh, L., 1981
McKenzie, D. P., 1968
McKenzie, S. L., 1975
McKusick, V. A., 1966, 1968
McPherson, A., 1973
McQuillen, K., 1959
Mears, K., 1958
Medawar, P. B., 1953, 1960
Mendel, G., 1856, 1865, 1866
Mendlewicz, J., 1972
Merril, C. R., 1971
Mertz, E. T., 1964
Meselson, M., 1958, 1961, 1975
Meyer, G., 1968
Miele, E. A., 1983
Miescher, F., 1866
Miller, C. O., 1956
Miller, J. F. A. P., 1962
Miller, J. R., 1977
Miller, O. L., 1969
Mills, D. R., 1965, 1967, 1973, 1983
Milner, R. J., 1982
Milstein, C., 1975, 1984
Mintz, B., 1967, 1975
Mirsky, A. E., 1968
Mitchell, H. K., 1948, 1974
Mohr, J., 1951
Monier, R., 1963
Monod, J., 1961, 1963, 1965
Montgomery, T. H., 1901
Moor, H., 1961
Morata, G., 1973, 1975
Morgan, J., 1968
Morgan, L. V., 1922
Morgan, T. H., 1910, 1911, 1912, 1919, 1933
Moses, M. J., 1956
Muhlenthaler, K., 1961
Mulder, G. J., 1838
Muller, H. J., 1916, 1918, 1927, 1946
Müller-Hill, B., 1966
Murray, K. M., 1974
Murray, N. E., 1974

Nakamoto, T., 1961
Nass, M. M. K., 1966
Nasse, C. F., 1820

Nathans, D., 1971, 1978
Neel, J. V., 1949
Nelson, O. E., 1964
Nicholson, G. L., 1972
Nilsson Ehle, H., 1909
Nirenberg, M. W., 1961, 1968
Nitsch, C., 1973
Noll, H., 1962, 1963
Noll, M., 1974
Nowell, P., 1960

O'Connell, C., 1978
Ochoa, S., 1956, 1959
Ojida, A., 1963
Okada, Y., 1966
Okamoto, M., 1958
Okamoto, T., 1963
Okazaki, R. T., 1968
Old, R. W., 1977
Olins, A. L., 1974
Olins, D. E., 1974
Olsson, L., 1980
Ono, T., 1923
O'Riordan, M. L., 1971
Ouchterlony, O., 1948
Oudin, J., 1946
Owen, R. D., 1945, 1951

Pace, N., 1983
Painter, T. S., 1933
Pak, W. L., 1969
Palade, G. E., 1952, 1956, 1960, 1965, 1974
Pardon, J. F., 1977
Pardue, M. L., 1969, 1970, 1972
Parker, R. C., 1981
Pasteur, L., 1861
Patterson, B. M., 1973
Pauling, L., 1949, 1962
Pearson, K., 1900
Pelc, S. R., 1953
Pelham, H. R. B., 1976
Pelling, C., 1959
Perutz, M. F., 1962
Peterson, R. L., 1967
Petricciani, J. C., 1971
Pigott, G. H., 1972
Pirie, N. W., 1937
Plaut, W., 1962
Plotkin, D. J., 1980
Polani, P. E., 1959
Pollack, W., 1964
Pontecorvo, G., 1952
Porter, K. R., 1953
Porter, R. R., 1962, 1972
Prescott, D. M., 1967
Pribnow, D., 1975

Pritchard, R. H., 1955
Ptashne, M., 1966
Puck, T. T., 1956
Punnett, R. C., 1906

Quigley, G. J., 1973
Quon, D., 1978

Ramanis, Z., 1970
Rambousek, F., 1912
Randolph, L. F., 1928
Rao, P. N., 1970
Raper, J. R., 1958
Rapoport, J. A., 1946
Raspail, F. V., 1825
Raymer, W. B., 1967
Reddy, E. P., 1982
Reddy, V. B., 1978
Redi, F., 1668
Reich, E., 1964
Renwick, J. H., 1968
Reszelbach, R., 1971
Reynolds, R. K., 1982
Rhoades, M. M., 1938
Rich, A., 1962, 1973, 1979
Richards, O. C., 1971
Richardson, C. C., 1966
Riggs, A. D., 1968
Riley, R., 1958
Ripoll, P., 1973
Ris, H., 1962, 1969
Ritossa, F. M., 1962, 1965
Roberts, B. E., 1973
Roberts, R. B., 1959
Robertson, W. R. B., 1911
Robinson, J. A., 1971
Robson, J. M., 1946
Rodgers, D. A., 1962
Röller, H., 1966
Roper, J. A., 1952
Rossenbeck, H., 1923
Rosset, R., 1963
Roth, J. R., 1970
Rothman, R., 1965
Rous, P., 1910, 1966
Roux, W., 1883
Rubin, G. M., 1978, 1982
Ruddle, F., 1969, 1980
Rusconi, S., 1981
Ruska, E., 1932
Russell, L. B., 1961, 1963
Russell, R., 1969

Sabatini, D. D., 1965
Sager, R., 1963, 1970

Sanger, F., 1952, 1958, 1965, 1975, 1977, 1980, 1981
Santos, E., 1982
Santos, J. K., 1923
Sarabhai, A. S., 1964
Saunders, G. F., 1981
Scangos, G. A., 1980
Schaffner, S., 1981
Schell, J., 1974
Schildkraut, C., 1960
Schleiden, M. J., 1838–39
Schlesinger, M., 1934
Schneider, A., 1873
Schoenheimer, R., 1942
Schramm, G., 1956
Schultz, J., 1936
Schwann, T., 1838–39
Schwartz, R. M., 1978
Scileppi, K., 1971
Scott, G., 1982
Scott-Moncrieff, R., 1936
Selander, R. K., 1973
Setlow, R. B., 1964
Seton, A., 1822–24
Shade, R. O., 1979
Shine, J., 1974
Shull, G. H., 1909
Siekevitz, P., 1956, 1960
Silber, R., 1972
Sim, G. K., 1978
Siminovitch, L., 1950
Sinclair, J. H., 1966
Singer, S. J., 1949, 1959, 1972
Sinsheimer, R. L., 1959, 1967
Sjöstrand, F. S., 1953
Slizynska, H., 1938
Slonimski, P. P., 1974
Smith, H. H., 1972
Smith, H. O., 1968, 1978
Smith, J. D., 1969
Smithies, O., 1955
Sneden, D., 1973
Snell, G. D., 1942, 1948, 1953, 1980
Sonneborn, T. M., 1937, 1938
Sorieul, S., 1960
Southern, E. M., 1973, 1975
Spallanzani, L., 1769
Spemann, H., 1918, 1935
Spiegelman, S., 1961, 1965, 1967, 1973
Spiteller, G., 1965
Spradling, A. C., 1982
Sprunt, A. D., 1915
Stadler, L. J., 1928, 1942
Staehelin, T., 1962, 1963
Stahl, F. W., 1958
Stanley, W. M., 1935, 1946
Stark, G. R., 1977

Steffensen, D. M., 1970
Steinbeck, K. E.,, 1981
Steinmann, E., 1953
Stern, C., 1931, 1936
Stern, H., 1971
Sternberg, P. W., 1983
Stetson, R. E., 1939
Stevens, B. J., 1966
Steward, F. C., 1958
Stormont, C., 1951
Strasburger, E., 1875
Streisinger, G., 1966
Stretton, A. O. W., 1964, 1965
Strong, J. A., 1959
Sturtevant, A. H., 1913, 1923, 1925, 1926, 1936
Suddath, F. L., 1973
Sulston, J., 1977, 1981
Sumner, J. B., 1926, 1946
Sutcliffe, J. G., 1979, 1982
Sutton, W. S., 1902
Suzuki, D. T., 1971
Suzuki, Y., 1972
Svedberg, T., 1923
Sweely, C. C., 1966
Symons, R. H., 1972
Synge, R. L. M., 1941
Szybalski, W., 1967, 1969

Takanami, M., 1963
Tanaka, Y., 1913
Tashiro, Y., 1965
Tatum, E. L., 1941, 1944, 1946, 1958
Taylor, J. H., 1957
Taylor, K., 1967
Teissier, G., 1934
Temin, H. M., 1970, 1975
Templeton, A. R., 1980
Terzaghi, E., 1966
Teuchy, H., 1974
Thomas, D. Y., 1968
Tiselius, A. W. K., 1933, 1939
Tissieres, A., 1974
Tjio, J. H., 1956
Todaro, G. I., 1969
Todd, A., 1952, 1957
Tokunaga, C., 1961
Tonegawa, S., 1976
Tracy, U. M., 1974
Travers, A. A., 1969
Trost, B. M., 1966
Tryon, R. C., 1929
Tschermak, E., 1900
Tsugita, A., 1966
Turpin, R., 1959
Twort, F. W., 1915
Tyler, S. A., 1954

Uber, F. M., 1942

van Beneden, E., 1883
Van Bruggen, E. F. J., 1966
van Larebeke, N., 1974
van Leeuwenhoek, A., 1677
van Montagu, M., 1974
Varmus, H. E., 1981
Vendrely, C., 1948
Vendrely, R., 1948
Virchow, R., 1855
Vogt, P. K., 1975
von Baer, K. E., 1827
von Behring, E., 1890, 1901
von Ehrenstein, G., 1961

Wagner, T. E., 1981
Wald, G., 1967
Waldeyer, W., 1888, 1903
Waldner, H., 1961
Wallace, H., 1966
Waller, J. P., 1961
Wang, A., 1979
Wang, L. H., 1975
Waring, M., 1966
Warner, J. R., 1962
Warner, N. L., 1962
Watkins, J. F., 1965
Watson, J. D., 1953, 1961, 1962
Watts-Tobin, R. J., 1961
Wegener, A., 1912
Weigel, J. J., 1961
Weill, L., 1974
Weinberg, E. S., 1972
Weinberg, W., 1908, 1910
Weintraub, H., 1977
Weinzierl, J., 1973
Weismann, A., 1887

Weiss, B., 1966
Weiss, M. C., 1967
Weiss, S. B., 1961
Wells, I. C., 1949
Westmoreland, B. C., 1969
Wettstein, F. O., 1963
White, R. L., 1977
Whittaker, R. H., 1959
Wilcox, K. W., 1968
Wilkie, D., 1968
Wilkins, M. H. F., 1962
Williams, C. A., 1955
Williams, R. C., 1955
Williamson, R., 1971
Wilson, E. B., 1896
Wimber, D. E., 1970
Winge, O., 1917, 1923
Wollman, E. L., 1956, 1958
Wood, W. B., 1966
Woods, P. S., 1957
Wright, S., 1931

Yalow, R. S., 1957, 1977
Yanofsky, C., 1962, 1964
Young, M. W., 1978
Yourno, J., 1970
Ysebaert, M., 1973

Zaenen, I., 1974
Zakharov, A. F., 1972
Zamecnik, P. C., 1958
Zaug, A. J., 1981
Zech, L., 1970
Zernicke, F., 1935
Zimm, B. H., 1973
Zinder, N. D., 1952
Zubay, G. L., 1960
Zuckerkandl, E., 1962

*In situations where a research contribution made at a later time is referred to in an earlier entry, both years are given. Thus Benda, 1890 (1898) indicates that his 1898 publication is referred to in a 1890 entry (on Altmann's bioblasts).

464

Bibliography

Adelberg, E. A., editor 1960 *Papers on Bacterial Genetics*. Little, Brown, Boston.

Babcock, E. B. 1950 *The Development of Fundamental Concepts in the Science of Genetics*. American Genetics Assoc., Washington, D. C.

de Beer, G. 1964 *Charles Darwin*. Doubleday, New York.

Boyer, S. H., editor 1963 *Papers in Human Genetics*. Prentice-Hall, Englewood Cliffs, N.J.

Brock, T., editor 1961 *Milestones in Microbiology*. Prentice-Hall, Englewood Cliffs, N.J.

Brosseau, G. E., Jr. 1967 *Evolution*. W. C. Brown, Dubuque, Iowa. (A collection of classic papers on evolution)

Carlson, E. A. 1966 *The Gene: A Critical History*. W. B. Saunders, Philadelphia.

Carter, G. S. 1957 *A Hundred Years of Evolution*. Macmillan, New York.

Corwin, H. O., and J. B. Jenkins, editors 1976 *Conceptual Foundations of Genetics: Selected Readings*. Houghton Mifflin Company, Boston.

Dampier, W. C. 1943 *A History of Science*. Cambridge University Press, Cambridge.

Darwin, C. 1959 *The Origin of Species*. Edited and introduced by J. W. Burrow (1968) from the first edition. Penguin Press, New York.

Darwin, C. 1962 *The Voyage of the Beagle*. Doubleday, Garden City, N.Y.

Dawes, B. 1952 *A Hundred Years of Biology*. Duckworth, London.

Dunn, L. C., editor 1951 *Genetics in the 20th Century*. Macmillan, New York.

Dunn, L. C. 1965 *A Short History of Genetics*. McGraw-Hill, New York.

Fruton, J. S. 1972 *Molecules and Life. Historical Essays on the Interplay of Chemistry and Biology*. Wiley-Interscience, New York.

Gabriel, M. L., and S. Fogel, editors 1955 *Great Experiments in Biology*. Prentice-Hall, Englewood Cliffs, N.J.

Gall, J. G., K. R. Porter, and P. Siekevitz, editors 1981 *Discovery in Cell Biology*. Rockefeller University Press, New York.

Gardner, E. J. 1960 *History of Life Science*. Burgess Publ. Co., Minneapolis.

Glass, B., O. Temkin, and W. L. Straus, Jr. 1959 *Forerunners of Darwin, 1745–1859*. Johns Hopkins Press, Baltimore.

Goldschmidt, R. B. 1956 *Portraits from Memory*. University of Washington Press, Seattle.

Goldstein, L. 1966 *Cell Biology*. W. C. Brown, Dubuque, Iowa. (A collection of classic papers on cytology)

Grant, V. 1956 The Development of a Theory of Heredity. *Am. Sci. 44*:158–179.

Hsu, T. C. 1979 *Human and Mammalian Cytogenetics: An Historical Perspective*. Springer Verlag, New York.

Hughes, A. 1959 *A History of Cytology*. Abelard-Schuman, New York.

Irvine, W. 1955 *Apes, Angels, and Victorians*. McGraw-Hill, New York.

Keller, E. F. 1983 *A Feeling for the Organism: The Life and Work of Barbara McClintock*. W. H. Freeman, San Francisco.

Levine, L. editor 1971 *Papers on Genetics: A Book of Readings*. C. V. Mosby, St. Louis.

Lewis, E. B., editor 1961 *Selected Papers of A. H. Sturtevant on Genetics and Evolution*. W. H. Freeman, San Francisco.

Lwoff, A., and A. Ullman 1979 *Origins of Molecular Biology: A Tribute to Jacques Monod*. Academic Press, New York.

Manger, L. N. 1979 *A History of the Life Sciences*. Marcel Dekker, New York.

Moore, J. A. editor 1972 *Readings in Heredity and Development*. Oxford University Press, New York.

Moore, R. 1961 *The Coil of Life. The Story of the Great Discoveries in Life Sciences*. Knopf, New York.

Muller, H. J. 1962 *Studies in Genetics*. Indiana University Press, Bloomington.

National Academy of Sciences of the USA; Biographical Memoirs. Columbia University Press, New York. Published at periodic intervals (Vol. 37 appeared in 1964).

The Nobel Prize Lectures in Physiology/Medicine, 1942–1962, 1922–1941. (2 vols), American Elsevier, New York.

Nordenskjöld, E. 1928 *The History of Biology, A Survey*. Knopf, New York.

Oppenheimer, J. 1963 Theodor Boveri: The Cell Biologists' Embryologist. *Quart. Rev. Biol. 38*:245–249.

Peters, J. A., editor 1961 *Classical Papers in Genetics.* Prentice-Hall, Englewood Cliffs, N.J.

Provine, W. B. 1971 *The Origins of Theoretical Population Genetics.* University of Chicago Press, Chicago.

Punnett, R. C. 1950 The Early Days of Genetics. *Heredity 4*:1–10.

The Royal Society, Biographical Memoirs of Fellows of the Royal Society. Burlington House, Picadilly, London, W 1. Published at periodic intervals (Vol. 7 appeared in 1961).

Singer, C. 1959 *A History of Biology.* Abelard-Schuman, New York.

Stent, G. S., editor 1960 *Papers on Bacterial Viruses.* Little, Brown, Boston.

Stent, G. S. editor 1981 *The Double Helix: A Personal Account of the Discovery of the Structure of DNA by James D. Watson. Text, Commentary, Reviews and Original Papers.* W. W. Norton, New York.

Stern, C., and E. R. Sherwood 1966 *The Origin of Genetics.* W. H. Freeman, San Francisco.

Stubbe, H. 1972 *History of Genetics from Prehistoric Times to the Rediscovery of Mendel's Laws.* Translated from the revised second German edition of 1965 by T. R. W. Walters. The MIT Press, Cambridge, Mass.

Sturtevant, A. H. 1965 *A History of Genetics.* Harper and Row, New York.

Taylor, J. H. editor 1965 *Selected Papers on Molecular Genetics.* Academic Press, New York.

Thomas, L. 1975 *The Lives of a Cell: Notes of a Biology Watcher.* Bantam Books, New York.

Voeller, B. R. editor 1968 *The Chromosome Theory of Inheritance: Classic Papers in Development and Heredity.* Appleton-Century-Crofts, New York.

Ward, H. 1927 *Charles Darwin: The Man and His Welfare.* Bobbs-Merrill, Indianapolis.

Williams, T. I. editor 1982 *A Biographical Dictionary of Scientists,* 3rd edition. John Wiley and Sons, New York.

Willier, B. J., and J. M. Oppenheimer, editors 1964 *Foundations of Experimental Embryology.* Prentice-Hall, Englewood Cliffs, N.J.

Wilson, E. B. 1925 *The Cell in Development and Heredity,* 3rd edition. Macmillan, New York.

Zubay, G. L., and J. Marmur, editors 1973 *Papers in Biochemical Genetics,* 2nd edition. Holt, Rinehart and Winston, New York.

Appendix D
Periodicals Cited in the Literature of Genetics, Cytology, and Molecular Biology

A listing of titles and addresses of publishers or distributors. The addresses of those firms publishing or distributing more than one journal are placed at the end of this list.

Acta Cytologica
    International Academy of Cytology
    2 Jaclynn Court
    St. Louis, Mo. 83132
Acta Geneticae Medicae et Gemellologiae
    Alan R. Liss, Inc.
Acta Genetica et Statistica Medica
    superseded by *Human Heredity*
Acta Medica Auxologica
    Centro Axologico Italiano De
    Piancavallo
    Corso Magenta, 42
    Milano 20123, Italy
Acta Paediatrica Scandinavica
    Almqvist and Wiksell Periodical Co.
    P.O. Box 62
    S101 20 Stockholm, Sweden
Acta Virologica
    Academic Press
Advances in Agronomy
    Academic Press
Advances in Applied Microbiology
    Academic Press
Advances in Biological and Medical Physics
    Academic Press
Advances in Biophysics
    Univerity of Tokyo Press
    7-3-1, Hongo
    Bunkyo-ku
    Tokyo, Japan
Advances in Cell and Molecular Biology
    Academic Press
Advances in Cell Biology
    Appleton-Century-Crofts

Advances in Clinical Chemistry
    Academic Press
Advances in Enzymology
    John Wiley and Sons
Advances in Experimental Medicine and Biology
    Plenum Publishing Co.
Advances in Genetics
    Academic Press
Advances in Human Genetics
    Plenum Publishing Co.
Advances in Immunology
    Academic Press
Advances in Metabolic Disorders
    Academic Press
Advances in Microbial Physiology
    Academic Press
Advances in Morphogenesis
    Academic Press
Advances in Pediatrics
    Year Book Medical
    Publishers
Advances in Protein Chemistry
    Academic Press
Advances in Radiation Biology
    Academic Press
Advances in Teratology
    Academic Press
Advances in Viral Oncology
    Raven Press
Advances in Virus Research
    Academic Press
Agri Hortique Genetica
    Plant Breeding Institute

Weibullsholm
Landskrona, Sweden
Agronomy Abstracts
American Society of Agronomy
Agronomy Journal
(formerly *Journal of the American Society of Agronomy*)
American Journal of Diseases of Children
American Medical Association
American Journal of Human Genetics
University of Chicago Press
American Journal of Medical Genetics
Alan R. Liss, Inc.
American Journal of Mental Deficiency
Boyd Printing Company
49 Sheridan Avenue
Albany, New York 12216
American Naturalist
University of Chicago Press
Animal Blood Groups and Biochemical Genetics
Dr. W. Junk, Publishers
Animal Breeding Abstracts
Commonwealth Agricultural Bureaux
Animal Production
Longman Publishers
Annales de Génétique
Expansion Scientifique Francaise
15 Rue Saint-Benoit
75278 Paris Cedex 06, France
Annales de Génétique et Sélection Animal
Institut National de la Recherche
Agronomique
Annales d'Embryologie et de Morphogenese
Centre National de la Recherche
Scientifique
15 Quai Anatole France
Paris (7e), France
Annales de l'Institut Pasteur
Masson et Cie.
Annals of Eugenics
superseded by the *Annals of Human Genetics*
Annals of Human Biology
Taylor and Francis, Ltd.
4 John Street
London WC IN 2ET, England
Annals of Human Genetics
Cambridge University Press
Annals Paediatrici
superseded by *Pediatric Research*
Annual Review of Biochemistry
Annual Reviews
Annual Review of Entomology
Annual Reviews
Annual Review of Genetics
Annual Reviews

Annual Review of Medicine
Annual Reviews
Annual Review of Microbiology
Annual Reviews
Arabidopsis Information Service
Dr. G. Röbbelen
Institute of Agronomy
University of Göttingen
Göttingen 3H, Germany
Archiv für Genetik
Orell Fussli Arts Graphiques S. A.
Imprimeries "Au Froschauer"
Dietzingerstrasse 3
8022 Zürich, Switzerland
Archiv der Julius Klaus Stiftung für Vererbungsforschung, Sozial-anthropologie and Rassenhygiene
superseded by Archiv für Genetik
Archives d'Anatomie Microscopique et de Morphologie Experimentale
Masson et Cie
Archiv für Microbiologie
Springer-Verlag
Archiv für die gesamte Virusforschung
Springer-Verlag
Archiv für Mikroskopische Anatomie und Entwicklungsmechanik
Springer-Verlag
Archives of Biochemistry and Biophysics
Academic Press
Archives of Disease in Childhood
British Medical Association
Archives of Pediatrics
superceded by *Clinical Pediatrics*
Aspergillus Newsletter
Dr. J. A. Roper
Department of Genetics
University of Sheffield
Sheffield S102 Tn, England
Atti Associazioni Genetica Italiana
Instituto Sperimentale per la
Cerealicoltura
via Cassia 176
Rome 00191, Italy
Australian Journal of Agricultural Research
Commonwealth Scientific and
Industrial Research Organization
Australian Journal of Biological Sciences
Commonwealth Scientific and
Industrial Research Organization
Bacteriological Proceedings
American Society for Microbiology
Bacteriological Reviews
American Society for Microbiology
Barley Newsletter
Dr. D. R. Metcalfe
Canada Agriculture Research Station

25 Dafue Road
Winnipeg 19
Manitoba, Canada

Behavior Genetics
Plenum Publishing Co.

Bibliographia Genetica
Martinus Nijhoff

Biken Journal
Osaka University
Research Institute for Microbial
Diseases
3 Dojima Nishimachi
Kita ku
Osaka, Japan

Biochemical and Biophysical Research Communications
Academic Press

Biochemical Genetics
Plenum Publishing Co.

Biochemical Medicine
Academic Press

Biochimica et Biophysica Acta
Elsevier Publishing Co.

Biocytologia
Masson et Cie.

Biological Bulletin (Woods Hole)
Woods Hole Oceanographic
Institution
Woods Hole, Mass. 02543

Biological Reviews of the Cambridge Philosophical Society
Cambridge University Press

Biometrics
Biometrical Society
Suite 621
806 15th Street N.W.
Washington, DC. 20005

Biophysical Journal
Rockefeller University Press

Biophysik
Springer-Verlag

Biopolymers
John Wiley and Sons, Inc.

Blood
Grune and Stratton

Blut
Springer-Verlag

Botanical Review
New York Botanical Garden

British Journal of Haematology
Blackwell Scientific Publishers

British Medical Bulletin
Longman Group Ltd.

British Poultry Science
Longman Group Ltd.

Brookhaven Symposia in Biology
Brookhaven National Laboratory

Biology Department
Upton, New York 11973

Canadian Journal of Genetics and Cytology
Genetics Society of Canada
Suite 907
151 Slater St.
Ottawa, Ontario, Canada K1PSHA

Canadian Journal of Microbiology
National Research Council of Canada

Cancer Genetics and Cytogenetics
Elsevier/North Holland

Carlsburg Laboratoriet, Comptes Rendus des Travaux
Danish Science Press
133 Gothersgade
1123 Copenhagen K, Denmark

Carnegie Institution of Washington
1530 P Street N.W.
Washington, D.C.

Carnivore Genetics Newsletter
Carnivore Genetics Research Center
Post Office Box 5
Newtonville, Mass. 02160

Caryologia
Journal of Cytology, Cytosystematics, and Cytogenetics
Via Lamarmora 4
Firenze, Italy

Cell
The MIT Press
28 Carleton Street
Cambridge, Mass. 02142

Cell and Tissue Kinetics
Blackwell Scientific Publishers

Cell and Tissue Research
(formerly *Zeitschrift für Zellforschung und mikroskopische Anatomie*)
Springer-Verlag

Cell Biology Communications (Amsterdam)
superseded by *Currents in Modern Biology*

Cell Biology International Reports
Academic Press

Cell Differentiation
Elsevier/North-Holland

Cellular and Molecular Biology
Pergamon Press

Cellular Immunology
Academic Press

la Cellule
S. A. Vander
Munstraat 10, B-3000
Louvain, Belgium

Chromosomes Today
Plenum Publishing Co.

Chromosoma
Springer-Verlag

Clinical Genetics
Munksgaard Forlag
Clinical Immunology and Immunochemistry
Academic Press
Clinical Pediatrics
(formerly *Archives of Pediatrics*)
J. B. Lippincott
Cold Spring Harbor Symposia on Quantitative Biology
Cold Spring Harbor Laboratory
Box 100
Cold Spring Harbor
Long Island, New York 11724
Comparative Biochemistry and Physiology
Pergamon Press
CRC Critical Reviews in Biochemistry
CRC Critical Reviews in Immunology
CRC Critical Reviews in Microbiology
Chemical Rubber Company Press
2000 NW 24th Street
Boca Raton, Florida 33431
Crop Science
Crop Science Society of America
677 South Segoe Road
Madison, Wisconsin 53711
Current Genetics
Springer-Verlag
Current Problems in Pediatrics
Year Book Medical Publishers
Currents in Modern Biology
(formerly *Cell Biology Communications, Amsterdam*)
North-Holland Publishing Co.
Current Topics in Cellular Regulation
Academic Press
Current Topics in Developmental Biology
Academic Press
Current Topics in Microbiology and Immunology
Springer-Verlag
Current Topics in Radiation Research
Elsevier/North-Holland
Cytobiologie
(superseded by the *European Journal of Cell Biology*)
Cytobios
Faculty Press
88 Regent St.
Cambridge, England
Cytogenetics
(superseded by *Cytogenetics and Cell Genetics*)
Cytogenetics and Cell Genetics
S. Karger AG
Cytologia (Tokyo)
Botanical Society of Japan
c/o The Toyo Bunko

Honkomagome 2-chome
28-21, Bunkyo-ku
Tokyo 113, Japan
Cytology and Genetics (USSR)
See *Tsitologiya i Genetika*
Developmental and Comparative Immunology
Pergamon Press
Developmental and Cell Biology
Cambridge University Press
Developmental Biology
Academic Press
Developmental Genetics
Alan R. Liss, Inc.
Development, Growth and Differentiation
(formerly *Embryologia*)
Maruzen Co. Ltd.
PO Box 605
Tokyo Central
Tokyo, Japan
Differentiation
Springer-Verlag
Dissertation Abstracts
University Microfilms
Xerox Company
300 North Zeeb Road
Ann Arbor, Michigan 48106
Drosophila Information Service
Dr. P. W. Hedrick
Division of Biological Sciences
University of Kansas
Lawrence, Kansas 66045
Egyptian Journal of Genetics and Cytology
Egyptian Society of Genetics
Department of Genetics
Faculty of Agriculture
Alexandria University
Alexandria, Egypt
Embryologia
superseded by *Development, Growth, Differentiation*
Endeavor
Pergamon Press
Environmental Mutagenesis
Alan R. Liss, Inc.
Environmental Mutagen Society Newsletter
Dr. F. J. deSerres
P.O. Box Y
Oak Ridge National Laboratory
Oak Ridge, Tenn. 37830
Enzyme
(formerly *Enzymologia, Biologica et Clinica*)
S. Karger AG
Enzymologia
superseded by *Molecular and Cellular Biochemistry*
Enzymologia Biologica et Clinica
superseded by *Enzyme*

Eugenics Quarterly
superseded by *Social Biology*
Eugenics Review
superseded by the *Journal of Biosocial Science*
Euphytica
Netherlands Journal of Plant Breeding
Lawickse Allee 166
Wageningen, Holland
European Journal of Biochemistry
Springer-Verlag
European Journal of Cell Biology
(formerly *Cytobiologie*)
Wissenschaftliche Verlagsgesellschaft
European Journal of Immunology
Academic Press
Evolution
Society for the Study of Evolution
Entomology Department
University of Kansas
Lawrence, Kansas 66044
Evolutionary Biology
Appleton-Century-Crofts
Experimental and Molecular Pathology
Academic Press
Experimental Cell Biology
S. Karger AG
Experimental Cell Research
Academic Press
Federation of European Biochemical Societies, Symposia
Elsevier/North-Holland
Federation Proceedings
Federation of American Societies
of Experimental Biology
9650 Rockville Pike
Bethesda, Maryland 20014
Gamete Research
Alan R. Liss, Inc.
Gene
Elsevier/North-Holland
Genen en Phaenen
Journal of the Dutch Genetical Society
Institute of Genetics
State University of Utrecht
Opaalweg 20
Utrecht, Holland
Genetica Agraria
Instituo di Zoologia
Universita
Pavia, Italy
Genética Ibérica
Sectión de Distribución de
Publicaciones
C.S.I.C.
Vitrubio, 8
Madrid 6, Spain

Genetical Research
Cambridge University Press
Genetica Polonica
Institute Plant Genetics
Polish Academy of Science
ul, Wojsha Polskiego 71c
Poznan, Poland
Genetica (The Hague)
Dr. W. Junk, Publishers
Genetic Engineering
Academic Press
Genetic Engineering: Principles and Methods
Plenum Press
Genetic Epidemiology
Alan R. Liss, Inc.
Genetics
Genetics
P.O. Drawer U
University Station
Austin, Texas 78712
Genetic Stocks Inventory
National Seed Storage Laboratory
Fort Collins, Colorado 70521
Genetik
Gustav Fischer Verlag
Genetika
See *Soviet Genetics*
Genetika i Selekcija
Academy of Agricultural Sciences
of Bulgaria
Bul. Dragon Cankov 6
Sofia, Bulgaria
Harvey Lectures
Academic Press
Hereditas
J. D. Törnqvists Bokhandel AB
Landskrona, Sweden
Heredity
Longman Group Ltd.
Histochemical Journal
Chapman and Hall
11 New Fetter Lane
London EC4, England
Histochemie
Springer-Verlag
Histochemistry
Springer-Verlag
Hoja Genetica
Sociedad Rioplatense de Genetica
Colonia, Uruguay
Human Biology
Wayne State University Press
5980 Cass Avenue
Detroit, Michigan 48202
Human Genetics
(formerly *Humangenetik*)
Springer-Verlag

Humangenetik
(formerly *Zeitschift für menschliche
Vererbungs-und Konstitutionslehre,*
renamed *Human Genetics*)
Human Heredity
(formerly *Acta Genetica et Statistica
Medica*)
S. Karger AG
Immunobiology
Gustav Fischer Verlag
Immunochemistry
Pergamon Press
Immunogenetics
Springer-Verlag
Immunological Communications
Marcel Dekker, Inc.
270 Madison Avenue
New York, New York 10016
Immunological Reviews
Munksgaard Förlag
Immunology
Blackwell Scientific Publications
Immunology Letters
Elsevier/North-Holland
Immunology Today
Elsevier/North-Holland
Indian Journal of Agricultural Science
Indian Council of Agricultural
Research
Business Manager
Krishi Bhavan
New Delhi, India
Indian Journal of Biochemistry and Biophysics
Council of Scientific and Industrial
Research
Public Information Directorate
Hillside Road
New Delhi 12, India
Indian Journal of Genetics and Plant Breeding
Indian Agricultural Research Institute
New Delhi 12, India
Indian Journal of Poultry Science
Indian Poultry Science Association
Krishi Bhavan
New Delhi, India
Infection and Immunity
American Society for Microbiology
*Information Newsletter of Somatic Cell
Genetics*
Dr. R. A. Roosa
Wistar Institute
36th Street at Spruce
Philadelphia, Penn. 19101
*International Journal of Insect Morphology
and Embryology*
Pergamon Press

*International Journal of Invertebrate
Reproduction*
Elsevier/North-Holland
*International Journal of Peptide and Protein
Research*
Munksgaard Förlag
International Journal of Radiation Biology
Taylor and Francis Ltd.
18 Red Lion Court
Fleet Street
London, E.C.4, England
International Review of Cytology
Academic Press
*International Society for Cell Biology,
Symposia*
Academic Press
In Vitro
Williams and Wilkins
Jackson Laboratory, Annual Report
The Jackson Laboratory
Bar Harbor, Maine 04609
Japanese Journal of Breeding
Japanese Society of Breeding
c/o Faculty of Agriculture
University of Tokyo
Bunkyo-ku
Tokyo 113, Japan
Japanese Journal of Developmental Biology
Yamashiro Publishing Co. Ltd.
Ogawa-Nishiiru
Teranouchidori, Kamikyo-ku
Kyoto, Japan
Japanese Journal of Genetics
Japan Publications Trading Co.
Japanese Journal of Human Genetics
Institute of Medical Genetics
Tokyo Medical and Dental University
Yushima 1-5, Bunkyo-ku
Tokyo, Japan
Japanese Journal of Microbiology
Japan Bacteriologists and Virologists
Department of Microbiology
Keio University School of Medicine
Shinamachi, Shinjuku
Tokyo, Japan
Journal de Génétique Humaine
Institute of Medical Genetics
8, Chemin Thury
1205 Genéve, Switzerland
Journal de Microscopie
Société Française de Microscopie
Electronique
Ecole Normale Supérieure
Laboratorie de Botanique
4, Rue Lhomond
Paris 5e, France

Journal of Agricultural Research
U.S. Department of Agriculture
14th Street & Independence Avenue
S.W.
Washington, D.C. 20250
Journal of Agricultural Science
Cambridge University Press
Journal of Applied Bacteriology
Academic Press
Journal of Bacteriology
American Society for Microbiology
Journal of Biochemistry
Japanese Biochemical Society
Department of Biochemistry
Faculty of Medicine
University of Tokyo
Tokyo, Japan
Journal of Biological Chemistry
Williams and Wilkins
Journal of Biophysical and Biochemical Cytology
superseded by *Journal of Cell Biology*
Journal of Biosocial Science
(formerly *Eugenics Review*)
Blackwell Scientific Publishers
Journal of Cell Biology
(formerly *Journal of Biophysical and Biochemical Cytology*)
Rockefeller University Press
Journal of Cell Science
(formerly *Quarterly Journal of Microscopical Science*)
Cambridge University Press
Journal of Cellular and Comparative Physiology
superceded by *Journal of Cellular Physiology*
Journal of Chronic Diseases
Pergamon Press
Journal of Cytology and Genetics
Banaras Hindu University
Varanasi, India
Journal of Dairy Research
Cambridge University Press
Journal of Dairy Science
American Dairy Science Association
113 North Neil Street
Champaign, Illinois 61820
Journal of Embryology and Experimental Morphology
Cambridge University Press
Journal of Experimental Biology
Cambridge University Press
Journal of Experimental Zoology
Alan R. Liss, Inc.

Journal of General Microbiology
Cambridge University Press
Journal of General Virology
Cambridge University Press
Journal of Genetic Psychology
Journal Press
Journal of Genetics
Habshiguda-16
Hyderabad-7
India
Journal of General and Applied Microbiology
Journal Press
Journal of Heredity
American Genetic Association
1507 M. Street N.W.
Washington, D.C. 20005
Journal of Histochemistry and Cytochemistry
Elsevier/North-Holland
Journal of Horitcultural Science
Headley Brothers Ltd.
The Invicta Press
Ashford, Kent, England
Journal of Human Evolution
Academic Press
Journal of Immunogenetics
Blackwell Scientific Publications
Journal of Immunological Methods
Elsevier/North-Holland
Journal of Immunology
Williams and Wilkins
Journal of Insect Physiology
Pergamon Press
Journal of Medical Genetics
Publishing Department
British Medical Association House
Tavistock Square
London WC1H9JR, England
Journal of Medical Virology
Alan R. Liss, Inc.
Journal of Mental Deficiency Research
National Society for Mentally Handicapped Children
117 Golden Lane
London EC1Y ORT, England
Journal of Molecular and Applied Genetics
Raven Press
Journal of Molecular Biology
Academic Press
Journal of Molecular and Cellular Immunology
Springer-Verlag
Journal of Molecular Evolution
Springer-Verlag
Journal of Morphology
Alan R. Liss, Inc.

Journal of the National Cancer Institute
National Cancer Institute
Bethesda, Maryland 22014
Journal of Neurogenetics
Elsevier/North-Holland
Journal of Obstetrics and Gynaecology of the British Commonwealth
Royal College of Obstetricians and Gynecologists
27 Sussex Place
Regents Park
London N.W. 1, 4RG
England
Journal of Pediatrics
C. V. Mosby Co.
Journal of Submicroscopic Cytology
Editrice Compositori Bologna
Viale XII Giugno 3
40124 Bologna, Italy
Journal of the American Society of Agronomy
superseded by the *Agronomy Journal*
Journal of the Australian Institute of Agricultural Science
Australian Institute of Agricultural Science
191 Royal Parade
Parkville, Victoria 3052
Australia
Journal of the European Society for Animal Blood Group Research
listed under *Animal Blood Groups and Biochemical Genetics*
Journal of Theoretical Biology
Academic Press
Journal of the Sericultural Society of Japan
Sericultural Experiment Station
3-55-30 Wada
Suginami-ku
Tokyo, Japan 166
Journal of Ultrastructure Research
Academic Press
Journal of Virology
American Society for Microbiology
Kihara Institute for Biological Research, Reports
Kihara Institute for Biological Research
Mutsukawa, Minami-ku
Yokahama, Japan
Laboratory Primate Newsletter
Dr. A. M. Schrier
Psychology Department
Brown University
Providence, Rhode Island 02912
Lancet
Little, Brown & Co.

34 Beacon St.
Boston, Massachusetts 02106
Life Sciences
Pergamon Press
Maize Genetics Corporation Newsletter
Dr. M. M. Rhoades
Botany Department
Indiana University
Bloomington, Indiana 47401
Mammalian Chromosome Newsletter
D. T. C. Hsu
Section of Cell Biology
M.D. Anderson Hospital
Houston, Texas 77025
Mankind Quarterly
American Philosophical Society
Library
105 South 5th St.
Philadelphia, Penn. 19106
Medical Genetics
Pergamon Press
Metabolism
Grune and Stratton
Methods in Cell Biology
Academic Press
Methods in Cell Physiology
Academic Press
Methods In Enzymology
Academic Press
Methods in Immunology and Immunochemistry
Academic Press
Methods in Medical Research
Year Book Medical Publishers
Methods in Virology
Academic Press
Microbial Genetics Bulletin
Dr. H. Adler
P.O. Box Y
Oak Ridge National Laboratory
Oak Ridge, Tenn. 37830
Mikrobiologiya
Mezhdunarodnaya Kniga
Molecular and Cellular Biochemistry
(formerly *Enzymologia*)
Dr. W. Junk, Publishers
Molecular and General Genetics
(formerly *Zeitschrift für Vererbungslehre*)
Springer-Verlag
Molecular Biology and Evolution
University of Chicago Press
Molecular Biology, Biochemistry and Biophysics
Springer-Verlag
Molecular Biology Reports
Dr. W. Junk, Publishers

Molecular Biology SSSR
Mezhdunarodnaya Kniga
Molecular Immunology
Pergamon Press
Molecular Pharmacology
Academic Press
Molekulyarnaya Biologiya
See *Molecular Biology SSSR.*
Monatsschrift für Chemie
Springer-Verlag
Monatsschrift für Kinderheilkunde
Springer-Verlag
Monographs in Human Genetics
S. Karger SG
Mosquito News
American Mosquito Control
Association
Box 278
Selma, California 93662
Mouse Newsletter
Dr. A. G. Searle, editor
M. R. C. Radiobiology Unit
Harwell, Didcot
Berkshire, England
Mutation Research
Elsevier/North-Holland
Mycologia
New York Botanical Garden
National Cancer Institute Monographs
U.S. Department of Health,
Education and Welfare
National Cancer Institute
Bethesda, Maryland 20011
*National Institute of Genetics (Mishima),
Annual Report*
National Institute of Genetics
Yata 1, 111 Mishima
Sizuoka-ken 411, Japan
Nature
Macmillan
Nature-New Biology
Macmillan
Naturwissenschaften
Springer-Verlag
Neurospora Newsletter
Dr. Barbara Bachman, editor
Department of Microbiology
Yale Medical School
New Haven, Connecticut
Nucleic Acids Research
Information Retrieval, Ltd.
1 Falconberg Court
London W1V5FG, England
The Nucleus
Cytogenetics Laboratory
Department of Botany
University of Calcutta

35, Ballygunge Circular Rd.
Calcutta 19, India
Oak Ridge National Laboratory Symposia
(supplements to the *Journal of Cell
Physiology*)
Wistar Institute Press
Oat Newsletter
Dr. J. A. Browning
Department of Botany
Iowa State University
Ames, Iowa 50010
Obstetrics and Gynecology
Elsevier/North-Holland
Pasteur Institute
(see *Annales de l'Institut Pasteur*)
Pathologia et Microbiologia
S. Karger AG
Pediatric Research
(formerly *Annales Paediatrici*)
Williams and Wilkins
Pediatrics
American Academy of Pediatrics
P.O. Box 1834
Evanston, Illinois 60204
Perspectives in Biology and Medicine
University of Chicago Press
*Philosophical Transactions of the Royal
Society of London, Series B, Biological
Sciences*
The Royal Society
6 Carlton House Terrace
London SW1Y 5AG, England
Photochemistry and Photobiology
Pergamon Press
Phytochemistry
Pergamon Press
Pisum Newsletter
Weibullsholm Plant Breeding Institute
Landskrona, Sweden
Planta
Springer-Verlag
Plant Breeding Abstracts
Commonwealth Agricultural Bureau
Plasmid
Academic Press
Poultry Science
Poultry Science Association
Texas A & M University
College Station, Texas 77843
*Proceedings of the National Academy of
Sciences of the United States of America*
National Academy of Sciences
2101 Constitution Avenue
Washington, D.C. 20025
*Proceedings of the Royal Society of Edinburgh,
Section B (Biological Sciences)*
Royal Society of Edinburgh

22 George St.
University of Edinburgh
Edinburgh EH 2-2 PQ, Scotland
*Proceedings of the Royal Society of London,
Series B, Biological Sciences*
Royal Society of London
6 Carlton House Terrace
London SW1Y SAG, England
Progress in Biophysics and Molecular Biology
Pergamon Press
Progress in Histochemistry and Cytochemistry
Gustav Fischer Verlag
Progress in Medical Genetics
Grune and Stratton
Progress in Medical Virology
S. Karger AG
*Progress in Nucleic Acid Research and
Molecular Biology*
Academic Press
Progress in Theoretical Biology
Academic Press
Protein, Nucleic Acid, and Enzyme
Kyoritsu Shippan Co., Ltd. Publishers
4-6-19 Kohinata
Bunkyo-ku
Tokyo, Japan
Protoplasma
Springer-Verlag
Quarterly Journal of Microscopical Science
superceded by the *Journal of Cell
Science*
Quarterly Review of Biology
Stony Brook Foundation
State University of New York
Stony Brook, New York 11790
Quarterly Reviews of Biophysics
Cambridge University Press
Radiation Botany
Pergamon Press
Radiation Research
Academic Press
Reports of the Tomato Genetics Cooperative
c/o Professor C. M. Rich
Department of Vegetable Crops
University of California
Davis, California 95616
Resumptio Genetica
Martinus Nijhoff
Revista Brasileira de Genetica
Sociedade Brasileira de Genetica
Departamento de Genetica
Faculdade de Medicina de Ribeirao
Preto
14100 Ribeirao Preto, S.P.
Brazil
Revue Suisse de Zoologie
Revue Suisse de Zoologie

Muséum d'Histoire Naturelle
Genève, Switzerland
Science
American Association for the
Advancement of Science
1515 Massachusetts Ave. N.W.
Washington, D.C. 20005
Scientia Genetica
Rosenberg & Sellier
Turin, Italy
Scientific Agriculture
Agricultural Institute of Canada
151 Slater Street
Ottawa, Ontario, Canada
Scientific American
415 Madison Ave.
New York, New York 10017
Seiken Ziho
Kihara Seibutsugaku Kenkyusho
Mazume, Muko-machi
Otokunigun, Kyoto-fu
Japan
Silvae Genetica
(formerly *Zeitschrift für Forstgenetik
und Forstpflanzen-Züchtung*)
J. P. Sauerlander's Verlag
Fruhenofstrasse 21
6 Frankfurt am Main,
West Germany
Social Biology
(formerly *Eugenics Quarterly*)
University of Chicago Press
Society for Developmental Biology, Symposia
Princeton University Press
Princeton, New Jersey 08540
Society for Experimental Biology, Symposia
Cambridge University Press
Society for General Microbiology, Symposia
Cambridge University Press
Somatic Cell Genetics
Plenum Publishing Company
Soviet Genetics
(an English translation of *Genetika*)
Plenum Publishing Co.
Stadler Genetics Symposia
University of Missouri Press
Columbia, Missouri 65201
Stain Technology
Williams and Wilkins
Studies in Drosophila Genetics
See *University of Texas Publications.*
Sub-Cellular Biochemistry
Plenum Press
Symposia Genetica
Instituto de Zoologia
Piazza Botta
Pavia, Italy

Theoretical and Applied Genetics
(formerly *der Züchter*)
Springer-Verlag
Theoretical Population Biology
Academic Press
Tissue and Cell
Longmans Group, Ltd.
Tissue Antigens
Munksgaard Förlag
Transactions of the British Mycological Society
Cambridge University Press
Transactions of the New York Academy of Sciences
New York Academy of Sciences
Transplantation
Williams and Wilkins
Transplantation Proceedings
Grune & Stratton, Inc.
Transplantation Reviews
Williams and Wilkins
Tribolium Information Bulletin
Dr. A. Sokoloff
Natural Sciences Division
California State College
San Bernadino, Cal. 92407
Trudy Instituta Genetiki i Selektsii
Biblioteka Akademia Nauk SSR
Birgevaja Lenaja 1
Lenningrad, B-164, U.S.S.R.
Tsitologiya
Mezhdunarodnaya Kniga
Tsitologiya i Genetika
Akademiia Nauk URSU, Kiev.
Institut Botaniki
Ul. Repina 4
Kiev, U.S.S.R.
Ultrastructure in Biological Systems
Academic Press
University of Texas, M. D. Anderson Hospital and Tumor Institute—Symposia on Fundamental Cancer Research
Texas Medical Center
Houston, Texas 77025
University of Texas Publications (Studies in the Genetics of Drosophila, Studies in Genetics)
University Station
Austin, Texas 78712
Vector Genetics Information Service
Vector Biology and Control Unit
World Health Organization
Avenue Appia
Geneva, Switzerland

Virology
Academic Press
Wheat Newsletter
Dr. E. G. Heyne
Agronomy Dept.
Kansas State University
Manhattan, Kansas 66502
Wilhelm Roux's Archives of Developmental Biology
Springer-Verlag
World's Poultry Science Journal
National Poultry Tests Ltd.
Eaton, Godalming
Surrey, England
Yearbook of Obstetrics and Gynecology
Year Book Medical Publishers
Yearbook of Pediatrics
Year Book Medical Publishers
Yeast
Dr. Herman J. Phaff, editor
University of California
Davis, Cal. 95616
Zeitschrift für Forstgenetik und Forstpflanzen-Züchtung
superseded by *Silvae Genetica*
Zeitschrift für Immunitätsforschung
Gustav Fischer Verlag
Zeitschrift für induktive Abstammungs-und Verebungslehre
superseded by *Zeitschrift für Verebungslehre*
Zeitschrift für menschliche Vererbungs-und Konstitutionslehre
superseded by *Humangenetik*
Zeitschrift für Naturforschung, Section B
Verlag der Zeitschrift für Naturforschung
Uhlandstrasse 11, P.O. Box 2645
D7400 Tübingen, West Germany
Zeitschrift für Vererbungslehre
(formerly *Z.f. induktive Abstammungs-und Vererbungslehre*)
superseded by *Molecular and General Genetics*
Zeitschrift für Zellforschung und mikroskopische Anatomie
superseded by *Cell and Tissue Research*
der Züchter
superseded by *Theoretical and Applied Genetics*

Multijournal Publishers

Academic Press, Inc., 111 Fifth Avenue, New York, New York 10003
Alan R. Liss, Inc., 41 East 11th Street, New York, New York 10011
American Medical Association, 535 North Dearborn Street, Chicago, Illinois 60610
American Society of Agronomy, 677 South Segoe Road, Madison, Wisconsin 53711
American Society for Microbiology, 1913 I Street N. W., Washington, D.C. 20006
Annual Reviews, Inc., 419 E. Camino Way, Palo Alto, California 94306
Appleton-Century-Crofts, 440 Park Ave. South, New York, New York 10016
Blackwell Scientific Publications, Osney Mead, Oxford OX2OEL, England
British Medical Association, Tavistock Square, London WC1H 9JR, England
Cambridge University Press, 32 East 57th St., New York, New York 10022
Commonwealth Agricultural Bureaux, Farnham Royal, Bucks, England
Commonwealth Scientific and Industrial Research Organization, 314 East Albert St., East
    Melbourne C. 2, Victoria, Australia
C. V. Mosby Co., 11830 Westline Industrial Dr., St. Louis, Missouri 63141
Dr. W. Junk, Publishers, PO Box 13713, 2501ES, The Hague, Netherlands
Elsevier/North-Holland, 52 Vanderbilt Ave., New York, New York 10017
Grune & Stratton Inc., 111 5th Ave., New York, New York 10003
Gustav Fischer Verlag, PO Box 7- 20143, Stuttgart, Germany
Harper & Row Publishers, 2350 Virginia Ave., Hagerstown, MD 21740
Institut National de la Recherche Agronomique, Service des Publication, Route de Saint Cyr, 78-
    Versailles, France
Japan Publications Trading Co., Ltd., 175 5th Avenue, New York, New York 10010
J. B. Lippincott Co., East Washington Square, Philadelphia, Pennsylvania 19105
John Wiley and Sons, 605 3rd Avenue, New York, New York 10010
Journal Press, 2 Commercial Street, Provincetown, Massachusetts 02657
Longman Group Ltd., 43-45 Annandale Street, Edinburgh, Scotland
Macmillan Journals Ltd., 4 Little Essex Street, London, WC2R 3LF, England
Martinus Nijhoff, P.O. Box 269, The Hague, Holland
Masson et Cie., 120 Boulevard Saint Germain, F75280, Paris 6e, France
Mezhdunarodnaya Kniga, Moscow G-200, USSR
Munksgaard Förlag, 35 Norre Sogade, DK 1370, Copenhagen K, Denmark
National Research Council of Canada, Acquisitions Section Library, Ottawa 2, Ontario, Canada
New York Botanical Garden, Bronx, New York 10458
Oliver and Boyd, Ltd., Tweeddale Court, 14 High St., Edinburgh EH1 1YL, Scotland
Pergamon Press, Maxwell House, Fairview Park, Elmsford, New York 10523
Plenum Publishing Co., 233 Spring St., New York, New York 10013
Raven Press, 1140 Avenue of the Americas, New York, New York 10036
Rockefeller University Press, 66th and York Avenue, New York, New York 10021
S. Karger AG, Arnold Böcklin Strasse 25, 4000 Basel 11, Switzerland
Springer-Verlag, 175 Fifth Avenue, New York, New York 10010
University of Chicago Press, 5750 Ellis Avenue, Chicago, Illinois 60637
Williams and Wilkins Co., East Preston St., Baltimore, Maryland 21202
Wissenschaftliche Verlagesgesellschaft MBH, P.O. Box 40, 7 Stuttgart 1, Germany
Wistar Institute Press, 3631 Spruce St., Philadelphia, Pennsylvania 19104
Year Book Medical Publishers, Inc., 35 East Wacker Drive, Chicago, Illinois 60601

Foreign Words Commonly Found in Scientific Titles

Abbildung -*G*- figure
Abhandlung -*G*- dissertation, transaction, treatise, paper
Abstammungslehre -*G*- theory of descent, origin of species
Abteil, Abteilung -*G*- division
Acta -*L*- chronicle
allgemein -*G*- general
angewandt -*G*- applied
Annalen -*G*- annals
Anzeiger -*G*- informer
Arbeiten -*G*- work
Atti -*I*- proceedings
Band -*G*- volume
Beiheft -*G*- supplement
Bericht -*G*- report
Bokhandel -*Sw*- bookstore
Boktryckeri -*Sw*- press
Bunko -*J*- library
Bunkyo -*J*- education
Buchbesprechung -*G*- book review
Comptes Rendus -*F*- proceedings
Daigaku -*J*- university
Doklady -*R*- proceedings
Entwicklungsmechanik -*G*- embryology
Ergänzungshefte -*G*- supplement
Ergebnis -*G*- conclusion
Folia -*L*- leaflet, pamphlet, journal
Förlag -*Sw*- publisher
Forschung -*G*- research
Fortbildung -*G*- construction
Forstgenetik -*G*- forestry genetics
Fortschritt -*G*- advance, progress
gesamt -*G*- general
Gesellschaft -*G*- association, society
hebdomadaire -*F*- weekly
Hefte -*G*- number (of a periodical)
Helvetica -*L*- Swiss
Hoja -*Sp*- paper, pamphlet, record, journal
Iberica -*L*- referring to Spain and Portugal
Idengaku -*J*- genetics
Inhalt -*G*- contents
Jahrbuch -*G*- yearbook, annual
Kenkyusho -*J*- research institute
Kniga -*R*- book
Kunde -*G*- science
Lebensmittel -*G*- nutrition
Lehrbuch -*G*- textbook

Mezhdunarodnaya -*R*- international
Monatsblätter -*G*- monthly journal
Nachrichten -*G*- news
Naturwissenschaft -*G*- natural science
Nauk -*R*- science
Österreich -*G*- Austria
Planches -*F*- plates
real -*Sp*- royal
Recueil -*F*- collection
Rendiconti -*I*- account
Resumptio -*D*- review
Revista -*Sp*- review
Rundschau -*G*- overview or survey
Sammlung -*G*- collection
Säugetier -*G*- mammal
Schriften -*G*- publication
Schweizerische -*G*- Swiss
Scripta -*I*- writing
Séance -*F*- session, meeting
Seibutsugaku -*J*- biology
Seiken -*J*- biological institute
Shokubutsugaku -*J*- botany
Silvae -*L*- forest
sperimentale -*I*- experimental
Shuppan -*J*- publication
Teil -*G*- part
Tierärtliche Medizin -*G*- veterinary medicine
Tijdschrift -*D*- magazine, periodical
Tome -*F*- volume
Toyo -*J*- East, Orient
Travaux -*F*- work
Untersuchungen -*G*- research
Vererbungslehre -*G*- genetics
vergleichen -*G*- comparative
Verhandlung -*G*- proceeding, transaction
Verlag -*G*- publishing house
Verslag -*D*- report, account
Vorbericht -*G*- preliminary report
Wissenschaft -*G*- science
Wochenschrift -*G*- weekly publication
Zasshi -*J*- magazine
Zeitschrift -*G*- periodical, journal, magazine
Zeitung -*G*- newspaper
Zellforschung -*G*- cytology
Zentralblatt -*G*- overview or survey
Ziho -*J*- journal
Züchtung -*G*- breeding, culturing, rearing

D = Dutch; *F* = French; *G* = German; *I* = Italian; *J* = Japanese, *L* = Latin; *R* = Russian; *Sp* = Spanish; *Sw* = Swedish.

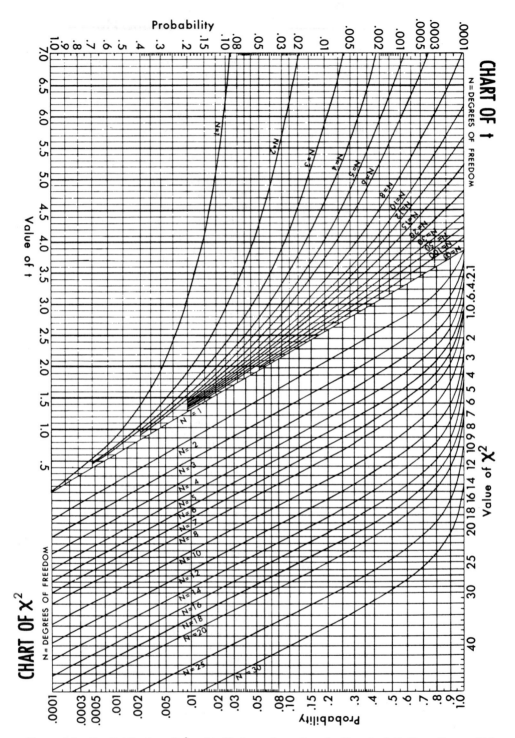

Charts giving the distribution of χ^2 and t. (Redrawn from *Genetics Notes* by J. F. Crow, Burgess Publishing Co.)